Green Energy and Technology

For further volumes:
http://www.springer.com/series/8059

Jorge de Brito · Nabajyoti Saikia

Recycled Aggregate in Concrete

Use of Industrial, Construction and Demolition Waste

Jorge de Brito
Department of Civil Engineering,
Architecture and Georesources
Instituto Superior Técnico (IST)
Lisboa
Portugal

Nabajyoti Saikia
Faculty of Science and Technology
Kaziranga University
Jorhat, Assam
India

All the figures presented in this book are adopted from the original figures

ISSN 1865-3529 ISSN 1865-3537 (electronic)
ISBN 978-1-4471-4539-4 ISBN 978-1-4471-4540-0 (eBook)
DOI 10.1007/978-1-4471-4540-0
Springer London Heidelberg New York Dordrecht

Library of Congress Control Number: 2012953118

Printed on acid-free paper

Springer is part of Springer Science+Business Media (www.springer.com)

Preface

Concrete is the most used man-made material in the world since its invention. Worldwide, about three tonnes of concrete are used annually per person. Concrete comprises three major fractions, aggregate: binder and water. The aggregate fraction in concrete is about 75 % of its total volume and therefore it plays a vital role in the overall performance of concrete. However, traditionally, more attention has been paid to develop novel binding phases of concrete as it is widely thought that the innovation in binder materials can help to develop innovative concrete materials. In fact, a significant improvement has been seen recently in this field such as the development of ultra-high strength concrete and self-compacting concrete.

It is common knowledge that the aggregates are the inert material in concrete, however, being their major constituents, their proper selection is very important to accomplish innovation in concrete production. In fact, the proper selection of aggregates and the manipulation of their size distribution are very important steps for the development of almost all types of special concrete. Moreover, the preparation of some types of concrete such as light and heavyweight concrete, concrete resistant to sound/vibration can only be achieved with proper selection of aggregates. They must not contain significant contents of deleterious components such as chlorides or sulphates, and they must also have proper shape and size to obtain a good quality concrete.

Another important recent developmental aspect in the field of cement and concrete science is the use of various types of recycled waste materials as fuel and raw material in cement production, as well as the use of these materials as aggregate in the production of various types of concrete. Cement and concrete production can consume a substantial percentage of the total generated waste materials, which can alleviate the acute environmental impact of these materials and also partly help to achieve the much needed sustainability in cement and concrete production. The use of waste materials as aggregate in concrete can consume vast amounts of them taking into account the scale of concrete production all over the world as well as the percentage of aggregate in the overall concrete volume.

Recycled Aggregate in Concrete is a recent development in the use of various types of waste materials in concrete production. The information that is scattered in various journals and conference proceedings published up to the end of March 2012 has been taken into consideration. The comprehensive information presented in the book will be helpful to graduate students, researchers and concrete technologists. It is also expected that the data presented in this book will be an essential reference for practicing engineers who face several problems concerning the use of these materials in concrete production.

The book can be divided into two parts: the compilation of varied literature data related to the use of various types of industrial waste as aggregates in concrete and the information related to the use of construction and demolition waste as aggregate in concrete. In the book, the properties of the aggregate and their effect on various concrete properties are presented separately. One chapter is devoted to describing a quantitative procedure to estimate the properties of concrete containing construction and demolition waste as aggregates. The current codes and practices developed in various countries to use construction and demolition waste as aggregates in concrete are discussed in the last chapter of the book. Moreover, several issues related to the sustainability of cement and concrete production are highlighted in the first chapter.

We would like to thank Mr. João Silvestre, Researcher, IST-Lisbon for his help during preparation of some of the figures and Ms. Grace Quinn, Editorial Assistance, Springer London, for her constant advice and help during the preparation of the manuscript. One of the authors (NJS) is also grateful to FCT, Portugal for providing financial assistance without which it would not have been possible to complete this work. Finally, we would like to thank Springer for publishing the book in excellent form.

Lisbon, Portugal, June 2012

Jorge de Brito
Nabajyoti Saikia

Contents

Chapter 1
Sustainable Development in Concrete Production

1.1 Introduction

Environmental issues such as climate change and associated global warming, depletion of natural resources and biodiversity, water and soil pollutions, generation of huge amounts of waste materials and their disposal are some of the great challenges faced by present-day civilisation. The emission of large amounts of particulate materials and various noxious gases including CO_2, the major greenhouse gas, into the atmosphere, due to rapid industrial and population expansions, is a major environmental concern and urgent action is necessary to control it. Each of these issues creates serious crisis to the future development of humankind if they are not tackled properly. The evaluation of the impact of the current developments on the environment is therefore an important agenda for present-day policy-makers and several initiatives have already been taken to tackle the problems related to these issues. Thus the term "sustainable development" was developed, which proposes a developing society, where people will live in a healthy environment with improved economic and social conditions.

The term "sustainable development" gains much attention after a United Nations report, published in 1987 (UN report 1987). It gains further momentum after a declaration published in a United Nations conference held in 1992 (Rio Summit 1992) and after the world summit on sustainable development held in Johannesburg in 2002 (World Summit 2002). According to the UN report, the term "sustainable development" is defined as the development that meets the needs of the present without compromising the ability of future generations to meet their needs (UN report 1987). The promotion of harmony among human beings and between humanity and nature is the main aim of sustainable development. However, several environmental, social and economic factors need to be considered to attain sustainable development. In this chapter, the environmental impacts of construction industry, more precisely the impacts of concrete production, on the environment will be focused.

J. de Brito and N. Saikia, *Recycled Aggregate in Concrete*,
Green Energy and Technology, DOI: 10.1007/978-1-4471-4540-0_1,

1.2 Sustainability in Construction Materials

The construction industry, one of the largest industries in the world, is notorious for having a major role on the emission of CO_2 into the atmosphere. Nowadays, the pace of development of this industry is increasing enormously all over the world especially in the developing countries due to rapid economic and industrial developments and consequent development of infrastructures and standard of living. As an example, currently, the construction industry is the second largest industry in India and the total investments in this sector account for nearly 11 % of the total gross domestic product (Construction Industry in India 2008). Similarly, the construction industry in China has been experiencing consistent growth for a long time and each year China spends nearly 16 % of its gross domestic product in this sector (China construction industry no date). By the end of 2001, about 36.69 million people in China worked on the construction sector. The construction industry is responsible for 7 % of total employment in the European Union (EU) and in the EU, the US and Japan combined, it employs more than 40 million people (OECD 2008).

The residential sector consumes huge amounts of energy all over the world. The energy used in the construction sector comprises direct use at the construction site and indirect energy used in the manufacture of the building materials. In the EU, about 40 % of total final energy is consumed by the residential and tertiary building sectors (Koukkari et al. 2007). According to Joseph and Tretsiakova-McNally (2010), building construction in the world consumes around 25 % of the global annual wood harvest; 40 % of stone, sand and gravel; and 16 % of water and also generates 50 % of the global output of greenhouse gases and agents of acid rains. The rapid expansion of this sector is creating a huge environmental problem all over the world and therefore recently several initiatives have been taken to tackle such problems.

To evaluate the environmental impact of construction materials, several issues need to be considered, namely collection, treatment and production of raw materials, construction, service life and demolition and disposal. In the whole process of construction, service life of the building and its demolition, not only huge amounts of energy of all sectors are consumed but also huge amounts of CO_2 emissions are created. These activities also consume huge amounts of most non-energy-related resources, create high volumes of waste and are responsible for enormous pollution in the atmosphere, soil and water. The uses of energy and the emission of CO_2 take place at various steps, such as raw material extractions, transportation, manufacture, demolition, service life and waste processing. Table 1.1 shows a typical example of calculation of the emitted amount of CO_2 into the atmosphere at the various steps of a building life cycle (BIS 2010).

Thus, by considering the size, importance, resources use and environmental impact of the construction industry, it is necessary to produce sustainable construction materials with increasing service life but minimum maintenance future needs. Sustainability in construction is also inevitable due to stringent regulations

Table 1.1 CO_2 emitted into atmosphere at the various steps of a building life cycle (BIS 2010)

Steps	Amount of CO_2 emitted	
	Quantity (Mt)	%
Design	1.3	0.5
Manufacture	45.2	15
Distribution	2.8	1
Operation on-site	2.6	1
In use	246.4	83
Refurbishment/demolition	1.3	0.4
Total	298.4	100

that have been adopted all over the world on the emission of greenhouse gases including CO_2 into the atmosphere to limit the rise of the global average temperature. As for example, EU is currently promoting a goal of 30 % reduction in greenhouse gas emissions by 2020 compared to 1990 levels in developed countries (Koukkari et al. 2007). The targets for various measures up to 2020 in the EU include (Koukkari et al. 2007):

- 20 % improvement of energy efficiency of cars, buildings and appliances and especially:
- 30 % reduction of final energy use of buildings;
- 20 % share of renewable energy on average;
- 10 % share of biofuels;
- Nearly 0 % emissions of new power plants.

Several factors such as energy saving methodologies and techniques, improved use of materials, increasing service life of products, further reuse/recycle of materials, eco-designing and emission control need to be considered for the development of sustainable construction materials. The durability of construction material is another factor that needs to be considered seriously for sustainable construction. A durable building material has a technically better and longer service life and therefore reduces the cost and amount of materials used in repair and in new constructions in a particular time period.

Material efficiency is one of the most important components of sustainable construction materials. Correct selection of materials by taking into account their complete service lifetime and by choosing products with minimal environmental impacts can reduce CO_2 emissions by up to 30 % (González and Navarro 2006). Without compromising on the quality of the end product, the use of locally produced materials as well as of renewable and recycled sources should be encouraged. In this way, transportation costs and problems associated with the disposal of other industrial waste can also be reduced. The recycling/reusability of construction products at the end of their service life should also be considered during the selection of materials. Higher recycling/reusability of construction products after their service life can reduce the generated amount of waste and associated disposal problems.

Other factors that greatly affect the selection of building materials are their costs and social requirements such as thermal comfort, good mechanical properties (strength and durability), aesthetic characteristics, health effect and the ability to build quickly. For example, the use of some building materials such as paints, treated wood or foams can have a toxic effect on the occupants of a building and therefore should be considered carefully. Ideally, the combination of all environmental, economic and social factors can give a clear description of a material and thus helps in a decision-making process regarding the selection of the materials suitable for buildings (Abeysundara et al. 2009). According to Calkins (2009), the materials that reduce the use of resources, minimise environmental impacts, pose no or low human health risks during their handling and service life, assist with sustainable site design strategies can be considered as sustainable construction materials.

Several codes and policies have been developed for environmentally efficient, carbon neutral, eco-designed building constructions. For example, the European commission developed a policy that takes into consideration the whole life cycle of the product, comprising three main phases: environmental impact of the products, environmental improvement of the products and policy implications. However, several problems still exist in addressing the issues related to sustainability in construction such as lack of innovation or inadequate level of skills. In the following section, sustainability in concrete production, the major construction material, will be briefly highlighted.

1.3 Sustainability in Concrete Production

Concrete is the major construction material and plays a vital rule in the development of current civilisation. It is the most used man-made material in the world since its invention. Worldwide, about three tonnes of concrete are used annually per person (Cement Concrete Aggregate Australia no date). The consumption of concrete as construction material in the world is over twice the total consumption of all other building materials including wood, steel, plastic and aluminium. It is reported that the total annual concrete production in the world is more than 10 billion tonnes (Meyer 2009). More than 0.9, 5 and 0.6 billion tonnes of Portland cement, aggregate and potable water, respectively, are necessary for the production of such an amount of concrete. The massive use of concrete as a construction material is due to its versatile properties. Properties such as strength, durability, affordability and abundance of raw materials make concrete the first choice material for most construction purposes. However, concrete production has several negative impacts on the environment, such as the emission of CO_2 and other greenhouse gases and the use of non-renewable natural resources like natural stone and water, and therefore a lot of attention has been paid recently to tackling the environmental issues related to their use in concrete preparation.

Concrete comprises various constituents and therefore the environmental impact of concrete production is a complex mechanism partly governed by the individual impacts from each of these constituents and partly governed by the combined effect of the constituents when they are mixed together. Therefore, sustainability issues related to concrete production need to be addressed by considering the individual as well as the combined effects of these constituents. On the other hand, improvement in concrete design, mechanical and durability properties and service life of concrete also need to be considered seriously as these factors also influence the environmental impact of concrete. These points will be presented briefly in the following sections. Table 1.2 outlines some topics related to the sustainability of concrete production (Eco-Serve).

1.4 Sustainability in Concrete by Improving Properties and Service Life of Concrete

The reduction of the environmental impact of concrete structures to a minimum without compromising on their performance is one of the major concerns for future sustainable development of the concrete industry. Sustainability in concrete production can be achieved by improving current practices, e.g. improvement or innovation in concrete mix and product design approaches (Khokhar et al. 2010, Joseph and Tretsiakova-McNally 2010), improvement of the performance of concrete-based products in their service lives. The improvement of mechanical and durability performances of concrete in their service life can indirectly reduce the CO_2 emission by increasing their service life and reducing the requirements of materials for repairing. In a report, it was estimated that reducing the volume of concrete by improving its mechanical strength can decrease the emissions of CO_2 by around 30 % (Habert and Roussel 2009). The use of innovative types of concrete such as high and ultra-high strength concrete and self-compacting concrete can also increase the sustainability in concrete production by giving flexibility in product design and by increasing material performance (Joseph and Tretsiakova-McNally 2010). Recent developments in self-healing concrete, which can repair cracks automatically, is an important step towards gaining sustainability in the concrete industry (Dry 2000). The use of innovative approaches in designing concrete and in using innovative materials in residential and commercial building sectors, one of the major users of concrete materials, can also reduce the amount of energy used during its service life. For example, phase change materials (PCM) can be used to increase the energy storage capacity of buildings and also to control the room temperature of buildings in summer and winter (Benz and Turpin 2007).

The use of polymeric materials as admixtures and for repair purposes and the application of nanotechnology are some recent innovations in concrete preparation, which also gives several economic and technical benefits towards obtaining sustainability. The addition of polymeric admixtures can improve several

Table 1.2 Sustainability issues in concrete production (Eco-Serve)

Environmental impact category	Societal issues	Economical issues	Some solutions
1 Land-use and exploitation of natural resources (excavations, quarrying, ground water and lime stone), mainly connected with the production of concrete constituents	Recreation versus industry. Planning of land-use. Utilisation of scarce resources.	Transportation cost. Use of local materials versus imported materials.	Innovation in production and processing of raw materials
2 Waste products from concrete production (washing/mixing water, cement slurry, form oil, rejected concrete and excess production)	Land-filling with the risk of leaching of heavy metals and hydrocarbons. Sorting and reusing.	Landfill taxes. Recycling into production. Demand from other industries.	Reusing waste materials generated within concrete production as well as from other industries
3 Emissions and energy consumption (CO_2, SO_2, embodied energy throughout production, transport and construction)	Commitment to reduce greenhouse effect and to behave in an energy conscious manner.	Energy taxes. Up-to-date production equipment and methods.	Reducing clinker content by blending other materials, optimising concrete mix design
4 Working environment (noise, vibrations, dust, accidents etc.)	Health problems.	Expenses for hospitalization and sick leave. Automated production equipment and methods.	By improving automation in concrete production, use of innovative concrete product such as fibre reinforced concrete and self-compacting concrete.

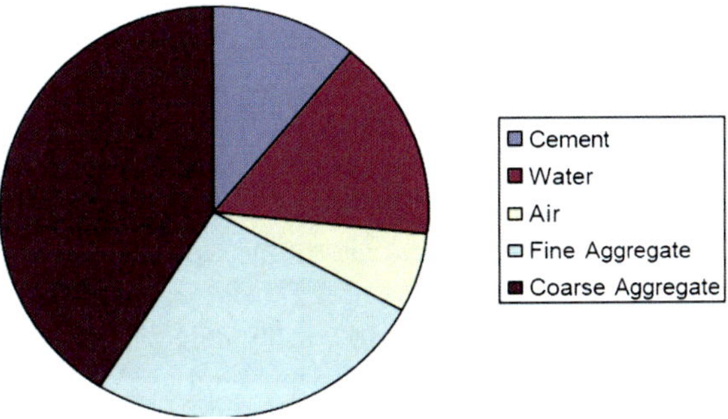

Fig. 1.1 Typical concrete mix

mechanical and durability properties of concrete and also indirectly reduce the emission of CO_2. Nanotechnology can provide huge opportunities towards gaining sustainability in concrete preparation. Improved understanding of nanostructure of cement hydration product, the only binding phases in concrete using nanotechnological tools, the use of nanomaterials such as nano silica, nano alumina and nanofibres in concrete preparation, the use of photo-catalysts such as nano TiO_2 for self-cleaning of concrete products, and the use of nanotechnology to monitor the performance during service life of concrete, are some of the recent inventions, which definitely decrease the environmental impact of concrete (Mukhopadhyay 2011). However, the toxic effect of nano-based products in human health during manufacturing and the service life of the resulting concrete products must thoroughly be investigated before their applications.

1.5 Sustainability in Concrete by Innovation in Concrete Constituents

Concrete mainly consists of at least three constituents: cement as a binding material, aggregates, the major part of concrete (normally accounting for 70–75 % of its volume) and water. A typical concrete composition is shown in Fig. 1.1. Each constituents of concrete has its own environmental impact; however, the sustainability of concrete as a material is strongly influenced by the cement and aggregate industries. The sustainability in water use in concrete has also become a big issue recently due to the huge consumptions of potable water during concrete preparation as well as the scarcity of potable water faced all over the world. Sustainability in water, cement and aggregate use in concrete preparation is highlighted in the remaining sections.

1.5.1 Sustainability in Water Use

The production of concrete needs huge quantities of potable water. About 15–18 % of the total volume of structural concrete mix is water. The concrete industry uses around 1 trillion gallons of water per year worldwide (Meyer 2005). Recent scarcity of water in many parts of the world requires the sustainable use of water in concrete production. Therefore, searching for alternative sources of concrete mixing water is necessary for sustainable growth of the concrete industry. Several types of waste or non-potable water can be considered, after treatment, as mixing water and some information is available in the literature to evaluate the acceptability of water used in concrete mixing (Abrams 1942; Steinour 1960; Kuhl 1928a, b; Neville 1997; Lobo and Mullings 2003; Cebeci and Saatci 1989). The waste water generated in concrete production can be a good option and in this way concrete production units can reach complete sustainability in terms of materials use with zero discharge. The waste water generated from sewage treatment and after domestic, agricultural and industrial use can also be considered for concrete mixing if these water samples meet some specific criteria in terms of concentrations of deleterious constituents. Seawater cannot be considered as a source of concrete mixing water due to presence of large amounts of chlorides.

Standards such as European EN 1008 (2002) and American ASTM C 1602-06 (ASTM C1602) regulate the quality of concrete mixing water, i.e. impose the restriction on the amount of deleterious components in water and allow using some type of recycled water in concrete mixing. It can be stated that some alternative sources of water can be considered as mixing water after treatment; however, while searching for an alternative source, the effect of chemical contaminants present in water on the properties of concrete produced and the health effect of chemical and biological constituents during handling must be considered seriously.

The preparation of concrete using less water by innovating in concrete mixing methodology and the use of water reducing admixtures in concrete are two good options that can help to achieve sustainability in water use in concrete preparation. The addition of chemical admixtures can reduce up to 20 % of water (Cement Concrete Aggregate Australia 2010).

1.5.2 Sustainability in Cement Production

Cement is one of the major constituents of concrete and therefore huge amounts of cement are produced all over the world. According to a report (EPA 2004), the world total annual production of hydraulic cement was about 2 billion metric tonnes (Gt) and this quantity of cement was sufficient to produce about 14–18 Gt/year of concrete (including mortars), and makes concrete the most abundant of all manufactured solid materials. In a recent report (US Geological Survey 2011), it was stated that the world annual production of cement in 2010 was 3,300 million

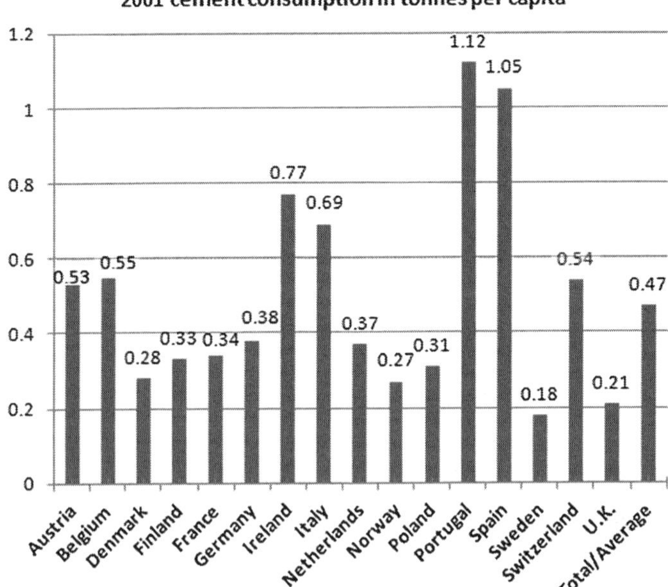

Fig. 1.2 Total annual cement consumption per capita in some European countries (Eco-Serve no date)

Table 1.3 CO_2 emissions from cement and concrete production (Wilson 1993)

	Amount of CO_2 emitted in production per		% of total CO_2
	Ton of cement (lbs)	Cubic yard of concrete (lbs)	
Source of CO_2 emission from energy use	1,410	381	60
Source of CO_2 emission due to limestone calcining	947	250	40
Total CO_2 emission	2,410	631	100

tonnes. Figure 1.2 shows the total annual consumption of cement per capita in some European countries in 2001.

The production of cement poses several sustainability issues that need to be handled properly to lessen the environmental impacts. The production of cement is a highly energy consuming process. The formation of cement clinker generally occurs at about 1,450 °C and limestone is the major source of raw material. According to Getting the Numbers Right (GNR) data for the year 2006 (CSI Report 2009), the thermal energy consumption for the production of one tonne of cement clinker was 3,690 MJ.

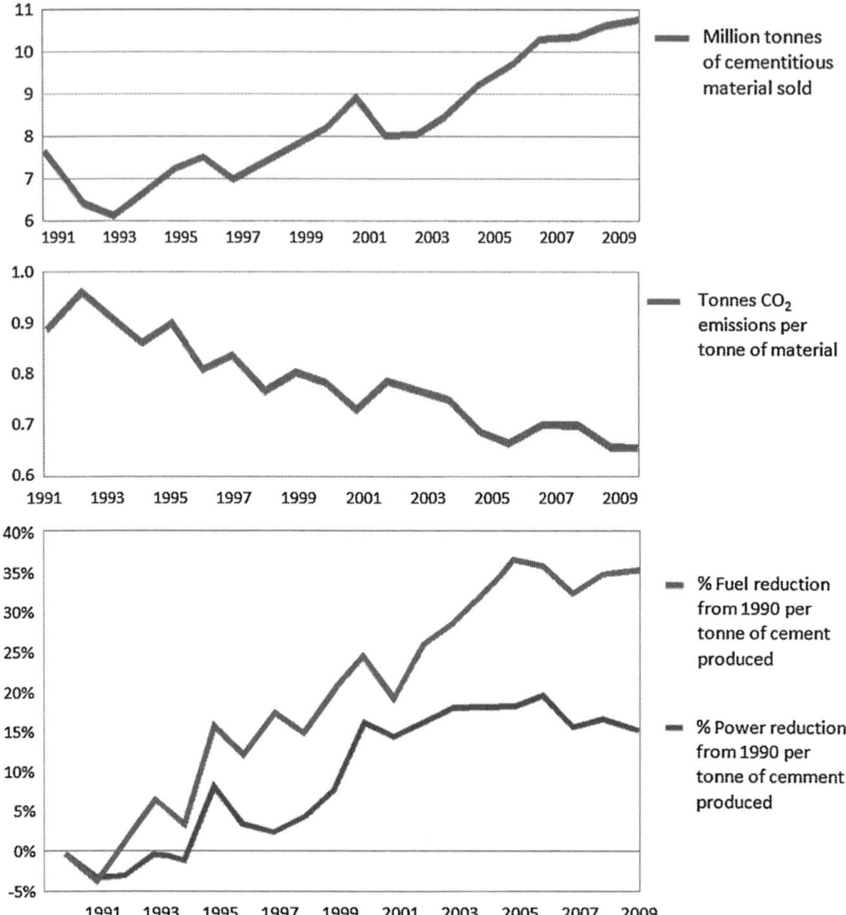

Fig. 1.3 Reduction in CO_2 emission and increase in fuel and power efficiency in Australian cement industry due to sustainability initiative (Cement Concrete Aggregate Australia, 2010)

After the power sector, the cement industry is one of the major CO_2 emitting sectors. The use of huge amounts of fuel as well as de-carbonation of limestone emits massive amounts of CO_2 and other gases into the atmosphere. It is widely accepted that the production of one tonne of cement roughly emits 1 tonne of CO_2. Table 1.3 shows the amount of CO_2 that is emitted during the production of cement and concrete. Therefore, the major focus to achieve sustainability in the cement industry is the reduction of greenhouse gases including CO_2 emissions into the atmosphere.

For a long period, cement industry has been working steadily to increase processing efficiency and decrease energy consumption due to significant consumption of energy as well as the emissions of toxic gases and particulate matters including CO_2 during cement manufacturing. These efforts also reduce the negative environmental impact of the cement industry.

For example, due to improvements in fuel efficiency as well as in power utilisation technology in Australian cement manufacturing industries, a reduction of about 23 % in emission of CO_2 per tonne of cement production was observed in 2009, in comparison to that observed in 1990, which is depicted in Fig. 1.3 (Cement Concrete Aggregate Australia 2010). However, the application of modern technology to reduce CO_2 emissions and improving the fineness of cement clinker for getting better technical properties can increase the thermal and electrical energy consumptions. Research is going on to develop nano-catalyst to reduce the clinkering temperature which will subsequently reduce the emission of CO_2 (Sobolev et al. 2006).

One recent global effort is the "The Cement Sustainability Initiative (CSI)". A total of 24 major cement producers, which account for about one-third of the world's total cement production with operations in more than 100 countries, got together to reach the goal of sustainability in cement industry. In this initiative, four points were identified to control the emission of greenhouse gases: thermal and electric energy efficiency, alternative fuels, clinker substitution, carbon capture and storage (CCS) (CSI Report 2009). Except for the last point, which is still at a demonstration stage, positive impacts of the other three points can already be seen.

Carbon capture and storage (CCS) is not yet a fully developed technology and additional research and demonstration are necessary to get benefits from this technology. However, several feasibility studies were already conducted and gave promising results. Post combustion capture techniques such as chemical absorption, membrane technology, oxy-fuel technology and carbonate-looping technology are some promising technologies that can provide solutions to control CO_2 emission. Moreover, technological, societal and economical aspects of these technologies must properly be addressed before their application.

In the next two sections, two common practices used to reduce global fuel consumption and emissions of CO_2 into the atmosphere of the cement industry, namely the use of alternative fuels and waste materials, will be briefly described.

1.5.2.1 Use of Waste Materials in Cement Kiln and Clinker Production

Several waste materials are nowadays used in cement kiln either as alternative fuel or as raw materials. According to a GNR report in 2006 (CSI Report 2009), globally 7 % of total fuel energy consumption in manufacturing of cement came from alternative fuels comprising biomass and energetic waste materials. Biomass has a great potential to be used as alternative fuel in cement kiln. Pure biomass such as animal meal, waste wood, saw dust and sewage sludge can be used to replace large amounts of fossil fuels and has the potential to reduce the emitted amounts of CO_2. Cement kilns can burn some waste materials such as used motor oil, spent solvents, printing inks, paint residue, cleaning fluids, waste textiles, papers and plastics, scrap tires, relatively more safely than an MSW incinerator because the extremely high temperatures in cement kilns result in very complete combustion with very low pollutant emissions (Knuttgen and Muench 2009; Wilson 1993). The use of waste materials in cement production can reduce the problem associated with their incineration or land-filling. Land-filling of these wastes could emit another greenhouse gas, methane, which has 21 times higher global warming potential than that posed by CO_2.

Some waste materials contain raw materials used in cement clinker manufacturing and therefore these materials can be used to replace some of the raw materials in cement clinker production too. The high temperature processing of waste in the production of cement clinker can destroy toxic organic compounds without the formation of dioxins, fix metals in the product and use mineral content as a constituent of clinker. Using some types of waste in clinker production may lower CO_2 emissions if the source of the calcium is different from $CaCO_3$. Extensive literature is available on the use of waste such as red mud, MSWI ash, steel mill scale, leather scraps and shavings, construction and demolition waste and various sludges in the production of clinker (Caponero and Tenorio 2000; Espinosa and Tenorio 2000; Galbenis and Tsimas 2006; Monshi and Asgarani 1999; Saikia et al. 2007; Trezza and Scian 2007; Vangelatos et al. 2009). The increasing production of belite-based cement can also reduce fuel requirements, which has also been an active research area for a long time (Odler 2000).

1.5.2.2 Blended Cement: Reduction of Clinker Content in Cement

A significant proportion of the total cement used in the world was of blended cement, produced by replacing a given amount of normal cement by supplementary cementing materials (SCM). According to GNR data (Geragthy no date), the global ratio of clinker to cement was 78 % in the year of 2006. Thus, about 400 million tons of clinkers in 2,400 million tons of cement produced on 2006 were replaced by other materials. Several natural materials such as natural pozzolans, ground limestone and waste materials such as blast furnace slag, coal fly ash and silica fume are extensively used as SCM in the preparation of blended cement. The impact of these materials on the properties of concrete is already well-known and reviewed extensively (Taylor 1997). The use of any type of SCM in cement production is location specific depending upon its availability. Similarly the whole amount of a waste material cannot be used in concrete preparation. For example, fly ash is produced in coal fired power plant and fly ash containing high amount of carbonaceous materials cannot be used as SCM. Fig. 1.4 shows the annual use of some SCM and gypsum by world-wide GNR participants to produce blended cements and Portland cement as reported in a CSI GNR Data (2010).

The reduction of the amount of cement used in the production of cement mortar and concrete by the use of natural and waste materials as SCM lowers the atmospheric emission of CO_2, reduces energy consumption, improves several concrete properties with increased service life and conveniently reduces the

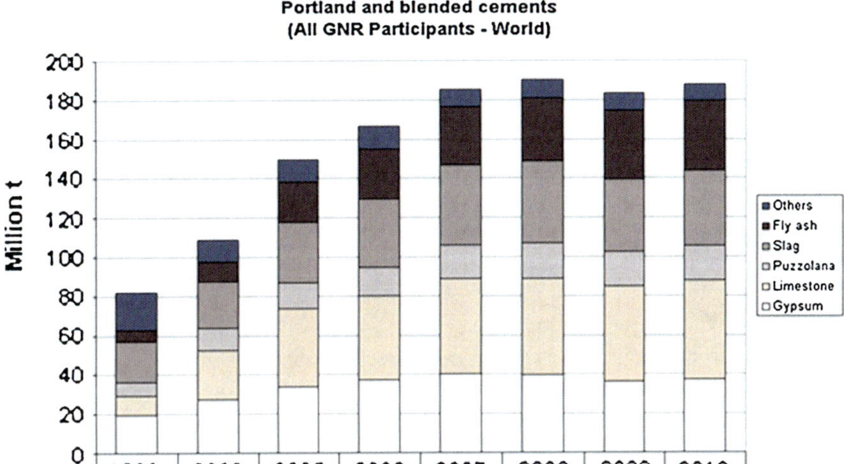

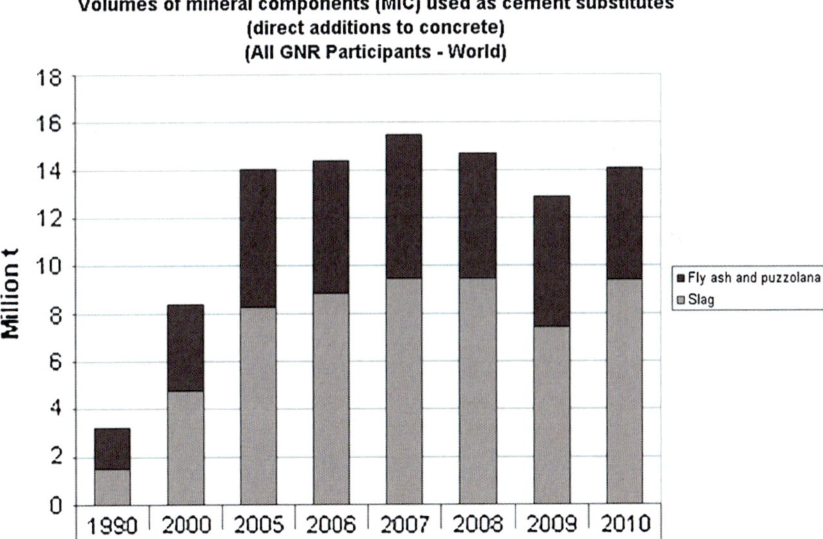

Fig. 1.4 Annual use of supplementary cement materials (MIC) in the preparation of blended cement and Portland cement by the cement industry (CSI GNR Data 2010)

problems associated with the disposal of these waste materials (Roskovic and Bjegovic 2005; Anand et al. 2006, Taylor 1997). Roskovic and Bjegovic (2005) observed around 25 and 29 % reduction in CO_2 emissions due to the substitution of 25 and 30 % of cement clinker by fly ash and slag respectively.

1.5.3 Sustainability in the Aggregate Industry

It is estimated that the global demand for aggregates used in construction is growing 4.7 % annually and in 2011 the global demand was 26.8 billion metric tons with a cost of $201 billion (Indian Concrete Journal 2008). Aggregates typically account for 70–80 % of the concrete volume and nearly for 92–96 % of asphalt pavement. Therefore, they play a substantial role in concrete properties such as workability, strength, dimensional stability and durability. Conventional concrete contains sand as fine aggregate and gravel in various sizes and shapes as coarse aggregate. Aggregates are one of the most abundantly used materials due to being major constituents of concrete.

Both fractions of aggregates (fine and coarse) are normally collected by mining. Sand and gravel are mined in two major techniques: in-stream extraction and land mining. Mined aggregates and rock are obtained by various ways such as blasting and dredging. Aggregates can be used at the size produced by nature due to weathering or after crushing larger stone. Washing, blending to grading requirements are generally done after extraction and processing of aggregates. As the fuel, labor and maintenance costs are the major expenses of the aggregate industry, aggregates are normally mined near the intended market because the cost of transportation is the major expense in this industry (Meador and Layher 1998, Ayenagbo et al. 2011).

In comparison to the environmental impact from cement production, aggregate mining or production has little impact as only simple extraction without fundamental alteration of material is necessary to obtain aggregates. However, recently, the mining of aggregates and rock is becoming an ecological problem in many parts of the world as the demand for sand and gravel is increasing rapidly due to rapid infrastructure activities all over the world. Aggregate mining is now creating ecological imbalance in several ways: damaging biodiversity of nearby areas, causing erosion in the coastal and river bank, polluting water by increasing turbidity and suspended solid mater, destroying livelihood of the peoples that rely on fishing, increasing flood, noise and dust pollutions, damaging landscape and generating waste in mining as well as in the processing sites. The mining, transportation and processing of aggregates also consume energy and therefore those processes also emit CO_2 into the atmosphere, although not so significant as that observed in cement production.

However, these problems are region specific and can be overcome by proper planning and policy implementation. The reclamation and stabilisation of pits, surface quarries and underground mines that result from aggregate mining should also be done. Another way that can make the aggregates industry more sustainable is through increased efficiency and improved technology in extraction and processing of aggregates. Reduction in dust during the processing stages is also important.

Sustainability in concrete production can be achieved by innovation in aggregate use too. The scarcity of high quality aggregates in construction sites can be solved by proper engineering of local aggregates to produce quality concrete, which will help to overcome CO_2 emission problem. The use of rejected aggregates in aggregate processing plants should also be considered in the future. The use of waste materials as aggregate in concrete is another good option to meet the sustainability goal in concrete production.

1.6 Use of Waste Materials as Aggregate in Concrete

The production of waste materials is an unavoidable stage of all industrial and human activities. This waste is now creating big environmental and economic problems all over the world. The management and treatment of industrial solid waste and municipal waste has recently been gaining importance worldwide. This waste ranges from relatively inert, e.g. glass bottles, excavated soil, construction and demolition waste, to hazardous waste with high concentrations of heavy metals and toxic organic compounds. Several discussions and initiatives were already taken to decrease the amount of waste production and its recycling/reusing. Several benefits can be achieved by recycling waste materials in other processes, such as decrease energy consumption, solve disposal problems, reduce deforestation and natural resources and reduce the health risks on human and other biotic components.

In 2004, 2006 and 2008, the total generation of waste in the 27 countries in European Union amounted to 2.68, 2.73 and 2.68 billion tonnes, respectively (Eurostat 2011). According to this statistics, each European Union citizen produced on average about 5.2 tonnes of waste in 2008 of which 196 kg were hazardous. Construction (859 million tonnes or 32.9 % of the total) and mining (727 million tonnes or 27.8 % to the total) are the major economic sectors that generated the greater part of wastes in 2008 (Fig. 1.5a). Out of the total waste generated from these two sectors, 97 % of waste was mineral waste or soils (excavated earth, road construction waste, demolition waste, dredging spoil, waste rocks, tailings etc.). The share of mineral waste and soils in relation to total waste and total hazardous waste produced was 65 % (Fig. 1.5b) and 41 %, respectively.

Substantial amounts of waste materials are also recovered, as seen in Fig. 1.6a. In the case of non-hazardous mineral waste originating mainly from construction and mining activities amounted to 754 million tonnes and represented 69 % of the total waste recovered (Fig. 1.6a). The recoveries of all types of waste also gradually increased from 2004 to 2008. For example, the recovery of mineral waste and animal and vegetable waste increased from 2004 to 2008 by 177 million tonnes or 31 % and 15 million tonnes or 30 %, respectively. The recovery rates of some waste materials over time are presented in Fig. 1.6b.

Sustainability in the construction sector can also be achieved by reuse/recycling of waste produced in several other industrial processes as raw materials or as secondary energy sources in construction material production; nowadays, cement and concrete production consumes huge amounts of these materials. Waste

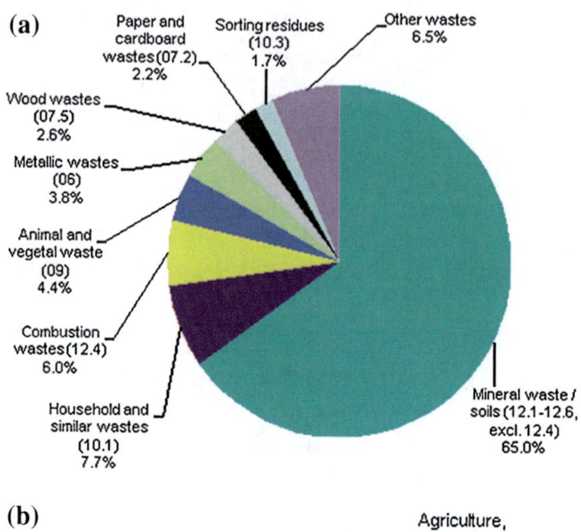

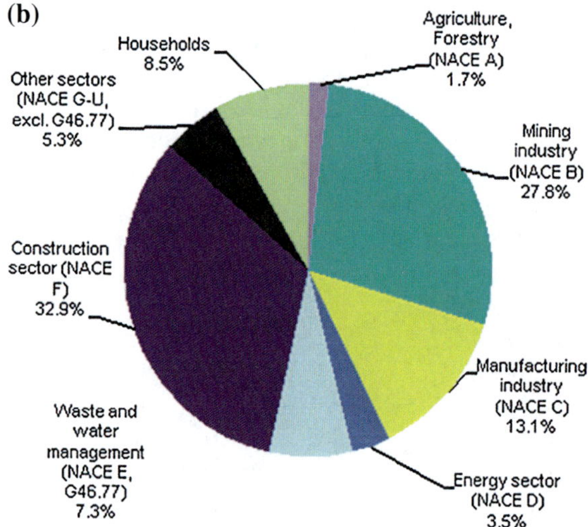

Fig. 1.5 The generation of total waste in European Union according to: **a** economic activity, **b** waste categories (Eurostat, 2011)

materials can be used as fuel in cement kilns or in brick preparation, as raw material for cement clinker/brick, as mineral additions to cement, or as granular material in cement mortar and concrete production.

Although vast amounts of waste material can be consumed in the production of cement clinker and blended cement, that consumption can be increased substantially if waste is used as aggregate in cement mortar and concrete. The use of waste material in this way can also solve problems of shortage of aggregates in construction sites, reduce environmental problems related to aggregate mining and, in some cases, reduce the cost of concrete production. The interest in using waste materials as aggregates is rapidly growing all over the world, and significant research is underway on the use of construction and demolition waste, granulated

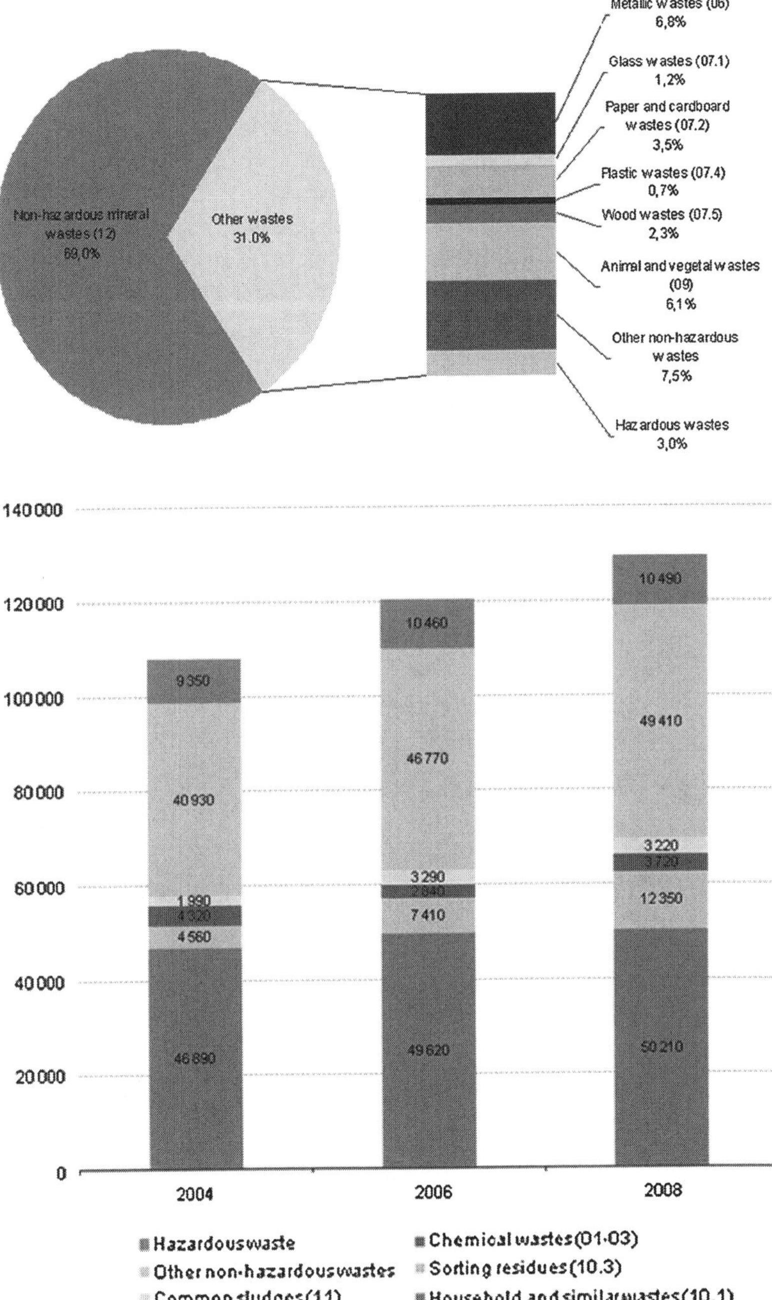

Fig. 1.6 Recovery of various waste materials in European Union in 2008 (Eurostat 2011)

coal ash, blast furnace slag, waste glass, waste plastics, rubber waste, sintered sludge pellets and other materials as replacement for traditional aggregates.

Depending on its properties, waste can be used in the cement and concrete industry without treatment or after treatment. Some types of waste have to be treated because they contain detrimental components or perform poorly. Preliminary physico-chemical and mineralogical characterisation of waste is therefore necessary before using the material in cement-based systems. Waste materials may be mixed with other materials to prepare aggregates too. Most industrial and municipal waste contains quite large amounts of toxic constituents, and therefore waste that contains contaminants should also be assessed for environmental impact as part of a proper evaluation of any waste materials that may be used as constituents in concrete. This indicates that chemical, civil, material and environmental engineering approaches are necessary before waste can be used safely and effectively in the construction industry.

However, before application of waste materials in construction, the cost factor needs to be considered seriously as it plays a vital rule particularly in the replacement of low-cost materials such as aggregates. To analyse the feasibility of a waste material as aggregate in concrete production, it is necessary to check factors such as possible environmental impact of waste if it remains unused for a long time, existence of other cost effective technology to recycle waste, comparison of cost of waste material with natural aggregates and technical feasibility of end-product. The identification of special properties inherent to recycled materials, which can show the advantages of waste material-based product over natural aggregates-based product, can be beneficial to reach the technical as well as economic viability of the use of waste materials (Meyer 2005).

1.7 Overview of the Book

The contents of this book describe comprehensively the use of several waste materials as aggregates in concrete production. The book is divided into a total of seven chapters including introduction. The second and third chapters describe the properties of industrial waste and construction and demolition waste used as aggregates respectively. The fourth and fifth chapters describe the properties of concrete containing aggregates generated from industrial waste and construction and demolition waste respectively. The sixth chapter discusses a quantitative procedure developed by the authors to estimate the properties of concrete containing construction and demolition waste as aggregates. In Chap. 7 current codes and practices developed in various countries to use construction and demolition waste as aggregates in concrete preparation are discussed.

The most promising waste that can be used as aggregate in the production of new concrete is that generated from demolition of construction materials. The waste generated from old concrete structures is one of the largest single components of solid waste and a promising material that can be used as aggregate in new concrete

Table 1.4 Produced and recycled amounts of concrete and asphalt pavement wastes in three European countries (Oikonomou 2005)

Country	Year of reporting	Type of waste	Amounts in millions metric ton	
			Produced	Used
Sweden	1999	Asphalt pavement	0.80	0.76
Denmark	1997	Demolition waste	1.5–2.0	Small quantities
		Concrete	1.06	0.90
		Asphalt pavement	0.82	0.82
		Ceramics (bricks, etc.)	0.48	0.33
Germany	1999	Asphalt pavement	12	6.0
		Other road materials	22	11
		Demolition waste	23	4.0
		Building and demolition waste	9.2	9.2

preparation. This waste is produced in the demolition of old structures and in destructions due to natural calamities. In a report in 1996, it was reported that each person of the European Union generated annually, on average, 500 kg of construction rubble and demolition waste (Oikonomou 2005). So, by considering a similar situation in other parts of world, it can be seen that a huge amount of concrete rubble and other construction waste is generated annually all over the world. Recycling of this waste should be encouraged because of reasons such as sustainable development of construction and concrete industries, protection of natural resources from further depletion and overcoming disposal problems. Therefore, recycling of construction waste as aggregate in concrete production is widely practiced in several European countries and in Japan. Table 1.4 shows generated and recycled amounts of some construction wastes in three countries in Europe.

The use of construction rubble as aggregates in new concrete production is a promising option to deal with the costs involved in the disposal of this waste, the scarcity of natural aggregates and the cost necessary for transportation of natural aggregates. However, several technical challenges need to be overcome to produce high-quality waste construction-based aggregates that can be used in all confidence to replace natural aggregates. If the economics related to recycling allow, it is also necessary to determine the possible application of this type of aggregate where the quality of the aggregates is less important. Several standards and codes of practice for using concrete rubble as aggregates in concrete preparation were developed and implemented in several countries, which is helping to achieve a much needed sustainability in the construction sector.

1.8 Conclusions

In conclusion, it can be stated that the implementation of stringent rules and regulations to overcome the impact of present development on the environment as a result of increasing global consensus on environmental issues require looking for sustainable development in various industrial sectors including the construction industry. Subsequently several steps have been initiated to minimise the impact of construction industry on the environment and to adjust on time to the changing political and societal scenario. Sustainability in construction is a complex issue that can be achieved by considering several individual factors that need to be addressed properly, some of which are briefly discussed in this chapter. Vast amounts of waste materials, an inevitable subproduct of human activities, are creating several environmental and economic problems, but are presently used or about to be used for construction purposes. The use of this waste will enhance the environmental suitability of construction industry and help in attaining the sustainability in construction sector. To use various waste materials as aggregates in concrete preparation, changes in the existing specifications and codes should be made without compromising the quality of the product.

References

Abeysundara UG, Babel S, Gheewala S (2009) A matrix in life cycle perspective for selecting sustainable materials for buildings in Sri Lanka. Build Environ 44(5):997–1004

Abrams DA (1942) Tests of impure waters for mixing concrete. Am Concr Inst J 20:442–486

Anand S, Vrat P, Dahiya RP (2006) Application of a system dynamics approach for assessment and mitigation of CO_2 emissions from the cement industry. J Environ Manage 79(4):383–398

ASTM C1602 (2004) ASTM C 1602-06, Standard specification for mixing water used in the production of hydraulic cement concrete, USA

Ayenagbo K, Kimatu JN, Gondwe J, Rongcheng W (2011) The transportation and marketing implications of sand and gravel and its environmental impact in Lome—Togo. J Econ Int Fin 3(3):125–138

Benz DP, Turpin R (2007) Potential applications of phase change materials in concrete technology. Cement Concr Compos 29(7):527–532

BIS (2010) Estimating the amount of CO_2 emissions that the construction industry can influence, supporting material for the low carbon IGT report, Autumn. http://www.bis.gov.uk/assets/biscore/business-sectors/docs/e/10-1316-estimating-co2-emissions-supporting-low-carbon-igt-report. Accessed May 2012)

Calkins M (2009) Materials for sustainable sites: a complete guide to the evaluation, selection and use of sustainable construction materials. Wiley, Hoboken

Caponero J, Tenorio JAS (2000) Laboratory testing of the use of phosphate-coating sludge in cement clinker. Resour Conserv Recycl 29(3):169–179

Cebeci OZ, Saatci AM (1989) Domestic sewage as mixing water in concrete. ACI Mater J 86(5):503–506

Cement Concrete Aggregate Australia (2010) Sustainable concrete materials, cement concrete and aggregates, Australia, pp 1–8. www.ccaa.com.au/sustainability/document2.pdf. Accessed May 2012

Cement Concrete Aggregate Australia (no date) Concrete—The responsible choice, cement concrete and aggregates, Australia, www.ccaa.com.au/sustainability/document1.pdf. Accessed on May 2012

China construction industry (no date) China construction industry. http://www.chinaorbit.com/china-economy/china-industry-sectors/china-construction.html. Accessed May 2012

Construction Industry in India (2008) Overview of the construction industry in India, The Indo-Italian Chambers of Commerce and Industry, Mumbai, India. http://www.centroesteroveneto.com/pdf/

Osservatorio %20Mercati/India/Ricerche %20di %20Mercato/2009/Construction %20Sector.pdf. Accessed Apr 2012

CSI GNR Data (2010) Cement sustainability initiative, getting the numbers right database (online). http://www.wbcsdcement.org/GNR-2010/index.html. Accessed May 2012

CSI Report (2009) Development of state of the art—techniques in cement manufacturing: trying to look ahead, cement sustainability initiative, CSI/ECRA-technology papers, Duesseldorf, Geneva

Dry CM (2000) Three designs for the internal release of sealants, adhesives, and waterproof chemicals into concrete to reduce permeability. Cem Concr Res 30(12):1969–1977

Eco-Serve (no date) Environmental impacts—sustainability, European construction in service of society. http://www.eco-serve.net/publish/cat_index_74.shtml. Accessed May 2012

EN 1008 (2002) European standard, EN 1008: mixing water for concrete—specification for sampling, testing and assessing the suitability of water, including water recovered from processes in the concrete industry, as mixing water for concrete, European Union

EPA (2004) Overview of Portland cement and concrete, EPA report, environmental protection agency, USA. http://www.epa.gov/osw/conserve/tools/cpg/pdf/rtc/app-a.pdf. Accessed May 2012

Espinosa DCR, Tenorio JAS (2000) Laboratory study of galvanic sludge's influence on the clinkerization process. Resour Conserv Recycl 31(1):71–82

Eurostat (2011) Waste statistics, European commission. http://epp.eurostat.ec.europa.eu/ statistics_explained/index.php/Waste_statistics. Accessed May 2012

Galbenis CT, Tsimas S (2006) Use of construction and demolition waste as raw materials in cement clinker production. China Particuol 4(2):83–85

Geragthy E (no date) Getting the numbers right. A database for the cement industry, cement sustainability initiative. http://www.wbcsdcement.org/pdf/houston/breakout_gnr.pdf. Accessed May 2012

González MJ, Navarro JG (2006) Assessment of the decrease of CO_2 emissions in the construction field through the selection of materials: practical case study of three houses of low environmental impact. Build Environ 41(7):902–909

Habert G, Roussel N (2009) Study of two concrete mix-design strategies to reach carbon mitigation objectives. Cem Concr Compos 31(6):397–402

Indian Concrete Journal (2008) Forecast for construction materials—cement and aggregates (News). Indian Concr J July. http://www.icjonline.com/news_july2008.htm. Accessed May 2012

Joseph P, Tretsiakova-McNally S (2010) Sustainable non-metallic building materials. Sustainability 2(2):400–427. doi:10.3390/su2020400, ISSN 2071-1050

Khokhar MIA, Roziere E, Turcry P, Grondin F, Loukili A (2010) Mix design of concrete with high content of mineral additions: optimisation to improve early age strength. Cement Concr Compos 32(5):377–385

Knuttgen R, Muench S (2009) Cement production—example operation—Lafarge Seattle. http://www.pavementinteractive.org/article/cement-production. Accessed May 2012

Koukkari H, Kuhnhenne M, Braganca L (2007) Energy in the sustainable European construction sector. In: sustainability of constructions—integrated approach to life-time structural engineering, 1st workshop, "sustainability of constructions", COST action C25, Lisbon, pp 0.23-0-34

Kuhl H (1928a) The chemistry of high early strength cement. Concrete (Cement Mill Edition) 33(1):109–111

Kuhl H (1928b) The chemistry of high early strength cement. Concrete (Cement Mill Edition) 33(2): 103–105

Lobo C, Mullings GM (2003) Recycled water in ready mixed concrete operations. Concr Focus 2(1):17–26, Spring

Meador MR, Layher AO (1998) In stream sand and gravel mining: environmental issues and regulatory process in the United States. Fisheries 23(11):6–13

Meyer C (2005) Concrete as a green building material, third international conference on construction materials, CONMAT'05, Vancouver

Meyer C (2009) The greening of the concrete industry. Cement Concr Compos 31(8):601–605

Monshi A, Asgarani MK (1999) Producing Portland cement from iron and steel slags and limestone. Cem Concr Res 29(9):1373–1377

Mukhopadhyay AK (2011) Next-generation nano-based concrete construction products: a review. In: Gopalakrishnan K, Birgisson B, Taylor P, Attoh-Okine N (eds) Nano-technology in civil infrastructure: a paradigm shift. Springer, Berlin, pp 207–223

Neville AM (1997) Properties of concrete, Fourth edn. Longman, England, p 734

Odler I (2000) Special inorganic cements. Modern concrete technology series, No. 8, E & FN SPON, Taylor & Francis Group, London

OECD (2008) Competition in the construction industry, policy roundtable: construction industry, organization for economic co-operation and development, DAF/COMP(2008)36. http://www.oecd.org/dataoecd/32/55/41765075.pdf. Accessed May 2012

Oikonomou ND (2005) Recycled concrete aggregates. Cem Concr Aggreg 27(2):315–318

Rio Summit (1992) Rio declaration on environment and development, United Nations conference on environment and development, Rio de Janeiro, Brazil. http://www.un-documents.net/rio-dec.htm. Accessed May 2012

Roskovic R, Bjegovic D (2005) Role of mineral additions in reducing CO_2 emission. Cem Concr Res 35(5):974–978

Saikia N, Kato S, Kojima T (2007) Production of cement clinkers from municipal solid waste incineration (MSWI) fly ash. Waste Manag 27(9):1178–1189

Sobolev K, Vivian IF, Ferrada-Gutiérrez M (2006) Nanotechnology of concrete, Universidad Autonoma de Nuevo Leon, Mexico. http://www.voyle.net/2006 %20Research/ Research-06-046.htm. Accessed May 2012

Steinour HH (1960) Concrete mix water—how impure can it be? Portland Cem Assoc J Res Dev Lab 3(3):32–50

Taylor HFW (1997) Cement chemistry, 2nd edn. Thomas Telford Publishing, London

Trezza MA, Scian AN (2007) Waste with chrome in the Portland cement clinker production. J Hazard Mater 147(1–2):188–196

US Geological Survey (2011) USGS mineral program cement report, United States Geological Survey

UN report (1987) Our common future, Chapter 2: towards sustainable development, report of the world commission on environment and development. http://www.un-documents.net/ocf-02.htm. Accessed May 2012

Vangelatos I, Angelopoulos GN, Boufounos D (2009) Utilization of ferroalumina as raw material in the production of ordinary Portland cement. J Hazard Mater 168(1):473–478

Wilson A (1993) Cement and concrete: environmental considerations. Environ Build News 2(2). http://www.buildinggreen.com/ auth/article.cfm/1993/3/1/Cement-and-Concrete-Environmental-Considerations. Accessed May 2012

World Summit (2002) Johannesburg declaration on sustainable development, world summit on sustainable development, Johannesburg, South Africa. http://www.un-documents.net/jburgdec.htm. Accessed May 2012

Chapter 2
Industrial Waste Aggregates

2.1 Introduction

The aggregates typically account for 70–80 % of the concrete volume and play a substantial role in different concrete properties such as workability, strength, dimensional stability and durability. Conventional concrete consists of sand as fine aggregate and gravel, limestone or granite in various sizes and shapes as coarse aggregate. There is a growing interest in using waste materials as alternative aggregate materials and significant research is made on the use of many different materials as aggregate substitutes such as coal ash, blast furnace slag, fibre glass waste materials, waste plastics, rubber waste, sintered sludge pellets and others. The consumption of waste materials can be increased manifold if these are used as aggregate into cement mortar and concrete. This type of use of a waste material can solve problems of lack of aggregate in various construction sites and reduce environmental problems related to aggregate mining and waste disposal. The use of waste aggregates can also reduce the cost of the concrete production. As the aggregates can significantly control the properties of concrete, the properties of the aggregates have a great importance. Therefore a thorough evaluation is necessary before using any waste material as aggregate in concrete. Significant work has been done on the use of several types of waste materials as an aggregate in preparation of cement mortar and concrete. In this section, various properties of some waste materials used as aggregate will be presented.

2.2 Types of Industrial Waste Aggregates

The properties of waste aggregates that will be highlighted in this section are: 1. Plastics wastes; 2. Coal ash; 3. Rubber tyre; 4. Slags; 5. Waste from food and agricultural industries; 6. Pulp and paper mill waste; 7. Leather waste; 8. Industrial sludge; 9. Mining industry waste.

J. de Brito and N. Saikia, *Recycled Aggregate in Concrete*,
Green Energy and Technology, DOI: 10.1007/978-1-4471-4540-0_2,

Depending on their generation, wastes can be separated into two types: those that directly result from industry as industrial by-products and those that can be named recycled wastes. The first type includes coal ash, various slags from metal industries, industrial sludge, waste from industries like pulp and paper mills, mine tailings, food and agriculture, and leather. The second type includes different plastic and rubber wastes.

A broad classification of industrial waste aggregate can be made depending on the chemical nature of wastes. Some waste aggregates come from production and use of organic materials. Plastics, rubber, leather and some food industries wastes are organic wastes. On the other hand, industrial slags, mining wastes, coal industry wastes and others are inorganic wastes. Glass reinforced plastics and some industrial sludge may contain both organic and inorganic materials.

Another classification of industrial waste aggregate can be done depending on the weight of waste aggregates. Some aggregates are lightweight by nature. Plastics, rubber, most food and agricultural industries wastes and coal bottom ash are of this kind. On the other hand, most of the industrial slags are heavier than conventional aggregates.

2.3 Coal Ash as an Aggregate in Concrete

Burning of coal generates two types of waste materials: fly ash and bottom ash. There are two types of bottom ashes, wet bottom boiler slag and dry bottom ash depending on both the boiler type and its design.

Coal fly ash, also known as pulverised fuel ash, is the finest fraction of these ashes, which are released from combustion chamber and transported by flue gases. Fly ash contains the non-combustible matter in coal along with a small amount of carbon that remains from incomplete coal combustion. Fly ash consists mostly of silt-sized and clay-sized glassy spheres. When pulverised coal is burned in a dry bottom boiler about 80 per cent of the unburned material or ash is entrained in the flue gas and is captured and recovered as fly ash. The remaining 20 % ash that is collected from the bottom of furnaces is called coal bottom ash (CBA), which is a coarse, incombustible by-product with a grain size similar to that of fine and coarse sized natural aggregates. Bottom ash is produced as a granular material and removed from the bottom of dry boilers.

Boiler slag, a coarse grained product, is produced from two types of wet bottom boilers, slag-tap and cyclone boilers. The slag-tap boiler burns pulverised coal while the cyclone boiler burns crushed coal. Both boiler types have a solid base with an orifice that can be opened to allow molten ash to flow into a hopper, which contains quenching water. When the molten slag comes in contact with the quenching water, the ash fractures instantly, crystallises, and forms pellets. High-pressure water jets wash the boiler slag from the hopper into a sluiceway, which then transmits the ash to collection basins for dewatering and further processing. Boiler slag is a coarse, angular, glassy, black material. When pulverised coal is burned in a slag-tap furnace, as much as 50 % of the ash is retained in the furnace as boiler slag. In a cyclone furnace, which burns crushed coal, 70–85 % of the ash is retained as boiler slag.

Properties of coal ash depend on coal type, pulverising system, combustion conditions, temperature, type of furnace, minerals in coals and milling system. Though a significant number of references are available on the properties and use of fly ash as a mineral addition in normal Portland cement, not much literature exists on the use of fly ash, coal bottom ash (CBA) and boiler slag as a granular additive into concrete. Again compared to CBA, very little work has been done on the use of other two ashes as aggregate in concrete. The properties of these ashes will be discussed separately.

2.3.1 Bottom Ash

2.3.1.1 Dry Bottom Ash

Some of physical properties of ash generated from dry bottom boiler (henceforth called coal bottom ash, CBA) as aggregate are presented in Table 2.1. These properties depend on the burning efficiency, the method by which the CBA is obtained and the type of combustion and thus the physical properties of bottom ash greatly vary in reported works. The density parameters of CBA are considerably lower than those of natural sand ($\sim$2.6 g/cm^3) and therefore CBA can be used as lightweight aggregate in concrete. CBA with relatively low density or specific gravity is often indicative of the presence of porous particles. Bottom ash with relatively high specific gravity (above 3.0) may indicate the presence of high amounts of iron. The bottom ash is a porous material and generally has high water absorption capacity. However, variation in water absorption capacity in different CBA is quite large, with a range of 2–32 %. The moisture content of bottom ash used by Andrade et al. (2009) as a fine aggregate in concrete is about 55 %. On the other hand, CBA is more brittle than natural sand and has a greater resemblance to cement clinker (Rogbeck and Knutz 1999).

The porosity and void content of CBA are generally higher than in natural aggregate and thus CBA can accommodate a high amount of water in a concrete mix. Therefore there is some difficulty in determining the exact water/cement ratio of concrete mixes containing CBA. However, the ability to incorporate high amount of water by CBA can be used as a reservoir of water for future hydration of cement. This behaviour, commonly known as internal curing is particularly useful for high strength concrete, where less water is used to make concrete. In this type of concrete, the hydration process leads to shortage of water in the cement paste. At this stage, the water content in CBA can promote a supply of water internally to the concrete for continuous hydration. These hydration products fill the pores or micro-cracks and improve concrete properties.

Table 2.1 Physical properties of coal bottom ash reported as aggregate in concrete

Reference	Properties						
	Unit weight, kg/m^3	Specific gravity/ density, kg/m^3	Water absorption (%)	Fineness modulus	Moisture content (%)	Maximum diameter (mm)	Porosity #Void content (%)
Kou and Poon (2009)		2,190 (SSD)	28.9 (1 h)	1.83			
Andrade et al. (2007)		1.674 (sp. gr.)		1.55	55		
Andrade et al. (2007)		1.67 (sp. gr.)		1.60		2.4	
Bai et al. (2005)		1.50 (SSD, sp. gr.)	30.4 (1 h)				
Bai et al. (2005)		1.58	32.2 (1 h)			5.0	
Kim and Lee (2011)		1.77 (OD) 1.87 (SSD)	5.45 (ASTM)	2.36		2.4	10.19[a]
Kim and Lee (2011)		1.64 (OD) 1.77 (SSD)	8.14 (ASTM)			13.2	13.34[a]
Yuksel et al.	620 (loose) 660 (dense)	1.39	6.10 (TS)			4.0	
Ghafoori and Bucholc (1996)		2.33 (OD) 2.47 (SSD)	7.0	2.80			
Lee et al. (2010)	1,268	1.84 (OD) 1.90 (SSD)		1.59		2	#33.3
	988	1.98 (OD) 2.00 (SSD)		3.81		8	#50.4
	1,040	1.78(OD) 1.90 (SSD)				11	#45.1
Park et al. (2009)	1,275	2,410	2.43			13	
Kou and Poon (2009)	1,270	2,480	2.11			20	

SSD saturated surface-dry; *sp. gr.* specific gravity; *TS* Turkish standard, *TS* EN-12620; [a] water accessible porosity

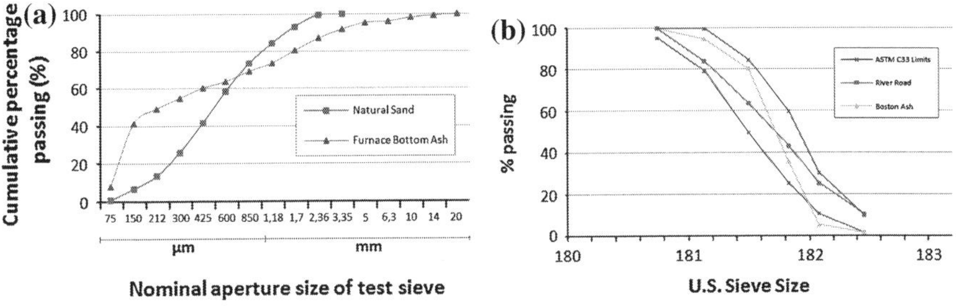

Fig. 2.1 Particle size distribution of a CBA and of natural sand **a** Bai et al. (2005); **b** Ghafoori and Bucholc (1996)

The particle size of CBA also depends on the factors indicated earlier. In literature, the use of CBA as fine or coarse aggregate is reported (Table 2.1). The CBA used in Ghafoori and Bucholc (1996) study was an well-graded fine aggregate with a fineness modulus (FM) of 2.80 (ASTM's recommendation of FM range for fine aggregate is 2.3–3.1). Figure 2.1 shows the particle size distribution of two typical CBA used as partial substitution of fine aggregate in concrete.

Compared to natural sand aggregate, which is dense, normally smooth in texture, and round in shape. The particles of CBA aggregate are porous and angular in shape. It has rough surface texture, large numbers of micro-pores (circular holes with a diameter in a range of 0.5–5 μm) and internal pores. Some spherical shaped fly ash particles having a few micrometer diameters can be deposited on the surfaces of CBA (Kim and Lee 2011). The shape and porous structure of CBA makes it necessary to use a high amount of water during preparation of the concrete mix. A typical pore-size distribution curve of bottom ash (presented in Fig. 2.2) indicates that the nanostructure of bottom ash is quite dense (Kim and Lee 2011).

The chemical properties of CBA are normally controlled by the properties of coal (its origin). Table 2.2 shows oxide composition of some CBA samples. CBA is composed primarily of silica (SiO_2), ferric oxide (Fe_2O_3) and alumina (Al_2O_3), with smaller quantities of calcium oxide (CaO), potassium oxide (K_2O), sodium oxide (Na_2O), magnesium oxide (MgO), titanium oxide (TiO_2), phosphorous pentoxide (P_2O_5) and sulphur trioxide (SO_3).

The minerals identified in different CBA used as aggregate in concrete are presented in Table 2.2. Figure 2.3 shows the X-ray diffraction (XRD) pattern of typical low calcium CBA. The major minerals found in CBA are quartz, mullite and a non-crystalline glassy phase. In some CBA, iron containing minerals like hematite and magnetite may also be present. The fused and glassy texture of CBA normally would make an ideal substitute for the aggregate fraction of concrete.

Unlike its companion—pulverised fuel ash (PFA) or coal fly ash, CBA usually has low pozzolanicity, which makes it unsuitable to be used as a mineral addition in cement. It may contain higher concentrations of unburned carbon. Some power plants recover coal mill rejects with bottom ash and therefore CBA may contain pyrites that come from mill rejects, which can cause expansion in concrete.

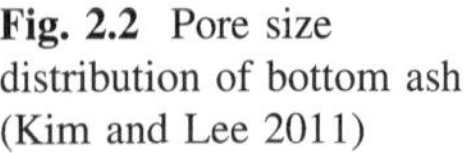

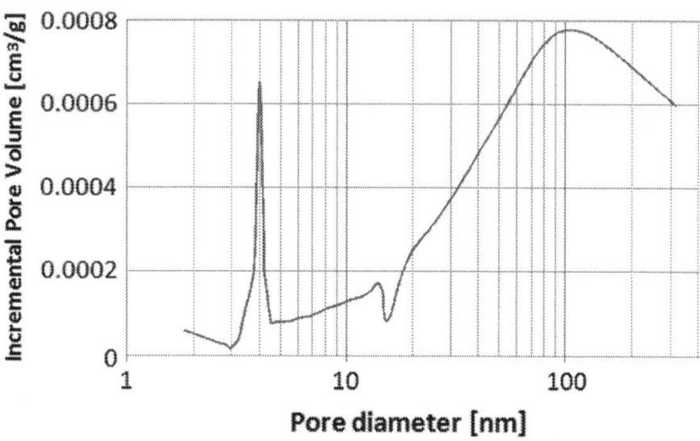

Fig. 2.2 Pore size distribution of bottom ash (Kim and Lee 2011)

2.3.1.2 Boiler Slag or Wet Bottom Ash

Boiler slag is a vitrified material, which is a very durable and environmentally stable form that permanently immobilises its chemical constituents into the glassy amorphous structure. Boiler slag is made of porous, glassy, angular, uniform sized smooth granular particles. The quenched slag becomes somewhat vesicular or porous if gases are trapped in the slag. Boiler slag generated from burning of lignite or sub-bituminous coal tends to be more porous than that of bituminous coals (Lovell and Te-Chih 1992).

The boiler slag primarily comprises particles, which can be regarded as single-sized coarse to fine sand with 90–100 % passing a 4.75 mm mesh sieve, 40–60 % passing a 2.0 mm mesh, 10 % or less passing a 0.42 mm mesh and 5 % or less passing a 0.075 mm mesh (Majizadeh et al. 1979). Boiler slag is black in colour, hard, and durable with a resistance to surface wear.

Boiler slag typically contains 40–60 % SiO_2, 18–38 % Al_2O_3, 2–7 % Fe_2O_3, 1–4 % CaO, 0.5–3.0 % MgO and 0.5–2.0 % TiO_2. The chemical composition of boiler slag is also governed by the coal source. Boiler slag exhibits less abrasion and soundness loss than bottom ash as a result of its glassy surface texture and lower porosity. The predominate minerals present in boiler slags are mullite, quartz, calcium silicate and quicklime. The specific gravity of boiler slag usually ranges from 2.3 to 2.9. The dry unit weight of boiler slag usually ranges from 960 to 1,440 kg/m^3. Occasionally, the dry unit weight of boiler slag may reach 1,760 kg/m^3.

Deleterious materials, such as soluble sulphates or coal pyrites, should be removed from the bottom ash and boiler slag before attempting to use these materials as an aggregate. Pyrites can be removed from the coal before it is burned using sink-float techniques, or from the bottom ash or boiler slag using magnetic separation. Due to salt content (soluble chlorides and sulphates) and low pH and electrical resistivity, bottom ash and boiler slag may be potentially corrosive and therefore evaluation of the corrosive nature of the bottom ash being used should be investigated. Corrosivity indicator tests normally used to evaluate bottom ash are pH, electrical resistivity, soluble chloride content and soluble sulphate content.

Table 2.2 Chemical and physical characteristics of bottom ash

Content (%)	Andrade et al. (2007)	Bai et al. (2005)	Cheriaf et al. (1999)	Kou and Poon (2009)	Lee et al. (2010)	Ozkan et al. (2007)	Kurama et al. (2009)	Park et al. (2009)	Ghafoori and Bucholc (1996)
Country	Brazil	UK	French	China	SKorea	Turkey	Turkey	SKorea	USA
Origin	nr	nr	Low-grade coal	nr	nr	nr	nr	nr	Lignite
Chemical analysis									
SiO_2	50.46	61.80	56.00	60.7	44.2, 48.0	59.53	54.5	47.9	41.70
Al_2O_3	28.35	17.80	26.70	18.3	31.5, 31.3	20.12	15.4	25.94	17.10
Fe_2O_3	10.69	6.97	5.80	6.56	8.87, 8.26	13.08	11.16	4.76	6.63
Na_2O	nr	0.95	0.20	0.89	nr	–	nr	1.38	1.38
K_2O	3.81	2.00	2.60	2.12	4.04, 4.14	0.06	1.34	0.67	0.40
CaO	2.07	3.19	0.80	3.25	2.00, 1.41	2.02	4.69	2.48	22.50
TiO_2	1.57	0.88	1.30	0.95	2.38, 2.29	nr	nr	0.86	nr
SO_3	0.34	0.79	0.1 (as S)	0.82	1.3, 1.2	Trace	1.30	1.03	0.42
ZrO_2	0.18	nr	nr	nr	nr	nr	nr	nr	nr
V_2O_5	0.09	nr	nr	nr	0.07, 0.09	nr	nr	nr	nr
MnO	0.07	nr	nr	nr	0.07, 0.06	nr	nr	nr	nr
ZnO	0.03	nr	nr	nr	nr	nr	nr	nr	nr
Y_2O_3	0.03	nr	nr	nr	nr	nr	nr	nr	nr
SrO	0.03	nr	nr	nr	nr	nr	nr	nr	nr
MgO	nr	1.34	0.60	1.28	2.6, 2.0	3.20	4.26	1.10	4.91
Cr_2O_3	nr	nr	nr	nr	0.03, 0.05	nr	nr	nr	nr
P_2O_5	nr	0.20	nr	nr	0.3, 0.33	nr	nr	nr	nr
Others	nr	0.49	nr	1.00	0.32, 0.40	nr	nr	nr	3.83
Cl	nr	nr	nr	nr	2.32, 0.47	nr	nr	0.18	nr
Loss on	ignition	2.30	3.61	4.60	4.13	nr	9.81	8.90	3.63
1.13									
Minerals									
nr	Quartz, mullite, glassy phase	nr	Mullite, quartz, glassy phase	nr	nr	nr	Quartz, magnetite, hematite, glassy phase	nr	nr

nr not reported

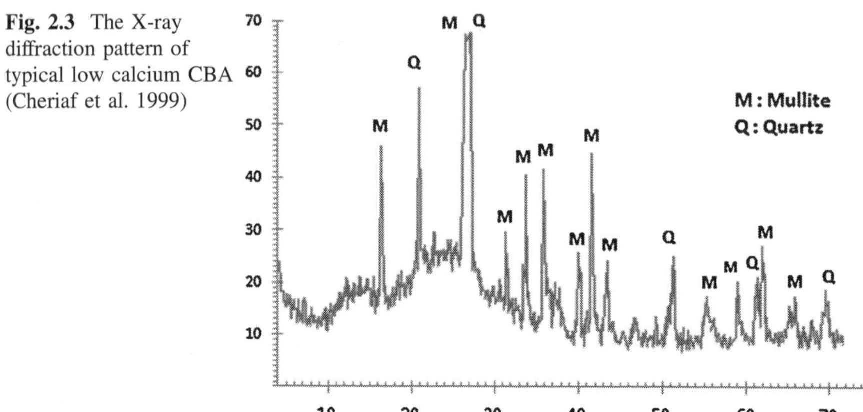

Fig. 2.3 The X-ray diffraction pattern of typical low calcium CBA (Cheriaf et al. 1999)

2.3.1.3 Fly Ash

The use of fly ash (FA) as partial replacement of normal Portland cement in concrete is very common nowadays and around 15–25 % of cement is generally replaced by FA in normal structural concrete mixes. However, the overall percentage utilisation remains very low in many countries, and most of the fly ash is dumped at landfills (Siddique 2003a, b; Ravina 1997). Much higher quantities of FA can be used in concrete if fly ash can partially replace the fine sand fraction in concrete mix. This replacement can be made by low quality fly ash too, which has low pozzolanic properties. Although many references are available on the use of fly ash as a supplementary cementing material in concrete, the number of available references on the use of fly ash as partial replacement of fine aggregates is not very large.

American Society for Testing and Materials (ASTM) C618-03 (Standard specification for coal fly ash and raw or calcinated natural pozzolan for use in concrete), classifies fly ash into two categories—Class F (low calcium) and Class C (high calcium) fly ash. Combustion of bituminous or anthracite coal normally produces Class F (low calcium) fly ash and combustion of lignite or sub-bituminous coal normally produces Class C (high calcium) fly ash.

The chemical composition of fly ash used as aggregate in concrete is presented in Table 2.3. The chemical composition of fly ash depends on the type of coal. More references are available on the use of low calcium fly ash as a replacement of fine aggregate in concrete than high calcium fly ash. High calcium fly ash contains large amounts of free lime and sulphite than that of low calcium fly ash. Due to the presence of undesirable chemical components in high calcium fly ash, the use of this material is much limited compared to low calcium fly ash.

The mineralogical composition of fly ash is very complex. Mineralogically, fly ashes are a heterogeneous mixture of mineral phases and amorphous glassy phases with small amount of unburned carbon. The glassy phase of low calcium fly ash is alumino-silicate type whereas that of high calcium fly ash is a mixture of calcium aluminate and ferrous alumino-silicate (Das and Yudhbir 2006).

Table 2.3 Chemical composition of some fly ashes used as a replacement of sand fraction in concrete

Constituents/ properties	Siddique (2003a , b)	Maslehuddin et al. (1989)	Rajamane et al. (2007)	Pofale and Deo (2010)	Papadakis (1999)	Papadakis (2000)
SiO_2	55.3	60.5	59	55.5	53.50	39.21
Al_2O_3	25.7	23.0	nr	31.3	20.40	16.22
Fe_2O_3	5.3	7.5	nr	6.4	8.66	6.58
CaO	5.6	2.0	1.02	1.02	3.38 (free: 0.36)	22.78 (free: 5.18)
MgO	2.1	1.0	0.30	0.21	–	–
TiO_2	1.3	nr		2.70	–	–
SO_3	1.4	0.3	–	0.44	0.63	4.30
Na_2O	0.4	nr	0.54 (total alkalis)	–	–	
K_2O	0.6	nr	0.25	–	–	
LOI	1.9	1.4	1.08	0.74	2.20	2.10
Moisture content	0.3	nr	nr	Nr	–	–
$SiO_2 + Al_2O_3 + Fe_2O_3$	86.3	91	–	93.2	82.16	62.01

Depending on the fly ash type, the mineral matter present in fly ash varies significantly. More than 188 minerals have been identified in fly ash, most of them trace minerals. High calcium fly ash contains large amounts of calcium-bearing minerals like lime, anhydrite, gypsum, tricalcium aluminate, alite, gehlenite, akermanite, portlandite and larnite. Some other minerals like quartz, hematite and magnetite are also present in high calcium fly ash. On the other hand, low calcium fly ash mainly contains quartz, mullite, hematite, magnetite and small amounts of calcite.

Silicon and aluminium are mainly present in a glassy phase, with small amounts of quartz and mullite included. Iron appears partly as oxides (magnetite and hematite), with the rest in a glassy phase.

The specific gravity of fly ash may vary from 1.3 to 4.8 (Joshi and Lohtia 1997). FA mainly consists of clay and silt-sized particles (particle diameter <45 micrometer), which are generally spherical in shape. There is a wide variation in the particle size distribution of fly ashes irrespective of the type of fly ash. The particle size distribution of fly ash mainly depends upon the initial grinding of the coal and the efficiency of the thermal power plant and even fluctuations in power generation (Lee et al. 2010). The shape of the FA may vary depending on the various physical and chemical factors. The glassy phase is mainly spherical in shape. However, large sized or irregular shaped particles can also be formed from the fusion of smaller fragments and incomplete melting. The colour of coal ash depends on its chemical and mineral constituents. High lime containing FA is normally tan and light in colour. Iron containing FA is brownish in colour. The dark grey to black colour of FA indicates the presence of a high amount of unburned carbon.

2.4 Industrial Slag

Slag is a partially vitreous by-product of smelting ore due to separating of the metal fraction from the worthless fraction. It can be considered a mixture of metal oxides; however, slags can contain metal sulphides and metal atoms in the elemental form.

2.4.1 Ferrous Slag

Ferrous slag is produced during the production of iron using blast furnace (blast furnace slag) as well as in the separation of the molten steel from impurities in steel-making furnaces (steel slag) (Fig. 2.4).

2.4.1.1 Steel Slag

Steel slag is produced during the separation of molten steel from impurities in steel furnaces. The slag occurs as a molten liquid and is a complex solution of silicates and oxides that solidifies upon cooling. There are several different types of steel slag produced during the steel-making process out of which basic oxygen furnace steel slag (BOF slag), electric arc furnace slag (EAF-slag) and ladle furnace slag (LDF-slag) or refining slag are important.

An electric arc furnace produces steel by melting recycled steel scrap, using heat generated by an arc, created by a large electric current. The slag is formed through the addition of lime, which is designed to remove impurities from within the steel. Slag has a lower density than steel and therefore floats on top of the molten bath of steel.

In the basic oxygen process, hot liquid blast furnace metal, scrap and fluxes, which consist of lime (CaO) and dolomitic lime (CaO.MgO), are charged into the furnace. The oxygen, injected into the furnace, combines with and removes the impurities in the charge. These impurities consist of carbon as gaseous carbon monoxide, and silicon, manganese, phosphorus and some iron as liquid oxides, which combine with lime and dolomitic lime to form the BOF steel slag.

After being tapped from the furnace, molten steel is transferred in a ladle for further refining to remove additional impurities still contained within the steel. This operation is called ladle refining because it is completed within the transfer ladle. During ladle refining, additional steel slags are generated by again adding fluxes to the ladle to melt.

Steel mill scale is produced during processing of iron in steel mills. During the processing of steel in steel mill, iron oxides, known as mill scale are formed on the surface of the metal during the continuous casting, reheating and hot rolling operations. The steel mill scale is removed by water sprays. The steel mill scale is somewhat similar to steel slag and therefore, like steel slag, it can be used in concrete production.

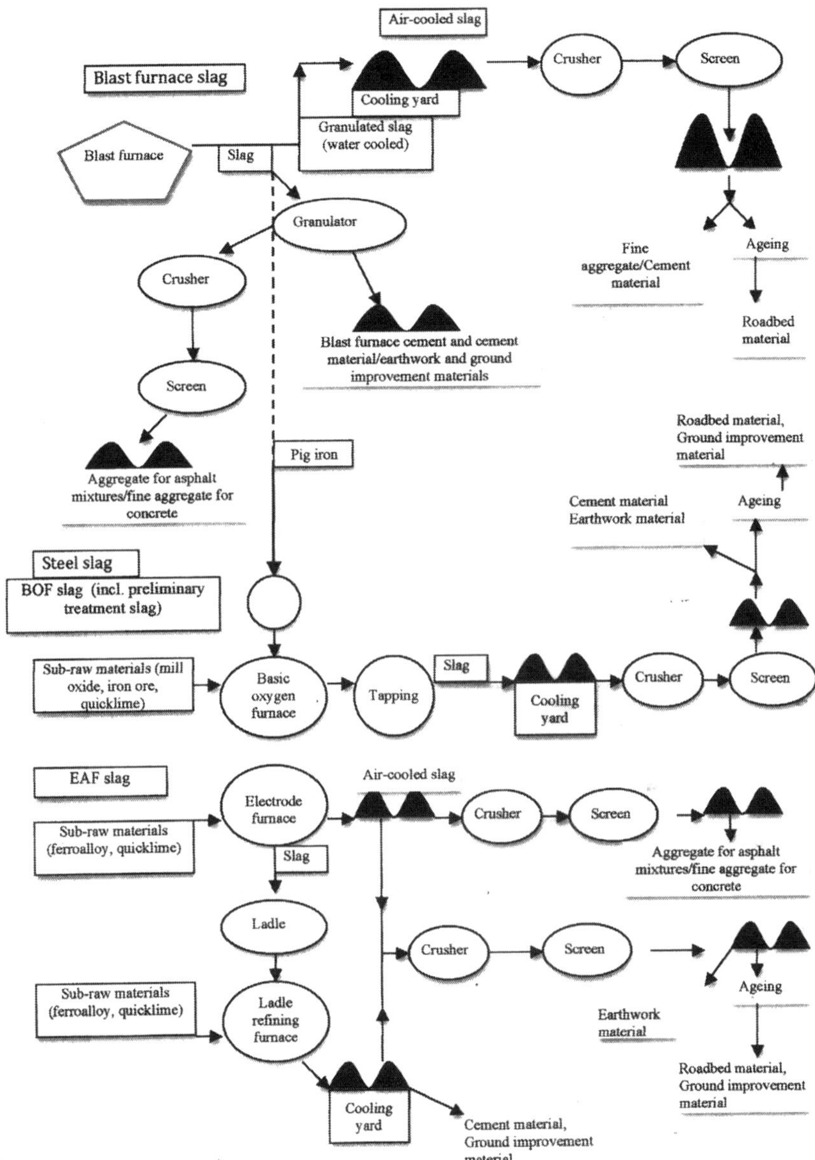

Fig. 2.4 Flowchart of production of ferrous slag (*courtesy* Nippon Slag Association)

The use of steel slag in the preparation of blended cement, or cement clinker, and as aggregate or hydraulic binder in road construction has been proposed by several authors for a long time (Mahieux et al. 2009; Monshi and Asgarani 1999; Shi and Qian 2000). However, not much is reported on the use of steel slag as aggregate in preparation of cement mortar and concrete. Compared to the use of

BOF-slag as aggregate in concrete, more references are available on EAF-slag. On the other hand, LDF-slag is mainly used as a mineral admixture in cement due to its particle size and possibly its mineralogy.

Steel slag is a crushed product having hard, dense, angular and roughly cubical particles (National Slag Association 2011). The EAF-slag used by Al-Negheimish et al. (1997) as a replacement of coarse aggregate in concrete was angular shaped with honeycombed surface texture. The angular shape of steel slag aggregate can help to develop very strong interlocking properties. The Flakiness Index (FI) of the steel slag aggregate is markedly lower than those for the dolerite and quartzite aggregates (Anastasiou and Papayianni 2006). The amount of clay lump and friable material content in steel slag is considerably lower than in natural aggregates (Almusallam et al. 2004).

Anastasiou and Papayianni (2006) evaluated several physical properties of crushed steel slag to be used as an aggregate in concrete. The authors concluded that the evaluated aggregate properties of steel slag are in between the limits of the standards and in the best categories. Some physical properties of the different types of steel slags used as aggregate in concrete or asphalt mix and reported by several researchers are presented in Table 2.4. Steel slag has higher abrasion resistance and lower crushing value than natural aggregates. On the other hand, the specific gravity and water absorption capacity of steel slag are higher than those of conventional aggregates. The porosity of a typical steel slag reported by Manso et al. (2004) is 10.5 %. EAF-steel slag aggregates have excellent resistance to fragmentation (Papayianni and Anastasiou 2010).

The surface texture of steel slag is rougher than that of limestone aggregate (Xue et al. 2006; Ahmedzade and Sengoz 2009). The scanning electron micrograph of EAF-slag indicates the presence of many pores on its surface. Steel slag has higher bulk density than natural aggregates (Al-Negheimish et al. 1997).

An adequate grading of steel slag is necessary to obtain better performance from concrete containing steel slag as aggregate (Manso et al. 2006). The particle size distribution curves of two typical EAF-slags used as coarse and fine aggregate in concrete are presented in Fig. 2.5. Depending on the cooling process, the particle size of steel slag may vary. Air-cooled steel slag consists of large sized granules and some powder (Wang et al. 2010).

The high content of free lime (free-CaO) and periclase (MgO) is the detrimental factor against using various steel slags as aggregate in concrete. The content of free lime (free-CaO) and periclase (MgO) in EAF-slag is considerably lower than in BOF-slag. As steel slag contains expansive materials like CaO and MgO, to be used in concrete slag is generally treated. Ageing or weathering of slag, steam and autoclave curing of slag are generally performed to reduce expansive oxide contents (Chen et al. 2007; Faraone et al. 2009; Lun et al. 2008; Pellegrino and Gaddo 2009). As the majority of the steel slag contains free CaO and MgO, experiments are generally performed to evaluate the free CaO and MgO content in the slag and soundness of steel slag aggregate.

Tables 2.5, 2.6 and 2.7 show the chemical compositions of BOF-, LDF- and EAF-slags, respectively. Steel slag is basic in nature (ratio of concentrations of

Table 2.4 Some physical properties of steel slag

Property	Steel slag	Type of steel slag	Natural aggregate	Type of natural aggregate	Reference
Specific gravity	3.51	EAF-slag	2.71	Quartzite	Almusallam et al. (2004).
Los Angeles abrasion (%)	11.6		19.2		
Water absorption (%)	0.85		1.60		
Specific gravity	3.64	EAF-slag	2.66	–	Al-Negheimish et al. (1997)
Water absorption (%)	0.54		1.67		
Specific gravity	3.35	EAF-slag			Manso et al. (2004)
Los Angeles abrasion (%)	~20				
Water absorption (%)	3.29				
Porosity (%)	10.5		–		
Specific gravity	3.02	EAF-slag	2.63	Limestone	Ahmedzade and Sengoz (2009)
Los Angeles abrasion (%)	20		29		
Shape (flat and long) (%)	<10	BOF-slag	10	Basalt	Xue et al. (2006)
LA abrasion (%)	13.1		14.9		
Crushing value (%)	12.0		12.9		
Water absorption (%)	1.18		0.70		
Specific gravity	3.48	BOF-slag	2.66	River crushed stone	Shen et al. (2009)
Water absorption (%)	2.2		1.3		
LA abrasion (%)	20.96		23.52		
Shape (flat and elongated) (%)	5.14		7.55		
Roundness index (%)	0.60		0.54		

$CaO + MgO$ to $SiO_2 + Al_2O_3$ greater than 1) (Tossavainen et al. 2007). The order of this ratio for LDF-, EAF- and BOF-slags can be arranged as EAF-slag $\leq$ LDF-slag $\leq$ BOF-slag. The reactivity and free CaO content of steel slag increases with its increasing basicity (Shi and Qian 2000). Compared to BOF- and LDF-slag, most EAF-slags contain less CaO and MgO in their chemical compositions. There are some aggregates reported as steel slags (Maslehuddin et al. 2003; Qasrawi et al. 2009), which contain more than 85 % of Fe_2O_3 and are therefore completely different from other EAF-slags (Table 2.8). This material can be considered as mill scale. The dominant minerals of EAF-slag can be classified into dicalcium silicate (C_2S), merwinite (C_3MS_2) or kirschsteinite (CFS), depending on the basicity of steel slag, of which the basicity of kirschsteinite is the lowest (Qian et al. 2002).

Although the concentrations of some toxic metals in steel slag are higher than those in normal soil, metals are strongly bound with the slag matrix and therefore are not readily leached. Therefore, steel slag cannot be considered as "characteristically hazardous" material (Proctor et al. 2000). However, in a recent report higher leachability values (above the standard limit for inert landfill) of Cr and Mo from EAF-slag were also reported (Tossavainen et al. 2007). As the chemical composition of steel slag is highly variable mineralogical composition of steel slag also varies. Table 2.9 shows the mineralogical compositions of various types of

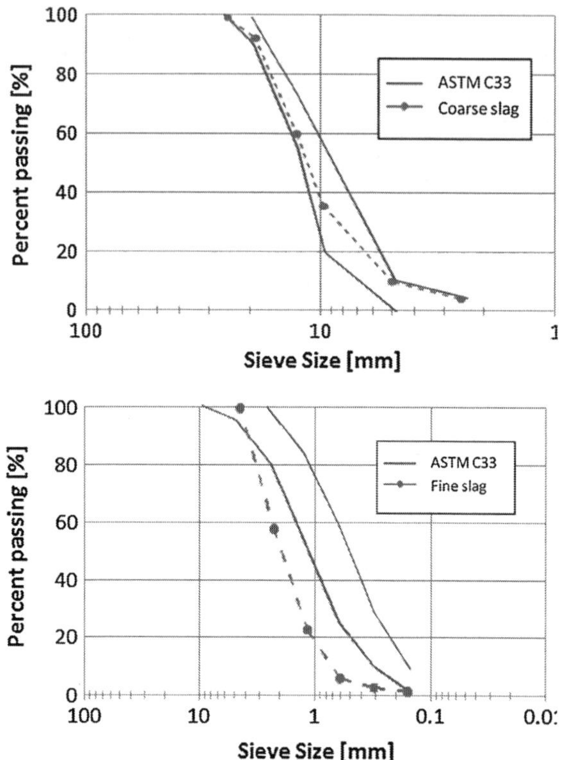

Fig. 2.5 Particle size distribution curves of two steel slags (Manso et al. 2004)

slags as reported by various authors. Different calcium silicates, solid solution of CaO-FeO-MnO-MgO and free-CaO are the common minerals in steel slag.

2.4.1.2 Blast Furnace Slag

When the blast furnace is tapped to release the molten iron, it flows from the furnace with molten slag floating on its upper surface. These two materials are separated using a weir, the molten iron being channelled to a holding vessel and the molten slag to a point where it is to be treated further. The final form of the blast furnace slag is dependent on the method of cooling. There are four main types of blast furnace slag: granulated; air-cooled; expanded and pelletised.

Chemically, the blast furnace slag contains mainly silica (30–35 %), calcium oxide (28–35 %), magnesium oxide (1–6 %), and Al_2O_3/Fe_2O_3 1.8–2.5 %. Due to its low iron content it can be safely used in the manufacturer of cement. Two types of blast furnace slag such as air-cooled slag and granulated slag are being generated from the steel plants. The specific gravity of the slag is approximately 2.90 with its bulk density varying in the range of 1,200–1,300 kg/m^3. The colour of granulated slag is whitish. The air-cooled slag is used as aggregate in road making

Table 2.5 Chemical compositions of BOF-slags

Author	Lun et al. (2008)	Xue et al. (2006)	Mahieux et al. (2009)	Tossavainen et al. (2007)	Altun and Yilmaz (2002)		Reddy et al. (2006)
Country	PR China	PR China	France	Sweden	Turkey		India
Type	Steel slag (from BOF)	BOF-slag	Weathered BOF-slag	BOF-slag	BOF-slag		BOF-slag
					S1	S2	
SiO_2 (%)	10.10	13.71	11.8	11.1	18.01	18.87	15.3
Al_2O_3 (%)	1.70	3.80	2.00	1.90	2.61	2.91	1.30
Fe and Fe_xO_y (%)	$Fe_2O_3 = 20.57$	$FeO = 21.85$ $Fe_2O_3 = 3.24$	$Fe_2O_3 = 22.6$	$FeO = 10.9$ $Fe_2O_3 = 10.7$ $Fe = 2.30$	14.10	11.73	Fe (total) = 16.2
CaO (%)	40.65	45.41	47.5	45.0	37.02	37.94	52.3
MnO (%)	0.06	3.27	1.90	3.10	7.52	7.72	0.39
MgO (%)	10.88	6.25	6.30	9.60	14.10	14.37	1.1
Free lime (%)	nr	nr	nr	nr	0.83	0.83	10
Basicity: (CaO + MgO)/(Al_2O_3 + SiO_2)	4.37	2.95	3.90	4.20	2.48	2.40	3.82

nr not reported

Table 2.6 Chemical compositions of LDF-slags

Author	Setien et al. (2009)			Tossavainen et al. (2007)	Shi and Qian (2000)	Rodriguez et al. (2009)
Country	Spain			Sweden	PR China	Spain
Type	1	2	3	LDF slag	LDF-slag fines	LDF-slag
SiO_2 (%)	15.0	19.8	12.6	14.2	26.4–26.9	17
Al_2O_3 (%)	12.5	4.3	18.6	22.9	4.3–5.2	11
Fe and Fe_xO_y (%)	$Fe_2O_3 = 2.1$	$Fe_2O_3 = 3.3$	$Fe_2O_3 = 1.6$	$Fe_2O_3 = 1.1$ $FeO = 0.5$ $Fe = 0.4$	$Fe_2O_3 = 1.0$–1.6	nr
CaO (%)	55.0	57.5	50.5	42.5	56.6–57.0	56
MnO (%)	0.36	0.42	0.52	0.2	0.5–1.0	
MgO (%)	7.5	11.6	11.9	12.6	3.2–4.2	10
Free lime (%)	19.0	3.5	9.5	nr	nr	nr
Free MgO (%)	3.0	10.0	8.0	nr	nr	nr
Basicity: $(CaO + MgO)/(Al_2O_3 + SiO_2)$	2.27	2.87	2.00	1.49	1.93	2.36

nr not reported

Table 2.7 Chemical compositions of EAF-slags

Author	Qian et al. (2002)	Luxan et al. (2000)		Manso et al. (2006)	Pellegrino and Gaddo (2009)		Tossavainen et al. (2007)	
Country	Singapore	Spain		Spain	Italy		Sweden	
Type	Kirschsteinite based EAF-slag	Black EAF-slag (fusion of scrap steel)		Black EAF-slag	EAF slag		EAF slag	
		S1	S2		Medium	Coarse	S1	S2
(contaminated by white slag)								
SiO_2 (%)	19.38	6.04	15.35	15.3	10.1	14.7	32.2	14.1
Al_2O_3 (%)	5.27	14.07	12.21	7.4	5.70	7.20	3.70	6.70
Fe and/or Fe_xO_y (%)	FeO = 30.54			Fe_2O_3 = 8.49			FeO = 27.41	FeO = 34.36
				FeO + Fe_2O_3 = 42.5	FeO = 37.2	FeO = 44.8	FeO = 3.30	
Fe_2O_3 = 1.00 Fe = 0.10	FeO = 5.60			Fe_2O_3 = 20.3 Fe = 0.60				
CaO (%)	22.48	29.11	24.40	23.9	24.2	29.5	45.5	38.8
MnO (%)	1.10	15.58	5.57	4.5	5.10	5.70	2.0	5.0
MgO (%)	9.51	3.35	2.91	5.1	1.90	4.60	5.2	3.9
Free lime (%)	nd	nr	nr	0.45	nr	nr	nr	nr
Free MgO (%)	nr	nr	nr	~1.0	nr	nr	nr	nr
Basicity:				(CaO + MgO)/ (Al_2O_3 + SiO_2)	1.30	1.61	0.99	1.28
1.65	1.60	1.41	2.05					

nd not detected; *nr* not reported

Table 2.8 Chemical composition of some steel slag described as steel slag or steel making slag and steel mill scale

Author	Faraone et al. (2009)	Al-Negheimish et al. (1997)	Chen et al. (2007)	Maslehuddinn et al. (2003)	Qasrawi et al. (2009)	Al-Otaibi (2008)
Country	Italy	Saudi Arabia	PR China	Saudi Arabia	Jordan	Kuwait
Type	Steel slag	Steel making slag (EAF-slag)	Steel slag	Steel slag	Unprocessed steel slag	Steel mill scale
SiO_2 (%)	13.2	17.43	13.40	1.00	0.80	1.37
Al_2O_3 (%)	6.80	7.85	1.04	nr	nr	0.10
Fe and/or Fe_xO_y (%)	$Fe_xO_y = 9.80$	Total Fe = 20.09 FeO = 18.33	FeO = 18.98 $Fe_xO_y = 8.45$	$Fe_xO_y = 89.0$	$Fe_xO_y = 97.05$	$Fe_xO_y = 94.61$
CaO (%)	37.9	30.53	51.51	$CaCO_3 = 10$	0.40	0.11
MnO (%)	13.5	2.66	1.65		1.07	1.03
MgO (%)	11.9	13.19	3.45		0.40	0.03
Basicity: (CaO + MgO)/ (Al_2O_3 + SiO_2)	2.49	1.73	3.81	–	–	–

nr not reported

Table 2.9 Mineralogical compositions of various slags

Authors	Country	Type of slag	Minerals identified
Rojas and Rojas (2004)	Spain	EAF-slag	Wustite/plustite, magnesioferrite/magnetite, hematite, larnite, bredigite/merwinite, gehlenite, birnessite/groutellite (manganese oxides)
Faraone et al. (2009)	Italy	EAF-slag	Fine fraction: glaucochroite $(CaMn)_2SiO_4$, iron manganese oxide (FeMnO) Medium: calcite ($CaCO_3$), glaucochroite ($(CaMn)_2SiO_4$), iron manganese oxide (FeMnO), magnetite (Fe_3O_4), portlandite ($Ca(OH)_2$)
Tossavainen et al. (2007)	Sweden	LDF-slag	Mayenite, ($Ca_{12}Al_{14}O_{33}$), free MgO. β-Ca_2SiO_4, γ-Ca_2SiO_4, $Ca_2Al_2SiO_7$
		BOF-slag	Larnite, β-Ca_2SiO_4, solid solution of (Fe,Mg,Mn)O, solid solution of (CaMg)O
		EAF-slag, S1	Merwinite, $Ca_3Mg(SiO_4)_2$, γ-Ca_2SiO_4, solid solution of spinel phase $(Mg, Mn)(Cr, Al)_2O_4$
		EAF-slag, S2	β-Ca_2SiO_4, wustite-type solid solution ((Fe, Mg, Mn)O), $Ca_2(Al, Fe)_2O_5$, Fe_2O_3
Qian et al. (2002)	Singapore	Kirschsteinite based EAF-slag	Kirschsteinite ($CaFe^{2+}(SiO_4)$), Mg-wustite
Luxan et al. (2000)	Spain	EAF-slag, S1 and S2	Major: gehlenite [Ca_2 $Al(Al,Si)_2$ O_7], larnite (Ca_2SiO_4) Minor: bredigite [$Ca_{14}Mg_2(SiO_4)_8$], manganese oxides (Mn_3O_4, MnO_2), magnesioferrite ($MgFe_2O_4$), magnetite (Fe_3O_4)

while the granulated slag is used for cement manufacturing. Its use as aggregate in concrete is not very usual although it has no behavioural problems, and there has been little research work done on the subject.

Because of their more porous structure, blast furnace slag aggregates have lower thermal conductivities than conventional aggregates. Their insulating value is of particular advantage in applications such as frost tapers (transition treatments in pavement sub-grades between frost susceptible and non-frost susceptible soils) or pavement base courses over frost-susceptible soils.

The granulated blast furnace slag is formed due to rapid quenching of molten slag, which converts it into a glassy state. Granulated slag possesses cementitious properties if it is ground finely. The size and physical properties of granulated blast furnace slag varies, depending on the chemical composition and method of production. Numerous studies are available on the properties of cement and concrete containing ground blast furnace slag as a latent hydraulic material. Slag cement is the hydraulic cement that results when molten slag from an iron blast furnace is rapidly quenched with water, dried and ground to a fine powder. Blast furnace slag sand was recognised as meeting JIS A 5012 standards in 1981, and has also been stipulated in guidelines for the Japan Society of Civil Engineers. It is also stipulated as JIS A 5308 (ready-mixed concrete).

Air-cooled blast furnace slag (air-cooled slag) is prepared by cooling the molten blast furnace slag slowly by ambient air. A small amount of water is generally used to spray the surface of the slag to assist in the cooling process. The air-cooled blast furnace slag is normally processed in a crushing and screening plant to manufacture products of particular maximum sizes and gradings. Crushed air-cooled slag is angular, roughly cubical, and has textures ranging from rough, vesicular surfaces to smooth glassy surfaces with conchoidal fractures. Processed air-cooled slag exhibits good abrasion resistance, good soundness characteristics and high bearing strength. The cementitious property of air-cooled slag is poorer than in other types of slag prepared by rapid quenching. Slag sand is very angular and coarse and therefore mixes containing this product require a high fine sand content or the use of a mix containing three sand types. In general, blast furnace slag processed for use as a concrete aggregate complies with the same requirements for naturally occurring dense aggregate. While complying with these requirements, air-cooled blast furnace slag aggregate differs from the range of naturally occurring dense aggregates in certain properties. The particle and bulk densities of air-cooled slag are slightly lower than those of natural aggregates. It has higher water absorption and Los Angeles value. Some typical properties of air-cooled slag are presented in Table 2.10.

If the molten slag is cooled and solidified by adding controlled quantities of water, air-, or steam, the resultant slag becomes a lightweight expanded or foamed type of product. Foamed blast furnace slag is distinguishable from air-cooled slag by its relatively high porosity and low bulk density. Crushed expanded slag is angular, roughly cubical in shape and has a texture that is rougher than that of air-cooled slag. The porosity of expanded blast furnace slag aggregates is higher than that of air-cooled slag. The bulk relative density of expanded slag is difficult to determine accurately, but it is approximately 70 % of that of air-cooled slag. Typical compacted unit weights for expanded blast furnace slag aggregates range from 800 to 1,040 kg/m^3.

The molten slag can be pelletised during cooling and solidification process. The produced pellets can be made more crystalline or more vitrified (glassy). Crystalline pellet can be used as aggregate. The pelletised blast furnace slag has smooth texture and round shape. Consequently, the porosity and water absorption are much lower than those of air-cooled or expanded slag. Pellet sizes range from 13 to 0.1 mm, with the bulk of the product in the 1.0–9.5 mm range. Pelletised blast furnace slag has a unit weight of about 840 kg/m^3.

Collins and Sanjayan (1999) reported the use of an air-cooled porous slag with maximum diameter of 14 mm as a coarse aggregate in preparation of alkali activated slag concrete. The specific gravity of the slag aggregate was 2.71, which was slightly lower than used basalt aggregate (2.95). The water absorption capacity of slag was 4.4 % and considerably higher than basalt aggregate (1.2 %).

Etxeberria et al. (2010) reported the use of blast furnace slag of size range of 4.75–25 mm as a complete or partial replacement of coarse aggregate in preparation of concrete. The physical properties of the aggregate were determined according to British standard, BS EN 1097-6:2000. The oven dry and saturated surface dry densities of the aggregate were 2.27 and 2.37 g/cm^3 respectively , which are lower than the used natural aggregate (about 2.7 g/cm^3). On the other

Table 2.10 Typical mechanical properties of air-cooled blast furnace slag (FHWA-RD-97-148)

Property	Value
Los Angeles abrasion (ASTM C131)	35–45 %
Sodium sulphate soundness loss (ASTM C88)	12 %
Angle of internal friction	40–45
Hardness (measured by Mohr's scale of mineral hardness)[a]	5–6
California bearing ratio (CBR), top size 19 mm (3/4 in)[b]	up to 250 %

[a] Hardness of dolomite measured on same scale is 3–4
[b] Typical CBR value for crushed limestone is 100 %

hand, the water absorption capacity of slag aggregate was 4.10 % and lower than that of natural aggregate (2.94 %).

Yuksel et al. (2011) used granulated blast furnace slag (GBFS) with maximum diameter of 4 mm as partial replacement of fine aggregate in concrete. The various properties of GBFS aggregate along with a natural fine aggregate are presented in Table 2.11.

Escalante-García et al. (2009) reported the use of a GBFS as fine aggregate in cement mortar. The GBFS employed as an aggregate was used as received and only a fraction of large sized particles was removed; 90 % was retained in the 0.420 mm sieve and 28 % was retained in the 1 mm sieve. The used GBFS has sharp edges and some particles have several surface pores.

Leshchinsky (2004) reported the use of air-cooled slag sand with maximum size of 6.3 mm as a fine aggregate in ready-mixed concrete. The moisture content of the used slag sand is within a range of 2–4 % above the saturated surface dry (SSD) content. According to the author, complete saturation is crucial for slag aggregate and for slag sand to achieve the required concrete properties.

2.4.2 Non-Ferrous Slag

Non-ferrous metallurgical slags are generated during refining of various metals such as Cu, Cr, Zn and treatment of waste such as Pb-acid batteries. Some physical properties of a few non-ferrous slags are presented in Table 2.12.

2.4.2.1 Copper Slag

Copper slag is a by-product obtained during the matte (molten copper sulphide) smelting and refining of copper (Biswas and Davenport 1976). Major constituents of a smelting charge are sulphides and oxides of iron and copper. The charge also contains oxides such as Al_2O_3, CaO, MgO, and principally SiO_2, which are either present in the original concentrate or added as a flux. As a result, copper-rich matte (sulphides) and copper slag (oxides) are formed as two separate liquid phases. The

Table 2.11 Physical properties of granulated blast furnace slag (Yuksel et al. 2011)

Property	Type of aggregate	
	GBFS	Natural
Loose bulk density (kg/m^3)	1,052	1,930
Dense bulk density (kg/m^3)	1,236	1,950
Specific gravity	2.08	2.60
Water absorption (%)	8.30	2.30
Amount of clay (%)	1.0	4.0
Loss on ignition (%)	1.8	5.0
Lightweight particles (%)	3.0	4.0

Table 2.12 Some physical properties of various non-ferrous slags (Publication number: FHWA-RD-97-148)

Property	Nickel slag	Copper slag	Phosphorus slag	Lead, lead–zinc, and zinc slags
Appearance	Reddish brown to brown-black	Black	Black to dark grey	Black to red
Texture	Massive, angular, amorphous texture	Glassy, more vesicular when granulated	Air-cooled is flat and elongated but granulated is uniform, angular	Glassy, sharp angular (cubical) particles
Unit weight, (kg/m^3)	3,500	2,800–3,800	Air-cooled: 1,360–1,440 Expanded: 880–100	<2,500–3,600
Absorption (%)	0.37	0.13	1.0–1.5	5.0

molten slag is discharged from the furnace at 1,000–1,300 °C. When liquid slag is cooled slowly, it forms a dense, hard crystalline product where a quick solidification by pouring molten slag into water provides amorphous granulated slag. Production of one tonne of copper generates approximately 2.2–3 tonnes of copper slag. In the United States the amount of copper slag produced is about four million tonnes (Collins and Ciesielski 1994), in Japan it is about two million tonnes per year (Ayano and Sakata 2000), and approximately 360,000, 244,000 and 60,000 tonnes of copper slag are produced every year in Iran, Brazil and Oman, respectively (Behnood 2005; Moura et al. 1999; Taeb and Faghihi 2002).

Recycling, recovering of metal, production of value added products and disposal in slag dumps or stockpiles are the options for management of copper slag. It has been widely used for abrasive tools, roofing granules, cutting tools, abrasive, tiles, glass, road-base construction, railroad ballast, asphalt pavements, cement clinker and blended cement production (Shi et al. 2008). Many researchers have investigated the use of copper slag as fine or coarse aggregate in the preparation of cement mortar and concrete.

The main concern in large-scale use of copper slags including in construction is the content of toxic elements in copper slag and the consequent leaching from slag and slag based products. However, copper slag has been excluded from the listed hazardous waste category of the United States Environmental Protection Agency

(USEPA). The United Nations (UN) Basel Convention on the Transboundary Movement of Hazardous Waste and its Disposal also ruled that copper slag is not a hazardous waste (Alter 2005).

The chemical composition of copper slag depends on the type of furnace, the metallurgical production process, and the composition of the extracted ore. The range of percentage of the main oxides of copper slag can vary as follows—Fe_2O_3: 35–60 %, SiO_2: 25–40 %, CaO: 2–10 %, Al_2O_3: 3–15 %, CuO: 0.3–2.1 %, MgO: 0.7–3.5 %. The density of copper slag varies between 3.16 and 3.87 g/cm^3, the average specific gravity of copper slag is about 3.5 g/cm^3 and the average water absorption of copper slag is 0.15–0.55. Water-cooled copper slag has a higher water absorption and lower unit weight than air-cooled copper slag due to its more porous texture (Shi et al. 2008). Air-cooled copper slag has a black colour and glassy appearance. The specific gravity varies with iron content, from 2.8 to 3.8. The unit weight of copper slag is somewhat higher than that of conventional aggregates. The absorption capacity of the material is typically very low (0.13 %). Granulated copper slag is more porous and therefore has lower specific gravity and higher absorption capacity than air-cooled copper slag. The granulated copper slag is made up of regularly shaped, angular particles, mostly between 4.75 and 0.075 mm in size (Emery 1995; Hughes and Halliburton 1973).

Air-cooled and granulated copper slag has a number of favourable mechanical properties for aggregate use, including excellent soundness characteristics, good abrasion resistance, and good stability. It has high friction angle due to sharp angular shape. However, the slag tends to be vitreous or 'glassy,' which adversely affects their frictional properties (skid resistance), a potential problem if used in pavement surfaces.

Khanzadi and Behnood (2009) reported the use of copper slag as a coarse aggregate in concrete, which meet the grading requirements of ASTM C 33 for 12.5–4.75 mm size aggregates. The physical properties of the copper slag aggregates along with a coarse natural limestone aggregate are presented in Table 2.13. The used copper slag has significantly higher specific gravity than the limestone aggregate. The high specific gravity and the glass-like smooth surface properties of irregular grain shape of copper slag aggregates can increase bleeding. The water absorption capacity of copper-slag is 0.4 % and lower than limestone aggregate, which is 0.6 %. Copper slag aggregate is harder than natural limestone aggregate.

Wu et al.(2009) reported the use of copper slag as a fine aggregate in concrete. The copper slag used in this investigation had a density of 3,660 kg/m^3, in contrast with the density of sand of 2,640 kg/m^3. The fineness modulus of the copper slag was 1.78, which was finer than sand, with a fineness modulus of 2.91. The particle size of the copper slag was well distributed within the range of 0.1–1 mm. The shape of sand is irregular with rounded edges at 50 × magnification while that of copper slag is angular with sharp edges. The surface texture of sand is rougher than that of copper slag at 200 × magnification. Under observation at 1,000 × magnification the presence of moisture on sand surface is visible, but the surface of copper slag is glassy and dry.

Al-Jabri et al. (2011) reported the use of a copper slag as a fine aggregate in the preparation of cement mortar and concrete. Before using the slag, the material was

Table 2.13 Some physical properties of copper slag and limestone aggregate (Khanzadi and Behnood 2009)

Properties	Copper slag aggregate	Limestone aggregate
Specific gravity, g/cm^3	3.59	2.65
Aggregate crushing value, %	10–21	23
Aggregate impact value, %	8.2–16	11
Water absorption, %	0.4	0.6
Particle size, mm	4.75–12.5	4.75–12.5

grinded in the laboratory into a fine powder to the required size. The particle size distribution of the ground slag along with normal sand is presented in Fig. 2.6.

Copper slag has high concentrations of silica, alumina and iron oxides. Results from specific gravity and water absorption tests revealed that copper slag has a specific gravity of 3.4 which is higher than that of sand (2.77), whereas the water absorption values for copper slag and sand were about 0.2 and 1.4 %, respectively.

Brinda et al. (2010) used a copper slag as partial replacement of sand and cement in concrete preparation. The physical properties of copper slag are presented in Table 2.14. The slag was made of black glassy particles and granular in nature and has a particle size range similar to sand. The specific gravity of the slag is 3.91. The bulk density of granulated copper slag varied from 1.9 to 2.15 kg/m^3, which is almost similar to the bulk density of conventional fine aggregate. The hardness of the slag lies between 6 and 7 in the Mohr scale. This is almost equal to the hardness of gypsum. The pH of aqueous solution of aqueous extract as per IS 11127 varies from 6.6 to 7.2. The free moisture content present in slag was found to be less than 0.5 %. The amount of silica in slag is about 26 %. The fineness of copper slag was calculated as 125 m^2/kg.

Ishimaru et al. (2005) reported the use of a copper slag as a replacement of sand in the preparation of concrete. The density, water absorption and fineness modulus of used copper slag are respectively 3.46 g/cm^3, 0.65 and 2.58 %.

2.4.2.2 Other Non-ferrous Slags

The use of other non-ferrous industrial slags as aggregate in concrete and cement mortar is also reported. Lead slag generated from lead smelting and from recycling of secondary lead batteries may be used as fine and coarse aggregate in the preparation of cement mortar and concrete.

Atzeni et al. (1996) reported the use of granulated slag generated during smelting of lead and zinc by two different processes (named as Imperial Smelting and Kivcet slags), as a partial and total replacement of the sand fraction in cement mortar and concrete. The slags were used as received from granulation plants. Both types of slags were essentially vitreous with small amount of gehlinite, crystobalite and iron oxides as minor crystalline components. Both slags contain FeO, SiO_2, CaO and Al_2O_3 as the major oxides with significant amounts of zinc

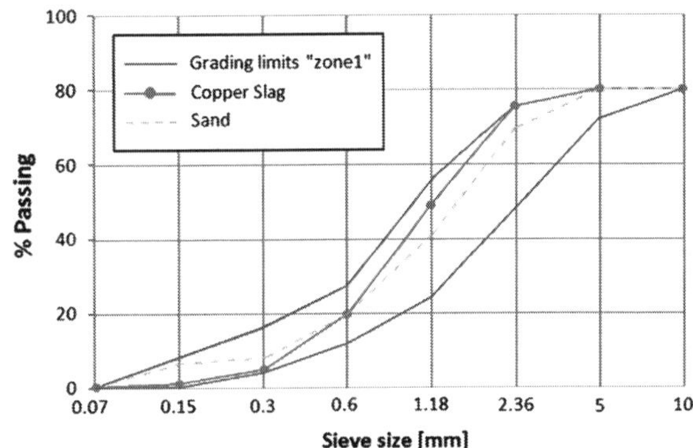

Fig. 2.6 Sieve analysis of a typical copper slag along with normal sand aggregate (Al-Jabori et al. 2011)

Table 2.14 Some physical properties of copper slag (Brinda et al. 2010)

Particle shape	Irregular
Appearance	Black and glassy
Specific gravity	3.91
Percentage of voids	43.20 %
Bulk density	2.08 g/cm^3
Fineness modulus	3.47
Angle of internal friction	51°20′

and lead. Most slag grains are rounded and others are polygonal as a result of conchoidal fracture caused by thermal shock induced by granulation. Imperial Smelting slag is generally less regular than Kivcet slag. The grain size distribution of both slags is similar to that of normal sand. The predominant fraction by mass is the 0.15–2 mm size range. The grains have very low porosity, on average 1 % by volume for the K slag and 3 % by volume for the IS slag.

Penpolcharoen (2005) reported the use of a secondary lead slag as an admixture and/or as an aggregate in the production of concrete blocks. The slag was produced from battery smelting using $CaCO_3$ as flux. The fine fraction of slag, which passed through ASTM sieve No. 200, was used as a partial substituent of ordinary Portland cement, while the coarse fraction of slag, which passed through ASTM sieve No. 4, was used to partially replace crushed limestone aggregate. The CaO content in slag is 6.2 times less than that in OPC, whereas its iron content, as FeO, is 15.1 times higher. The Fe in slag is mainly contained as FeO and therefore weight gain was observed during determination of loss on ignition test. The physical properties of fine and coarse fractions of slags are presented in Table 2.15 along with OPC and crushed stone aggregate.

Sorlini et al. (2004) reported the use of two different types of slag that contained very high amounts of Zn as aggregate in the preparation of concrete. A fresh slag obtained after water cooling and a cured slag after storage of about 6 months in landfill were used in this study. The slag is generated as a by-product of the conversion of electric arc furnace (EAF) dusts (with zinc concentration of 18–35 %) into an impure zinc oxide, called Waelz oxide (with zinc concentration

Table 2.15 Some physical properties of secondary lead slag (Penpolcharoen 2005)

Property	OPC	Fine slag	Crushed stone	Coarse slag
Specific gravity	3.15	3.62	2.71	3.62
Blain fineness	3,380	3,333	–	–
Fineness modulus	–	–	4.74	4.70

of 55–65 %), that can be reprocessed in metallurgical plants. This slag is classified as dangerous waste. Both slags contain very high concentrations of calcium (43 %). Zinc and lead concentrations are about 6 and 1 % respectively; other metals, like manganese, cadmium, copper, chromium and arsenic, are also present in considerable amounts. However, the leachability of these toxic elements in concrete blocks with this slag as aggregate is within the specified limit.

Zelic (2005) investigated the use of a high-carbon ferrochromium slag as aggregate in the preparation of concrete. Before using the slag as aggregate, the original air-cooled slag was crushed to a grain size in the range of 0–16 mm, first by a hammer crusher and then by a cone beaker. The metallic globules of ferro-chromium metal were then removed from the crushed slag by the Remer jig treatment by means of difference in specific gravities of slag and metallic fraction. The specific gravity of slag and FeCr metal is 3.2 and 7.1 g/cm^3, respectively. The refined slag (0–16 mm) fraction contains up to the 11 % Cr.

The part of the Cr_2O_3 component in the slag is 8.3 % in mass smaller after the jig treatment than in the original slag before the treatment, which was the actual aim of the applied process of metal concentration and its removal from the slag. No chlorides were observed in the slag and the sulphur SO_3 content of 0.50 % in mass is much lower than the allowed value of 1.00 % in mass, indicating that the slag does not contain harmful components that may be found in aggregates. The major minerals present in slag are: forsterite, $(Mg, Fe)_2SiO_4$, common-spinel ($MgAl_2O_4$), chrome-spinel ($(Mg, Fe)(Cr,Al)_2O_4$) as unaltered chromium ore and also of small amounts of enstatite ($MgSiO_3$). The slag exhibits better performance than normal limestone aggregate during the Loss Angeles and Bohme wear tests. The fineness modulus of the unfractionated slag varied from 4.0 to 5.1 suggesting a too uneven size distribution of the slag, as the upper acceptable limit is 3.6. The volume mass of ferrochromium slag particles is about 3,250–3,310 kg/m^3 and that of the limestone ones about 2,700 kg/m^3. The Faury coefficients, k, obtained for both the (4–8 mm) and the (8–16 mm) fractions of slag are 0.17 and 0.22, respectively. The results of the mortar bar expansion test indicate the slag aggregate as non-reactive in terms of alkali-silica reaction. The frost resistance of slag is comparable to natural limestone aggregate and well within the standard specifications.

Morrison et al. (2003) reported the use of ferro-silicate slag, generated during the production of zinc in Imperial Smelting Furnace (ISF-slag) was as aggregate in concrete road construction (Morrison et al. 2003, Morrison and Richardson 2004). The slag is glassy and granular in nature and has a particle size range similar to sand, indicating that it could be used as a replacement for sand present in cementitious mixes. The density of this slag (3,900 kg/m^3) is higher than that of traditional aggregates, suggesting it may have advantages over these aggregate materials in certain applications, such as in noise barriers. The X-ray diffraction (XRD) showed the presence of inclusions of metallic iron, although the principal crystalline components detected were FeO and ZnS.

Metwally et al. (2005) reported the use of a slag produced from recycling of spent lead-batteries as aggregate in concrete production. Recycled-lead slag (RLS) was used as both fine and coarse aggregate in concrete manufacture after crushing into the desired gradation by using a roller mill. The physical and mechanical properties of the used fine aggregate (sand and fine lead-slag) and coarse aggregate (gravel and coarse lead-slag) are given in Table 2.16. The coarse lead-slag were washed carefully and dried before mixing to remove any impurities and organic matters, which may weaken its bond with the cement paste. Mixing water was clean tap water free from impurities and organic matters.

Pereira et al. (2000) reported the use of a salt cake slag, produced from aluminium scrap re-melted as partial replacements of either sand or cement in the preparation of cement mortar production after washing. Non-washed slags cannot be mixed with cement due to the volumetric expansion observed during setting, high chloride content and release of high concentrations of noxious gases. As the concentrations of some toxic elements in the majority of these slags are very high, before application the environmental suitability as well as the long-term mechanical and durability performance of concrete containing slag must be evaluated for the application of these materials as aggregate in concrete.

2.5 Other Waste as Aggregate

The meat and bone mill bottom ash (MBM-BA) used by Cyr and Ludmann (2006) as a fine aggregate in cement mortar preparation has a grading size between 0 and 2 mm and a mean diameter of 0.4 mm (Fig. 2.7). The physical properties of MBM-BA are presented in Table 2.17. The bulk density of MBM-BA is around 900 kg/m^3 and much lower than that of normal sand (1,500 kg/m^3). The average density of the ash is 2,900 kg/m^3. The external specific surface area, calculated from the density and the particle size distribution (considering cylindrical particles), is about 3 m^2/g. The BET method gives a specific surface area of 3,000 m^2/kg, which is a thousand times higher than the value calculated using the particle size distribution. This significant difference is related to a large open porosity of the grains, leading to water absorption of 11 %, a very high value compared to normalised siliceous sand (less than 1 %).

The strength of MBM-BA is evaluated using a friability coefficient, defined as the ratio between the mass of crushed particles (less than 0.1 mm) and the whole mass of material before crushing. The friability coefficient of MBM-BA, measured as 37 %, is well within the specified value, which is 40 % for 60 MPa concrete intended for building construction.

Table 2.16 Physical properties of fine and coarse lead slags along with natural aggregates (Metwally et al. 2005)

Property	Natural sand	Natural gravel	Fine lead slag	Coarse lead slag
Specific gravity	2.57	2.51	4.28	3.79
Density (kg/m^3)				
Loose	1,600	1,530	2,280	1,990
Dense	1,869	1,640	2,820	2,120
Voids (%)				
Loose	38	39	47	47
Dense	27	35	34	44
Water absorption (%)	0.42	0.35	3.95	2.35
Fineness modulus	2.49	6.90	3.35	6.95
Crushing value (%)	–	13.5	–	29.3
Impact value (%)	–	6.82	–	13.4
Loss Angeles value (%)	–	16.7	–	47.0

The major minerals present in MBM-BA are hydroxylapatite $Ca_5(PO_4)_3(OH)$ and whitlockite $Ca_3(PO_4)_2$. Calcium hydroxylapatite is the major inorganic constituent of bones and teeth. Trace amounts of some minerals such as quartz, hematite and magnetite are also present in MBM-BA, which probably came from other waste used as co-combustion materials. Although MBM-BA contains small amounts of trace elements, the leachability of these meets the regulation set by French Government for using in road construction.

Öztürk and Bayrakl (2005) investigated the possibilities of the use of tobacco waste, a by-product in a cigarette factory as aggregate in lightweight concrete production. The organic and water content in this aggregate are 66.21 and 25.45 % respectively.

Yellishetty et al. (2008) evaluated the recycling potential of four different iron ore mineral wastes as aggregate in the preparation of concrete. The waste was collected randomly, from different fresh waste dumps as recommended ASTM D75; C702 sampling of aggregates. It was mixed to form a representative homogeneous mixture. The sieve analyses of typical particle overburden material from the waste dumps of the corresponding iron ore mines are presented in Fig. 2.8. A volume equivalent of approximately 50 % of the total is within the usable range as aggregates for making of concrete cubes. Of the whole volume, 20 % is in the range of sand–silt–clay material. Variability in size distribution is an indication of inherent geological properties of mine waste material and of the crushing and compaction that waste undergoes while breaking/loosening.

Various physico-mechanical properties of the mine waste were measured according to ASTM methods and presented in Table 2.18. In aggregates from different mines, the fineness modulus (FM) was within the range of IS2386 specifications prescribed for aggregate for civil constructions. Most of the physical properties of the mine aggregates, belonging to different mines, were within ±5 % range of standard specifications and were in conformity with the concurrent Indian Standards for aggregates.

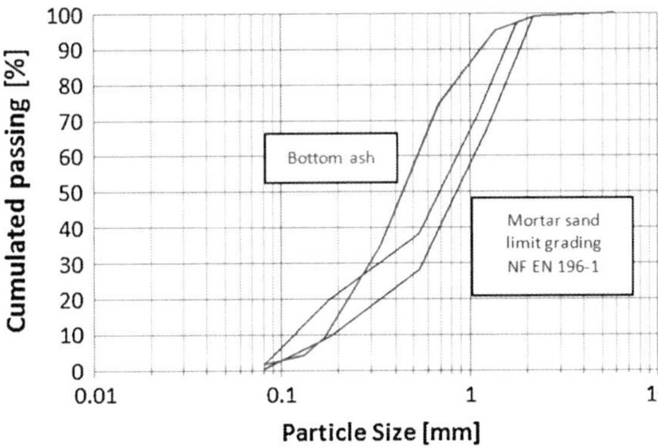

Fig. 2.7 Particle size distribution of MBM-BA, in comparison with the limits for sand used for standard mortar (Cyr and Ludmann 2006)

Table 2.17 Physical properties of meat and bone mill ash (Cyr and Ludmann 2006)

Density (kg/m^3)	2,900
Bulk density (kg/m^3)	900
Specific surface area, BET (m^2/kg)	3,000
Particle size range (mm)	0–2
Mean diameter (mm)	0.4
Morphology	Irregular particles
Adsorption coefficient (water) (%)	11
Friability coefficient (%)	37

Gallardo and Adajar (2006) reported the use of paper mill sludge as a partial replacement of fine aggregate in concrete preparation. The composition of paper sludge is generally a function of the raw material, manufacturing process, chemicals used, final products and wastewater treatment techniques. Sludge from pulp and paper mills are mainly cellulose fibres generated at the end of the pulping process prior to entering the paper machines. Paper sludge obtained directly from a mill wastewater treatment plant is composed generally of 50–75 % organics (cellulose fibres and tissues) and 30–50 % kaolinite clay. The paper mill sludge used in this study was air-dried for several days and facilitated by an artificial method using a blow drier. The dry sludge was then crushed and sieved using sieve No. 4 to separate the fine particles to be used as fine aggregates in the mix. The physical properties of paper sludge along with fine aggregates (sand) are presented in Table 2.19. Sludge was lighter than normal sand and it has very high water absorption capacity. The large absorption percentage of sludge is indicative of its high porosity or entrapped air content and that it absorbs water faster than sand. The high absorption rate is due to hydrogen bonding in the property of sludge. There was no silica content in paper sludge while calcium and iron is of lesser amount as compared to fine aggregates. Some other element contents in paper sludge are lower than the normal sand.

Ahmadi and Al-Khaja (2001) investigated the use of paper sludge, obtained from a tissue paper manufacturing facility as partial replacement of fine aggregate in the preparation of concrete. The sieve analysis of sludge and normal sea sand indicates

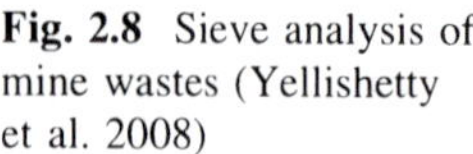

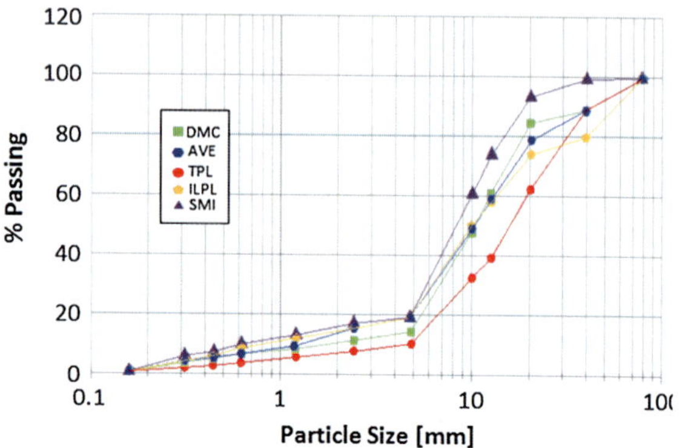

Fig. 2.8 Sieve analysis of mine wastes (Yellishetty et al. 2008)

that the used sludge is coarser than the sea sand. About 50 % of the particles of sludge are coarser than 2 mm compared with sea sand in which 100 % of the particles pass the 2.36 mm sieve. The chemical analysis of the waste shows that it has a low pH value with acceptable levels of other chemical parameters and falls within the permissible limits established in Bahrain for use of waste in construction.

Kuo et al. (2007) reported the use of petroleum reservoir sludge as fine aggregate in the preparation of cement mortar. The composition of sludge is presented in Table 2.20. The sludge contains very high amount of smectite clay (>60 %), which can cause a detrimental expansion to some extent, when it is mixed with water. Hence, reservoir sludge cannot be directly used as fine aggregates in concrete and therefore, before using it as aggregate, hydrophilic sludge is converted into hydrophobic by using a cationic surfactant.

The sieve analysis of modified sludge organically modified reservoir sludge (OMRS) along with normal sand is presented in Table 2.21. More than 80 % of the OMRS particles are smaller than 0.6 mm but larger than 0.075 mm. From the thermal analysis of reservoir sludge before and after the treatment, the authors confirmed that the OMRS particles have been organically modified and became hydrophobic before they were mixed with water, quartz sands and cement.

Kinuthia et al. (2009) reported the use of dark coloured colliery spoil, generated during mining of coal as replacement of fine and coarse aggregate in the preparation of medium strength concrete. It was obtained as two materials, a fine fraction of low plasticity, and a coarser non-plastic fraction. The two fractions were blended in equal proportions to produce a well-graded colliery spoil material, which is presented in Fig. 2.9 along with other constituents of prepared concrete. The specific gravity of both types of spoils was 1.8 and lighter than normal aggregate.

Ilangovana et al. (2008) reported the use of quarry rock dust as fine aggregate in the preparation of concrete. The authors defined the quarry rock dust as residue, tailing or other non-valuable waste material after the extraction and processing of rocks to form fine particles less than 4.75 mm. The level of utilisation of this dust in nations like Australia, France, Germany and UK has reached more than 60 % of its total production. The physical properties of the used dust are presented in Table 2.22.

Table 2.18 Physical properties of various mine waste aggregates (Yellishetty et al. 2008)

Properties	Sample 1	Sample 2	Sample 3	Sample 4
Fineness modulus: coarse	7.5	8.0	7.0	8.0
Fineness modulus: fine	3.3	3.4	3.2	3.5
Flakiness index, %	10	13	14	11
Elongation index, %	11	15	14	11
Impact value, %	19.09	29.00	15.75	21.94
Specific gravity	2.5	2.3	2.6	2.5
Water absorption, %	11.00	13.10	5.59	8.13
Bulk density: rodded, kg/l	1.39	1.25	1.40	1.30
Bulk density: loose, kg/l	1.36	1.22	1.30	1.20
Void ratio: rodded, %	44.40	46.80	46.00	48.00
Void ratio: loose, %	45.60	48.08	47.00	52.00
Crushed value, %	30.0	30.0	28.6	29.9
Abrasion value, %	29.42	30.00	30.00	29.00
Angularity number	3	3	1	7

Table 2.19 Properties of normal sand and paper sludge (Gallardo and Adajar 2006)

Properties	Sand	Sludge
Moisture content (%)	10.00	25.77
Specific gravity	2.60	1.57
Water absorption (%)	4.00	22.35
Silica (%)	18.91	0.00

The major oxide composition of the quarry rock dust and fine sand is presented in Table 2.23. Rock dust contains lesser amounts of silica and higher amounts of alumina, iron oxide, calcium oxide, magnesium oxide and titanium oxide than natural sand. On the other hand, the total alkali content in these materials is almost the same.

2.6 Ceramic Industry Waste

In the ceramic industry several types of waste are generated. According to Pacheco-Torgal and Jalali (2010), ceramic waste can be separated into two categories in accordance with the source of raw materials. In each category, the fired ceramic waste was classified according to the production process. This classification is presented in Fig. 2.10. These were produced by using red and white ceramic pastes. However, the use of white paste is more frequent and much higher in volume.

It has been estimated that about 30 % of the daily production in the ceramic industry goes to waste (Senthamarai and Devadas 2005). In Europe, the amount of waste in the different production stages of the ceramic industry reaches 3–7 % of

Table 2.20 Composition of reservoir sludge (Kuo et al. 2007)

Physical properties	
Gravel content, %	0
Sand content %	0
Silt content	31.9
Clay content	68.1
Specific gravity	2.72
Chemical composition	
Silica	53.03
Alumina	22.32
Iron oxide	8.56
Others	16.09

Table 2.21 Sieve analysis of sludge (OMRS) and quartz sands (Kuo et al. 2007)

OMRS		Quartz sand	
Sieve size	Cumulative passing (%)	Sieve size	Cumulative passing (%)
4.75	100	4.75	100
2.36	100	2.36	100
1.18	100	1.18	98.6
0.600	100	0.600	26.2
0.300	86.8	0.300	3.4
0.150	55.2	0.150	1.4
0.075	19.3	0.075	0.2

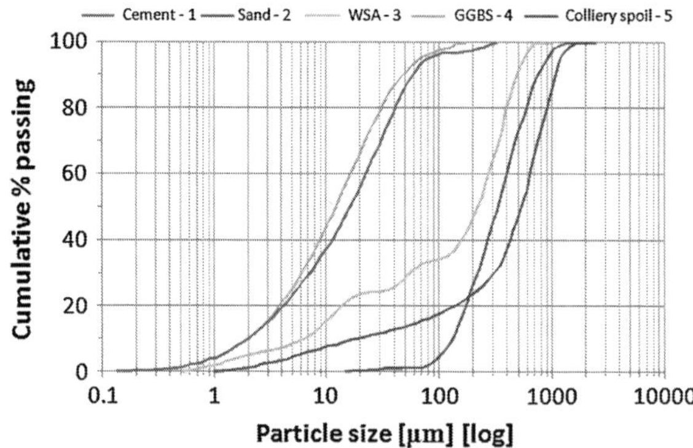

Fig. 2.9 Particle size distributions of colliery waste along with other materials (Kinuthia et al. 2009)

its global production (Fernandes et al. 2004). Although the reutilisation of ceramic waste has been practiced, the amount of waste reused in that way is still negligible (Pacheco-Torgal and Jalali 2010). Ceramic waste can be used safely in the production of concrete due to some of its favourable properties. Ceramic waste is durable, hard and highly resistant to biological, chemical and physical degradation

Table 2.22 Physical properties of quarry rock dust and natural sand (Ilangovana et al. 2008)

Properties	Quarry rock dust	Fine sand
Specific gravity	2.54–2.60	2.60
Bulk relative density (kg/m^3)	1,720–1,810	1,460
Water absorption (%)	1.2–1.5	Nil
Moisture content (%)	Nil	1.5
Particles finer than 0.075 mm (%)	12–15	6
Sieve analysis (Indian standard specification)	Zone II	Zone II

Table 2.23 Oxide composition of quarry rock dust and natural sand (Ilangovana et al. 2008)

Oxide constituents	Quarry rock dust (%)	Natural sand (%)
SiO_2	62.48	80.78
Al_2O_3	18.72	10.52
Fe_2O_3	6.54	1.75
CaO	4.83	3.21
MgO	2.56	0.77
Na_2O	Nil	1.37
K_2O	3.18	1.23
TiO_2	1.21	Nil
Loss on ignition	0.48	0.37

forces. In the following paragraphs, the aggregate properties of some ceramic waste will be discussed.

The waste used by Senthamarai and Devadas (2005) as coarse aggregate in the preparation of normal concrete was collected from a ceramic electrical insulator industry. The material was too big to be fed into a crushing machine and therefore it was broken into small pieces of about 100–150 mm with a hammer. The surface was also deglazed manually by chisel and hammer. These small pieces were then fed into a jaw crusher to reach the required 20 mm size.

The various physical properties of waste as well as those of natural aggregates are presented in Table 2.24. According to the same authors, the properties of ceramic waste aggregate were similar to those of natural crushed stone aggregate. The specific gravity and fineness modulus of aggregate were 2.45 and 6.88 respectively. The surface texture of the ceramic waste aggregate was found to be smoother than that of crushed stone aggregate. In the soundness test, after 30 cycles, the weight loss of ceramic waste aggregate was 51 % less than that of conventional crushed stone aggregate, since ceramics are more resistant to all chemicals.

Binici (2007) used ceramic industry waste as a partial replacement of fine aggregate (40–60 %) in the preparation of normal concrete. The bigger waste pieces were processed into the 4 mm or less size range by a procedure similar to that adopted by Senthamarai and Devadas (2005). The size grading of the waste ceramic fine aggregate was suitable for concrete production. The various physical properties of the waste aggregate are given in Table 2.25 along with those of conventional fine aggregate. Like the coarse aggregate properties of ceramic waste

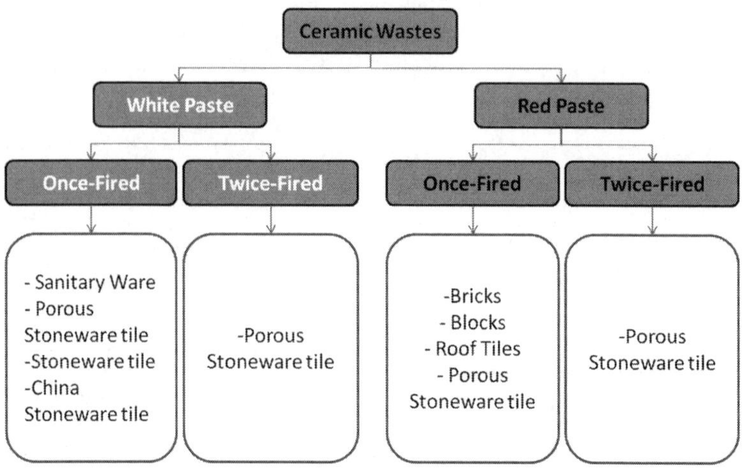

Fig. 2.10 Classification of ceramic wastes according to Pacheco-Torgal and Jalali (2010)

reported by Senthamarai and Devadas (2005), the properties of fine waste aggregates are similar to those of fine conventional aggregate. Chemically the ceramic waste fine aggregate is composed of silica (88.4 %) with 7.3 % of aluminium oxide.

Suzuki et al. (2009) used a porous waste ceramic coarse aggregate in the preparation of high-performance concrete. The aggregate was recovered from the waste of a local ceramic production plant. The aggregate was used in saturated surface-dry conditions.

Table 2.26 summarises the main characteristics of the porous ceramic waste aggregate along with natural coarse and fine aggregates and a commercial lightweight coarse aggregate that is used for internal wet-curing purposes.

The particles of the waste ceramic aggregate are coarser than those of the natural coarse aggregate (Fig. 2.11).

Topcu and Canbaz (2007) reported the use of crushed tile waste as aggregate in the preparation of concrete. Tile is produced from natural materials sintered at high temperatures and it does not contain any harmful chemicals. The tile waste was crushed into 4–16 mm and 16–31.5 mm sizes before using. The unit weight, specific weight and water absorption of crushed tiles were 925, 1,904 kg/m^3 and 11.56 % respectively. Abrasion losses were 21 and 82 % for the Los Angeles abrasion tests at 100 and 500 cycles, respectively. The abrasion of crushed tile was quite high. The maximum amount of abrasion was 50 % at 500 cycles.

Guerra et al. (2009) investigated the use of a crushed recycled sanitary porcelain ceramic ranging from 5 to 20 mm as a partial substitution of natural coarse aggregate. The composition of the clay from which sanitary porcelain was made was: quartz 30 %, feldspar 26 %, kaolin 26 %, clay 18 % and lastly, glazes and metal oxides. The waste was used without cleaning due to the purity of the materials. However, the porcelain was crushed and sieved in the laboratory to get the required size range. The density and moisture content in the porcelain aggregate were 2.36 g/cm^3 and

Table 2.24 Physical properties of coarse ceramic aggregate (Senthamarai and Devadas 2005)

Properties	Ceramic waste aggregate	Crushed stone aggregate
Specific gravity	2.45	2.68
Maximum size (mm)	20	20
Fineness modulus	6.88	6.95
Water absorption, 24 h (%)	0.72	1.20
Bulk density (kg/m^3)		
Loose	1,200	1,350
Compacted	1,325	1,566
Voids (%)		
Loose	50	48
Compacted	45	44
Crushing value (%)	27	24
Impact value (%)	21	17
Abrasion value (%)	28	20
Soundness test: weight loss after 30 cycles (%)	3.3	6.8

Table 2.25 Physical properties of coarse ceramic aggregate (Binici 2007)

Properties	Ceramic waste aggregate	Fine aggregate
Specific gravity	2.44	2.65
Maximum size (mm)	4	4
Fineness modulus	2.68	2.68
Water absorption, 24 h (%)	0.71	0.75
Bulk density (kg/m^3)	1,395	1,695
Voids (%)	44.20	46.20
Abrasion value (%)	28	–
Soundness test: weight loss after 30 cycles (%)	4.2	–

0.11 % respectively. The sieve analyses of this type of aggregate and of normal coarse aggregate (gravel) are presented in Table 2.27.

Torkittikul and Chaipanich (2010) reported the use of earthenware ceramic waste collected from ceramic industry as fine aggregate in the preparation of normal concrete. The larger ceramic waste pieces were broken with a hammer into smaller pieces ($\leq$10 mm) and then they were crushed using a jaw crusher until the percentage passing sieve mesh No. 4 (opening 4.75 mm) was 100 %. The particle size distribution of ceramic waste used in this investigation was kept the same as that of sand by using sieves of mesh Nos. 4, 8, 16, 30, 50 and 100. The maximum particle size, water absorption and specific gravity of ceramic waste were 4 mm, 1.25 % by mass and 2.31, respectively. The particle shape of the crushed and sieved ceramic aggregate as observed by optical microscope (OM) was more angular than that of natural sand. The surface texture of ceramic waste as observed by scanning electron microscope (SEM) was found to be rougher than that of sand.

Pacheco-Torgal and Jalali (2010) reported the use of four different types of ceramic waste as coarse and aggregate as well as replacement of cement in the

Table 2.26 Physical properties of porous ceramic waste aggregate, natural coarse aggregate and a commercial lightweight aggregate (Suzuki et al. 2009)

Properties	Ceramic waste aggregate	Natural coarse aggregate	Natural fine aggregate	Commercial lightweight aggregate
Specific gravity	2.27	2.92	2.62	1.27
Water absorption capacity (%)	9.31	0.88	2.41	12.2–22.3
Fineness modulus	6.66	6.51	3.21	6.47
Crushing rate	21.4	7.86	–	37.0

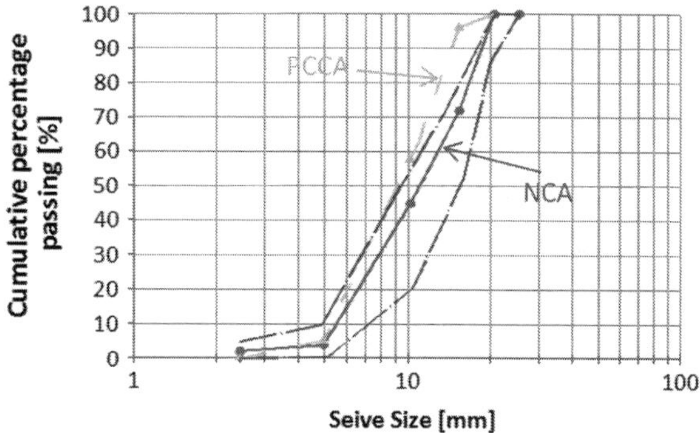

Fig. 2.11 Grading curve of porous coarse ceramic aggregate (PCCA) and of natural coarse aggregate (Suzuki et al. 2009)

production of concrete. These are: ceramic bricks; white stoneware once fired; sanitary ware; white stoneware twice-fired. The major oxide constituents present in these waste types are silica and alumina. The major mineral phases present in all the waste types are quartz and feldspars. All the waste types were crushed with a jaw crusher to make the ceramic aggregate. The coarse and fine sized aggregates and ceramic powder were obtained after sieving. The densities of sand and coarse sized ceramic waste were 2,210 and 2,263 kg/m^3 respectively. The water absorption capacities of sand and coarse sized ceramic aggregates were 6.1 and 6.0 % respectively.

De Brito et al. (2005) reported the use of ceramic hollow bricks fragments from the making of partition walls as a coarse aggregate for the production of non-structural concrete for pavement slabs. To characterise ceramic waste aggregate along with natural aggregate, the volume index of the different size fractions, the compacted oven dry bulk density of both types of aggregates, the compacted air-dried and water saturated bulk densities of waste ceramic aggregate, the specific densities of both types of aggregates at dry and saturated surface dry conditions, and the water absorption capacity were measured. The results are presented in Table 2.28.

The volume index indicates the shape of the particles: aggregates nearly spherical have an index near 1, compared to elongated ones with a smaller index. From the results of volume indices of various size fractions, the authors conclude that the grinding process of the recycled aggregates is a critical parameter since it strongly affects the volume index and therefore deserves further study. From the

Table 2.27 Sieve analysis of waste porcelain aggregate and natural gravel (Guerra et al. 2009)

Sieve mesh size (mm)	Amount retained after sieving (%)	
	Waste porcelain	Gravel
20	95.08	100
10	23.32	25.02
5	1.32	4.10
2.5	0.06	2.18
1.25	0.04	2.04
0.63	0.03	1.95
0.32	0.03	1.82
0.16	0.03	1.59

Table 2.28 Some physical properties of waste ceramic aggregate and of natural limestone aggregate (de Brito et al. 2005)

Properties	Waste ceramic aggregate	Limestone aggregate
Volume index of size fractions (mm)		
6.35–9.52	0.202	0.162
4.76–6.35	0.144	0.149
2.38–4.76	0.153	0.239
Bulk density (kg/m^3):		
Oven dry	1,159	1542
Air-dry	1,167	–
Saturated surface dry	1,265	–
Specific density (kg/m^3)		
Dry	2,029	2,626
Saturated surface dry	2,273	2,657
Water absorption (%)	12	1

water absorption value, the authors conclude that pre-saturation of ceramic aggregate is necessary and the weight of water absorbed by the ceramic aggregates should be measured by deducting the total weight of the aggregates just before they are mixed from their weight before saturation.

Guney et al. (2010) investigated the use of waste foundry sand as aggregate in the preparation of high-performance concrete. Foundry sand is high-quality silica sand and is a by-product from the production of both ferrous and non-ferrous metal casting. The raw sand used to cast metal is normally of higher quality than a typical bank run or the natural sand used in construction. However a small amount of clay (bentonite or kaolinite) is used as a binding material, binding the sand, and this mix is referred as green sand. Thus, foundry sand or green sand consists of high-quality 85–95 % silica sand, 7–10 % bentonite or kaolinite clay, 2–5 % water and about 5 % sea coal. Chemical binders, such as phenolic urethane, are also used to create sand cores (American Foundrymen's Society 2004). In the casting process, moulding sand is recycled and reused several times and therefore the recycled sand degrades to a point that it cannot be reused in the casting process. At this

Table 2.29 Physical properties of waste foundry sand (Guney et al. 2010)

Properties	Waste foundry sand
Specific gravity	2.45
Co-efficient of uniformity	5.50
Fines content (<74 μm), %	24
Active clay content (<2 μm), %	5
Moisture content (%)	3.25
pH	9.1
Organic content (%)	4.3

point, the old sand is removed from the cycle as a by-product, new base sand is introduced, and the cycle begins again.

The waste foundry green sand used by Guney et al. (2010) was used as foundry sand in steel and metal moulding facilities for the production of metal-steel parts at high temperatures (1,500 °C) for about 8–10 times until losing its moulding properties. The grain size distribution of waste foundry sand was uniform, with 100 % of the material under 1 mm, 10 % of foundry sand greater than 0.5 mm and 5 % smaller than 0.125 mm; it is black in colour due to the sea-coal organic binder used in the foundry sand. The used waste foundry sand is composed mainly of silica (98 %) with very small amounts of other oxides. The physical properties of waste foundry sand are presented in Table 2.29. The waste foundry sand was generally sub-angular to round in shape and have rough surface texture.

2.7 Use of Plastic Waste as Aggregate

Many references are available on the use of waste plastic as aggregate, filler or fibre in the preparation of cement mortar and concrete (Siddique et al. 2008). In this section, only the cases where plastic is used as aggregate replacement are presented. The various aggregate properties of the different types of plastic waste are presented. Finally, the possible future studies on plastic waste as aggregate in cement mortar and concrete are evaluated.

2.7.1 Types of Plastic Waste Used in the Preparation of Cement Mortar and Concrete

Different types of plastic waste such as polyethylene terephthalate (PET) bottle (Akcaozoglu et al. 2010; Albano et al. 2009; Choi et al. 2005, 2009; Kim et al. 2010; Marzouk et al. 2007; Yesilata et al. 2009), polyvinyl chloride (PVC) pipe (Kou et al. 2009), high density polyethylene (HDPE) (Naik et al. 1996), thermo-setting plastics (Panyakapo and Panyakapo 2008), shredded plastic waste (Al-Manaseer and Dalal 1997; Ismail and Al-Hashmi 2008), expanded polystyrene foam (EPS) (Kan and Demirboga 2009), glass reinforced plastic (GRP) (Asokan et al. 2010) have been used as aggregate and filler in the preparation of cement mortar and concrete. Here, only the properties of those waste plastics which are used as aggregate will be discussed.

2.7.2 Sources and Preparations of Plastic Aggregate

The majority of plastic aggregate used in different studies came from plastic bottles or containers waste. In general plastic bottles are grinded at the laboratory by using a grinding machine and then sieved to get the suitable size fraction. However, in some studies, plastic waste with suitable sizes is collected from a plastic waste treatment plant. In this case, sieving into suitable size range was done at the laboratory. In some of the studies, treatment of plastic waste was done by heating, melting followed by mixing with other materials or other techniques.

Akcaozoglu et al. (2010) used granules from shredded PET bottle waste as aggregate, which were supplied from a commercial company. The bottles were obtained by picking up waste PET and then washing and mechanically crushing them into granules.

The waste PET (WPET) aggregate used in the Frigione (2010) study was manufactured from PET-bottle waste, unwashed and not separated on the basis of colour. To prepare WPET aggregate, PET-bottle waste, with a thickness of 1–1.5 mm, was grinded in a blade mill to the size of 0.1–5 mm. Then, the resulting particles were separated, through sieves, into a similar size grading to that of natural sand.

In the Batayneh et al. (2007) study the plastic waste aggregate was prepared by grinding original waste plastics into small sized particles. The size distribution of the prepared plastic aggregate was within the fine aggregate specified gradation limit as recommended in BS882:1992.

Ismail and AL-Hashmi (2008) studied the behaviour of concrete containing fibre-shaped plastic waste, which represents the discarded waste from plastic containers collected from plastic manufacturing plants. It consists of approximately 80 % polyethylene and 20 % polystyrene. After collection of plastic waste it was crushed into suitable size range.

In the Marzouk et al. (2007) investigation the polyethylene terephthalate (PET) waste was obtained from drinking water bottles that were first separated, washed and shredded. The shredded plastics were shredded once again, using a propeller crusher with grids of differently-sized meshes, at room temperature and moisture, in order to yield differently-sized aggregate. Plastic aggregate with three different size ranges, i.e. 50, 20 and 10 mm, were used in this study.

In the Remadania et al. (2009) study PET aggregate was prepared from drinking water bottles. For this purpose, PET-bottles were first separated, washed and shredded. The particles thus derived were then shredded once again, using a propeller crusher in order to control granular limit with crushing and to facilitate matrix-aggregate adhesion due to their irregular shape and rough surface texture.

Kou et al. (2009) produced PVC plastics granules by grinding scraped PVC pipes into small granules with about 95 % passing the 5 mm sieve.

Panyakapo and Panyakapo (2008) prepared an aggregate by grinding melamine waste. In this study, the ground melamine waste, retained by ASTM sieve numbers 10–40, was used.

In the Hannawi et al. (2010) investigation polyethylene terephthalate (PET) and polycarbonate (PC) waste was obtained from an industry.

Fraj et al. (2010) used the coarse rigid polyurethane foam waste with size range of 8–20 mm as coarse aggregate, which came from the destruction of insulation panels used in building industry. In order to maintain a comparable aggregate size distribution in the various concrete compositions this waste was sieved into five different size ranges. Mounanga et al. (2008) reported the behaviour of lightweight cement mortar containing rigid polyurethane foam waste with 0–10 mm size range as aggregate, which also came from the destruction of insulation panels used in building industry.

Laukaitis et al. (2005) used crumbled recycled foam polystyrene waste as well as spherical large and fine blown polystyrene waste in his investigation. Polystyrene granules of three types: blown (large + fine) and crumbled were used in this study. The crumbled granules were produced by mechanically disintegrating unusable or poor quality polystyrene slabs and from recycled polystyrene foam plastic. The foam was beaten for 5 min in a horizontal beater, which expanded the foam volume by 40 times. The hydrophilisated polystyrene granules were prepared by soaking in water and under water saturated condition in vacuum desiccators.

Choi et al. (2009) prepared an aggregate by mixing granulated waste PET-bottle with powdered river sand at 250 °C. After air-cooling the mixture, the prepared aggregate and remaining powdered sand fraction was screened by using a 0.15 mm sieve. Choi et al. (2005) also prepared another type of plastic-based aggregate by mixing powdered blast furnace slag with granulated waste PET-bottle at 250 °C. The schematic diagram to produce PET aggregate according to Choi et al. (2009) is presented in Fig. 2.12.

Kan and Demirboga (2009) prepared an aggregate from waste-expanded polystyrene (EPS) foams. This modified waste-expanded polystyrene (MEPS) aggregate was prepared by melting EPS foam waste in a hot air oven at 130 °C for 15 min. The aggregate was separated into two size fractions similar to those of natural aggregate: 0–4 mm (fine aggregate) and 4–16 mm (coarse aggregate).

2.7.3 Evaluation of Properties of Plastic Aggregate

The major property, evaluated in almost all waste plastic aggregate related studies, was their size grading that was generally done by standard sieving methods (Batayneh et al. 2007; Frigione 2010; Ismail and Al-Hashmi 2008; Kou et al. 2009; Panyakapo and Panyakapo 2008; Marzouk et al. 2007). However Albano et al. (2009) adopted a different approach to estimate the size distribution of plastic aggregate. In his approach, sizes of the plastic aggregate were measured by means

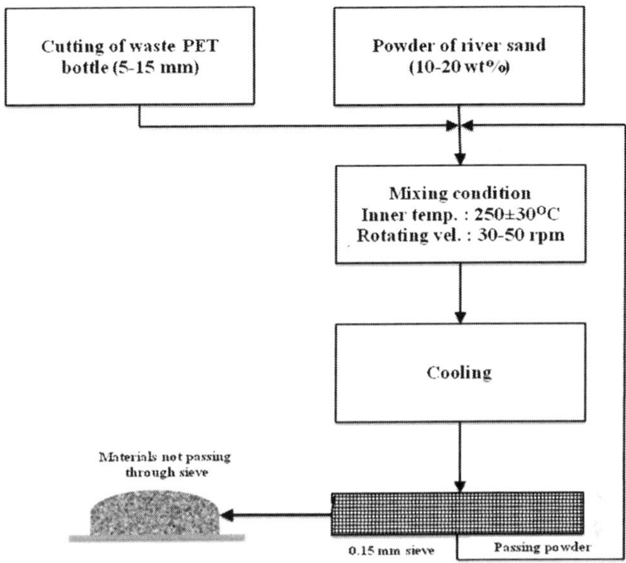

Fig. 2.12 Manufacturing process of sand coated PET aggregate (Choi et al. 2009)

of an electronic magnifying glass and the average particle size was determined using a software application. In the Mounanga et al. investigation (2008) the fine and coarse fractions of rigid polyurethane foam were analysed by laser size grading and sieve analysis method, respectively.

Other aggregate properties, such as bulk density, specific gravity, and water absorption, were also evaluated. Although the adopted procedure to evaluate these parameters was not described, standard procedures used for natural aggregate can be used. On the other hand, some other properties of plastic aggregate, such as tensile and compressive strengths, elasticity modulus, and decomposition temperature, were also reported. However, details of the experimental methods adopted to evaluate these properties were not provided. In some studies, parameters such as melting and initial degradation temperatures and melt flow index (MFI), which may be interesting for the evaluation of the fire behaviour of concrete containing plastic waste, heat capacity and thermal conductivity, were also determined.

2.7.4 Types and Ratio of Substitution of Natural Aggregate by Plastic Aggregate in Cement Mortar/Concrete Mixes

Plastic aggregate is generally produced from big sized waste plastic samples like drinking water bottles, other plastic containers and pipes. Therefore, both coarse and fine sized natural aggregate can be replaced by plastic aggregate. Both partial and full substitutions of natural aggregate by plastic aggregate were reported in various references (Table 2.30).

2.7.5 Properties of Plastic Aggregate

In different studies, several types of plastic waste were used as aggregate. As the origins of this aggregate were completely different from that of natural aggregate i.e. one is organic and the other is inorganic, a big difference in properties was generally observed. The properties of the types of plastic used as aggregate in concrete are presented in Table 2.31.

The use of polyethylene terephthalate (PET) as aggregate was studied extensively compared to other types of plastic aggregate, namely the replacement of fine ($\leq$ 4 mm) and coarse ($\geq$4 mm) natural aggregate with similar size fractions of PET. However, in several studies fine natural aggregate of cement mortar and concrete was replaced with coarse sized PET aggregate too (Albano et al. 2009; Ismail and Al-Hashmi 2008).

Polyethylene terephthalate (PET) is used in various purposes such as in the preparation of beverage, food and other containers, thermoforming applications and synthetic fibre. PET may exist both as an amorphous and a semi-crystalline polymer (with particle sizes in nanometre to micrometre ranges) depending on its processing and thermal history. PET consists of polymerised units of the monomer ethylene terephthalate, with repeating $C_{10}H_8O_4$ units. The molecular formula of PET can be represented as $(C_{10}H_8O_4)_n$ and the molecular structure is presented in Fig. 2.13.

The degradation of PET is possible in highly alkaline solution like concrete pore fluid (Silva et al. 2005). The ions present in pore fluid, Ca^{2+}, Na^+, K^+, and OH^-, can attack the C–O bonds of PET and split the polymer into two groups: the group of the aromatic ring and that of aliphatic ester. The alkali ions can interact with aromatic rings and form Ca, Na, and K-terephthalates. On the other hand, hydroxyl ion can form ethylene glycol by reacting with aliphatic ester group.

Polyethylene terephthalate is thermoplastic polyester with Young's modulus in the range of 1,700–2,500 MPa and tensile strength up to 75 MPa. It has excellent chemical resistance properties. It is a semi-crystalline polymer, with a melting point of about 260 °C and initial degradation temperature of about 412 °C. In several references, the specific gravity and bulk density of the plastic aggregate were reported is the ranges of 1.24–1.36 g/cm^3 and 326–547 kg/m^3 respectively. The PET has very low thermal conductivity (0.13–0.24 W m^{-1} K^{-1}) compared to limestone (1.26–1.33 W m^{-1} K^{-1}) and sandstone (1.7 W m^{-1} K^{-1}). Again the specific heat capacity of PET (1.0–1.1 kJ kg^{-1}K^{-1}) is higher than that of limestone (0.84 kJ kg^{-1}K^{-1}) and sandstone (0.92 kJ kg^{-1}K^{-1}).

Table 2.30 Types of substitution of natural aggregate by plastic aggregate in cement mortar/concrete

Reference	Types of composite	Types and amounts of substitution	Origin of plastic waste
Albano et al. (2009)	Concrete	Fine aggregate 10 and 20 vol.%	PET-bottle
Batayneh et al. (2007)	Concrete	Fine aggregate 5, 10, 15, 20 vol.%	–
Ismail and Al-Hashmi (2008)	Concrete	Fine aggregate 10, 15, 20 wt%	Plastic containers (80 % polyethylene and 20 % polystyrene)
Kou et al. (2009)	Lightweight aggregate concrete	Fine aggregate 5, 15, 30, 45 vol.%	PVC pipe
Marzouk et al. (2007)	Mortar	Fine aggregate 2, 5, 10, 15, 20, 30, 50, 70, 100 vol.%	PET-bottle
Panyakapo and Panyakapo (2008)	Non-load-bearing lightweight concrete	With sand fraction in aerated concrete	Melamine waste
Fraj et al. (2010)	Concrete	Coarse aggregate 34, 35, 45 vol.% of concrete	Waste polyurethane foam collected after destruction of insulation panels used in the building industry
Frigione (2010)	Concrete	Fine aggregate 5 wt%	PET-bottle
Remadnia et al. (2009)	Mortar	Fine aggregate 30, 50, 70 vol.%	PET-bottle
Hannawi et al. (2010)	Mortar	Fine aggregate 3, 10, 20 50 vol.%	A mixture of polyethylene terephthalate (PET) and polycarbonate industrial waste

(continued)

Table 2.30 (continued)

Reference	Types of composite	Types and amounts of substitution	Origin of plastic waste
Kan and Demirboğa (2009)	Concrete	Fine and coarse aggregate 25, 50, 75, 100 vol.%	Waste packaging materials composed of expanded polystyrene foams
Mounanga et al. (2008)	Mortar	Fine aggregate 13.1–33.7 vol.% of concrete	Waste polyurethane foam collected after destruction of insulation panels used in the building industry
Akcaozoglu et al. (2010)	Mortar	Fine aggregate 50 and 100 wt%	PET-bottle
Choi et al. (2005, 2009)	Mortar and concrete	Fine aggregate 25, 50, 75, 100 vol. %	PET-bottle

Table 2.31 Properties of the types of plastic used as aggregate in concrete

Reference	Type of plastic	Particle size/shape	Density/specific gravity*/ apparent bulk density#	Water absorption	Other properties
Albano et al. (2009)	PET	0.26 and 1.14 cm (average size of two fractions)	–	–	MP: 248 °C Initial degradation temperature: 412 °C MFI: 70 g/10 min
Batayneh et al. (2007)	Waste plastic (PET?)	0.15–4.75 mm	–	–	–
Ismail and Al-Hashmi (2008)	80 % polyethylene + 20 % polystyrene	Length: 0.15–12 mm; Width: 0.15–4 mm; Fibre-shaped	#386.7 kg/m^3	0.02 %	CS: poor; TS: 5,000 psi
Marzouk et al. (2007)	PET	Type A: ≤0.5 cm; Type C: ≤0.2 cm; Type D: ≤0.1 cm	Type A:#326 kg/m^3; Type C: #345 kg/m^3; Type D: #408 kg/m^3	–	–
Remadnia et al. (2009)	PET	≤4 mm/Thin	#327 kg/m^3	0	TS: 75 MPa MP: 249-271 °C TC: 0.13 W/m K MHC: 1.1–1.3 kJ/kg K
Akcaozoglu et al. (2010)	PET	0.25–4 mm	*1.27 g/cm^3		
Hannawi et al. (2010)	PET	1.6–10 mm	#547 kg/m^3/*1.36		Colour: white; MP: 255; YM: 1,700–2,510 MPa
Hannawi et al. (2010)	PC	≤5 mm	#646 kg/m^3/*1.24	–	Colour: transparent; MP: 230–250; YM: 2,700 MPa
Choi et al. (2005)	PET (coated with slag)	Round and smooth	1.39/#844 kg/m^3	0	FM: 4.11
Choi et al. (2009)	PET (coated with sand)	0.15–4.75 mm/round and smooth	1.39/#844 kg/m^3	–	FM: 4.11

(continued)

Table 2.31 (continued)

Reference	Type of plastic	Particle size/shape	Density/specific gravity*/ apparent bulk density#	Water absorption	Other properties
Frigione (2010)	PET	Thickness: 1–1.5 mm Size: 0.1–5 mm	–	–	–
Kou et al. (2009)	PVC	≤5 mm/granular	1,400/#546 kg/m^3	–	CS: 65 MPa
Panyakapo and Panyakapo (2008)	Melamine waste	<10 mm	*1.48	5.6 %	TS: 60 MPa temperature resistance: 300 °C
Fraj et al. (2010)	Rigid polyurethane foam	8–20 mm	*45/#21 kg/m^3	13.9 %	Porosity: 98 %; CS: 174 kPa; YM: 5.6 MPa
Mounanga et al. (2008)	Rigid polyurethane foam	<10 mm	45 ± 2 kg/m^3 (apparent density)/2,191 kg/m^3 (density without porosity)	–	Porosity: 98 %
Kan and Demirboğa (2009)	Treated polystyrene foam	Coarse: 4–8 mm; Fine: 0–4 mm	#Coarse: 220 kg/m^3; #Fine: 162 kg/m^3;	4.1 % (w/w)	CS: 1.76–8.22 MPa; TC: 0.037–0.052 W/m K; SGF: 0.22–0.24 (coarse) 0.31–0.34 (fine)

MP melting point; *MFI* melt flow index; *CS* compressive strength; *TS* tensile strength; *TC* thermal conductivity; *MHC* mass heat capacity; *YM* Young modulus; *FM* fineness modulus; *SGF* specific gravity factor
*: specific gravity; #: apparent bulk density

Fig. 2.13 Molecular formula of polyethylene terephthalate (*C* carbon; *H* hydrogen; *O* oxygen; *n* nos. of monomers)

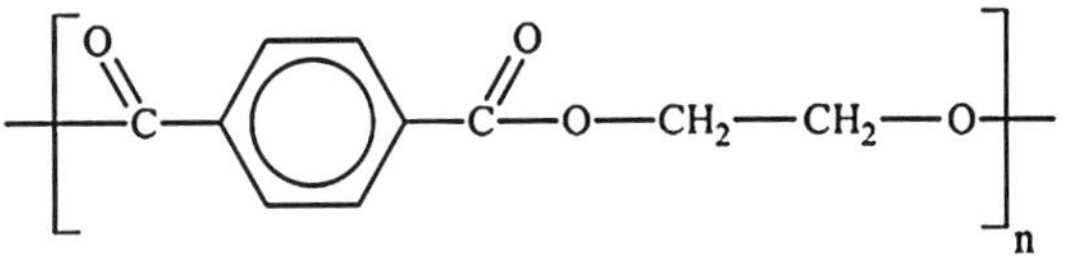

[C: carbon H: hydrogen O: oxygen; n: nos. of monomers]

2.8 Waste Tyre Rubber

The tremendous growth of automobile industry and the increasing use of car as the main means of transportation have increased its production, thus generating huge amounts of tyre rubber wastes. Extensive research works have been carried out to find the way to utilise the rubber tyre waste in various applications. Like plastic wastes, rubber tyre is non-degradable in nature at ambient conditions. This has generated massive stockpiles of used tyre and is creating huge environmental problems including fire hazards. Recently many countries have forbidden land filling of scrap tyres and therefore recycling of this material in the production of other products has immense importance. Out of several management options, the use of waste scrape tyre in the production of cement mortar and concrete is a promising path.

About 300 million tyres were generated in the USA in 2005 and the total number of scrap tyres consumed in end-use markets reached approximately 260 million (Rubber Manufacturer Association 2006). About 190 million scrap tyres remained in stockpile at the end of 2005 in the USA (Rubber Manufacturer Association 2006). This is a simple example and if this scenario is considered for the whole world, the amount of rubber tyres remaining as waste will be increased manifold.

Ganjian et al. (2009) classified tyre rubber into two classes according to the type of vehicles that use the tyre. The natural and synthetic rubber contents in car tyres are different from those in truck tyres (Table 2.32).

According to the use of tyre rubber in concrete preparation, it can be separated into three types (Ganjian et al. 2009):

1. Shredded or chipped rubber is used to replace gravel. By shredding the rubber pieces, particles about 13–76 mm big are produced.
2. Crumb rubber is used to replace sand with size range 0.425–4.75 mm and is manufactured by using special mills. The size of rubber particles depends on the type of mill used and the generated temperature.
3. Ground rubber can be used as a filler material to replace cement. The tyres are subjected to two stages of magnetic separation and screening to produce this size of rubber particles. In micro-milling process, the rubber particles made are in the range of 0.075–0.475 mm.

Several studies were made to evaluate the properties of concrete incorporating recycled tyre rubber as aggregate or filler material. The size, proportion in concrete mix, and surface texture of rubber particles affect the strength of concrete

Table 2.32 Compositions of car and truck tyres (Ganjian et al. 2009)

Constituents	Content in	
	Car	Truck
Natural rubber	14	27
Synthetic rubber	27	14
Black carbon	28	28
Fabric, filler accelerators, anti-ozonants	16–17	16–17
Steel	14–15	14–15

containing used tyre rubber. In the following sections, a literature survey on the aggregate properties of tyre rubber will be presented.

Ganjian et al. (2009) used two types of tyre waste as replacement of coarse aggregate and of cement in the preparation of concrete. The chipped rubber to replace coarse aggregates in normal concrete was prepared from big pieces of tyre rubber in the laboratory using scissors. The grading size of the rubber aggregate was similar to that of coarse natural aggregates (Fig. 2.14).

The relative density of chipped rubber as reported by these authors was 1.3. They found very low values of various strength properties of concrete containing rubber aggregate compared to concrete containing natural aggregate due to lack of proper bonding between rubber aggregates and the cement paste as compared to cement paste and natural aggregates. This is due to the organic nature of the rubber aggregate, which does have any interaction with cement paste.

Khaloo et al. (2008) reported the use of two types of scrap tyre rubber particles as aggregate in concrete preparation. The used crumb rubber was a fine material with grading close to that of sand and the coarse tyre chips used as coarse aggregate were produced by mechanical shredding. According to Neville (1995) tyre particles finer than 0.15 mm can disturb the cement paste reaction and therefore in this study, these particles were removed from the tyre aggregate source.

The various physical properties of tyre rubber aggregate along with those of fine and coarse aggregates were determined according to ASTM standard test methods and presented in Table 2.33. The low specific gravity and unit weight of rubber aggregate reduces the unit weight of concrete, which is prepared by replacing heavy natural aggregates with rubber aggregate. Due to the non-polar nature of rubber particles and their tendency to entrap air on their rough surfaces, concrete containing rubber aggregate has a higher air content than normal concrete. The modulus of elasticity of rubber aggregate with respect to mineral aggregates is very low and therefore rubber aggregates act as large pores, and do not significantly contribute to the resistance to externally applied loads.

The grading of tyre rubber materials was determined based on the ASTM C136 method and presented in Fig. 2.15. The grading curve of rubber materials was determined by using crushed stones in each sieve in order to provide adequate pressure on tyre rubber particles to pass the sieves.

Benazzouk et al. (2003) investigated the use of two types of rubber aggregates, compact rubber aggregates (CRA) and expanded rubber aggregates (ERA) as a

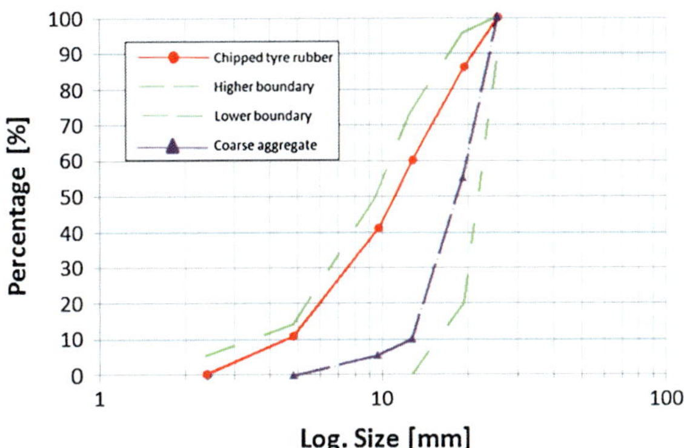

Fig. 2.14 Grading size of chipped rubber and natural coarse aggregate (Ganjian et al. 2009)

Table 2.33 Physical properties of different aggregates (Khaloo et al. 2008)

Aggregate type	Specific gravity	Water absorption (%)	Fineness modulus	Unit weight (kg/m^3)
Natural coarse aggregate	2.65	2.66	NA	1,701.3
Natural fine aggregate	2.67	5.01	5.34	1,716.8
Tyre rubber aggregate	1.16	–	NA	1,150

partial replacement of natural aggregate in the preparation of concrete. CRA has smooth surfaces with a water-accessible porosity of 0.3 %. The magnitude of the strain before fracture (strain is defined as the ratio of the length at failure and the initial length) is 85 %. ERA is a soft aggregate with alveolar surfaces. The magnitude of the strain before fracture and the water absorption of ERA are 200 and 3 % respectively.

The tyre rubber waste particles were reduced into three groups of 1–4, 4–8 and 8–12 mm size grading by means of mechanical grinding followed by sieving. According to author, the rubber aggregate differed from mineral aggregates in terms of both the strain magnitude and the non-brittle characteristic under loading.

The physical and mechanical properties of rubber aggregates are shown in Table 2.34. The hardness of the rubber aggregates was determined according to ASTMD 2240-75, where hardness is defined as the resistance offered by a specimen to the penetration of a hardened steel truncated cone.

Snelson et al. (2009) reported the preparation of concrete by using tyre rubber waste and ash. Rubber chips ranging from 15 to 20 mm were used to replace equal proportions of two different sized coarse limestone aggregate. The steel wires present in rubber chips were removed by an electromagnet during the shredding process.

Pierce and Blackwell (2003) reported the use of crumb tyre rubber as lightweight aggregate in the preparation of controlled low-strength material. The crumb rubber was produced from recapping truck tyres by using a sharp rotating disc. The tread was shaved off into 15 cm and smaller strips. The strips were then grinded down into crumb rubber. According to the authors, the production of crumb rubber

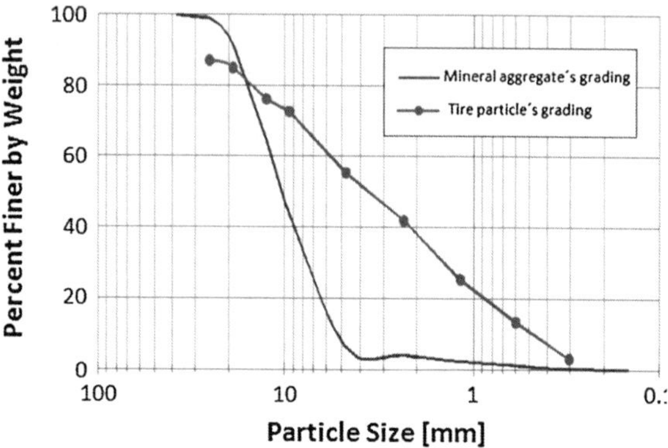

Fig. 2.15 Grading curves of natural and tyre aggregates (Khaloo et al. 2008)

Table 2.34 Properties of rubber aggregates (Benazzouk et al. 2003)

Properties	CRA	ERA
Unit weight (kg/m^3)	1,286	1,040
Hardness (shore)	85	35
Modulus of elasticity (Mpa)	68	12

from truck tyre recaps was less expensive than that from the whole tyre because the tread is free from any fibrous material. The used crumb rubber aggregate was coarser than the ASTM C 33-02A specified limit for concrete sand. The crumb rubber used in this study was in dry state. However used crumb rubber aggregate can absorb a small amount of water equal to 2.4 % of its dry weight in the saturated surface dry condition that was higher than that of concrete sand, which is normally about 0.5 % or less. The bulk specific gravity of the crumb rubber aggregate varied between 0.53 and 0.60, which is nearly five times less than sand. Crumb rubber can thus be considered a lightweight aggregate source due to its low specific gravity.

The authors compared some properties measured during the investigation including cost of crumb rubber aggregate with some lightweight aggregates available in the markets and they are presented in Table 2.35.

The authors concluded that the cost of crumb rubber aggregate compared favourably with that of other lightweight aggregates and were lower than the costs of microlite and perlite.

Sukontasukkul and Chaikaew (2006) reported the use of crumb tyre rubber as a partial replacement of aggregate to produce concrete paving blocks. The crumb rubber particles passing ASTM sieve No. 6 and ASTM sieve No. 20 were used separately as aggregate in concrete. The specific gravity and fineness modulus of both crumb rubber aggregates are presented in Table 2.36 and their particle size distributions were presented in Fig. 2.16. According to the authors, the higher water requirement of concrete mix containing rubber aggregate than that of a conventional concrete mix was due to the low specific gravity and high specific surface area of rubber crumb.

Table 2.35 Comparison of properties of crumb rubber aggregate with some commercially available lightweight aggregates (Pierce and Blackwell 2003)

Types of lightweight aggregates	Bulk specific gravity	Bulk dry density (kg/ m^3)	Cost per ton (US $)
Crumb rubber	0.53–0.63	0.4	200+
Specrete microlite®	0.40–0.45	0.1	500
Vermiculite	2.5 (unexpanded)	0.06–0.16 (unexpanded)	100–150
Perlite	2.2–2.4 (unexpanded)	0.03–0.40 (unexpanded)	320–400

Table 2.36 Some physical properties of crumb rubber aggregates (Sukontasukkul and Chaikaew 2006)

Properties	Crumb rubber No. 6	Crumb rubber No. 20
Average bulk specific gravity	0.97	0.88
Average bulk specific gravity (SSD)	0.98	0.89
Average apparent specific gravity	0.98	0.89
Average absorption (%)	1.01	1.70
Fineness modulus	4.98	2.62

2.9 Concluding Remarks

The aggregates typically account for 70–80 % of the concrete volume and play a substantial role in different concrete properties such as workability, strength, dimensional stability and durability. There is a growing interest in using waste materials as alternative aggregate materials and significant research is made on the use of many different materials as aggregate substitutes. The waste aggregates whose properties are highlighted in this section are: 1. Coal ash; 2. Ferrous and non-ferrous Slag; 3. Waste from food and agricultural industries; 4. Pulp and paper mill waste; 5. Leather waste; 6. Industrial sludge; 7. Mining industry waste; 8. Ceramic wastes; 9. Plastics wastes and 10. Rubber tyre.

Depending on the physic-chemical properties of aggregates, these can be classified in various ways. Out of several industrial waste aggregates, some waste types like coal ash aggregates, some types of slag and ceramic waste can be beneficially used as an aggregate in the preparation of concrete and cement mortars. On the other hand, some waste aggregates like coal bottom ash, plastic wastes and rubber tyre can be used as lightweight aggregates. Some types of waste like rubber waste and plastic waste are organic in nature and therefore they do not interact with cement pastes, ultimately reducing various mechanical properties of the resulting concrete composites.

The majority of the industrial waste aggregates have some special properties, which can be applied to develop some special purpose cement-based materials. For example, plastic and rubber aggregates have high toughness value, can absorb

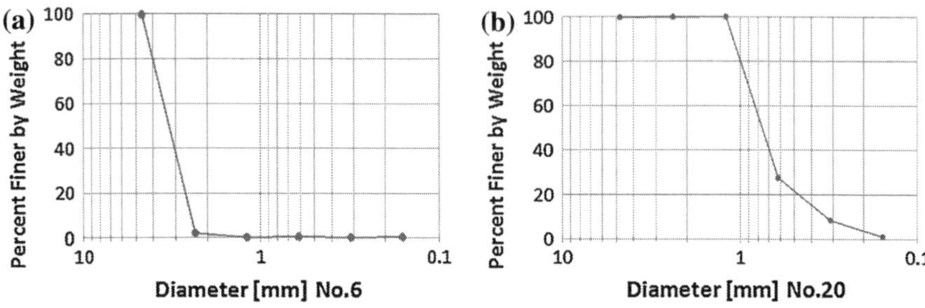

Fig. 2.16 Particle size distributions of crumb rubber aggregates (Sukontasukkul and Chaikaew 2006)

energy during failure and have better acoustic properties than normal aggregates. The majority of non-ferrous slags are heavier than normal aggregates, which can be useful to develop radiation resistant concrete.

The majority of industrial waste types may contain several toxic constituents like toxic elements and organics and therefore removal of these constituents from these aggregates as well as the fate of these constituents in the application of these aggregates and in cement composites are also studied. Research on the long-term behaviour of toxic constituents present in these aggregates and cement composites containing these aggregates needs to be addressed properly before application of these materials.

The treatment or modification of some industrial aggregates to improve its aggregate properties is also reported. For example, plastic waste is coated with inorganic powders and treated physic-mechanically to improve its interactions with cement paste. Similarly, tyre rubber can be treated by physic-chemical methods and petroleum sludge can be treated with surfactant to improve aggregate properties.

Though a significant number of studies are available for the use of some industrial waste types in concrete and mortar preparation, the evaluation of the properties of used aggregates has not been properly addressed, which limits the understanding of the properties of concrete and mortar containing these aggregates. Therefore, a thorough evaluation of properties of industrial waste before application in cement mortar and concrete preparation is an important step and in every study it should be addressed properly.

References

Ahmadi B, Al-Khaja W (2001) Utilization of paper waste sludge in the building construction industry. Resour Conserv Recycl 32(2):105–113

Ahmedzade P, Sengoz B (2009) Evaluation of steel slag coarse aggregate in hot mix asphalt concrete. J Hazard Mater 166(1–3):300–305

Akcaozoglu S, Atis CD, Akcaozoglu K (2010) An investigation on the use of shredded waste PET bottles as aggregate in lightweight concrete. Waste Manage 32(2):285–290

Albano C, Camacho N, Hernandez M, Matheus A, Gutierrez A (2009) Influence of content and particle size of waste pet bottles on concrete behaviour at different w/c ratios. Waste Manage (Oxf) 29(10):2707–2716

Al-Jabri KS, Al-Saidy AH, Taha R (2011) Effect of copper slag as a fine aggregate on the properties of cement mortars and concrete. Constr Build Mater 25(2):933–938

Al-Manaseer AA, Dalal TR (1997) Concrete containing plastic aggregates. Concr Int 19(8):47–52

Almusallam AA, Beshr H, Maslehuddin M, Al-Amoudi OSB (2004) Effect of silica fume on the mechanical properties of low quality coarse aggregate concrete. Cem Concr Compos 26(7):891–900

Al-Negheimish AI, Al-Sugair FH, Al-Zaid RZ (1997) Utilization of local steel making slag in concrete. J King Saud Univ Eng Sci 9(1):39–55

Al-Otaibi S (2008) Recycling steel mill scale as fine aggregate in cement mortars. Eur J Sci Res 24(3):332–338

Alter H (2005) The composition and environmental hazard of copper slags in the context of the Basel convention. Resour Conserv Recycl 43(4):353–360

Altun IA, Yılmaz I (2002) Study on steel furnace slags with high MgO as additive in Portland cement. Cem Concr Res 32(8):1247–1249

American Foundrymen's Society (2004) Foundry sand facts for civil engineers. Report No.: FHWA-IF-04-004 prepared by American Foundrymen's Society Inc. for Federal Highway Administration Environmental Protection Agency Washington, DC, USA, 80 p

Anastasiou F, Papayianni I (2006) Criteria for the use of steel slag aggregates in concrete. In: Konsta-Gdoutos MS (ed) Measuring, monitoring and modeling concrete properties. Springer, The Netherlands, pp 419–426

Andrade LB, Rocha JC, Cheriaf M (2007) Evaluation of concrete incorporating bottom ash as a natural aggregates replacement. Waste Manage (Oxf) 27(9):1190–1199

Andrade LB, Rocha JC, Cheriaf M (2009) Influence of coal bottom ash as fine aggregate on fresh properties of concrete. Constr Build Mater 23(2):609–614

Asokan P, Osmani M, Price ADF (2010) Improvement of the mechanical properties of glass fibre reinforced plastic waste powder filled concrete. Constr Build Mater 24(4):448–460

Atzeni C, Massidda L, Sanna U (1996) Use of granulated slag from lead and zinc processing in concrete technology. Cem Concr Res 26(9):1381–1388

Ayano T, Sakata K (2000) Durability of concrete with copper slag fine aggregate, Fifth CANMET/ACI international conference on durability of concrete, SP-192. American Concrete Institute, Farmington Hills, pp 141–158

Bai Y, Darcy F, Basheer PAM (2005) Strength and drying shrinkage properties of concrete containing furnace bottom ash as fine aggregate. Constr Build Mater 19(9):691–697

Batayneh M, Marie I, Ibrahim A (2007) Use of selected waste materials in concrete mixes. Waste Manage (Oxf) 27(12):1870–1876

Behnood A (2005) Effects of high temperatures on high-strength concrete incorporating copper slag aggregates. In: Seventh international symposium on high-performance concrete, SP-228-66, Washington, USA, pp 1063–1075

Benazzouk A, Mezreb K, Doyen G, Goullieux A, Queneudec M (2003) Effect of rubber aggregates on the physico-mechanical behaviour of cement–rubber composites-influence of the alveolar texture of rubber aggregates. Cem Concr Compos 25(7):711–720

Binici H (2007) Effect of crushed ceramic and basaltic pumice as fine aggregates on concrete mortars properties. Constr Build Mater 21(6):1191–1197

Biswas AK, Davenport WG (1976) Extractive metallurgy of copper, 1st edn. Pergamon Press, Oxford

Brinda D, Baskaran T, Nagan S (2010) Assessment of corrosion and durability characteristics of copper slag admixed concrete. Int J Civil Struct Eng 1(2):192–211

Chen M, Zhou M, Wu S (2007) Optimization of blended mortars using steel slag sand. J Wuhan Univ Technol 22(4):741–744

Cheriaf M, Rocha JC, Pera J (1999) Pozzolanic properties of pulverized coal combustion bottom ash. Cem Concr Res 29(9):1387–1391

Choi YW, Moon DJ, Chung JS, Cho SK (2005) Effects of waste PET bottles aggregate on the properties of concrete. Cem Concr Res 35(4):776–781

Choi YW, Moon DJ, Kim YJ, Lachemi M (2009) Characteristics of mortar and concrete containing fine aggregate manufactured from recycled waste polyethylene terephthalate bottles. Constr Build Mater 23(8):2829–2835

Collins RJ, Ciesielski SK (1994) Recycling and use of waste materials and by-products in highway construction, NCHRP (National Cooperative Highway Research Program, Synthesis of Highway Practice), Issue No. 199, Transportation Research Board, Washington, USA

Collins F, Sanjayan JG (1999) Strength and shrinkage properties of alkali-activated slag concrete containing porous coarse aggregate. Cem Concr Res 29(4):607–610

Cyr M, Ludmann C (2006) Low risk meat and bone meal (MBM) bottom ash in mortars as sand replacement. Cem Concr Res 36(3):469–480

Das SK, Yudhbir (2006) Geotechnical properties of low calcium and high calcium fly ash. Geotech Geol Eng 24(2):249–263

de Brito J, Pereira AS, Correia JR (2005) Mechanical behaviour of non-structural concrete made with recycled ceramic aggregates. Cem Concr Compos 27(4):429–433

Emery JJ (1995) Dominican Republic mega project uses hi-tech hot mix, Ontario Hot Mix Producers Association, OHMPA. Asphaltopics 8(2):23–56

Escalante-Garcia JI, Magallanes-Rivera RX, Gorokhovsky A (2009) Waste gypsum-blast furnace slag cement in mortars with granulated slag and silica sand as aggregates. Constr Build Mater 23(8):2851–2855

Etxeberria M, Pacheco C, Meneses JM, Berridi I (2010) Properties of concrete using metallurgical industrial by-products as aggregates. Constr Build Mater 24(9):1594–1600

Faraone N, Tonello G, Furlani E, Maschio S (2009) Steelmaking slag as aggregate for mortars: effects of particle dimension on compression strength. Chemosphere 77(8):1152–1156

Fernandes M, Sousa A, Dias A (2004) Environmental impact and emissions trade. Ceramic industry. A case study, Portuguese Association of Ceramic Industry APICER, Coimbra, Portugal

Fraj AB, Kismi M, Mounanga P (2010) Valorization of coarse rigid polyurethane foam waste in lightweight aggregate concrete. Constr Build Mater 24(6):1069–1077

Frigione M (2010) Recycling of PET bottles as fine aggregate in concrete. Waste Manage (Oxf) 30(6):1101–1106

Gallardo RS, Adajar MAQ (2006) Structural performance of concrete with paper sludge as fine aggregates partial replacement enhanced with admixtures. In: Symposium on infrastructure development and the environment 2006, 7–8 Dec 2006, SEAMEO-INNOTECH, University of the Philippines, Diliman, Quezon City, Philippines

Ganjian E, Khorami M, Maghsoudi AA (2009) Scrap-tyre-rubber replacement for aggregate and filler in concrete. Constr Build Mater 29(5):1828–1836

Ghafoori N, Bucholc J (1996) Investigation of lignite-based bottom ash for structural concrete. J Mater Civ Eng 8(3):128–137

Guerra I, Vivar I, Llamas B, Juan A, Moran J (2009) Eco-efficient concretes: the effects of using recycled ceramic material from sanitary installations on the mechanical properties of concrete. Waste Manage (Oxf) 29(2):643–646

Guney Y, Sari YD, Yalcin M, Tuncan A, Donmez S (2010) Reuses of waste foundry sand in high-strength concrete. Waste Manage (Oxf) 30(8–9):1705–1713

Hannawi K, Kamali-Bernard S, Prince W (2010) Physical and mechanical properties of mortars containing PET and PC waste aggregates, Waste Manage 30(11):2312–2320

Hughes ML, Halliburton TA (1973) Use of zinc smelter waste as highway construction material. Highw Res Rec 430:16–25

Ilangovana R, Mahendrana N, Nagamani N (2008) Strength and durability properties of concrete containing quarry rock dust as fine aggregate. ARPN J Eng Appl Sci 3(5):20–26

Ishimaru K, Mizuguchi H, Hashimoto C, Ueda T, Fujita K, Ohmi M (2005) Properties of concrete using copper slag and second class fly ash as a part of fine aggregate. J Soc Mater Sci 54(8):828–833 (in Japanese)

Ismail ZZ, Al-Hashmi EA (2008) Use of waste plastic in concrete mixture as aggregate replacement. Waste Manage (Oxf) 28(11):2041–2047

Joshi RC, Lohtia RP (1997) Fly ash in concrete production, properties and uses. Gordon and Breach Science Publishers, India

Kan A, Demirboga R (2009) A novel material for lightweight concrete production. Cem Concr Compos 31(7):489–495

Khaloo AR, Dehestani M, Rahmatabadi P (2008) Mechanical properties of concrete containing a high volume of tire-rubber particles. Waste Manage (Oxf) 28(12):2472–2482

Khanzadi M, Behnood A (2009) Mechanical properties of high-strength concrete in-corporating copper slag as coarse aggregate. Constr Build Mater 23(6):2183–2188

Kim HK, Lee HK (2011) Use of power plant bottom ash as fine and coarse aggregates in high-strength concrete. Constr Build Mater 25(2):1115–1122

Kim SB, Yi NH, Kim HY, Kim JHJ, Song YC (2010) Material and structural performance evaluation of recycled PET fibre reinforced concrete. Cem Concr Comp 32(3):232–240

Kinuthia J, Snelson D, Gailius A (2009) Sustainable medium-strength concrete (CS-concrete) from colliery spoil in South Wales UK. J Civil Eng Manage 15(2):149–157

Kou SC, Poon CS (2009) Properties of concrete prepared with crushed fine stone, furnace bottom ash and fine recycled aggregate as fine aggregates. Constr Build Mater 23(8):2877–2886

Kou SC, Lee G, Poon CS, Lai WL (2009) Properties of lightweight aggregate concrete prepared with PVC granules derived from scraped PVC pipes. Waste Manage (Oxf) 29(2):621–628

Kuo WY, Huang JS, Tan TE (2007) Organo-modified reservoir sludge as fine aggregates in cement mortars. Constr Build Mater 21(3):609–615

Kurama H, Topcu IB, Karakurt C (2009) Properties of the autoclaved aerated concrete produced from coal bottom ash. J Mater Process Technol 209(2):767–773

Laukaitis A, Zurauskas R, Keriene J (2005) The effect of foam polystyrene granules on cement composite properties. Cem Concr Compos 27(1):41–47

Lee HK, Kim HK, Hwang EA (2010) Utilization of power plant bottom ash as aggregates in fibre-reinforced cellular concrete. Waste Manage (Oxf) 30(2):274–284

Leshchinsky A (2004) Slag sand in ready-mixed concrete. Concrete 38(3):38–39

Lovell CW, Te-Chih K (1992) Corrosivity of Indian bottom ash, Transportation Research Record No. 1345, Transportation Research Board, Washington, DC, USA, 52 p

Lun Y, Zhou M, Cai X, Xu F (2008) Methods for improving volume stability of steel slag as fine aggregate. J Wuhan Univ Technol 23(5):737–742

Luxan MP, Sotolongo R, Dorrego F, Herreroh E (2000) Characteristics of the slags produced in the fusion of scrap steel by electric arc furnace. Cem Concr Res 34(4):517–519

Mahieux PY, Aubert JE, Escadeillas G (2009) Utilization of weathered basic oxygen furnace slag in the production of hydraulic road binders. Constr Build Mater 23(2):742–747

Majizadeh K, Bokowski G, El-Mitiny R (1979) Material characteristics of power plant bottom ashes and their performance in bituminous mixtures: a laboratory investigation, In: 5th international ash utilization symposium. Report No. METC/SP-79/10, Part 2, US Department of Energy, Morgantown, West Virginia

Manso JM, Gonzalez JJ, Polanco JA (2004) Electric arc furnace slag in concrete. J Mater Civ Eng 16(6):639–645

Manso JM, Polanco JA, Losanez M, Gonzalez JJ (2006) Durability of concrete made with EAF slag as aggregate. Cem Concr Compos 28(6):528–534

Marzouk OY, Dheilly RM, Queneudec M (2007) Valorisation of post-consumer waste plastic in cementitious concrete composites. Waste Manage (Oxf) 27(2):310–318

Maslehuddin M, Al Mana AI, Samim M, Saricimen H (1989) Effect of sand replacement on the early-age strength gain and corrosion-resisting characteristics of fly ash concrete. ACI Mater J 86(1):58–62

Maslehuddin M, Sharif AM, Shameem M, Ibrahim M, Barry MS (2003) Comparison of properties of steel slag and crushed limestone aggregate concretes. Constr Build Mater 17(2):105–112

Metwally MEA, Seleem MH, Balaha MM, Abd El-Rahman H (2005) Utilizing of slag produced from recycling of spent lead-batteries as concrete aggregate. Alex Eng J 44(6):883–892

Monshi A, Asgarani MK (1999) Producing Portland cement from iron and steel slags and limestone. Cem Concr Res 29(9):1373–1377

Morrison C, Richardson D (2004) Re-use of zinc smelting furnace slag in concrete. Eng Sustain 157(4):213–218

Morrison C, Hooper R, Lardner K (2003) The use of ferro-silicate slag from ISF zinc production as a sand replacement in concrete. Cem Concr Res 33(12):2085–2089

Mounanga P, Gbongbon W, Poullain P, Turcry P (2008) Proportioning and characterization of lightweight concrete mixtures made with rigid polyurethane foam wastes. Cem Concr Compos 30(9):806–814

Moura W, Masuero A, Dal Molin D, Vilela A (1999) Concrete performance with admixtures of electrical steel slag and copper concerning mechanical properties. In: 2nd CANMET/ACI international conference on high-performance concrete, SP-186 American Concrete Institute, Farmington Hills, MI, pp 81–100

Naik TR, Singh SS, Huber CO, Brodersen BS (1996) Use of post-consumer waste plastics in cement-based composites. Cem Concr Res 26(10):1489–1492

National Slag Association (2011) NSA product information: steel slag base and subbase aggregates, PI 207. National Slag Association, 25 Stevens Avenue, Building A, West Lawn, PA 19609. (http://www.nationalslag.org/archive/sf_prod_info_sheet.pdf. Accessed January 2011)

Neville AM (1995) Properties of concrete, 4th edn. Longman, London

Ozkan O, Yuksel I, Muratoglu O (2007) Strength properties of concrete incorporating coal bottom ash and granulated blast furnace slag. Waste Manage 27(2):161–167

Öztürk T, Bayrakl M (2005) The possibilities of using tobacco wastes in producing lightweight concrete. Agricultural Engineering International: the CIGR E-Journal, Vol. VII, Manuscript BC 05 006

Pacheco-Torgal F, Jalali S (2010) Reusing ceramic wastes in concrete. Constr Build Mater 24(5):832–838

Panyakapo P, Panyakapo M (2008) Reuse of thermosetting plastic waste for lightweight concrete. Waste Manage (Oxf) 28(9):1581–1588

Papadakis VG (1999) Effect of fly ash on Portland cement systems Part I. Low-calcium fly ash. Cem Concr Res 29(11):1727–1736

Papadakis VG (2000) Effect of fly ash on Portland cement systems Part II. High-calcium fly ash. Cem Concr Res 30(10):1647–1654

Papayianni I, Anastasiou E (2010) Production of high-strength concrete using high volume of industrial by-products. Constr Build Mater 24(8):1412–1417

Park SB, Jang YI, Lee J, Lee BJ (2009) An experimental study on the hazard assessment and mechanical properties of porous concrete utilizing coal bottom ash coarse aggregate in Korea. J Hazard Mater 166(1):348–355

Pellegrino C, Gaddo V (2009) Mechanical and durability characteristics of concrete containing EAF slag as aggregate. Cem Concr Compos 31(9):663–671

Penpolcharoen M (2005) Utilization of secondary lead slag as construction material. Cem Concr Res 35(6):1050–1055

Pereira DA, de Aguiar D, Castro F, Almeida MF, Labrincha JA (2000) Mechanical behaviour of Portland cement mortars with incorporation of Al-containing salt slags. Cem Concr Res 30(7):1131–1138

Piercea CE, Blackwell MC (2003) Potential of scrap tire rubber as lightweight aggregate in flowable fill. Waste Manage (Oxf) 23(3):197–208

Pofale AD, Deo SV (2010) Comparative long term study of concrete mix design procedure for fine aggregate replacement with fly ash by minimum voids method and maximum density method. KSCE J Civil Eng 14(5):759–764

Proctor DM, Fehling KA, Shay EC, Wittenborn JL, Green JJ, Avent C, Bigham RD, Connolly M, Lee B, Shepker TO, Zak MA (2000) Physical and chemical characteristics of blast furnace, basic oxygen furnace, and electric arc furnace steel industry slags. Environ Sci Technol 34(8):1576–1582

Qasrawi H, Shalabi F, Asi I (2009) Use of low CaO unprocessed steel slag in concrete as fine aggregate. Constr Build Mater 23(2):1118–1125

Qian G, Sun DD, Tay JH, Lai Z, Xu G (2002) Autoclave properties of kirschsteinite-based steel slag. Cem Concr Res 32(9):1377–1382

Rajamane NP, Annie Peter J, Ambily PS (2007) Prediction of compressive strength of concrete with fly ash as sand replacement material. Cem Concr Compos 29(3):218–223

Ravina D (1997) Properties of fresh concrete incorporating a high volume of fly ash as partial fine sand replacement. Mater Struct 30(8):473–479

Reddy AS, Pradhan RK, Chandra S (2006) Utilization of basic oxygen furnace (BOF) slag in the production of a hydraulic cement binder. Int J Miner Process 79(2):98–105

Remadnia A, Dheilly RM, Laidoudi B, Quéneudec M (2009) Use of animal proteins as foaming agent in cementitious concrete composites manufactured with recycled PET aggregates. Constr Build Mater 23(10):3118–3123

Rodriguez A, Manso JM, Aragon A, Gonzalez JJ (2009) Strength and workability of masonry mortars manufactured with ladle furnace slag. Resour Conserv Recycl 53(11):645–651

Rogbeck J, Knutz A (1999) Coal bottom ash as light fill material in construction. Waste Manage (Oxf) 16(1):125–128

Rojas MF, Sanchez de Rojas MI (2004) Chemical assessment of the electric arc furnace slag as construction material: expansive compounds. Cem Concr Res 34(10):1881–1888

Rubber Manufacturer's Association (2006) Scrap tire markets in the United States, 2005 edn. Nov 2006, Rubber Manufacturer's Association, 1400 K Street, NW, Washington DC 20005 (http://www.rma.org/scrap_tires. Accessed May 2011

Senthamarai RM, Devadas MP (2005) Concrete with ceramic waste aggregate. Cem Concr Compos 27(9–10):910–913

Setien J, Hernandez D, Gonzalez JJ (2009) Characterization of ladle furnace basic slag for use as a construction material. Constr Build Mater 23(5):1788–1794

Shen D-H, Wu C-M, Du J-C (2009) Laboratory investigation of basic oxygen furnace slag for substitution of aggregate in porous asphalt mixture. Constr Build Mater 23(1):453–461

Shi C, Qian J (2000) High performance cementing materials from industrial slag—a review. Resour Conserv Recycl 29(2):195–207

Shi C, Meyer C, Behnood A (2008) Utilization of copper slag in cement and concrete. Resour Conserv Recycl 52(11):1115–1120

Siddique R (2003a) Effect of fine aggregate replacement with Class F fly ash on the mechanical properties of concrete. Cem Concr Res 33(4):539–547

Siddique R (2003b) Effect of fine aggregate replacement with Class F fly ash on the abrasion resistance of concrete. Cem Concr Res 33(11):1877–1881

Siddique R, Khatib J, Kaur I (2008) Use of recycled plastic in concrete: a review. Waste Manage 28(10):1835–1852

Silva DA, Betioli AM, Gleize PJP, Roman HR, Gomez LA, Ribeiro JLD (2005) Degradation of recycled PET fibres in Portland cement-based materials. Cem Concr Res 35(9):1741–1746

Snelson DG, Kinuthia JM, Davies PA, Chang S-R (2009) Sustainable construction: composite use of tyres and ash in concrete. Waste Manage (Oxf) 29(1):360–367

Sorlini S, Collivignarelli C, Plizzari G, Foglie MD (2004) Reuse of Waelz slag as recycled aggregate for structural concrete. In: International RILEM conference on the use of recycled materials in building and structures, Barcelona, pp 1086–1094

Sukontasukkul P, Chaikaew C (2006) Properties of concrete pedestrian block mixed with crumb rubber. Constr Build Mater 20(7):450–457

Suzuki M, Meddah MS, Sato R (2009) Use of porous ceramic waste aggregates for internal curing of high-performance concrete. Cem Concr Res 39(5):373–381

Taeb A, Faghihi S (2002) Utilization of copper slag in the cement industry. ZKG Int 55(4):98–100

Topcu IB, Canbaz M (2007) Utilization of crushed tile as aggregate in concrete. Iran J Sci Technol Trans B Eng 31(5):561–565

Torkittikul P, Chaipanich A (2010) Utilization of ceramic waste as fine aggregate within Portland cement and fly ash concretes. Cem Concr Compos 32(6):440–449

Tossavainen M, Engstrom F, Yang Q, Menad N, Larsson ML, Bjorkman B (2007) Characteristics of steel slag under different cooling conditions. Waste Manage (Oxf) 27(7):1335–1344

Wang G, Wang Y, Gao Z (2010) Use of steel slag as a granular material: volume expansion prediction and usability criteria. J Hazard Mater 184(1–3):555–560

Wu W, Zhang W, Ma G (2009) Optimum content of copper slag as a fine aggregate in high strength concrete. Mater Des 31(6):2878–2883

Xue Y, Wu S, Hou H, Zha J (2006) Experimental investigation of basic oxygen furnace slag used as aggregate in asphalt mixture. J Hazard Mater B 138(2):261–268

Yellishetty M, Karpe V, Reddy EH, Subhash KN, Ranjith PG (2008) Reuse of iron ore mineral waste in civil engineering constructions: A case study. Resour Conserv Recycl 52(11):1283–1289

Yesilata B, Isıker Y, Turgut P (2009) Thermal insulation enhancement in concretes by adding waste PET and rubber pieces. Constr Build Mater 23(5):1878–1882

Yüksel I, Bilir T (2007) Usage of industrial by-products to produce plain concrete elements. Constr Build Mater 21(3):686–694

Yüksel I, Siddique R, Özkan O (2011) Influence of high temperature on the properties of concretes made with industrial by-products as fine aggregate replacement. Constr Build Mater 25(2):967–972

Zelic J (2005) Properties of concrete pavements prepared with ferrochromium slag as concrete aggregate. Cem Concr Res 35(12):2340–2349

Chapter 3
Construction and Demolition Waste Aggregates

3.1 Introduction

The growth of the world population, widespread urbanisation and the economic condition of developing countries has remarkably increased the pace of development of the construction industry. As a result of these activities, old constructions are being demolished to make new buildings. Due to these large-scale demolitions, a huge amount of debris is generated all over the world, which is causing serious environmental pollutions including a disposal problem. Recently, it was reported that about 850 million tonnes of construction and demolition waste (CDW) were generated in the EU per year, representing 31 % of the overall waste generation (Fisher and Werge 2009).

Using CDW as an aggregate in the preparation of new concrete has immense potential and it has been the object of investigation for a long time. Using CDW as aggregate can reduce the use of natural aggregates and the problem of mining them. However, in comparison to natural aggregate (NA) the quality of CDW aggregate is poor (which will be addressed thoroughly in subsequent sections), which restricts its use in varieties of construction applications. The cost of concrete production containing CDW aggregate also needs to be considered in terms of its large-scale application in the construction sector. Recent implementation of stringent rules and regulations on the disposal of several types of wastes including CDW all over the world, however, can also help in the large-scale application of CDW aggregate in productions of various types of concrete productions.

Recycling CDW in the making of new construction also increases the life cycle of construction materials. To improve the recycling amount of CDW, some countries ratified some governmental laws and specific regulations. Due to these initiatives, the recycling level of this material in some countries reached about 90 % of the total generated amount.

As CDW can be used as aggregate in new concrete preparation, like NA, this type of aggregate is characterised using methods similar to those used for NA charac-

J. de Brito and N. Saikia, *Recycled Aggregate in Concrete*,
Green Energy and Technology, DOI: 10.1007/978-1-4471-4540-0_3,

terisation. In fact, aggregate occupies more than 75 % volume of concrete mix and therefore the characterisation of properties of a new material to be used as aggregate should be evaluated properly, namely because it is produced from different types of materials. To classify the different types of CDW aggregates, BS 8500 (2002) defines two types of CDW aggregates. The concrete aggregate containing a minimum of 95 % crushed concrete is defined as recycled concrete aggregate (RCA) and 100 % crushed masonry-based aggregate is defined as recycled aggregate (RC). However, here, all types of aggregates will be considered as CDW aggregates.

The grain size, specific gravity/density, water absorption, Los Angeles abrasion and crushing value of CDW aggregates are studied in detail. As CDW contains several contaminants, such as wood, plastics and gypsum, these contaminating materials should be removed before the application of CDW as an aggregate in concrete. In this section, the properties of CDW as aggregate will be discussed in detail from the existing literature data.

3.2 Preparation of CDW Aggregate

In several countries recycling plants have been established to produce CDW aggregate. For the preparation of aggregates, in general, waste concrete elements are mechanically broken into small-sized pieces. The small pieces are further crushed into small-sized pieces using crushers. After crushing, different sized fractions are screened using a sieving device and used as aggregates. Rubble from demolished concrete buildings is generally contaminated with mortar paste, gypsum and minor quantities of other substances such as wood, plastics, metals and glass. These impurities have several deleterious effects and therefore are unsuitable for concrete production. Consequently, in most of the cases, the impurities present in CDW must be separated during the process. The concrete produced in the laboratory is also used to prepare CDW aggregate. The use of laboratory produced concrete allows control of its production which can then be characterised thoroughly (Fonseca et al. 2011). A schematic diagram of two typical plants that recycle CDW as aggregate is presented in Fig. 3.1.

The production process of CDW aggregate affects its quality and composition. The original concrete used to make CDW aggregate plays a very important role in the aggregate properties. Further processing and higher quality source concrete result in better quality aggregates (Nagataki et al. 2004).

3.3 Composition of CDW Aggregates

In the literature, two types of CDW aggregates are reported. Some CDW aggregates contain natural aggregates with adhered mortar. These are produced from recycled precast concrete and test specimens. On the other hand, in some CDW aggregates,

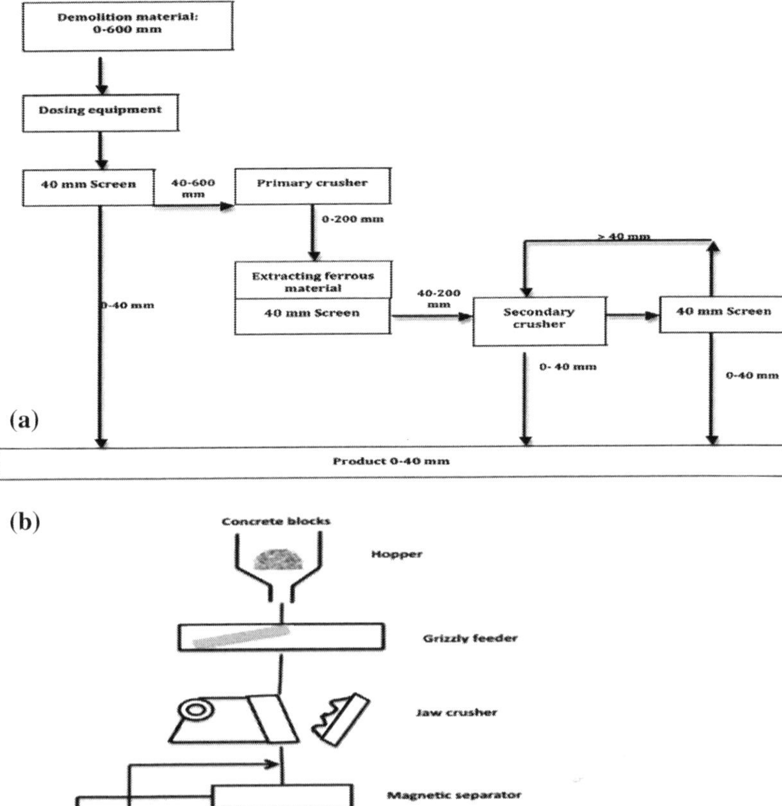

Fig. 3.1 A typical CDW recycle plant **a** Gonzalez-Fonteboa and Martinez-Abella (2008), **b** Eguchi et al. (2007)

several types of contaminants such as bitumen mixtures, plastics, bricks and tiles are present in minor amounts with natural aggregate and adhered mortar. Distributions of materials in a few construction demolition wastes reported in various published works and summarised by Coelho and de Brito (2011) are presented in Table 3.1.

The composition of CDW aggregate depends on the type of construction demolition waste used to prepare this type of aggregate. A typical CDW aggregate prepared from normal concrete block contains 65–70 % of coarse and fine sized normal aggregates and 30–35 % of cement paste (Poon et al. 2004a, b). The CDW

Table 3.1 Distributions of materials in construction demolition wastes (Coelho and de Brito 2011)

Materials	Pereira (2002)	Costa and Ursella (2003)	Reixach et al. (2000)	Franklin Associates (1998)
	Amount in %			
Concrete and ceramics	58.3	84.3	85.0	24.0
Metals	8.3	0.08	1.8	2.0
Wood	8.3		11.2	42.0
Plastics	0.83		0.20	32.0
Bituminous concrete	10.0	6.9		
Other waste	14.2	8.8	1.8	
Total	100	100	100	100

Table 3.2 Composition of coarse CDW aggregates (particle fraction: 4–16 mm) (Limbachiya et al. 2007)

Constituents	Proportions in % (m/m)		
	CDWA 1	CDWA 2	CDWA 3
Concrete	92.4	92.1	85.5
Masonry	1.9	1.6	5.3
Asphalt	4.9	1.4	3.3
Lightweight material[a]	0.0	0.6	0.5
Fines	0.2	3.4	4.4
Miscellaneous materials[b]	0.5	0.9	1.0

[a] density 1,000 kg/m^3 ; [b] glass, timber, plastic, metal, etc.

aggregate prepared from concrete used for bituminous road construction may contain an organic part. Coarse CDW aggregate with particle size of 4–16 mm as reported by Limbachiya et al. (2007) contain different amounts of asphalt in its composition (Table 3.2).

Gonzalez-Fonteboa and Martinez-Abella (2008) reported the use of CDW material obtained from real demolition debris in Spain as aggregate in the preparation of concrete. CDW with size range 0–40 mm was separated into two particle size fractions with ranges of 10–25 and 4–12 mm. The composition of these fractions is presented in Fig. 3.2.

There may be small variations of composition of CDW aggregate for different size fractions. Table 3.3 shows the composition of two size fractions of CDW aggregate generated in a recycling plant (Sani et al. 2005).

The presence of other crushed materials such as ceramic brick and tiles in CDW aggregate is also reported depending on the source of CDW. Corinaldesi and Moriconi (2009a) reported the use of a CDW aggregate collected from a recycling plant with an average composition of 70 % old concrete, 27 % bricks and tiles and 3 % miscellaneous materials (asphalt, glass, wood, paper and other similar construction debris).

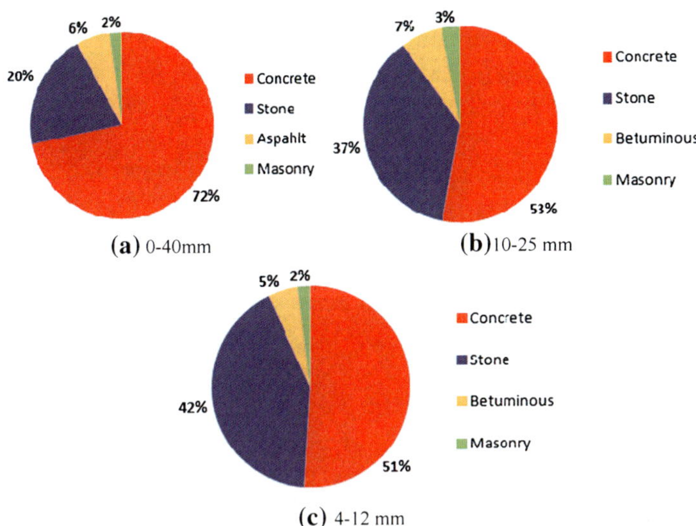

Fig. 3.2 Compositions of CDW with different size range Gonzalez-Fonteboa and Martinez-Abella (2008)

Table 3.3 Composition of two size fractions of CDW aggregate (Sani et al. 2005)

Constituents	Amount (%) in	
	Sand (0–5 mm)	Gravel (5–15 mm)
Masonry	32	25
Inert	30	29
Concrete	35	45
Bitumen	2	0.5
Wood, glass, plastic etc.	1	0.5

CDW aggregates may comprise three types of particles: some particles of natural coarse aggregates are held together and surrounded fully or partially by a layer of mortar and some lumps of mortar embedded with varying proportions of smaller natural aggregates (Akbarnezhad et al. 2011).

3.4 Attached Mortar Contents in CDW Aggregate and Methods of Evaluation

The properties of CDW aggregate are dependent on the amount of mortar content in the CDW aggregate. The amount of mortar content in CDW is dependent on the number of crushing processes in the production plants. The attached mortar content in the CDW aggregate can be reduced by increasing the number of crushing processes. However, increasing the number of crushing processes increases the

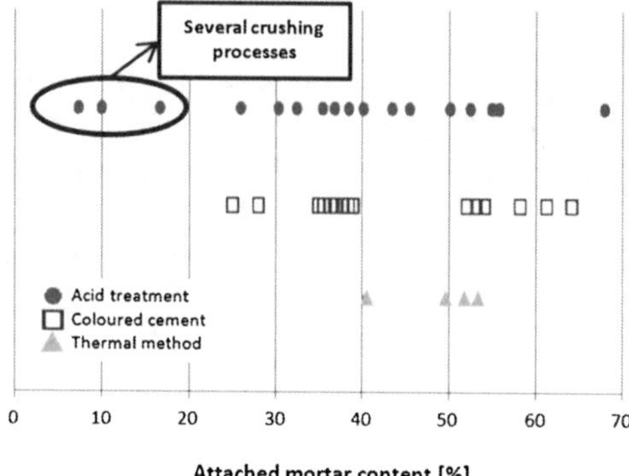

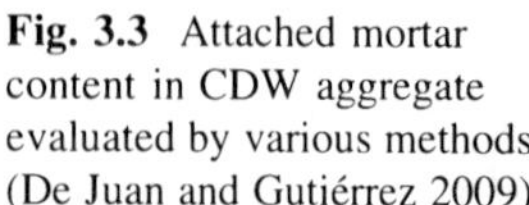
Fig. 3.3 Attached mortar content in CDW aggregate evaluated by various methods (De Juan and Gutiérrez 2009)

production costs of aggregates and therefore an optimisation is necessary to get high-quality CDW aggregate with minimal production costs. Some authors claim that the type of crusher used in the crushing process also affects the adhered mortar content in the CDW aggregate (Corinaldesi and Moriconi 2009b; Etxeberria et al. 2007a).

De Juan and Gutiérrez (2009) reported the presence of 33–55 % and 23–44 % mortar contents in 4–8 and 8–16 mm fractions of a CDW aggregate, respectively. The authors used a thermal method to determine the attached mortar content in CDW aggregate. Several authors (Etxeberria et al. 2007a; Katz 2003; Zaharieva et al. 2003) also found an increasing cement and mortar content in CDW aggregate with decreasing size range. CDW aggregate prepared from low strength concrete contains less adhered mortar than CDW aggregate prepared from high strength concrete (Etxeberria et al. 2007a; Padmini et al. 2009). The amount of mortar content in CDW aggregate has profound effects on its aggregate properties, which will be discussed later.

In the literature, four different methods are described to determine attached mortar contents (Abbas et al. 2009). In one method, the attached mortar content is determined by using a solution of dilute hydrochloric acid, where dilute acid dissolves the cement pastes without affecting the remaining aggregate fractions (Nagataki et al. 2000). However, this method is not suitable for CDW containing some types of aggregates like limestone aggregate as acid can also dissolve such types of aggregates. In the second method, a new concrete is produced using CDW aggregate with a coloured cement. The mortar surface is easily detected in a slice specimen by means of the different colour between both natural aggregate and the new mortar (Ravindrarajah and Tam 1985). In yet another method, several cycles of soaking in water and heating were done to remove attached mortar content from the surfaces of the natural aggregate present in CDW aggregate (De Juan and Gutiérrez 2009). Abbas et al. (2009) developed a rapid analysis method in which CDW aggregate is subjected to a few daily cycles of freezing and thawing in a sodium sulphate solution. A comparison of the attached mortar content in CDW aggregate determined by three different methods in various references is presented in Fig. 3.3 (De Juan and Gutiérrez 2009), which shows that the thermal method gives the lowest value and the acid treatment method the highest one.

Apart from increasing the mechanical processing steps, several other methods are proposed to remove the mortar content and therefore improve the quality of CDW aggregate (Akbarnezhad et al. 2011). These are thermal treatment, mechanical treatment, thermal–mechanical treatment, acid soaking, chemical–mechanical treatment and microwave-assisted treatment.

3.5 Properties of CDW Aggregate

3.5.1 Density/Specific Gravity

Density is one of the fundamental parameters of aggregate and is important to design concrete mixes and control several properties of the resulting concrete. Table 3.4 shows the density of recycled aggregates reported in several works. The density of CDW aggregate is lower than that of natural aggregates. This is due to the existence of porous and less dense cement paste in the CDW aggregates. Due to their origin and size, CDW aggregates may have different densities depending on the amount of adhered mortar paste. The CDW aggregates with less mortar paste have higher density than the CDW with higher content of cement paste (Fig. 3.4).

Poon et al. (2004a, b) reported a lower density for CDW aggregate obtained from high performance concrete (HPC) than that of CDW aggregate made from normal strength concrete (NC). Both concrete types were prepared by using the same type of granite aggregate but the HPC contained fly ash and silica fume as mineral additions. Santos et al. (2002a) found slight variations in the density parameters along with other properties of two types of the CDW aggregates prepared from two concrete types with different 28-day compressive strengths of 45 and 56 MPa (Table 3.4). The mortar contents in the 56 MPa and 45 MPa concretes were, respectively, 36.3 and 49.4 %. Gomez-Soberon (2002) found that the density of CDW aggregate increased with increasing particle size. De Brito and Robles (2010) and De Brito and Alves (2010) reported that the density of the mixture of CDW and normal aggregates showed high correlation coefficients in the graphical analysis for the various hardened concrete properties.

The bulk density of CDW aggregate is also lower than that of normal aggregates (Table 3.5). The bulk density of CDW aggregates is generally in the range of 1,150–1,400 kg/m^3 with a few exceptions. According to Ferreira et al. (2011), the lesser bulk density of CDW aggregate compared to natural aggregate is due to the greater volume of voids between particles in CDW aggregate. Ferreira et al. (2011) found 48.8 and 50.4 % of bulk void contents in the natural and CDW aggregates, respectively.

Table 3.4 Density/specific gravity of CDW aggregates

Reference	Origin	Types of CDW aggregates	Size of CDW aggregate (mm)	Density (kg/m^3)/specific gravity	
				CDW aggregate	Natural aggregate
Watanabe et al. (2007)	–	Coarse		2,520	
		Sand		2,470	
Topçu and Guncan (1995)	Construction site	Coarse	8–31.5	2,450[a]	
Gonzalez-Fonteboa and Martinez-Abella (2008)	Clean concrete from demolished debris	Coarse	4–12	2,350	
				2,470[b]	
		Coarse	10–25	2,370	
				2,480[b]	
Courard et al. (2010)	Bituminous road pavements	–	2–20	2,634	
Poon et al. (2007)	Unwashed demolished concrete from construction sites	Coarse	10 (maximum diameter)	2,490	
		Coarse	20 (maximum diameter)	2,570	
Rao et al. (2011)	C&D debris obtained from culvert	Coarse	–	2,470 (SSD)	
Corinaldesi et al. (2002)	Processed rubble from recycling plant	Fine	–	2,150 (SSD)	
Corinaldesi and Moriconi (2009a, b)	Recycling plant	Fine	0–5	2,290	
Miranda and Selmo (2006)	Ceramic wall prepared in the laboratory	Fine	4.8 (maximum)	2,680	
	Mortar prepared in the laboratory		0.55 (maximum)	2,600	
	Concrete prepared in the laboratory		4.8 (maximum)	2,670	
Poon et al. (2004a, b)	CDW from normal concrete	Coarse	–	2,409	
	CDW from high performance concrete	Coarse	–	2,390	

(continued)

Table 3.4 (continued)

Reference	Origin	Types of CDW aggregates	Size of CDW aggregate (mm)	Density (kg/m^3)/ specific gravity	
				CDW aggregate	Natural aggregate
Poon and Chan (2007)	Recycled concrete	Fine	0–5	2,310 (SSD)	
				2,093 (dry)	
	Recycled tiles			2,199 (SSD)	
				1,882 (dry)	
	Recycled brick			2,042 (SSD)	
				1,560 (dry)	
Eguchi et al. (2007)	CDW prepared at the laboratory	Coarse	20 (maximum)	2,220–2280	
	CDW from a building in a power plant	Coarse	20 (maximum)	2,221–2,231	
Tangchirapat et al. (2008)	CDW prepared in laboratory	Fine	0–4.5	2,310	
		Coarse	4.5–25	2,450	
Courard et al. (2010)	Bituminous road pavements	Fine and coarse	2–20 mm with about 86.5 % 10–20 mm	2,634	
Topçu (1997)	Laboratory made C16 concrete	Coarse	–	2,450	

(continued)

Table 3.4 (continued)

Reference	Origin	Types of CDW aggregates	Size of CDW aggregate (mm)	Density (kg/m^3)/specific gravity	
				CDW aggregate	Natural aggregate
Gomez-Soberon	150-day cured laboratory made concrete and then ground by roller grinder	Coarse	10–20	2,280 (dry)	
				2,410 (SSD)	
			5–12	2,260 (dry)	
				2,420 (SSD)	
		Fine	0–5	2,170 (dry)	
				2,380 (SSD)	
Fonseca et al. (2011), Amorim et al. (2012)	Laboratory made concrete with 28-day strength of 39.6 MPa	Coarse	4–25.4	2,310 (OD)	2,550 (OD)
				2,450 (SSD)	2,580 (SSD)
Evangelista and de Brito (2007, 2010)	Laboratory made concrete with 28-day strength of 29.6 MPa	Fine	<4	1,913 (OD)	2,544 (OD)
				2,165 (SSD)	2,564 (SSD)

(continued)

Table 3.4 (continued)

Reference	Origin	Types of CDW aggregates	Size of CDW aggregate (mm)	Density (kg/m^3)/ specific gravity	
				CDW aggregate	Natural aggregate
Gomes and de Brito (2009)	35-day crushed laboratory made C30/37 concrete	Coarse	<25.4	2,448 (OD) 2,526 (SSD)	2,550 (OD) 2,573 (SSD)
	Mortar and brick mixture from a 6-month-old wall	Coarse	<25.4	2,160 (OD) 2,301 (SSD)	
Ferreira et al. (2011)	Collected from a mobile recycling plant of CDW waste	Coarse	4–25.4	2,300 (OD) 2,440 (SSD)	2,600 (OD) 2,640 (OD)
Santos et al. (2002a)	CDW waste from a demolished stadium	Coarse	4–25.4	2,460 (SSD)	2,660
Santos et al. (2002a)	Laboratory made concrete with two different strength classes: A: 56 MPa; B: 45 MPa	Coarse	19 (maximum)	A: 2,364 (OD) 2,480 (SSD) B: 2,329 (OD) 2,458 (SSD)	2,620 (OD) 2,657 (SSD)

[a] Specific gravity; [b] water saturated density; *SSD* saturated surface dry; *OD* oven dry

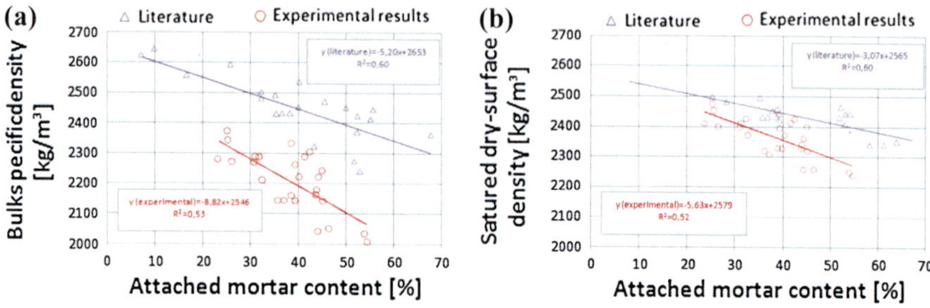

Fig. 3.4 Relationship between attached mortar content and density (De Juan and Gutiérrez 2009). **a** Bulk specific density; **b** Saturated surface dry density

3.5.2 Water Absorption

The water absorption capacity of CDW aggregate is higher than that of normal aggregate (which is less than 1 % for almost all current aggregates) as CDW aggregate is composed of cement paste, which is porous by nature and therefore can absorb high amounts of water. The water absorption capacity of various types of CDW aggregates is presented in Table 3.6. The variation of water absorption capacity reported in various references is due to the variation of cement paste content in this aggregate as well as the content of other components such as crushed clay brick and tiles, which have very high water absorption capacity. De Juan (2004) reports that water absorption of natural aggregate is between 0 and 4 %, while for adhered mortar it is between 16 and 17 %, to give water absorption of CDW aggregate in between 0.8 and 13 %, with an average of 5.6 %. Concrete rubble from most demolished buildings contain concrete materials along with crushed clay bricks (CCB), tiles and other materials, which is not only costly but also technically impossible to remove from concrete components.

Vieira et al. (2011) plotted the water absorption capacity of CDW aggregate as a function of time (Fig. 3.5). Their absorption graph reveals that the water absorption of CDW aggregate occurs mainly during the initial period of immersion, reaching about 80 % of its absorption potential after only 5 min of immersion. After this period, the increase in water absorption is much slower, tending to a value of 84 % after 30 min. This type of measurement can help to predict the behaviour of CDW aggregate during concrete mixing and therefore help to determine the amount of additional water to be introduced in the mix to compensate for the extra absorption of the CDW aggregates. De Juan (2004) also reported that the 70–90 % of water absorption potential of CDW was utilised after 10 min of submersion.

CDW aggregate obtained from lower W/C ratio concrete, and so with lower porosity and higher mechanical strength, has lower water absorption (Hansen and Narud 1983; Santos et al. 2002b). The particle size of CDW aggregate has a significant effect on its water absorption capacity: the fine CDW aggregate has higher water absorption capacity than the coarse CDW aggregates due to relatively

Table 3.5 Bulk densities of CDW aggregates

Reference	Origin	Types of aggregates	Aggregate size (mm)	Bulk density (kg/m^3)	
				CDW	Natural
Topçu and Guncan (1995)	Construction site	Coarse	8–31.5	1,161 (loose)	
Topçu (1997)	Laboratory made C16 concrete	–	–	1,161	
Rao et al. (2011)	C&D debris obtained from culvert	Coarse	–	1,340	
Miranda and Selmo (2006)	Ceramic wall prepared in the laboratory	Fine	4.8 (maximum)	1,270	
	Mortar (cement, lime and sand) prepared in the laboratory		0.55 (maximum)	1,370	
	Concrete prepared in the laboratory		4.8 (maximum)	1,530	
Eguchi et al. (2007)	CDW prepared in the laboratory	Coarse	20 (maximum)	1,260–1,310	
	CDW from a building in a power plant			1,280–1,370	
Park et al. (2005)	–	Coarse	5–13 mm	1,402	
Olorunsogo and Padayachee (2002)	Collected from local supplier	Coarse	19 mm (maximum)	1,397 (compact); 1,362 (loose)	1,458 (compact); 1,344 (loose)
Fonseca et al. (2011), Amorim et al. (2012)	Laboratory made concrete with 28-day strength of 39.6 MPa	Coarse	4–25.4	1,170	1,440
Evangelista and de Brito (2007, 2010)	A laboratory made concrete with 28-day strength of 29.6 MPa	Fine	<4	1,234	1,517
Ferreira et al. (2011)	Collected from a mobile recycling plant of CDW waste	Coarse	4–25.4	1,140	1,330
Santos et al. (2002a)	Laboratory made concrete with two different strength classes: A: 56 MPa; B: 45 MPa	Coarse	19 (maximum)	A: 1,372 B: 1,393	1,534

Table 3.6 Water absorption capacity and porosity of CDW aggregate

Reference	Origin	Types of aggregates	Aggregate size (mm)	Water absorption (%)	
				CDW aggregate	Natural aggregate
Topçu and Guncan, (1995)	Construction site	Coarse	8–31.5	7	
Gonzalez-Fonteboa and Martinez-Abella (2008)	Clean concrete from demolished debris	Coarse	4–12	4.82	
			10–25	4.59	
Courard et al. (2010)	Bituminous road pavements	Fine and coarse	2–20 mm with about 86.5 % 10–20 mm	4.58	
Poon et al., (2007)	Unwashed demolished concrete from construction sites	Coarse	10 (maximum diameter)	4.3	
			20 (maximum diameter)	3.5	
Rao et al. (2011)	C&D debris obtained from culvert	Coarse	–	3.1	
Corinaldesi et al. (2002)	Processed rubble from recycling plant	Fine	0–5	10	
Corinaldesi and Moriconi (2009a)	Rejected precast concrete	Fine	0–5	8.8	
	Processed rubble from recycling plant[a]			7.1	
	Crushed and processed red brick			16.2	
Miranda and Selmo (2006)	Ceramic wall prepared in the laboratory	Fine	4.8 (maximum)	11.5	
	Mortar (cement, lime and sand) prepared in the laboratory		0.55 (maximum)	1.0	
	Concrete prepared in the laboratory		4.8 (maximum)	2.0	
Poon and Chan (2007)	Recycled concrete	Fine	0–5	10.3	
	Recycled brick			30.9	
	Recycled tiles			16.9	
Eguchi et al. (2007)	CDW prepared at the laboratory	Coarse	20 (maximum)	5.69–7.92	
	CDW from a building in a power plant			6.22–7.66	
Tangchirapat et al. (2008)	CDW prepared in laboratory	Fine	0–4.5	11.91	
		Coarse	4.5–25	5.61	
Li et al. (2009)	Old concrete pavement	Coarse	5–31.5	8.7	

(continued)

Table 3.6 (continued)

Reference	Origin	Types of aggregates	Aggregate size (mm)	Water absorption (%)	
				CDW aggregate	Natural aggregate
Corinaldesi and Moriconi (2009b)	Processed rubble from recycling plant	Coarse	15 (maximum)	8	
		Fine	5 (maximum)	10	
Katz (2003)	CDW prepared in laboratory from concrete cured at different time periods	Coarse	>9.5	1 day:3.2 3 day:3.4 28 day:3.3	
		Medium	2.36– 9.5	1 day:9.7 3 day:8.1 28 day:8.0	
		Fine	<2.36	1 day:11.2 3 day:11.4 28 day:12.7	
Tam et al. (2007)	Recycling plant	Coarse	20	1.65	0.77
			10	2.63	0.57
Fonseca et al. (2011), Amorim et al. (2011)	Laboratory made concrete with 28-day strength of 39.6 MPa	Coarse	4–25.4	6.1	1.5
Evangelista and de Brito (2007, 2010)	Laboratory made concrete with 28-day strength of 29.6 MPa	Fine	<4	13.1	0.8
Gomes and de Brito (2009)	35-day crushed laboratory made C30/37 concrete	Coarse	<25.4	8.49	2.29
	Mortar and brick mixture from a 6-month-old wall	Coarse	<25.4	16.34	
Ferreira et al. (2011)	Collected from a mobile recycling plant of CDW waste	Coarse	4–25.4	5.8	1.2
Santos et al. (2002a)	CDW waste from a destructed stadium	Coarse	4–25.4	6.0	0.8
Santos et al. (2002a)	Laboratory made concrete with two different strength classes: A: 56 MPa; B: 45 MPa	Coarse	19 (maximum)	A: 4.9 B: 5.5	1.14

[a] With a composition of 72 % concrete, 25 % masonry and 3 % bitumen

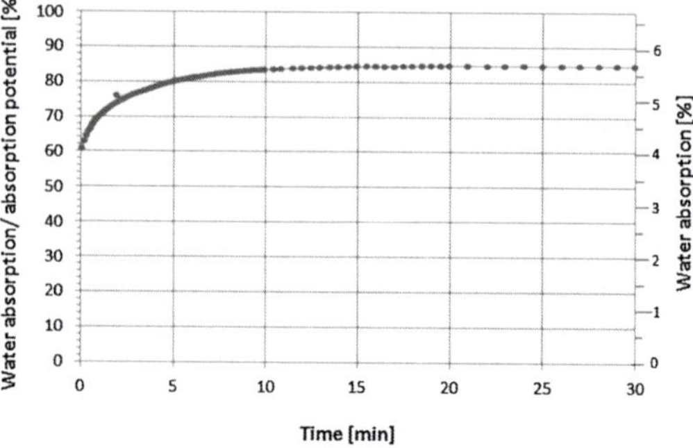

Fig. 3.5 Absorption of water by CDW aggregate with respect to increasing immersion time (Vieira et al. 2011)

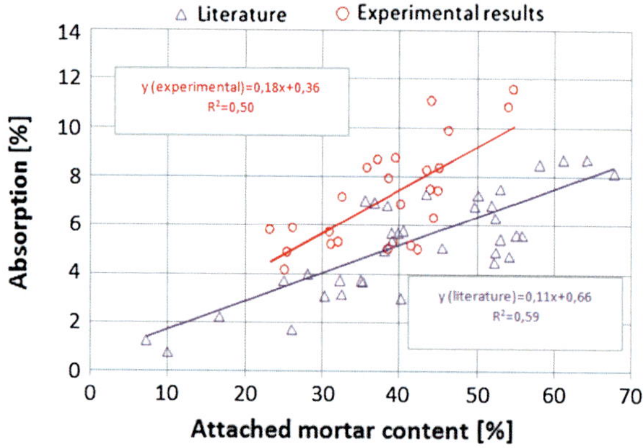

Fig. 3.6 Relationship between attached mortar content and water absorption capacity of CDW aggregate (De Juan and Gutiérrez 2009)

higher amounts of adhered mortar than that in coarse-sized CDW aggregate. On the other hand, CDW aggregates prepared with different curing ages have almost identical water absorption capacities (Katz 2003). According to Katz (2003), the difference between the size groups seem to be a result of the relative amounts of cement paste or adhered mortar content in the crushed material regardless of its age at crushing. The relationship between the attached mortar content and water absorption capacity of CDW aggregate is presented in Fig. 3.6. De Juan and Gutiérrez (2009) also established an inverse relationship between CDW aggregate density and water absorption capacity (Fig. 3.7).

The fast and higher water absorption capacity of CDW aggregate by comparison with that of natural aggregate implies a lower degree of workability for the same water/cement ratio of concrete containing CDW aggregate than that produced with natural aggregates and therefore a greater amount of additives are necessary in order to make up for the loss of workability of concrete prepared by using CDW.

Pre-soaking of CDW aggregate before preparation of concrete can prevent the suction of the mixing water (Zaharieva et al. 2003). However, complete saturation of CDW aggregates may increase the bleeding during preparation of concrete mix (Poon et al. 2004b) and affect the mechanical performance of resulting concrete

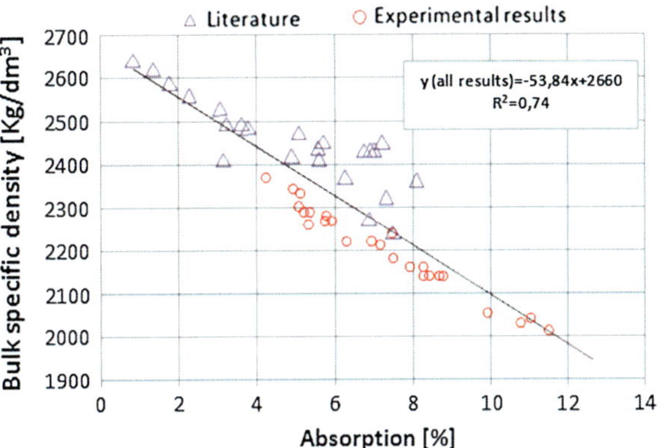

Fig. 3.7 Relationship between density and water absorption capacity of CDW aggregate (De Juan and Gutiérrez 2009)

due to the formation of a weak interfacial transition zone between the saturated recycled coarse aggregates and the new cement paste (Etxeberria et al. 2007a). Etxeberria et al. (2007a) recommended a humidity level of 80 % of the total absorption capacity of CDW aggregate for better performance of concrete containing CDW aggregate. Oliveira and Vazquez (1996) reported that the mechanical performance of concrete prepared by pre-saturated followed by 30 min air-dried CDW aggregate is better than the performances of concrete containing oven dry as well as saturated surface dry CDW aggregates. The humidity level of air-dried CDW aggregate was about 90 % with respect to saturated surface dry aggregate.

In some investigations, an extra amount of water is added to the concrete mix corresponding to the water absorbed by the CDW aggregate (Matias and de Brito 2004; Oliveira and Vázquez 1996; Santos et al. 2005). Ferreira et al. (2011) describe this method as mixing water compensation method. The amount of water added depends on the initial water content and effective absorption of CDW aggregate during the mixing period. Potential water absorption and absorption evolution with time should also be known in order to predict the water to cement ratio (W/C) after the mixing period. This ought to guarantee that the water added does indeed correspond to the amount of water absorbed by the CDW aggregate.

The mixing water compensation method has the advantage of making it possible to produce both concrete containing CDW aggregate and conventional concrete in a similar way. However, Oliveira and Vázquez (1996) note that the water absorption of CDW aggregate may not correspond to the free water absorption determined in the laboratory, since the pores of CDW aggregate are filled with cement paste during mixing, which may lead to an excess of water in the mix, and thus an undesirable increase in the effective water to cement ratio (W/C).

Ferreira et al. (2011) used a different approach to keep the effective water to cement ratio (W/C) constant in the different concrete compositions containing CDW aggregate to ensure compensation during the mixing process. A schematic diagram of the procedure is presented in Fig. 3.8. To achieve this goal, these authors added directly to the concrete mix an additional amount of water equivalent to the absorption expected from CDW aggregate during the mixing process,

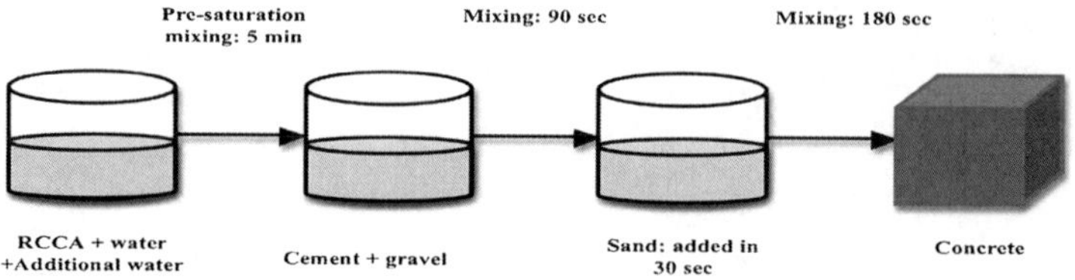

Fig. 3.8 A typical mixing procedure of concrete mix containing CDW aggregate (Ferreira et al. 2011)

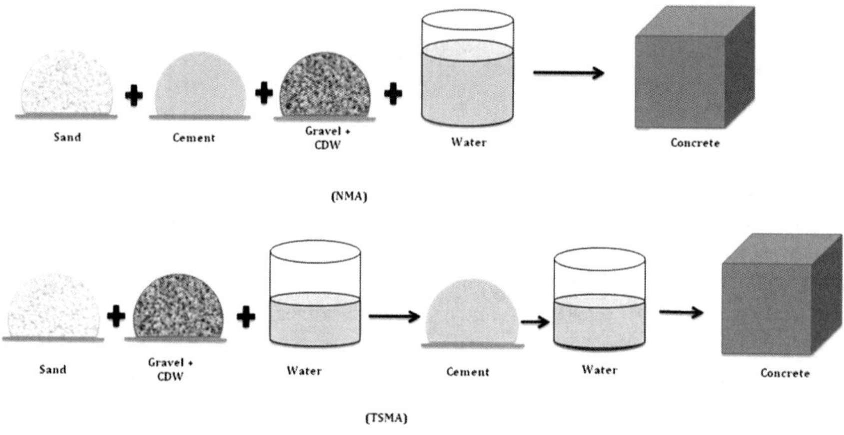

Fig. 3.9 Normal (NMA) and two-stage (TSMA) concrete mixing procedures adopted by Tam et al. (2008)

and created conditions for its absorption. This additional amount of water equivalent is called effective water absorption (Barra and Vázquez 1998). Santos et al. (2002b) found 1.70 and 1.82 % of effective water absorptions for two types of concrete. The test procedure consists in determining the evolution of water absorbed by the immersed CDW aggregate sample, previously oven dried, by means of a hydrostatic balance.

Tam et al. (2008) followed a two-stage mixing approach (TSMA) to compensate the higher water absorption capacity of CDW aggregate. The schematic diagram of this procedure along with normal concrete mixing approach (NMA) adopted in their study is presented in Fig. 3.9. According to the authors, the use of half of the required water for mixing leads to the formation of a thin layer of cement slurry on the surface of CDW aggregate, which permeates into the porous old cement mortar and fills the old cracks and voids (Tam et al. 2007). A stronger interfacial transition zone (ITZ) is thus developed by effectively developing some strength enhancing chemical products, namely etringite, portlandite and calcium silicate hydrate.

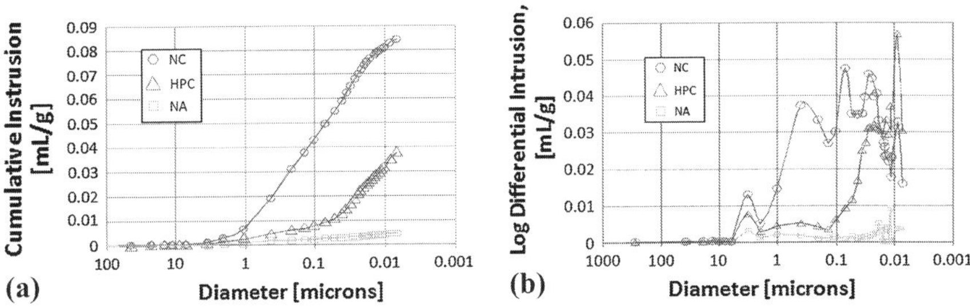

Fig. 3.10 Mercury intrusion porosity of CDW aggregate (Poon et al. 2004a, b). **a** Cumulative pore size distribution; **b** differential pore size distribution

3.5.3 Porosity

As the CDW aggregate has higher water absorption capacity than the normal aggregate, therefore it has higher water accessible porosity than normal aggregate. Poon et al. (2004a) evaluated the mercury intrusion porosity of natural coarse aggregate (NA), CDW aggregate prepared from normal concrete (NC) and CDW aggregate prepared from high-performance concrete (HPC), which are presented in Fig. 3.10. The authors reported that the porosities of NA, NC and HPC were 1.60, 16.81 and 7.86 %, respectively. According to the authors, the higher porosity of CDW aggregates is attributed to the adhered cement paste. The pores in the NC aggregate mainly distributed between 0.01 and 1 mm, whereas the majority of pores in the HPC aggregate located in the region of less than 0.1 mm (Fig. 3.10b). The finer pore size distribution of the recycled HPC was due to the use of pozzolanic admixtures, which substantially improved the microstructure of the cement paste and the paste–aggregate interfacial transition zone (ITZ) (Poon et al. 2004a).

3.5.4 Mechanical Properties of CDW Aggregate

3.5.4.1 Los Angeles Abrasion

The aggregate abrasion value is defined as the percentage loss in weight by abrasion, so that a high value denotes low resistance to abrasion (Neville 1981). Several tests are performed to evaluate the aggregates' abrasion value. Of these, the Los Angeles abrasion test is more commonly used all over the world and therefore the abrasion values obtained from this test reported in various references are considered in this section. According to ASTM C-33, "Standard specification for concrete aggregates," the Los Angeles abrasion value should be less than 50 % for aggregate used to produce concrete, and should be less than 40 % for aggregate used to make roads.

Table 3.7 Los Angeles abrasion value of CDW aggregate

Reference	Aggregate size (mm)	Los Angeles abrasion (%)	
		CDW aggregate	Normal aggregate
Gonzalez-Fonteboa and Martinez-Abella (2008)	5–40	39.65	–
Gonzalez-Fonteboa and Martinez-Abella (2007)	4–12	32	32
	10–25	34	27
Courard et al. (2010)	10–20	25	–
Rao et al. (2011)	4–20	37.1	21.56
Tangchirapat et al. (2008)	5–30	33.08	21.7
López-Gayarre et al. (2009)	4–20	37.2, 33.1	24, 26.4
Li et al. (2009)	–	20 (crushed by jaw crusher followed by an impact crusher),	–
		24.2 (crushed by jaw crusher)	–
Fonseca et al. (2011)	4–25.4	42.7	29.5
Gomes and de Brito (2009)	<25.4	37.96	28.52
		65.47	

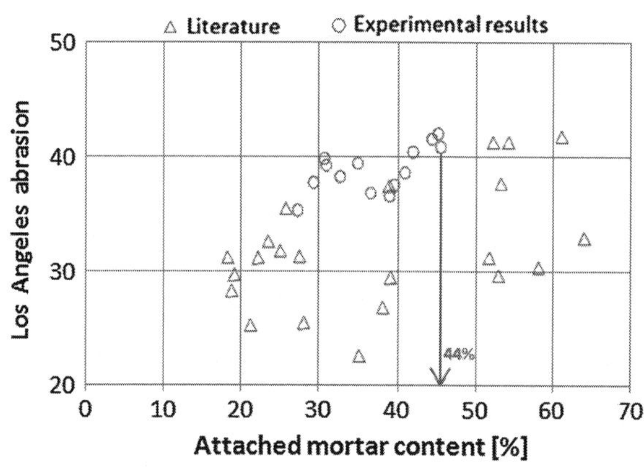

Fig. 3.11 Relationship between attached mortar content and Los Angeles abrasion value (De Juan and Gutiérrez 2009)

Table 3.7 shows some Los Angeles values for CDW as well as normal aggregate reported in various references. In general, CDW aggregate has a lower abrasion value than natural aggregate because of the presence of adhered mortar, which disintegrates during abrasion along with some parts of natural aggregate. However, this value for the majority of the CDW aggregate reported in the literature meets the various standard requirements for concrete and road constructions. De Juan and Gutierrez (2009) found an increasing trend of the Los Angeles value with increasing adhered mortar content, which is presented in Fig. 3.11. Although it is not clear, the particle size of CDW aggregate might have some effect on its Los Angeles abrasion value. Hansen and Narud (1983) found the abrasion value of 22.4 and 41.4 % for the same type of CDW aggregates with size ranges of 16–32 mm and 4–8 mm, respectively. On the other hand, Gonzalez-

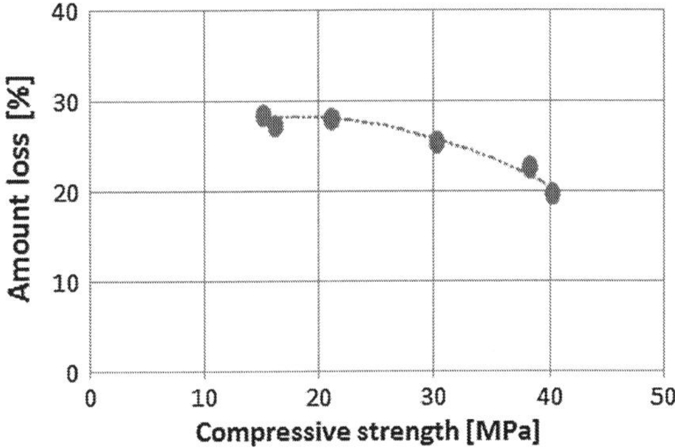

Fig. 3.12 Relationship between Los Angeles abrasion value and compressive strength (Topçu 1997)

Fonteboa and Martınez-Abella (2007) found lower abrasion values for coarse aggregates than that for fine aggregates.

The Los Angeles abrasion value of CDW aggregate prepared from high-strength concrete is higher than that prepared from low-strength concrete (Hasaba et al. 1981; De Juan and Gutiérrez 2009; Tabsh and Abdelfatah 2009). The change in Los Angeles value with respect to compressive strength value is presented in Fig. 3.12 (Topçu 1997). The Los Angeles abrasion value of CDW aggregate can be improved by changing the crushing procedure. Li et al. (2009) determined the Los Angeles abrasion value of 24.2 % for CDW aggregate crushed using a jaw crusher, and a value of 20 % when an impact crusher is used after crushing of CDW aggregate by a jaw crusher.

3.5.4.2 Strength and Toughness

The strength of aggregate affects the strength properties of the resulting concrete. The strength of aggregate depends on the composition, texture and structure of aggregate (Neville 1981). In some reports, the crushing strength of CDW aggregate is reported as a 10 % fines value. In the 10 % fines value determination test, a higher numerical result denotes a higher strength of aggregate (Neville 1981). According to BS 882: 1992, a minimum value of 150 kN is required for aggregate to be used in structural elements. On the other hand, a minimum of 100 kN is required for minor structural elements, pre-stressed concrete elements, road construction and insulation barriers (BS 882: 1992 and ASTM D448-03: 2000). According to BS 882:1992 and ASTM 2940-03:2000 specifications, a 10 % crushing value of 50 kN is sufficient for aggregate to be used in concrete prepared for applications such as non-structural element, base course, embankment and fill (Tam and Tam 2007).

The strength of CDW aggregate, reported as the 10 % crushing value, is presented in Table 3.8. The crushing value of CDW aggregate is in all cases lower than that of normal aggregate. However, in the majority of the cases, this value meets the 100 kN criteria for various applications specified in the BS and ASTM standards.

Table 3.8 10 % crushing value of CDW aggregate and normal aggregate

Reference	10 % crushing value (kN)	
	CDW aggregate	Normal aggregate
Poon et al. (2009)	117.0	159.0
Rao et al. (2011)	120.5	231.3
Tam and Tam (2007)	61–155.6	189.4
Poon et al. (2004a)	NC: 101.9; HPC: 123.8	159.7
Poon et al. (2009)	72, 88	–
Limbachiya et al. (2000)	160	289

NC CDW aggregate prepared from normal concrete
HPC CDW aggregate prepared from high performance concrete

The crushing value of CDW aggregate is also higher than that of normal aggregate. Limbachiya et al. (2000) reported a crushing value of 20 % for CDW aggregate with size 10–14 mm compared to a crushing value of 14 % for natural aggregate of similar size. A similar trend is observed by several researchers (Sagoe-Crentsil et al. 2001; Xiao et al. 2005). Like other parameters of CDW aggregate, the crushing value of CDW aggregate also varies depending on the crushing technique. Li et al. (2009) found a crushing value of 27 % for CDW aggregate crushed using a jaw crusher, and a crushing value of 23 % when an impact crusher is used after crushing CDW aggregate with a jaw crusher.

The toughness of the aggregate is determined using the aggregate impact value. Aggregate with high toughness exhibits a low impact value. An acceptable limit of this value is between 25 and 45 %. In general, 25 % is specified for heavy-duty concrete elements, 35 % for sub-base applications and 30 % for other low-grade applications (Tam and Tam 2007). The impact value of CDW aggregate is higher than that of normal coarse aggregate (Table 3.9). According to Rao et al. (2011), this is due to the separation and crushing of adhered mortars in CDW aggregate during testing.

3.5.5 Particle Shape and Texture

The surface of CDW aggregate is rough and porous due to the presence of mortar adhered (Rao et al. 2007; Domingo et al. 2010). According to Topçu (1997), the shape of CDW aggregate is angular because of debris. Limbachiya et al. (2000) used a CDW aggregate, which was found to be coarser, porous and rougher but equi-dimensional to that of natural gravel. Poon et al. (2004a) reported a CDW aggregate, which is more inhomogeneous, porous, less dense and weaker than crushed granite. From the optical and scanning electron microscopy (SEM) studies of CDW aggregate, Malhotra (1976) concluded that the particles of CDW aggregate tended to be more angular than those of natural aggregate (NA). According to Zaharieva

Table 3.9 Impact value of CDW aggregate along with normal aggregate

Reference	10 % crushing value (%)	
	CDW aggregate	Normal coarse aggregate
Rao et al. (2011)	35.0	17.37
Wong et al. (2007)	17.89	10.25
Limbachiya et al. (2000)	23.7	19.7

Table 3.10 Shape parameters of CDW and natural aggregate

Reference	Parameters	CDW aggregates	Natural aggregates
Etxeberria et al. 2007b	Shape index (%)	28	25
Gomez-Soberon (2002)	Shape coefficient	10–20 mm: 0.363 5–12 mm: 0.444	10–20 mm: 0.364 5–12 mm: 0.576
	Elongation index	10–20 mm: 6 5–12 mm: 8	10–20 mm: 15 5–12 mm: 19
Gonzalez-Fonteboa et al. (2011)	Flakiness index (%)	4–20 mm: 7	8–20 mm: 7 4–12 mm: 14
Gonzalez-Fonteboa and Martinez-Abella (2007)	Flakiness index (%)	4–12 mm: 9 10–25 mm: 7	4–12 mm: 25 10–25 mm: 11
Tam et al. (2008)	Flakiness index (%)	10 mm: 10.44–17.82 20 mm: 5.70–12.96	-
Vieira et al. (2011)	Shape index (%)	22.3	14.0–18.3
Fonseca et al. (2011)	Shape index (%)	24.3	11.1
Ferreira et al. (2011)	Flakiness index (%)	10	9

et al. (2003), CDW aggregate presents a cracked surface which contributes to an increase in water and air flows into the aggregates and between the cement paste and the aggregates. The natural and CDW aggregates used in Gonzalez-Fonteboa and Martinez-Abella (2008) study were angular with multiple cracking faces.

The shape and flakiness indices of various CDW aggregates reported in the literature are not much different from those of natural aggregates used in the same studies. Table 3.10 shows some shape parameters for CDW and natural aggregates reported in various studies. Chen et al. (2003) analysed the particle shape of CDW and natural coarse aggregates by measuring several types of axis lengths of both types of aggregates and found similar particle shapes for both types of aggregates. On the other hand, through the analysis of shape indices of natural and CDW aggregates, Vieira et al. (2011) concluded that CDW aggregate is sharper than the normal coarse aggregate since the shape index of CDW aggregate is about 34 % higher than that of the normal aggregate. Etxeberria et al. (2007b) reported the shape indices of CDW and conventional aggregates as 0.28 and 0.25, respectively. According to the author, the better shape of the CDW aggregates facilitated their use for concrete production.

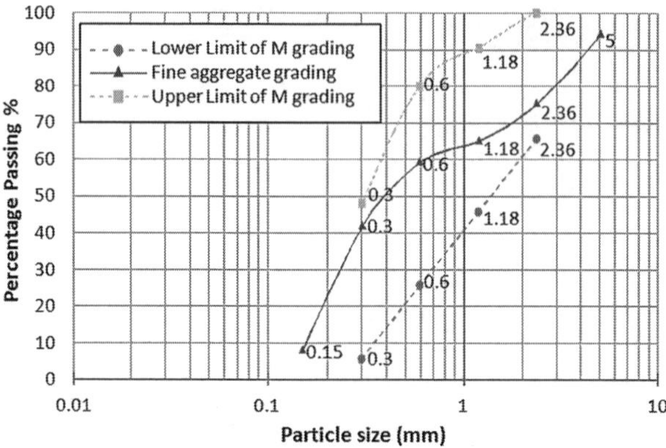

Fig. 3.13 Grading curve of fine CDW aggregate along with limits for M grading according to BS EN12620 (Yang et al. 2011)

3.5.6 Grading Size

The size distributions of coarse and fine CDW aggregates are generally different from the corresponding fractions of natural aggregates. However, in a production plant and also in several studies, CDW aggregate is produced by adopting crushing and screening processes and therefore CDW aggregate generally falls within the limits of mixing gradation for preparation of required types of concrete. The use of similar crushing technique with the same maximum size (or if the crusher is set a specific opening) generates CDW aggregates with almost similar grading behaviour (Katz 2003; Chen et al. 2003). Yang et al. (2011) compared the sieve analysis results of fine CDW aggregate with the grading limits from British standard, BS EN12620, and concluded that fine CDW aggregate used in their study could be categorised as medium class, as the passing percentage of 150 μm fraction of fine CDW aggregate fell in the region of medium grading limits (Fig. 3.13).

The particle sizes of the fine CDW aggregates used by Khatib (2005) are similar but coarser than those of natural class M sand. Tangchirapat et al. (2008) and Zaharieva et al. (2003) also reported that the CDW fine aggregate is generally coarser than normal sand. The fineness modulus of CDW and natural aggregates, reported in various references, are presented in Table 3.11.

Chen et al. (2003) found similar grade size distributions in the fine and coarse fractions of two types of CDW aggregates, which are slightly different in their compositions but processed by similar crushers with the same maximum size (Fig. 3.14).

Corinaldesi et al. (2002) and Vieira et al. (2011) found different grading curves for natural and CDW fine aggregates, which are presented in Fig. 3.15. For this reason, Vieira et al. (2011) separated various fractions of CDW aggregates so that the grading curve of CDW aggregate matched the grading curve of natural aggregate with similar fineness modulus. Although this type of procedure is difficult in practical terms, it enables comparisons between mix compositions with

Table 3.11 Fineness modulus of the CDW and natural aggregates reported in various references

Reference	Fineness modulus of	
	CDW aggregate (type of aggregate)	Natural aggregate (type of aggregate)
Chen et al. (2003)	2.61–2.68	2.95 (fine)
Nagataki et al. (2004), Gokce et al. (2004)	6.39–6.69	6.48 (coarse)
Tu et al. (2006)	6.35 (coarse); 2.74 (fine)	2.78 (fine)
Evangelista and de Brito (2007, 2010)	2.38 (fine)	2.38 (fine)
Tangchirapat et al. (2008)	3.55 (fine); 6.40 (coarse)	3.04 (fine); 6.04 (coarse)
Rao et al. (2011)	6.68 (coarse)	6.78 (coarse)
Lin et al. (2004)	6.75 (coarse); 3.10 (fine)	3.28 (fine)

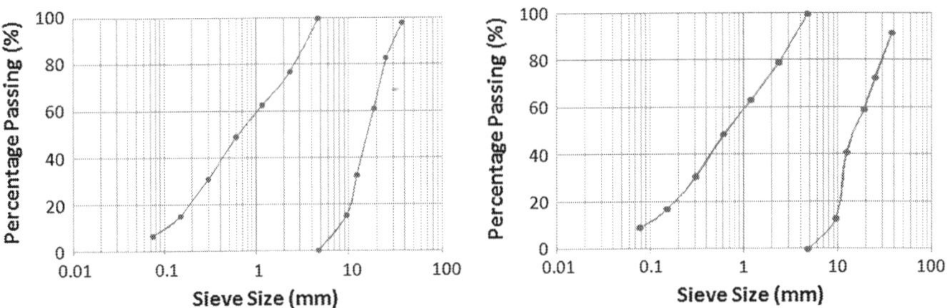

Fig. 3.14 Grading size distributions of fine and coarse fraction of two types of CDW aggregates (Chen et al. 2003)

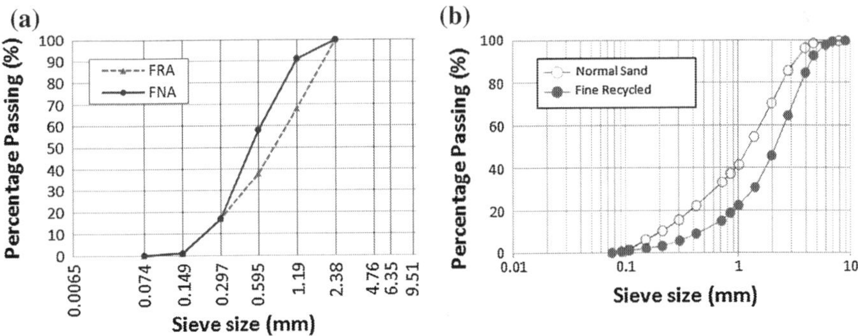

Fig. 3.15 Grading curve of fine natural (FNA) and fine CDW (FRA) aggregates: **a** Vieira et al. (2011) (FNA and FRA: fine natural and fine CDW aggregates); **b** Corinaldesi et al. (2002)

the same particle size distribution, even though the replacement ratios differ. The fine and coarse CDW aggregates used by Tu et al. (2006) also do not follow the ASTM C33 grading standard to use in concrete preparation and therefore the authors remixed these aggregates to meet the requirement. Gonzalez-Fonteboa

Table 3.12 Durability properties of CDW and natural aggregates

Reference	Sulphate soundness (%) (type of aggregate)		Frost resistance	
	CDW	Normal		
Zaharieva et al. (2003)	25.7 (fine); 26.4 (coarse)	3.8 (coarse)	26.7 (coarse)	5.6 (coarse)
Tabsh and Abdelfatah (2009)	9–14 (coarse)	<9	(Coarse)	
Lin et al. (2004)	17.9 (coarse); 10.8 (fine)	–		
Gokce et al. (2004)	9.1 (coarse); 2.6 (fine)			

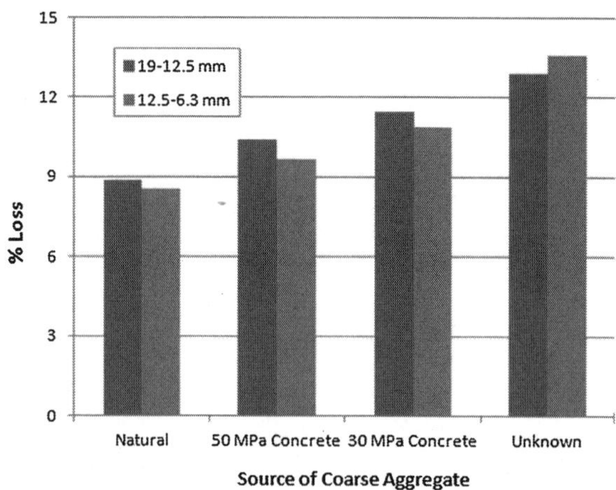

Fig. 3.16 Soundness of normal and CDW aggregate in sulphate solution (Tabsh and Abdelfatah 2009)

et al. (2011) mixed the two fractions of natural coarse aggregates to get the similar grading curve of coarse CDW aggregate. Similarly, Yang et al. (2011) remixed the coarse CDW aggregate to meet the required specifications.

3.5.7 Other Properties

Aggregate durability properties such as soundness, frost vulnerability are very important for the evaluation of aggregate to use concrete for various purposes. The durability performance of CDW aggregate is generally inferior to that of normal aggregate. Some durability properties reported in the literature for CDW and natural aggregates are presented in Table 3.12. The soundness of CDW aggregate in sulphate solution is considerably poorer than that of normal aggregate. However, some CDW aggregates meet the standard requirement (Tabsh and Abdelfatah 2009). Tabsh and Abdelfatah (2009) found better soundness of fine CDW aggregate than the coarse one for CDW aggregate produced in the laboratory. However, the fine fraction of CDW aggregate collected from a dumping ground exhibits better soundness behaviour than the coarse fraction (Fig. 3.16).

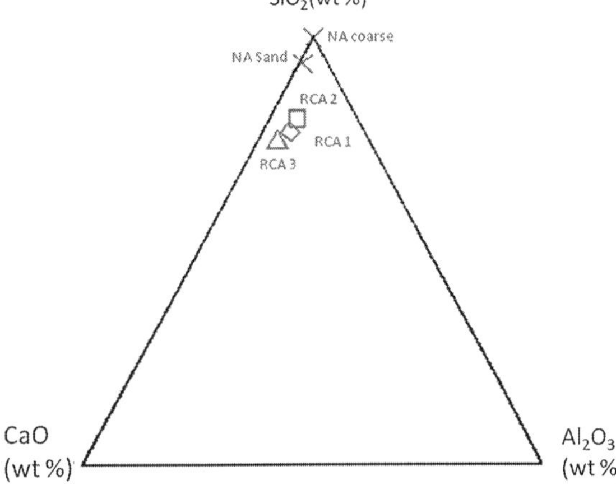

Fig. 3.17 Position of natural aggregate (NA) and CDW aggregates (RCA) in CaO-SiO_2-Al_2O_3 ternary diagram (Limbachiya et al. 2007)

Several reports indicate that CDW aggregate contains slightly higher amounts of chlorides and sulphates than normal aggregate. However, this amount is generally lower than the limits set for them in standard specifications. Tam and Tam (2007) determined chloride and sulphate contents in several CDW aggregates and compared the results with standard specifications. Their results suggest that sulphate and chloride content in all the aggregates, except one type, meet the standard specifications. Rahal (2007) also found slightly higher amounts of chloride (0.3 %) in CDW aggregate than in normal aggregate (0.14 %). Corinaldesi (2009) did not find any organic and alkali silica reactive materials in CDW fine aggregate. They also found low chloride and sulphate contents, which are lower than the standard specifications. To evaluate the potential surface activity of fine particles (clay, organic materials and ferrous hydroxides) present in CDW aggregate, Courard et al. (2010) analysed its fine fractions, those passing through the 80 μm sieve, by measuring the methylene blue value, according to the standard specified method described in Belgian standard, NBN B11-210. They found very low methylene blue value, which means that the aggregate is acceptable for use in concrete.

Chemically, CDW aggregate contains lower amounts of SiO_2 and higher amounts of CaO and Al_2O_3, the three major oxides, than the contents of those in natural fine and coarse aggregates. The CaO-SiO_2-Al_2O_3 ternary diagram for these three major oxides present in CDW and natural aggregate as reported by Limbachiya et al. (2007) is presented in Fig. 3.17. In comparison with CDW aggregates (RCA), the two natural aggregates are found to be at the top of the triangle, indicating the presence of higher amounts of SiO_2 than the CDW aggregate. The diagram also shows a comparable composition for all three CDA aggregates, obtained from three different C&D sources, with a richer composition in Al and Ca oxides when compared to coarse and fine NAs.

The mineralogical compositions of CDW aggregate are slightly different from those of natural aggregates due to the presence of different constituents from different sources such as hydrated cement, ceramics and natural aggregates (Bianchini et al. 2005; Limbachiya et al. 2007).

3.6 Concluding Remarks

Using CDW as an aggregate in the preparation of new concrete has been the subject of investigation for a long time. In general, the quality of CDW aggregate is poorer than that of natural aggregate, which restricts its use in various construction applications. Information gathered from this review can be summarised as follows:

1. The production process of CDW aggregate affects its quality and composition. Further processing and higher quality source concrete result in better quality aggregates;
2. Depending on its source, CDW aggregates may contain only natural aggregates with adhered mortar or several types of contaminants such as bitumen mixtures, plastics, bricks and tiles in minor amounts along with natural aggregate and adhered mortar contents;
3. Almost all properties of CDW aggregate are dependent on its mortar content. The adhered mortar may be present as lumps embedded with varying proportions of smaller natural aggregates or on the surface of the natural aggregate present in CDW aggregate or as a binder to two or more natural aggregate particles;
4. Several methods are proposed to determine adhered mortar content in CDW aggregates and also several beneficiation techniques can be applied to decrease mortar content in CDW aggregate;
5. All types of densities along with bulk density of CDW aggregate are lower than those of natural aggregates, due to the existence of porous and less dense cement paste/mortar in the CDW aggregates;
6. As adhered mortar in CDW aggregate is porous by nature, it can absorb high amounts of water. Thus the porosity and water absorption capacity of CDW aggregate is very high compared to that of natural aggregate. The variation of water absorption capacity in various CDW aggregates reported in various references is due to the variation of cement paste content in this aggregate as well as the content of other components such as crushed clay brick and tiles. The higher water absorption capacity of CDW aggregate substantially deteriorates the workability of resulting concrete, which finally affects the various properties of concrete;
7. Additional water is necessary to compensate for the extra absorption of the CDW aggregates during preparation of concrete mix. Several methods are adopted to improve the performance of CDW aggregate during concrete mixing;
8. CDW aggregate has a lower Los Angeles abrasion value than that of natural aggregate because of the presence of adhered mortar. However, this value for the majority of the CDW aggregate reported in the literature meets the various standard requirements for concrete and road constructions. The toughness of CDW aggregate is also lower than that of natural aggregate;

9. Compared to natural aggregate, the surface of CDW aggregate is rough and porous due to the presence of mortar adhered. The shape and flakiness indices of various CDW aggregates reported in the literature are comparable to those of natural aggregates used in those studies;
10. The size distributions of coarse and fine CDW aggregates are generally different from the corresponding fractions of natural aggregates. However, in a production plant and also in several studies, CDW aggregate is produced by adopting crushing and screening processes and therefore CDW aggregate generally falls within the limits of mixing gradation for the preparation of the required concrete types;
11. The durability performance of most of the CDW aggregates is considerably poorer than that of natural aggregate. The concentration of some deleterious chemical components in CDW aggregate is slightly higher than the natural aggregate, although the amount falls within the limits specified in various specifications;
12. Chemically, CDW aggregate contains lower amounts of SiO_2 and higher amounts of CaO and Al_2O_3, the three major oxides, than the content of those in natural fine and coarse aggregates. Similarly, the mineralogical compositions of CDW aggregate are slightly different from those of natural aggregate due to the presence of different constituents from different sources.

References

Abbas A, Fathifazl G, Fournier B, Isgor OB, Zavadil R, Razaqpur AG, Foo S (2009) Quantification of the residual mortar content in recycled concrete aggregates by image analysis. Mater Charact 60(7):716–728

Akbarnezhad A, Ong KCG, Zhang MH, Tam CT, Foo TWJ (2011) Microwave-assisted beneficiation of recycled concrete aggregates. Constr Build Mater 25(8):3469–3479

Amorim P, de Brito J, Evangelista L (2012) Concrete made with coarse concrete aggregate: influence of curing on durability. ACI Mater J 109(2):195–204

Barra M, Vázquez E (1998) Properties of concretes with recycled aggregates: influence of properties of the aggregates and their interpretation. Sustainable construction: use of recycled concrete aggregate, Dundee, pp 19–30

Bianchini G, Marrocchino E, Tassinari R, Vaccaro C (2005) Recycling of construction and demolition waste materials: a chemical–mineralogical appraisal. Waste Manage (Oxf) 25(2):149–159

Chen H-J, Yen T, Chen K-H (2003) Use of building rubbles as recycled aggregates. Cem Concr Res 33(1):125–132

Coelho A, de Brito J (2011) Distribution of materials in construction and demolition waste in Portugal. Waste Manage Res 29(8):843–853

Corinaldesi V (2009) Mechanical behavior of masonry assemblages manufactured with recycled-aggregate mortars. Cem Concr Comp 31(7):505–510

Corinaldesi V, Moriconi G (2009a) Behaviour of cementitious mortars containing different kinds of recycled aggregate. Constr Build Mater 23(1):289–294

Corinaldesi V, Moriconi G (2009b) Influence of mineral additions on the performance of 100 % recycled aggregate concrete. Constr Build Mater 23(8):2869–2876

Corinaldesi V, Giuggiolini M, Moriconi G (2002) Use of rubble from building demolition in mortars. Waste Manage (Oxf) 22(8):893–899

Costa U, Ursella P (2003) Construction and demolition waste recycling in Italy, WASCON 2003—Progress on the road to sustainability. San Sebastian, Spain, pp 231–239

Courard L, Michel F, Delhez P (2010) Use of concrete road recycled aggregates for roller compacted concrete. Constr Build Mater 24(3):390–395

De Brito J, Alves F (2010) Concrete with recycled aggregates: the Portuguese experimental research. Mater Struct 43(1):35–51

De Brito J, Robles R (2010) Recycled aggregate concrete (RAC): methodology for estimating its long-term properties. Indian J Eng Mater Sci 17(6):449–462

De Juan MS (2004) Study on the use of recycled aggregates in structural concrete production (in Spanish), PhD thesis, Polytechnic University of Madrid, Madrid

De Juan MS, Gutiérrez PA (2009) Study on the influence of attached mortar content on the properties of recycled concrete aggregate. Constr Build Mater 23(2):872–877

Domingo A, Lazaro C, Gayarre FL, Serrano MA, Lopez-Colina C (2010) Long term deformations by creep and shrinkage in recycled aggregate concrete. Mater Struct 43(8):1147–1160

Eguchi K, Teranishi K, Nakagome A, Kishimoto H, Shinozaki K, Narikawa M (2007) Application of recycled coarse aggregate by mixture to concrete construction. Constr Build Mater 21(7):1542–1551

Etxeberria M, Mari A, Vázquez E (2007a) Recycled aggregate concrete as structural material. Mater Struct 40(5):529–541

Etxeberria M, Vázquez E, Mari A, Barra M (2007b) Influence of amount of recycled coarse aggregates and production process on properties of recycled aggregate concrete. Cem Concr Res 37(5):735–742

Evangelista L, de Brito J (2007) Mechanical behaviour of concrete made with fine recycled concrete aggregates. Cem Concr Compos 29(5):397–401

Evangelista L, de Brito J (2010) Durability performance of concrete made with fine recycled concrete aggregates. Cem Concr Compos 32(1):9–14

Ferreira L, de Brito J, Barra M (2011) Influence of pre-saturation of recycled coarse concrete aggregates on structural concretes mecanical and durability properties. Mag Concr Res 63(8):617–627

Fisher C, Werge M (2009) EU as a recycling society. ETC/SCP working paper 2/2009, p 25. http://scp.eionet.europa.eu.int Accessed on May 2011

Fonseca N, de Brito J, Evangelista L (2011) The influence of curing conditions on the mechanical performance of concrete made with recycled concrete waste. Cem Concr Compos 33(6):637–643

Franklin Associates (ed) (1998) Characterization of building-related construction and demolition debris in the United States, Report No. EPA530-R-98-010, prepared for the U.S. Environmental Protection Agency, Municipal and Industrial Solid Waste Division, Office of Solid Waste, USA

Gokce A, Nagataki S, Saeki T, Hisada M (2004) Freezing and thawing resistance of air-entrained concrete incorporating recycled coarse aggregate: the role of air content in demolished concrete. Cem Concr Res 34(5):799–806

Gomes M, de Brito J (2009) Structural concrete with incorporation of coarse recycled concrete and ceramic aggregates: durability performance 42(5):663–675

Gomez-Soberon JMV (2002) Porosity of recycled concrete with substitution of recycled concrete aggregate: an experimental study. Cem Concr Res 32(8):1301–1311

Gonzalez-Fonteboa B, Martinez-Abella F (2007) Shear strength of recycled concrete beams. Constr Build Mater 21(4):887–893

Gonzalez-Fonteboa B, Martinez-Abella F (2008) Concretes with aggregates from demolition waste and silica fume: materials and mechanical properties. Build Environ 43(4):429–437

Gonzalez-Fonteboa B, Martinez-Abella F, Eiras-Lopez J, Seara-Paz S (2011) Effect of recycled coarse aggregate on damage of recycled concrete. Mater struct 44(10):1759–1771

Hansen TC, Narud H (1983) Strength of recycled concrete made from crushed concrete coarse aggregate. Concr Int 5(1):79–83

Hasaba S, Kawamura M, Toriik K (1981) Drying shrinkage and durability of concrete made of recycled concrete aggregates. Trans Jpn Concr Inst 3:55–60

Katz A (2003) Properties of concrete made with recycled aggregate from partially hydrated old concrete. Cem Concr Res 33(5):703–711

Khatib JM (2005) Properties of concrete incorporating fine recycled aggregate. Cem Concr Res 35(4):763–769

Li J, Xiao H, Zhou Y (2009) Influence of coating recycled aggregate surface with pozzolanic powder on properties of recycled aggregate concrete. Constr Build Mater 23(3):1287–1291

Limbachiya MC, Leelawat T, Dhir RK (2000) Use of recycled concrete aggregate in high-strength concrete. Mater Struct 33(9):574–580

Limbachiya MC, Marrocchin E, Koulouris A (2007) Chemical–mineralogical characterisation of coarse recycled concrete aggregate. Waste Manage (Oxf) 27(2):201–208

Lin Y-H, Tyan Y–Y, Chang T-P, Chang C-Y (2004) An assessment of optimal mixture for concrete made with recycled concrete aggregates. Cem Concr Res 34(8):1373–1380

López-Gayarre F, Serna P, Domingo-Cabo A, Serrano-López MA, López-Colina C (2009) Influence of recycled aggregate quality and proportioning criteria on recycled concrete properties. Waste Manage (Oxf) 29(12):3022–3028

Malhotra VM (1976) Testing hardened concrete: non-destructive method, American Concrete Institute Monograph 9. Detroit, Michigan

Matias D, de Brito J (2004) Influence of the shape of the coarse recycled concrete aggregates on the workability and strength of concrete (in Portuguese), Structural Concrete 2004. FEUP, Porto, pp 187–194

Miranda LFR, Selmo SMS (2006) CDW recycled aggregate renderings: Part I—analysis of the effect of materials finer than 75 lm on mortar properties. Constr Build Mater 20(9):615–624

Nagataki S, Gokce A, Saeki T (2000) Effects of recycled aggregate characteristics on the performance parameters of recycled aggregate concrete. In: Proceedings of the fifth CANMET/ACI international conference on durability of concrete, Barcelona, pp 51–71

Nagataki S, Gokce A, Saeki T, Hisada M (2004) Assessment of recycling process induced damage sensitivity of recycled concrete aggregates. Cem Concr Res 34(6):965–971

Neville AM (1981) Properties of concrete, 3rd edn. Pitman Books Limited, London

Oliveira MB, Vázquez E (1996) The influence of retained moisture in aggregates from recycling on the properties of new hardened concrete. Waste Manage (Oxf) 16(1–3):113–117

Olorunsogo FT, Padayachee N (2002) Performance of recycled aggregate concrete monitored by durability indexes. Cem Concr Res 32(8):179–185

Padmini AK, Ramamurthy K, Mathews MS (2009) Influence of parent concrete on the properties of recycled aggregate concrete. Constr Build Mater 23(2):829–836

Park SB, Seo DS, Lee J (2005) Studies on the sound absorption characteristics of porous concrete based on the content of recycled aggregate and target void ratio. Cem Concr Res 35(9):1846–1854

Pereira L (2002) Construction and demolition waste recycling: the case of the Portuguese northern region (in Portuguese). Master Thesis in Civil Engineering, Minho University, Braga, Portugal

Poon CS, Chan D (2007) Effects of contaminants on the properties of concrete paving blocks prepared with recycled concrete aggregates. Constr Build Mater 21(1):164–175

Poon CS, Shui ZH, Lam L (2004a) Effect of microstructure of ITZ on compressive strength of concrete prepared with recycled aggregates. Constr Build Mater 18(6):461–468

Poon CS, Shui ZH, Lam L, Fok H, Kou SC (2004b) Influence of moisture states of natural and recycled aggregates on the slump and compressive strength of concrete. Cem Concr Res 34(1):31–36

Poon CS, Kou SC, Lam L (2007) Influence of recycled aggregate on slump and bleeding of fresh concrete. Mater Struct 40(9):981–986

Poon CS, Kou SC, Wan H-W, Etxeberria M (2009) Properties of concrete blocks prepared with low grade recycled aggregates. Waste Manage (Oxf) 29(8):2369–2377

Rahal K (2007) Mechanical properties of concrete with recycled coarse aggregate. Build Environ 42(1):407–415

Rao A, Jha KN, Misra S (2007) Use of aggregates from recycled construction and demolition waste in concrete. Resour Conserv Recycl 50(1):71–81

Rao MC, Bhattacharyya SK, Barai SV (2011) Behaviour of recycled aggregate concrete under drop weight impact load. Constr Build Mater 25(1):69–80

Ravindrarajah RS, Tam CT (1985) Properties of concrete made with crushed concrete as coarse aggregate. Mag Concr Res 37(3):29–38

Reixach FM, Cuscó AS, Barroso JMG (2000) Situatión actual y perspectives de futuro de los resíduos de la construcción. Intitut de Tecnologia de la Construcció de Catalunya (IteC), Catalunya, Spain. (in Spanish)

Sagoe-Crentsil KK, Brown T, Taylor AH (2001) Performance of concrete made with commercially produced coarse recycled concrete aggregate. Cem Concr Res 31(5):707–712

Sani D, Moriconi G, Fava G, Corinaldesi V (2005) Leaching and mechanical behaviour of concrete manufactured with recycled aggregates. Waste Manage (Oxf) 25(2):177–182

Santos J, Branco FA, de Brito J (2002a) Compressive strength, modulus of elasticity and drying shrinkage of concrete with coarse recycled concrete. XXXIAHS World Congress on Housing, Coimbra, Portugal, pp 1685–1691

Santos J, Branco FA, de Brito J (2002b) The use of coarse recycled concrete aggregates in the production of new concrete (in Portuguese), Structures 2002. LNEC, Lisbon, pp 227–236

Santos J, Branco FA, de Brito J (2005) Reinforced concrete beams with recycled aggregates from demolished concrete of a stadium. SB05—3rd international conference on sustainable building, Tokyo, pp 2011–2018

Tabsh SW, Abdelfatah AS (2009) Influence of recycled concrete aggregates on strength properties of concrete. Constr Build Mater 23(2):1163–1167

Tam VWY, Tam CM (2007) Crushed aggregate production from centralized combined and individual waste sources in Hong Kong. Constr Build Mater 21(4):879–886

Tam VWY, Tam CM, Wang Y (2007) Optimization on proportion for recycled aggregate in concrete using two-stage mixing approach. Constr Build Mater 21(10):1928–1939

Tam VWY, Wang Y, Tam CM (2008) Assessing relationships among properties of demolished concrete, recycled aggregate and recycled aggregate concrete using regression analysis. J Hazard Mater 152(2):703–714

Tangchirapat W, Buranasing R, Jaturapitakkul C, Chindaprasirt P (2008) Influence of rice husk–bark ash on mechanical properties of concrete containing high amount of recycled aggregates. Constr Build Mater 22(8):1812–1819

Topçu IB (1997) Physical and mechanical properties of concretes produced with waste concrete. Cem Concr Res 27(12):1817–1823

Topçu IB, Guncan NF (1995) Using waste concrete as aggregate. Cem Concr Res 25(7):1385–1390

Tu T-Y, Chen Y–Y, Hwang C-L (2006) Properties of HPC with recycled aggregates. Cem Concr Res 36(5):943–950

Vieira JPB, Correia JR, de Brito J (2011) Post-fire residual mechanical properties of concrete made with recycled concrete coarse aggregates. Cem Concr Res 41(5):533–541

Watanabe T, Nishibata S, Hashimoto C, Ohtsu M (2007) Compressive failure in concrete of recycled aggregate by acoustic emission. Constr Build Mater 21(3):470–476

Wong YD, Sun DD, Lai D (2007) Value-added utilisation of recycled concrete in hot-mix asphalt. Waste Manage (Oxf) 27(2):294–301

Xiao J, Li J, Zhang C (2005) Mechanical properties of recycled aggregate concrete under uniaxial loading. Cem Concr Res 35(6):1187–1194

Yang J, Du Q, Bao Y (2011) Concrete with recycled concrete aggregate and crushed clay bricks. Constr Build Mater 25(4):1935–1945

Zaharieva R, Buyle-Bodin F, Skoczylas F, Wirquin E (2003) Assessment of the surface permeation properties of recycled aggregate concrete. Cem Concr Compos 25(2):223–232

Chapter 4
Use of Industrial Waste as Aggregate: Properties of Concrete

4.1 Introduction

As indicated in an earlier section, aggregates account for the largest part of the concrete volume and therefore play a substantial role in almost all concrete properties such as workability, strength, dimensional stability, and durability. Recently, several waste materials have been studied to be used as aggregate in concrete. The use of waste as aggregates can consume vast amounts of waste materials as this is the major component of cement mortar and concrete.

In this section, properties of concrete with various types of waste aggregates generated from various industries will be presented. The major focus will be given on the behaviour of concrete with various waste industrial aggregates; however, if information is not available on a particular property of concrete, the same or similar property of cement mortar with that particular waste will be considered.

4.2 Coal Bottom Ash

Many references are available on the properties and use of fly ash (FA) as a mineral addition in conventional Portland cement concrete. However, not much has been reported on the use of FA and coal bottom ash (CBA) as aggregate in concrete. CBA falls into the bottom of the furnace in modern large thermal power plants and constitutes about 20 % of total ash content of the coal fed into the boilers. The properties of CBA depend on the coal type, pulverising system, combustion conditions, temperature, type of furnace, minerals in coals and milling system and these are already presented in detail in Chap. 2. Here, the effect on CBA aggregate on the several concrete properties will be discussed.

J. de Brito and N. Saikia, *Recycled Aggregate in Concrete*,
Green Energy and Technology, DOI: 10.1007/978-1-4471-4540-0_4,

4.2.1 Fresh Concrete Properties

4.2.1.1 Workability/Slump Behaviour

The slump behaviour of concrete due to the incorporation of CBA is probably dependent on its shape, porosity and surface texture. Therefore, two parallel views exist concerning the slump behaviour of concrete with CBA.

Agarwal et al. (2007) measured the workability of concrete with CBA as replacement of fine aggregate using the compacting factor test, described in Indian standard, IS 1199-1959. They found that the workability of concrete decreased as the replacement level of the fine aggregates with CBA increased (Fig. 4.1a). The increase in the specific surface due to increased fineness of fine aggregate as well as a greater amount of water needed for the mix of ingredients to get closer packing result in decrease of the workability of the mix.

Kim and Lee (2011) reported contrasting slump values for concrete with fine and coarse CBA used as a partial or full replacement of natural fine and coarse aggregate, respectively. They concluded that the flow characteristics of fresh concrete were slightly reduced by the use of coarse CBA, whereas the effect of fine CBA can be neglected. They observed a 20.8 % reduction of slump of fresh concrete that contains coarse CBA only as coarse aggregate. More complicated shape and rougher surfaces of CBA than normal aggregate and a lowering of aggregate–cement paste lubrication effect due to absorption of some free cement paste and water by porous CBA are the major causes of this reduction. On the other hand, the porosity and water absorption capacity of fine CBA is lower than

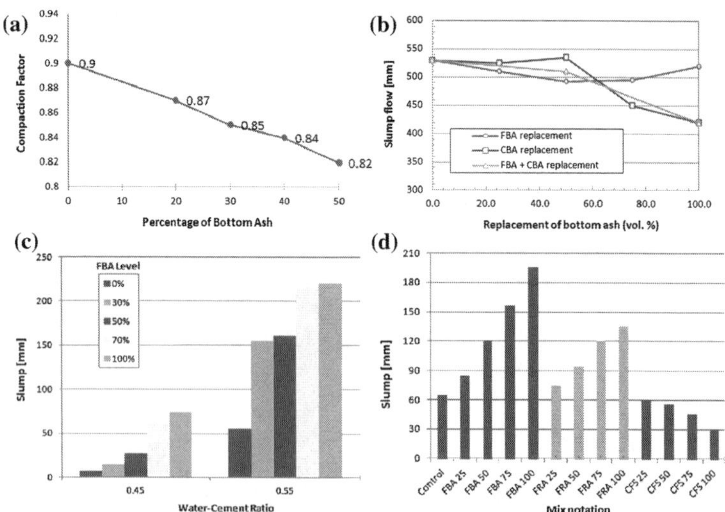

Fig. 4.1 Workability behaviour of concrete with coal bottom ash aggregate. **a** Aggarwal et al. 2007; **b** Kim and Lee 2011; **c** Bai et al. 2005; **d** Kou and Poon 2009

that of coarse CBA and therefore concrete with fine CBA absorbs negligible amounts of cement paste and water during mixing, which does not affect the slump value of the resulting concrete mix. Their results are presented in Fig. 4.1b.

Kasemchaisiri and Tangtermsirikul (2008) also observed reduction in slump value due to incorporation of CBA as a partial substitution of fine aggregate. According to these authors, this was due to an increase in frictional forces between aggregate particles as CBA is highly irregular in shape and it has rough surface texture.

Bai et al. (2005) reported an increase in slump due to the partial substitution of sand with fine CBA in the concrete mix (Fig. 4.1c). According to the authors, the presence of a "ball-bearing effect", due to the replacement of irregular shaped normal sand by spherical shaped fine CBA aggregate increases the slump value of the concrete mix.

Kou and Poon (2009) also reported increasing slump of fresh concrete mix due to the incorporation of fine CBA as partial or total replacement of normal sand (Fig. 4.1d). An increase in free water content in the concrete mix with fine CBA by comparison with that of the conventional concrete mix, due to the high water absorption capacity of CBA, increases slump.

4.2.1.2 Bleeding Behaviour

Andrade et al. (2009) observed bleeding of water during the preparation of a fresh concrete mix with CBA as aggregate. The authors reported that the concrete mix started to segregate (i.e. the aggregates and cement particles tended to occupy the bottom of the container) during the concrete preparation and moulding process due to the difference in weight of various constituents in the concrete mix. The water loss due to the addition of CBA as partial replacement of sand fraction in the concrete mix is presented in Fig. 4.2a. The bleeding of water increases with increasing content of CBA. Ghafoori and Bucholc (1996) observed a similar behaviour for concrete with fine CBA as partial substitution of sand (Fig. 4.2b). According to these authors, the higher bleeding of fresh concrete mix due to

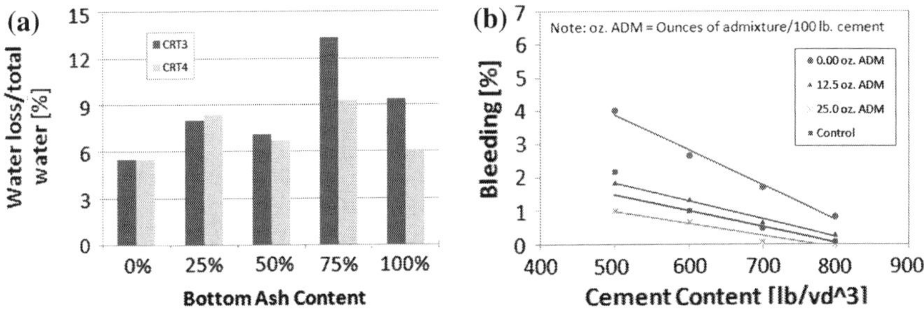

Fig. 4.2 Water loss due to bleeding of concrete mixes with various amounts of fine CBA aggregates. **a** Andrade et al. (2009); **b** Ghafoori and Bucholc (1996)

addition of CBA by comparison with conventional concrete is due to the increased demand of water during the mixing of concrete with CBA.

Andrade et al. (2007) reported that the water, absorbed during mixing, is desorbed at a later stage and increases bleeding. Decreases in water/cement (w/c) ratio and addition of air-entraining admixture can significantly decrease bleeding of concrete with CBA (Andrade et al. 2009; Ghafoori and Bucholc 1996).

4.2.1.3 Density

As the density of CBA is considerably lower than that of normal fine and coarse aggregates, the inclusion of CBA aggregate in concrete decreases its unit weight or density. Another factor that is pointed out in some of the studies is the higher w/c ratio of concrete with CBA than in conventional concrete, which introduces more air bubbles in the concrete mix. Figure 4.3 shows the density of concrete with two different size ranges (Lee et al. 2010). The size ranges of the CBA aggregate present in concrete mixes F1 and F2 are, respectively, 0–2 and 2–8 mm. A significant decrease in density was observed due to the incorporation of CBA aggregate in concrete. Yüksel et al. (2007) reported about 30 % reduction in fresh density of concrete briquette (block) with CBA used to replace 50 % (in volume) of 0–4 mm sand.

4.2.2 Hardened Concrete Properties

4.2.2.1 Density of Concrete

Just like for fresh-state density, the incorporation of CBA aggregate also decreases the dry density of hardened concrete due to the low bulk density of CBA aggregate. Experimental results of two different types of concrete are presented in Fig. 4.4.

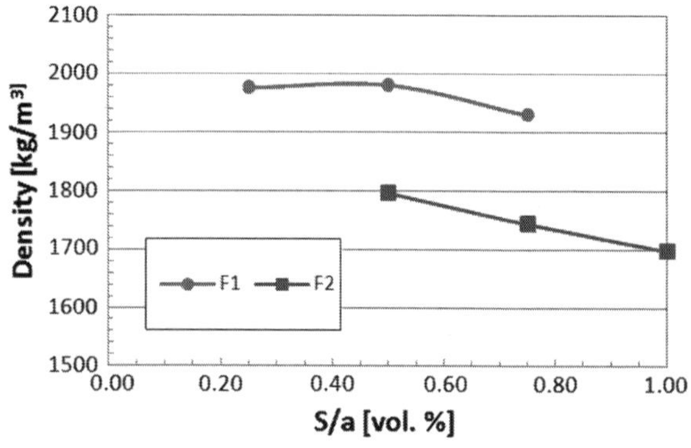

Fig. 4.3 Density of concrete with CBA aggregate (Lee et al. 2010)

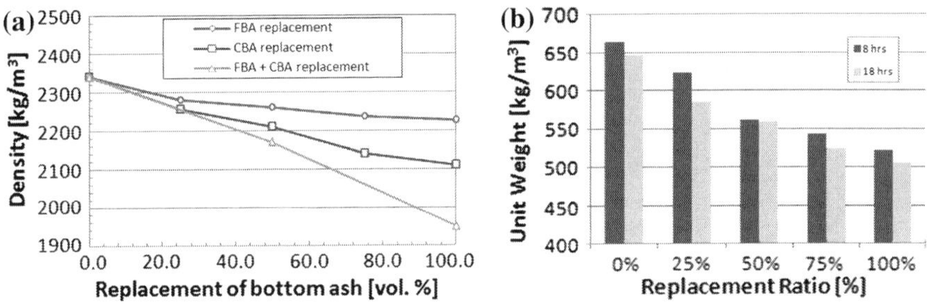

Fig. 4.4 Density of two different types of hardened concrete. **a** Normal (Kim and Lee 2011); **b** Autoclaved aerated (Kurama et al. 2009)

4.2.2.2 Compressive Strength

Variations in compressive strength of concrete due to the incorporation of CBA aggregate were observed depending on the method of preparation of concrete. Bai et al. (2005) observed lower compressive strength of concrete with CBA aggregate than that of conventional concrete at constant w/c ratio, while the two types of concrete exhibited almost similar compressive strength at constant slump (Fig. 4.5).

The same authors concluded that 30 % of natural sand could be replaced with CBA aggregate to produce concrete in the 40–60 N/mm^2 compressive strength range without detrimentally affecting the permeation and drying shrinkage properties of structural concrete. Yüksel et al. (2011) also observed a decreasing trend

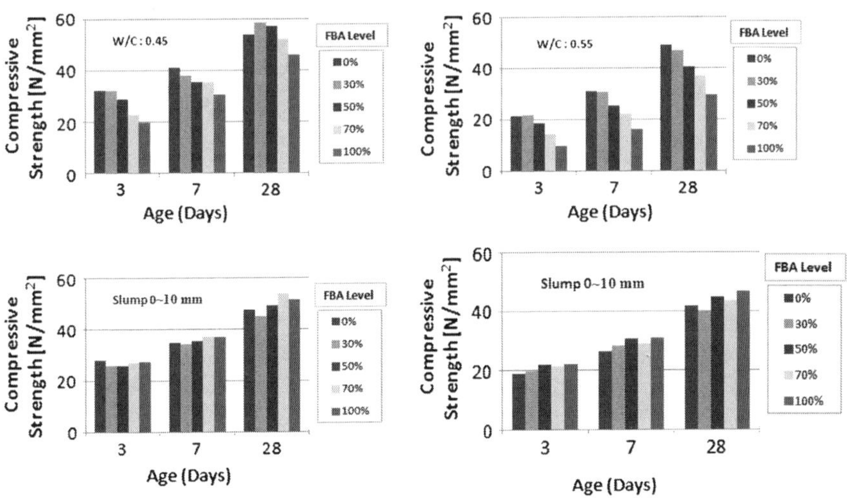

Fig. 4.5 Compressive strength of concrete with CBA at constant w/c and slump values (Bai et al. 2005)

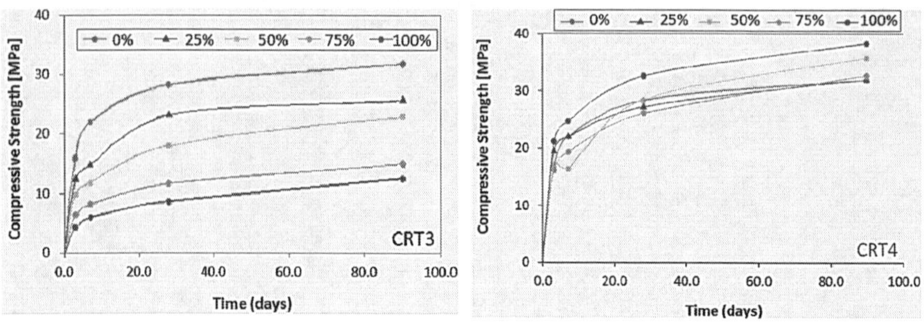

Fig. 4.6 Compressive strengths of concrete with CBA (Andrade et al. 2007)

in compressive strength of concrete due to increasing addition of CBA for constant w/c.

Kim and Lee (2011) did not observe any significant changes in the compressive strength of concrete with constant w/c value due to the incorporation of coarse and fine CBA as partial and full replacement of coarse and fine aggregate, respectively. According to the authors, this was due to the presence of higher amounts of cement paste than those observed in other studies. Kuruma et al. (2009) observed higher compressive strength in autoclaved aerated concrete with fine CBA replacing 50 % of natural sand than in similar conventional concrete, and compressive strength further decreased with increasing replacement level. This was mainly due to the pozzolanic activity of CBA, which increased at autoclaved aerated conditions and therefore forms additional amounts of products like calcium silicate hydrate gel, and strengthen the structure.

Andrade et al. (2007) reported that the consideration of water content in CBA during the preparation of a concrete mix has profound effect on the compressive strength of hardened concrete. Their results are presented in Fig. 4.6. The authors prepared two types of concrete: in one type the moisture content in CBA was not considered to determine the water amount in the mix (CRT3) and in the other type the moisture content in CBA was considered for that effect (CRT4).

As the CBA is slightly pozzolanic by nature, the strength development pattern with respect to elapsed time for concrete with CBA is different from that of conventional concrete. In most studies, it was reported that this type of concrete gains strength at a slower rate in the initial period of curing and grows faster at the latter stage of curing (Andrade et al. 2007; Agarwal et al. 2007; Ghafoori and Bacholc 1996). According to these authors, bottom ash takes parts in hydration reaction at the latter stages of curing and forms other products. In Fig. 4.6, the strength development behaviour of two types of concrete with CBA aggregate is presented. Park et al. (2009) reported that the failure of concrete with coarse CBA aggregate was predominantly by aggregate fracture instead of binder fracture and interface fracture, due to the lesser hardness of CBA versus that of normal aggregate.

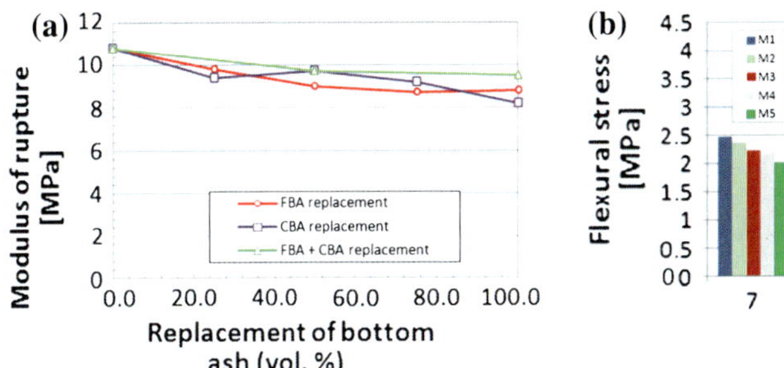

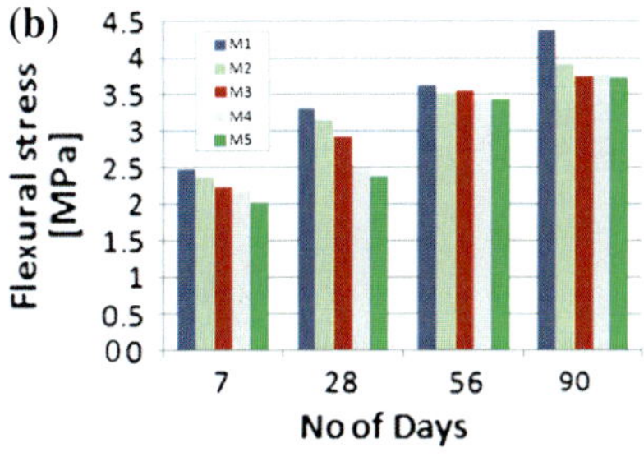

Fig. 4.7 Flexural strength of concrete with CBA aggregate (M1, M2, M3, M4, M5 in Fig. 4.7b represents concrete with CBA replacing 0, 20, 30, 40 and 50% by weight of natural sand). **a** Kim and Lee 2011; **b** Aggarwal et al. 2007

4.2.2.3 Flexural Strength

Some authors observed a significant reduction in the flexural strength of concrete due to the incorporation of CBA, as well as to increasing CBA content, even though the effect observed in compressive strength behaviour was insignificant (Kim and Lee 2011, Agarwal et al. 2007). Some experimental results are presented in Fig. 4.7. On the other hand, Triches et al. (2007) observed an increase in flexural strength of roller compacted concrete (RCC) due to the addition of CBA as partial substitution of natural sand because of the pozzolanic activity of CBA as well as improvements of aggregate arrangement in the concrete matrix. Kuruma et al. (2009) observed higher flexural strengths for concrete prepared by replacing 50 % (by weight) of natural sand with fine CBA.

Ghafoori and Bacholc (1996) reported that conventional concrete exhibited higher flexural strength at low content of cement than concrete with CBA fine aggregate. However, they were able to reduce this difference by increasing cement content in concrete mix as well as by adding chemical admixture in the concrete mix

Table 4.1 Flexural strength of concrete (psi) with natural sand and CBA as fine aggregates (Ghafoori and Bucholc 1996)

Curing age (day)	Cement content in concrete (lb/yd^3)				Cement content in concrete (lb/yd^3)			
	500	600	700	800	500	600	700	800
	Natural sand (C)				CBA aggregate (BA)			
7	505	646	716	818	376	501	641	752
28	595	722	797	881	481	622	748	916
90	688	830	945	985	573	707	815	925
	CBA aggregate + 12.5 oz ADM/100 lb cement (ADM1)				CBA aggregate + 25.0 oz ADM/100 lb cement (ADM2)			
7	469	588	743	809	452	675	809	951
28	674	788	840	926	636	830	967	1054
90	711	826	904	963	721	882	1021	1137

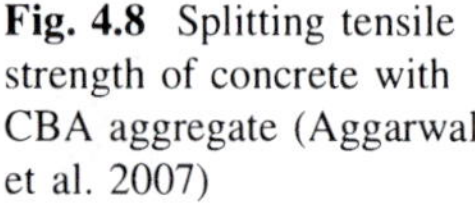

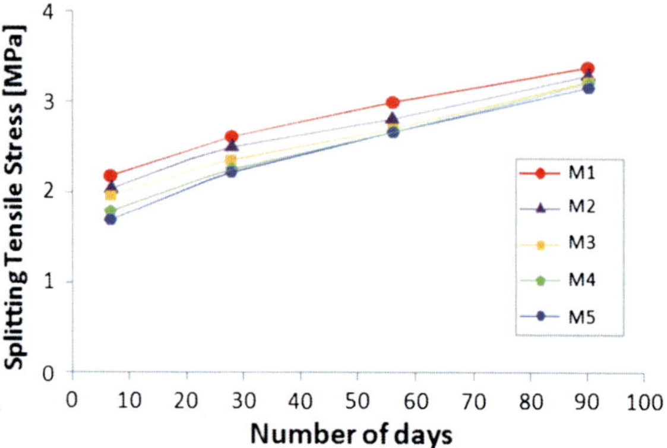

Fig. 4.8 Splitting tensile strength of concrete with CBA aggregate (Aggarwal et al. 2007)

with CBA. For a given content of admixture, the flexural strength of CBA concrete became higher than that of control concrete. Their results are presented in Table 4.1.

4.2.2.4 Splitting Tensile Strength

The splitting tensile strength of concrete decreases as replacement percentage of fine aggregate by CBA rises and increases with the curing age (Agarwal et al. 2007). The highest and lowest gains in splitting tensile strength were observed at 20 and 50 % replacement of fine aggregates with bottom ash, respectively (designated M2 and M5). Plain concrete reaches 64, 77 and 88 % of 90-day strength at 7, 28 and 56 days of curing, respectively, whereas these values for concrete with CBA at 20, 30, 40 and 50 % replacement levels were in the ranges of 62–86 %, 60–83 %, 56–83 % and 53–84 %, respectively. Their results are presented in Fig. 4.8.

Ghafoori and Bacholc (1996) reported that the inclusion of CBA in concrete had more influence on splitting tensile strength than on compressive strength. At low cement content (500 lb/yd^3 of concrete), the splitting tensile strength of concrete with CBA aggregate was lower than that of conventional concrete at the early ages of curing and was similar after 56 days of curing. However, the initial dormant period of CBA concrete can be overcome by adding admixtures to this type of concrete. On the other hand, the splitting tensile strength of both types of concrete was similar for the concrete mix with 600 lb/yd^3 of cement. These results are presented in Fig. 4.9.

4.2.2.5 Static Elastic Modulus

The static elastic modulus of concrete with CBA is significantly lower than that of conventional concrete. Ghafoori and Bacholc (1996) reported that the higher elastic modulus value of conventional concrete than that of CBA concrete was due to the lower paste porosity of conventional concrete than that of CBA concrete, as the w/c value of conventional concrete was lower than that of CBA concrete, as well as to the higher bulk density of natural sand aggregate than that of CBA

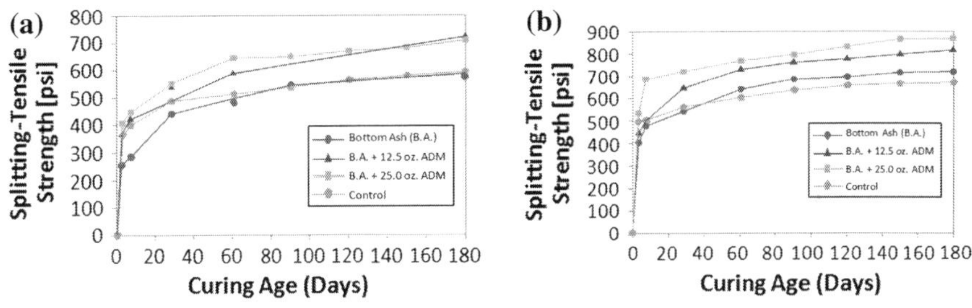

Fig. 4.9 Splitting tensile strength of normal and CBA concrete (Ghafoori and Bucholc 1996). **a** Cement content: 500 lb/yd^3; **b** Cement content: 600 lb/yd^3

Table 4.2 Modulus of elasticity of different types of concrete (Ghafoori and Bucholc 1996)

Cement content (lb/yd^3)	Modulus of elasticity, (psi) of concrete ($\times$ 10^6)			
	C	CBA	ADM1	ADM2
500	5.02	3.32	3.64	3.83
600	5.58	3.80	4.04	4.36
700	5.80	3.86	4.32	5.05
800	5.74	4.25	4.47	5.20

Details about concrete mix proportions are presented in Table 4.1

aggregate. Their results are presented in Table 4.2. The authors achieved a significant improvement of the modulus of elasticity for CBA concrete by using a higher amount of cement along with the addition of an admixture.

Kim and Lee (2011) also observed a similar modulus of elasticity of high-strength concrete (HSC) prepared by replacing 50 % by volume of aggregate with CBA, beyond which it dropped quickly (Fig. 4.10). The authors observed a higher reduction in the modulus of elasticity using coarse CBA aggregate than for fine CBA. The reduction in modulus of elasticity of concrete due to 100 % replacement of fine natural aggregate (NA) by fine CBA was about 15 %, whereas these values

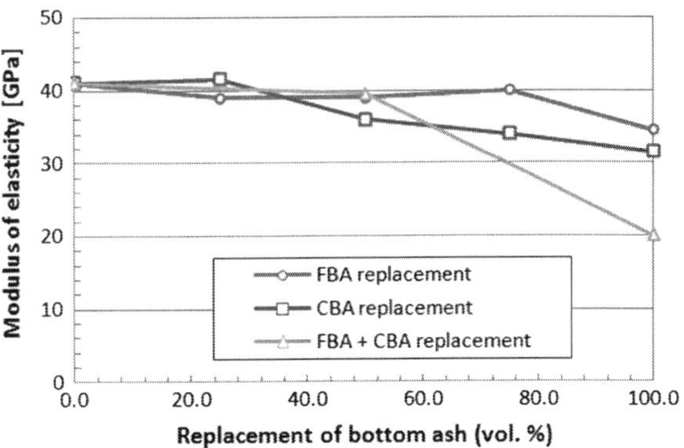

Fig. 4.10 Modulus of elasticity of concrete with various amounts of CBA aggregates (Kim and Lee 2011)

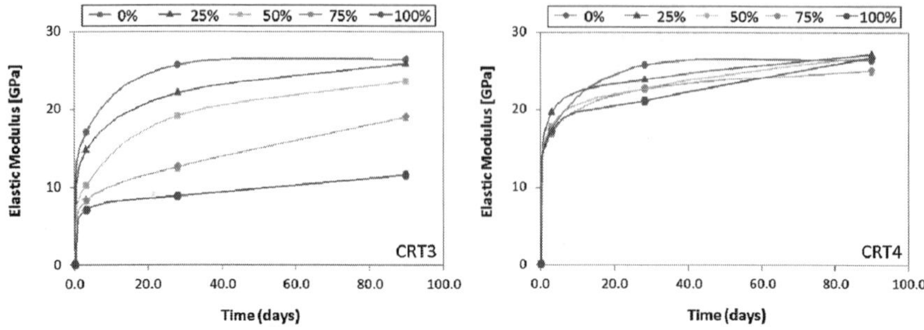

Fig. 4.11 Modulus of elasticity of concrete with CBA (Andrade et al. 2007)

for 100 % replacement by coarse CBA and 100 % replacement by a mixture of fine and coarse CBA were, respectively, 22.5 and 51 %.

Andrade et al. (2007) also observed significant reduction in elastic modulus value for concrete with fine CBA as partial or full replacement of fine aggregate fraction (Fig. 4.11). This reduction was prominent at the early stage of curing. However, at the latter stages of curing the pozzolanic reaction of CBA made the microstructure of concrete denser and improved the mechanical properties including the elasticity modulus. However, changes in water content during concrete the mix preparation by considering the water content in CBA can improve the elastic modulus behaviour of hardened concrete. The authors refer this fraction as CRT4 in Fig. 4.11.

The same authors plotted stress–strain curves for concrete with CBA aggregates. However, they did not find too much difference for CRT4 type concrete, but other types gave scattered results at all ages.

4.2.3 Durability Behaviour

4.2.3.1 Drying Shrinkage

The drying shrinkage of concrete is generally affected by the addition of CBA aggregates, as this material is porous by nature and therefore absorbs a large amount of water. Bai et al. (2005) reported that the concrete with CBA as a replacement of sand fraction at constant w/c value exhibited lower drying shrinkage than conventional concrete. This is due to the release of moisture absorbed by CBA during dry condition that keeps the mortar in a moist condition. On the other hand, for constant slump value, shrinkage increased with increasing content of CBA. However, in this condition, the authors found a comparable drying shrinkage of concrete with CBA replacing 30 % by weight of natural sand. Their results are presented in Fig. 4.12.

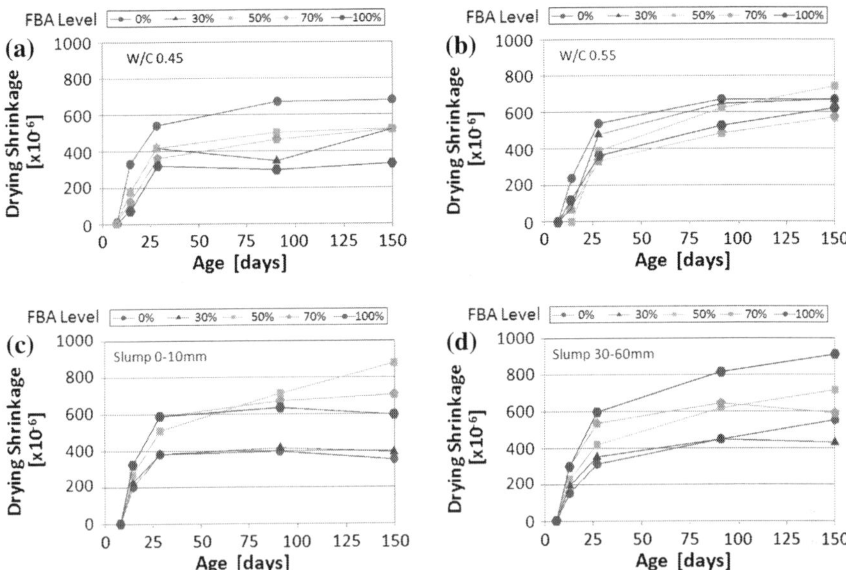

Fig. 4.12 Drying shrinkage of concrete with fine CBA aggregates (Bai et al. 2005). **a** W/C 0.45. **b** W/C 0.55. **c** Slump range 0–10 mm. **d** Slump range 30–60 mm

Kim and Lee (2011) observed similar results for concrete with CBA prepared at constant w/c value and slump value. The lower shrinkage value of concrete due to an increasing content of CBA aggregate at constant slump was due to the decrease in free water content in concrete with CBA. Ghafoori and Bacholc (1996) also reported lower drying shrinkage for CBA concrete than the conventional concrete despite the higher w/c value of the former concrete.

The plastic shrinkage (early volume change) of concrete is also affected by the inclusion of CBA aggregates. Ghafoori and Bacholc (1996) found about 35 % reduction in plastic shrinkage of concrete with fine CBA aggregates than that observed for normal concrete due to higher bleeding of the former type of concrete. Incorporating a low content of chemical admixture had little effect on the plastic shrinkage of CBA concrete but it could be increased considerably by adding a higher dosage of admixture. Bleeding water was significantly reduced for higher admixture content, and therefore increased the shrinkage value. Andrade et al. (2009) also found a reduction in the plastic shrinkage of concrete due to the incorporation of CBA aggregates. However, the shrinkage of concrete prepared by considering the moisture and free water content in CBA aggregates was higher than that of the reference concrete. The reduction in shrinkage was due to the higher content of bleeding water as well as to the absorption of water by CBA.

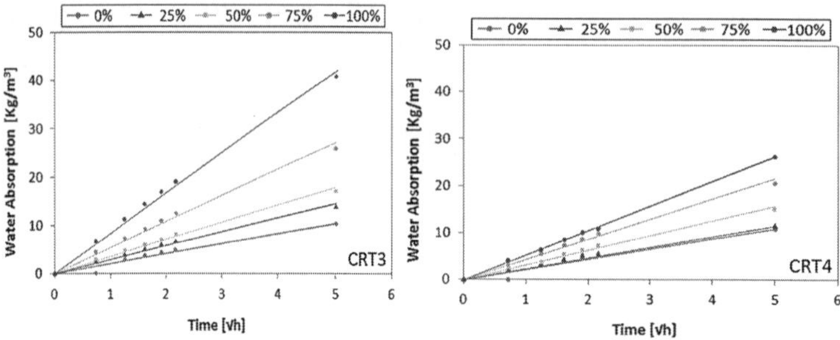

Fig. 4.13 Capillary water absorption of concrete with CBA aggregates (Andrade et al. 2009)

4.2.3.2 Capillary Water Absorption

Andrade et al. (2009) reported higher capillary water absorption values for concrete with various amounts of CBA as fine aggregates than for the reference concrete. According to the authors, the addition of porous CBA aggregate in concrete not only provides some free water due to bleeding but it also provides a pore system that is different from that of the reference concrete. However, capillary water can be reduced significantly if the water content in bottom ash is considered during concrete mixing. Their results are presented in Fig. 4.13. They also measured the sorptivity coefficient for different concrete mixes from capillary absorption data, which are presented in Table 4.3. Yüksel et al. (2007) also observed increasing capillary water absorption coefficients with increasing CBA content except for the 10 % replacement level of fine aggregate by volume. At this level, the pozzolanic activity of CBA decreases the porosity of concrete but at higher substitution levels the porosity of CBA increases the overall porosity of concrete and therefore increases the permeability.

4.2.3.3 Chloride Permeability

Ghafoori and Bacholc (1996) observed about 120 % higher current flow in CBA concrete than in conventional concrete in the rapid chloride permeability test. The authors also reported that the addition of an admixture reduced the chloride permeability of CBA concrete; in this concrete, the current flow was about 61 % higher than in the reference concrete. Kou and Poon (2009) reported contrasting

Table 4.3 Water sorptivity coefficient (kg m^{-2} $h^{0.5}$) of different type of concrete (Andrade et al. 2009)

Type of concrete	Amount of CBA in concrete (mass%)				
	0	25	50	75	100
CRT3	2.2	3.1	3.8	5.7	8.9
CRT4		2.3	3.2	4.4	5.3

behaviour of chloride permeability of concrete with CBA aggregate prepared at constant w/c value and at constant slump. At constant w/c value, the chloride permeability of CBA concrete was higher than that of conventional concrete and it increased with the CBA content, because of the looser microstructure of CBA concrete than that of the control concrete due to its higher free water content. On the other hand, at constant slump, the free water content in CBA concrete was not as high as in the reference concrete and therefore chloride permeability was reduced. Kasemchaisiri and Tangtermsirikul (2008) observed an increase in chloride permeability of 7-day cured self-compacting concrete with increasing CBA aggregate content. However, the authors observed a similar permeability for CBA concrete and the reference concrete at latter stages of curing, possibly due to the pozzolanic reaction of CBA.

4.2.3.4 Carbonation Depth

Kasemachaisiri and Tangtermsirikul (2008) measured the carbonation depth of 28- and 56-day cured concrete with CBA as a partial replacement of fine aggregates by using accelerated carbonation test. They found a slightly higher carbonation depth for concrete with 10 % CBA than that for the reference concrete. However, these values were much higher for concrete with 20 and 30 % CBA than for the reference concrete. The increase in porosity of concrete due to the addition of porous CBA in concrete led to deeper carbonation. However, the difference between the carbonation depth of the conventional concrete and the concrete with CBA aggregate decreased as curing time increased due to the pozzolanic activity of CBA.

4.2.3.5 Resistance to Chemical Attack

Kasemachaisiri and Tangtermsirikul (2008) reported higher resistance to sulphate attack of self-compacting concrete with CBA aggregates than of the reference concrete. The sulphate resistance of CBA concrete increased with increasing content of CBA. The observed improved performance of CBA concrete by comparison with the reference concrete was due to the predominance of sulphate enhanced pozzolanic activity of CBA aggregate over the porosity induced by porous CBA. Ghafoori and Bacholc (1996) did not observe any substantial differences between the expansions of CBA concrete and conventional concrete after 6-month exposure of concrete specimens in sulphate solution. The test was performed according to the ASTM C 1012 standard method.

4.2.3.6 Abrasion Resistance

Ghafoori and Bacholc (1996) observed higher depth of wear for concrete with CBA fine aggregates than for conventional concrete. The average depth of wear

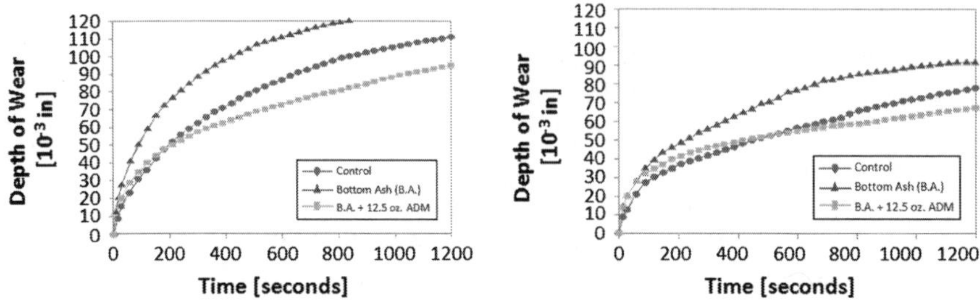

Fig. 4.14 Depth of wear of concrete with CBA and conventional concrete for two cement contents (Ghafoori and Bucholc 1996)

for CBA concrete was about 40 % higher that of the conventional concrete. However, a low content admixture can improve the abrasion resistance of CBA concrete and make it exhibit better performance than conventional concrete. These results are presented in Fig. 4.14. Yüksel et al. (2007) observed similar abrasion values in concrete with various percentage of CBA as fine aggregates replacement and conventional concrete.

4.2.3.7 Resistance to Freeze–Thaw and Dry–Wet Cycles

Yüksel et al. (2007) observed better performance of concrete with CBA aggregate than of conventional concrete subjected to freeze–thaw cycles. The strength loss for the concrete with CBA at 10 and 20 % replacement levels was almost the same but at higher substitution level the strength loss again increased and at 100 % replacement level it became similar to that of conventional concrete. They also observed further improvement of CBA concrete by mixing it with blast furnace slag aggregate (BFS). Their results are presented in Fig. 4.15. Ghafoori and Bucholc (1996) also observed better performance of concrete with CBA as fine aggregates than conventional concrete when the concrete specimens were intermittently subjected to freeze–thaw cycles.

Fig. 4.15 Freeze–thaw resistance of concrete with CBA (K-series), BFS concrete (C-series) and concrete with a mixture of CBA and BFS (CK-series) (Yüksel et al. 2007)

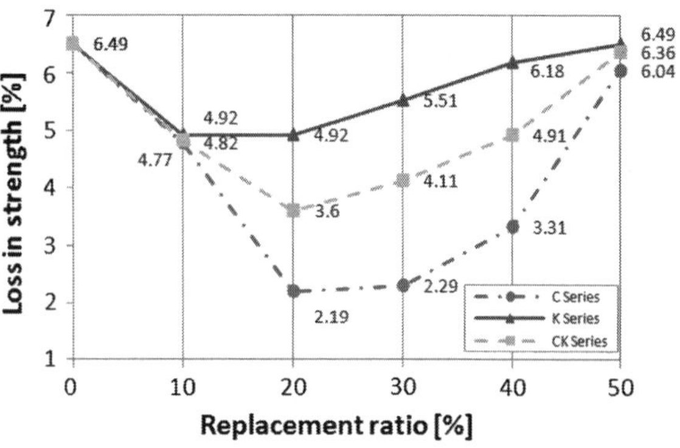

Yüksel et al. (2007) observed higher compressive strength losses for concrete with CBA aggregates than that observed for conventional concrete when both types of concrete were subjected to intermittent wet–dry cycles due to the increase in porosity of CBA concrete. This increase further increases the strength loss. Mixing BFS aggregate with CBA aggregates improved the performance of CBA concrete under these environmental conditions.

4.2.3.8 High Temperature Behaviour

The addition of CBA as a partial substitution of fine aggregates in concrete up to a certain level improved its high temperature performance (Yüksel et al. 2007, 2011). The percentage of residual compressive strength of CBA concrete at 20 % replacement level was the highest, then it gradually decreased at higher substitution rates and it was similar to that of conventional concrete at 40 % replacement level. Compared to the surface of the post-fired reference concrete, which had several randomly distributed cracks, the surfaces of post-fired CBA concrete contained very few cracks.

4.2.4 Coal Fly Ash

The use of FA as a pozzolan is well documented and many standard code of practice already recommend its use as pozzolanic material in concrete. However, limited studies are available on the use of FA as fine aggregates in concrete preparation. In this section, the concrete properties will be briefly highlighted.

As FA consists of very fine spherical particles, a concrete mix with FA as aggregates is more workable, cohesive, mobile, compactable and pumpable than conventional concrete (Ravina 1997; Pofale and Deo 2010). However, the water requirement of reference concrete to reach similar consistency to concrete with FA as partial replacement of sand depends on the size of the sand fraction (Ravina 1997). Ravina (1997) observed similar water requirements for concrete mixes with FA replacing fine sand and higher water requirements for concrete mixes where FA replaced relatively coarse sand. Siddique (2003a, b) observed a decreasing slump trend due to the incorporation of increasing replacement contents of fine aggregates by FA aggregates.

Siddique (2003a) also observed decreasing air-content values with increasing content of FA as partial replacement of fine aggregates. However, concrete fresh density increased with increasing FA contents. These results are presented in Table 4.4.

Bleeding of concrete due to the addition of FA aggregates was similar to that of conventional concrete. The addition of water reducers and retarders significantly increased bleeding and the addition of superplasticizers reduced bleeding (Ravina 1997).

Table 4.4 Fresh concrete properties of concrete with fly ash as partial replacement of fine aggregates (Siddique 2003a)

Properties	Amounts of sand replaced by fly ash (%)					
	0	10	20	30	40	50
Water/cement	0.47	0.48	0.49	0.49	0.49	0.50
Slump (mm)	100	90	65	40	30	20
Air content (%)	5.2	4.8	4.4	4.0	3.8	3.2
Density (kg/m^3)	2308	2310	2314	2314	2316	2319

The addition of class F FA as a partial replacement of fine aggregate improved the compressive, flexural and splitting tensile strengths as well as the modulus of elasticity of the resulting concrete and the more so the greater the curing time (Siddique 2003a). Similar compressive strength behaviour of concrete due to the addition of low calcium FA was also observed by Maslehuddin et al. (1989). This increase is due to the densification of microstructure due to the pozzolanic reaction of FA. Figure 4.16 shows the compressive, splitting tensile and flexural strengths of concrete with FA as partial substitution of fine aggregates. Papadakis (1999) observed higher compressive strength development of mortar after 14 days of curing when low calcium FA was used as partial replacement of fine aggregates. On the other hand, the strength development of concrete was observed only after 91 days when the same FA was used to replace cement. The same author observed higher compressive strength, bound water and total porosity of mortar when a high calcium FA was used to partially replace the fine aggregates, whereas the strength was almost similar during the experimental curing period when the same FA used to replace cement (Papadakis 2000).

The addition of FA as fine aggregates replacement also decreases the depth of wear during the abrasion resistance test (Fig. 4.16). The improvement of abrasion resistance of concrete is due to the increase in compressive strength due to the addition of FA (Siddique 2003b). Seo et al. (2010) observed similar drying shrinkage cracking of concrete when coal FA replaced part of cement or fine aggregates and slightly higher performance than that of conventional concrete. Maslehuddin et al. (1989) observed significantly higher chloride corrosion resistance of concrete with FA as fine aggregates than that of conventional concrete (Fig. 4.17). Hwang et al. (1998) reported that the addition of coal FA as aggregate in mortar improved the carbonation behaviour if the w/c ratio was properly maintained.

4.2.5 Other Coal Ash

Dhir et al. (2000) reported the use of pulverised fuel ash (PFA) as partial replacement of the sand fraction in concrete. The authors found lower slump for concrete with moist-cured PFA aggregates and the slump further decreased as the

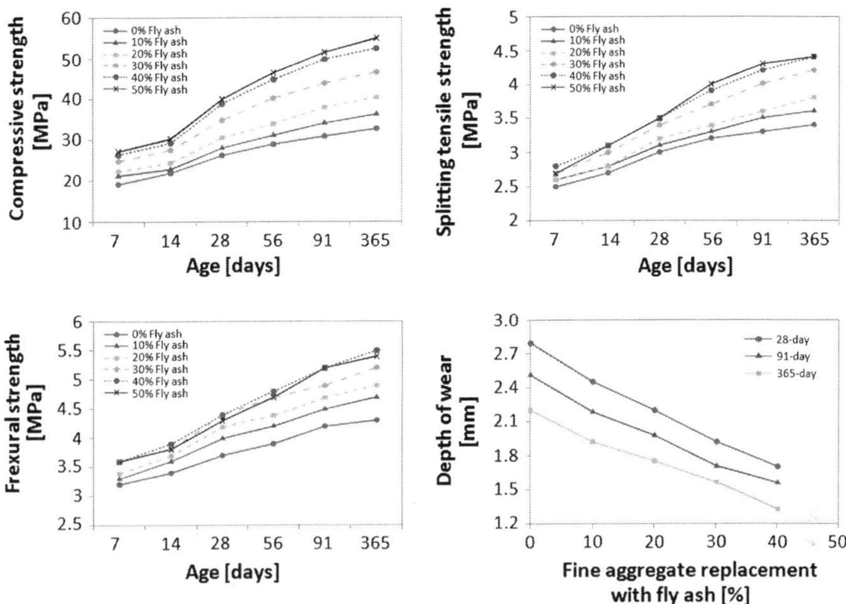

Fig. 4.16 Properties of concrete with fly ash aggregates (Siddique 2003a, b)

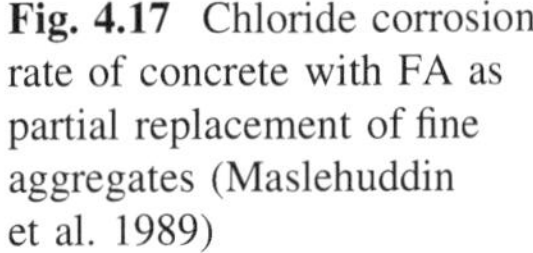

Fig. 4.17 Chloride corrosion rate of concrete with FA as partial replacement of fine aggregates (Maslehuddin et al. 1989)

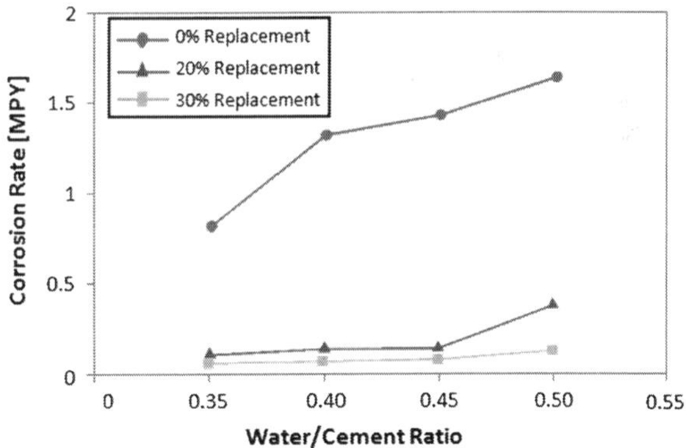

content of PFA rose due to the increase of the fine content in the concrete mix. For dry PFA, the slump was not significantly affected at PFA/sand ratio of 0.05; but slump was reduced due to the addition of a higher amount of dry PFA. However, the authors observed similar cohesion and finishability of the concrete with PFA aggregates and the conventional concrete mix. To control the slump of the concrete mix with PFA, the authors suggested using a superplasticizer. They observed better bleeding performance of concrete mix with both dry and moist PFA aggregates than that of the conventional concrete mix.

The compressive strength of concrete with PFA aggregates was higher than that of the conventional concrete and differences increased as the PFA content increased as well as the curing age. The authors observed a particle size effect of PFA on the compressive strength of PFA-based concrete; the compressive strength of concrete with fine PFA was higher than that observed for concrete with coarse PFA. According to the authors, the higher strength was due to the pozzolanic effect of PFA.

Nataraja et al. (2007) reported the use of a burnt coal cinder as full substitution of NA in concrete. The authors found a substantial reduction in compressive strength of concrete due to the addition of coal cinder. The strength of concrete was 27.6 MPa for 7 days for burnt coal cinder whereas with crushed NA it was 35.5 MPa. At 28 days of curing, the strength of burnt coal cinder concrete increased to 38.5 MPa with a corresponding value for crushed aggregate concrete of 55.0 MPa. Compared to natural coarse aggregate, burnt coal cinder had a low crushing value and therefore the failure of concrete with burnt coal cinder occurred mostly due to aggregate crushing which decreased the compressive strength.

4.3 Steel Slag

Several types of steel slag are generated in the steel making process. The generation of these slags and their properties are already presented in Chap. 2. The presence of free calcium and magnesium oxides in some types of slag restrict their use as an aggregate in concrete. In a recent study, it was concluded that the use of electric arc furnace (EAF) slag as coarse and fine aggregates in concrete preparation can be considered while ladle furnace slag cannot be used for this purpose due to controversial results obtained after durability tests (Polanco et al. 2011). Several studies also reported the presence of high amounts of free lime and magnesium oxide in basic oxygen furnace steel slag. In this section, the properties of concrete with EAF-slag or steel slag, which do not exert too deleterious effects on the durability performance of concrete, will be presented. Significant information is available on the behaviour of conventional as well as HSC with steel slag as aggregates, and it will be discussed systematically in this section.

4.3.1 Fresh Concrete Properties

Changes in workability behaviour (slump) of fresh concrete mixes with slag aggregates were observed due to the large variation in aggregate properties such as water absorption capacity, size and shape and surface texture of various steel slags that were reported in various studies.

A concrete mix with a small amount of EAF-slag had a similar slump value to the one of a conventional concrete; however, increasing the addition of EAF-slag significantly reduces slump (Etxeberria et al. 2010). Manso et al. (2004) reported that concrete mixes with EAF-slag as the only fine and coarse aggregates lacked cohesion, and therefore collapsed during mixing. The complete substitution of coarse aggregate EAF aggregates of similar size and the substitution of the 0–4 mm fraction of NA by a 1:1 mixture of EAF-slag and limestone filler with particle size <1 μm in the concrete mix can eliminate this problem. Qasrawi et al. (2009) also observed marginal reduction in slump for concrete mixes with steel slag replacing up to 50 % by weight of fine NA and concrete can be classified as having moderate slump. However, concrete with 100 % slag was sticky with slump almost nil. The increase in the fine content and angular particle content of the concrete mix due to the addition of slag as well as the slightly higher water absorption capacity of slag by comparison with that of natural sand were the causes of the observed slump loss. On the contrary, Al-Negheimish et al. (1997) did not observe any significant difference between the slump of concrete with steel slag as coarse aggregates and conventional concrete at equal w/c value.

The bulk density of the majority of steel slags is significantly higher than that of NA, and therefore the dry density of concrete with steel slag is generally higher than that of conventional concrete. According to Papayianni and Anastasiou (2010), heavyweight concrete with a density of 2750 kg/m^3 could be produced by using EAF-slag. Masleduddin et al. (2003) reported that the density of a fresh concrete mix with EAF-slag with a 3.51 specific gravity replacing 45–65 % by weight of crushed limestone aggregates were in the range of 2436–2769 kg/m^3, whereas the density of concrete with crushed limestone aggregates with a 2.54 specific gravity was 2330 kg/m^3. Al-Negheimish et al. (1997) observed a similar increase in density due to the replacement of natural coarse aggregate by steel slag. However, Qasrawi et al. (2009) observed a very slight increase (<5 %) in the density of concrete with steel slag, used to replace up to 50 % by weight of fine NA, and the resulting concrete was reported to be normal weight according to ASTM specifications.

4.3.2 Hardened Concrete Properties

4.3.2.1 Compressive Strength

In several references, it was reported that the compressive strength of concrete with EAF-slag as coarse and fine aggregates replacement was similar to or even higher than that of conventional aggregate. However, contrasting results are also available on compressive strength behaviour due to use of EAF-slag as aggregates in concrete.

Al-Negheimish et al. (1997) observed similar compressive strength behaviours in concrete with coarse conventional and with coarse EAF-slag aggregates as curing time increased and for three different curing conditions, namely moist curing at 21 °C, curing at 28 °C with 45 % humidity and curing at 55 °C and 10 %

humidity. However, curing conditions had a significant effect on the compressive strength of steel slag concrete. Standard moist curing of this type of concrete exhibited the highest compressive strength, followed by moderate and high temperature curing. Moreover, at the latter stages of curing, deterioration of compressive strength was observed in moist-cured samples possibly due to the formation of expansive products.

Pellegrino and Gaddo (2009) observed higher compressive strength in concrete with EAF-slag than in conventional concrete after 7, 28 and 74 days of curing. However, in this study, significantly higher amounts of fluidifying agent and slightly higher amounts of aerating agent were used during preparation of concrete with EAF-slag aggregate than those used in conventional concrete preparation. The compressive strength of conventional concrete stabilized after 28 days of curing while the compressive strength of concrete with EAF-slag increased with curing time up to 74 days.

The 28-day compressive strengths of conventional concrete and concrete with unprocessed steel slag in the Maslehuddin et al. (2003) study were 39.7 and 41.6 MPa, respectively. These concrete mixes were prepared using a similar composition with coarse aggregate to total aggregate ratio of 0.60. The 28-day compressive strength of concrete with slag aggregates with coarse aggregate to total aggregate ratios of 0.45, 0.50, 0.55 and 0.65 were, respectively, 31.4, 37.7, 37.6 and 42.7 MPa. The authors also concluded that a coarse aggregate to total aggregate proportion of 50 % may be adopted to minimise the weight effect of heavy steel slag aggregates.

Almusallam et al. (2004) and Beshr et al. (2003) compared the compressive strength of concrete with steel slag coarse aggregates to that of concrete with three types of limestone aggregates. Concrete was prepared with a w/c ratio of 0.35 and slump of 50–75 mm using a superplasticizer so that the compressive strength performance can be related with the mechanical properties of the aggregates. After 28 days of curing, the compressive strength of concrete specimens prepared with calcareous, dolomitic, and quartzitic limestone and steel slag aggregates were 43, 45, 47 and 54 MPa, respectively. According to the authors, for HSC the bulk of the compressive load is borne by the aggregate rather than the cement paste alone and therefore failure occurs through the aggregate. Thus, the compressive strength of HSC depends on the mechanical prosperities of the coarse aggregates. Since the steel slag aggregate had better mechanical properties than the other aggregates, the incorporation of steel slag in concrete improved its compressive strength.

Papayianni and Anastasiou (2010) determined a 28-day compressive strength of 64.2 and 70.3 MPa for HSC with crushed limestone aggregate (reference concrete) and concrete with coarse EAF aggregates, respectively. The compressive strength of concrete with EAF-slag as fine and coarse aggregates was 77.9 MPa and it was about 21.3 % higher than that of the reference concrete. The authors also observed a higher rate of strength gain for concrete with slag aggregates during the initial periods (0–7 days) of curing (89.2–92.2 % of 28-day strength) than that observed for the reference concrete (81.8 % of 28-day strength).

Etxeberria et al. (2010) observed lower compressive strength for concrete with EAF-slag as the only coarse aggregates than that of a conventional concrete due to a higher effective w/c value (0.69) than that of the conventional concrete (0.65). However, in comparison with conventional concrete, the authors observed slightly higher and almost equal compressive strength of concrete mixes prepared by replacing, respectively, 25 and 50 % by volume of coarse nature aggregates by slag. On the other hand, the compressive strength of concrete prepared at lower w/c value increased with higher content of EAF-slag used to replace 0, 25, 50 and 100 % by volume of coarse NA (w/c equal to 0.57, 0.58, 0.59, and 0.60, respectively).

Manso et al. (2006) observed low compressive strength of concrete with EAF-slag as fine and coarse aggregates, due to its very poor workability behaviour. But the compressive strength of concrete with EAF-slag was comparable to that of conventional concrete at latter stages of curing (6 months and 1 year) when the fine and coarse NA in concrete were replaced according to the following methods: (1) complete replacement of coarse NA by similar size fractions of EAF-slag; (2) complete replacement of coarse NA by similar size fractions of EAF-slag along with the replacement of an equal amount of fine limestone aggregates by EAF-slag fine aggregates. In the case of the second method, the grain size of limestone aggregates was below 1 mm and therefore they act as a filler material.

Qasrawi et al. (2009) observed higher compressive strength for three different types of concrete (with design cube strength of 25, 35 and 45 MPa) prepared by replacing 15 and 30 % by weight of fine aggregates by steel slag than that for conventional concrete. However, at the replacement ratios of 50 and 100 % by weight, the compressive strength for all concrete types with slag aggregate were lower than for conventional concrete. The increase in compressive strength of concrete with EAF-slag up to a certain replacement level was due to the higher angularity of steel slag aggregates compared to NA, which therefore increased the binding between cement paste and aggregates. However, for higher slag incorporation levels, the percentage of the 0.15 mm aggregates fraction in concrete increased due to the higher content of this fraction in slag (about 40 % of total content). Thus, less cement was available to coat the slag particles and therefore the paste–aggregate bonding decreased, which ultimately reduced the compressive strength.

4.3.2.2 Splitting Tensile Strength

Several authors reported that the incorporation of steel slag as aggregates in concrete increases the splitting tensile strength just like it does the compressive strength as discussed in the previous section. However, results are also available where improvements of compressive strength but deterioration of splitting tensile strength was observed.

Al-Negheimish et al. (1997) observed higher 28-day splitting strength for concrete with steel slag coarse aggregate than for conventional concrete at three different curing conditions. This difference was more significant for curing conditions with moderate and high temperatures along with dry environment than that for normal moist curing conditions. Almusallam et al. (2004) and Beshr et al. (2003) also observed higher splitting tensile strength for concrete with steel slag aggregates than that for three other types of concretes using calcareous, dolomitic and quartzitic coarse aggregates when the authors investigated the behaviour of coarse aggregate type on the mechanical performance of concrete. Papayianni and Anastasiou (2010) observed 28-day splitting strength of 5.20, 5.52 and 5.89 MPa for HSC with crushed limestone aggregates (reference concrete), concrete with coarse EAF-slag aggregates and concrete with EAF-slag as fine and coarse aggregates, respectively. Pellegrino and Gado (2009) observed higher splitting tensile strength for concrete with steel slag with 2–22.4 mm size range as aggregates than for concrete with NA.

Etxeberria et al. (2010) found lower splitting tensile strength for concrete with various amounts of steel slag as coarse and fine aggregates than for conventional concrete with effective w/c ratios of 0.55 and 0.50. However, in the same study, higher compressive strength for concrete with steel slag than for conventional concrete with w/c ratio of 0.50 was reported. A slight improvement of compressive strength while a slight deterioration of splitting tensile strength due to the addition of steel slag aggregates in concrete was also observed in the Maslehuddin et al. (2003) study. The author obtained compressive strength of 41.6 MPa and splitting tensile strength of 6.26 MPa for concrete with steel slag aggregates and compressive strength of 39.7 MPa and splitting tensile strength of 6.33 MPa for conventional concrete.

4.3.2.3 Flexural Strength

Al-Negheimish et al. (1997) reported a slightly higher 28-day flexural strength for concrete with steel slag as coarse aggregates than that for conventional concrete for various curing conditions. The authors also observed significant effect on the flexural behaviour of concrete with steel slag aggregates due to changes in curing conditions. Papayianni and Anastasiou (2010) found 28-day flexural strength of 8.30, 9.13 and 9.96 MPa, respectively, for HSC with crushed limestone aggregates (reference concrete), concrete with coarse EAF-slag aggregates and concrete with EAF-slag as partial replacement of fine NA and complete replacement of natural coarse aggregates. Maslehuddin et al. (2003) observed lower flexural strength for concrete with coarse steel slag aggregates that added up to 60 % of total aggregates than that for conventional concrete with coarse crushed limestone aggregates. However, a coarse steel slag content of 65 % of total aggregates in concrete gave a higher flexural value (4.21 MPa) than in concrete with coarse limestone aggregates (3.96 MPa).

Qasrawi et al. (2009) observed an increase in flexural strength in concrete with increasing replacement of natural sand by fine steel slag aggregates up to a 50 % ratio by weight. However, replacement of 100 % sand by slag was not beneficial when compared to the other replacement levels at the ages of 28 and 90 days. For all replacement ratios, concrete with slag aggregates exhibited higher flexural tensile strength than conventional concrete. This can be attributed to the better mechanical properties of steel slag in addition to its higher angularity by comparison with NA, which increased the bond between aggregates and paste.

4.3.2.4 Young's Modulus of Elasticity

Al-Negheimish et al. (1997) observed a significantly higher 28-day Young's modulus of elasticity for concrete with slag as coarse aggregates than for conventional concrete with natural gravel as coarse aggregates. In this study, the 28-day Young's modulus of elasticity for concrete with slag and conventional concrete was 34.3 and 27.9 GPa, respectively. This higher modulus of elasticity was caused by the increase in concrete weight associated with the higher bulk density and modulus of elasticity of slag aggregate in comparison with NA. Beshr et al. (2003) and Almusallam et al. (2004) observed higher Young's modulus of elasticity for concrete with steel slag as coarse aggregates when a comparison was made between this modified concrete and concrete with three types of limestone aggregates. The authors determined Young's modulus of elasticity values of 29.6, 21.6, 24.4 and 28.8 GPa for concrete with steel slag, calcareous limestone, dolomitic limestone and quartzitic limestone as a coarse aggregates, respectively. This was due to the better mechanical properties of steel slag aggregates than those of the other aggregates. The addition of silica fume to cement further increases the Young's modulus and for 15 % replacement level of Portland cement this value was 40.4 GPa for concrete with steel slag aggregates (Almusallam et al. 2004). According to the authors, the type of aggregates had a more significant effect on the modulus of elasticity than on the compressive strength of concrete. Etxeberria et al. (2010) observed similar modulus of elasticity for concrete with EAF-slag and conventional concrete, particularly at lower water to cement ratios. Two types of concrete were prepared by varying the amount of cement and w/c ratio (w/c values of 0.5 and 0.55). The EAF-slag was used to replace 25, 50 and 100 % by volume of coarse NA. Their results are presented in Table 4.5.

Table 4.5 Modulus of elasticity of concrete with EAF-slag aggregates (Etxeberria et al. 2010)

Concrete type	Amount of EAF-slag to replace coarse aggregate (in volume) (%)	Young's modulus of elasticity (GPa)	
		w/c = 0.50	w/c = 0.55
Ref	0	36.4	30.1
EAF25	25	35.6	30.3
EAF50	50	36.2	26.3
EAF100	100	36.2	23.5

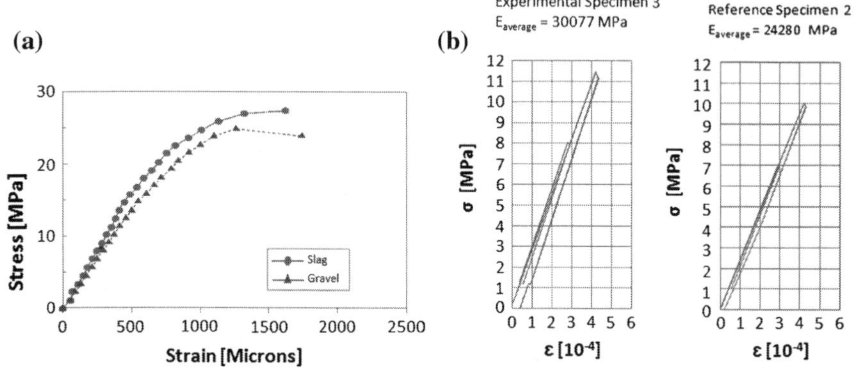

Fig. 4.18 Stress–strain curve of conventional concrete and concrete with steel slag as coarse aggregates. **a** Al-Negheimish et al. 1997; **b** Pellegrino and Gaddo 2009

Al-Negheimish et al. (1997) also observed an increase in the stiffness of concrete, which was clear in the plotted stress–strain curve (Fig. 4.18a). Pallegrino and Gaddo (2009) also plotted a stress–strain curve to determine the Young's modulus of elasticity (Fig. 4.18b). The value calculated from this graph for concrete with EAF-slag and conventional concrete was 30.7 and 24.1 GPa, respectively. The higher stiffness of EAF-slag concrete compared to conventional concrete was due to the higher density and roughness of steel slag aggregates compared to NA.

4.3.2.5 Abrasion Behaviour

Few data are available on the abrasion behaviour of concrete with steel slag aggregates. Papayianni and Anastasiou (2010) reported that concrete made with EAF-slag as coarse aggregates improved the abrasion resistance compared to reference concrete by 73.9 %. This improvement can be further improved to 77.4 % if the concrete is prepared by partially replacing fine aggregates and using EAF-slag as coarse aggregates.

4.3.3 Durability Properties

As steel slag contains some deleterious components like free lime and magnesia (periclase), the expansion of concrete with steel slag at various experimental conditions was reported to evaluate the effect of these components on the resulting concrete. In this section, those properties along with others will be discussed.

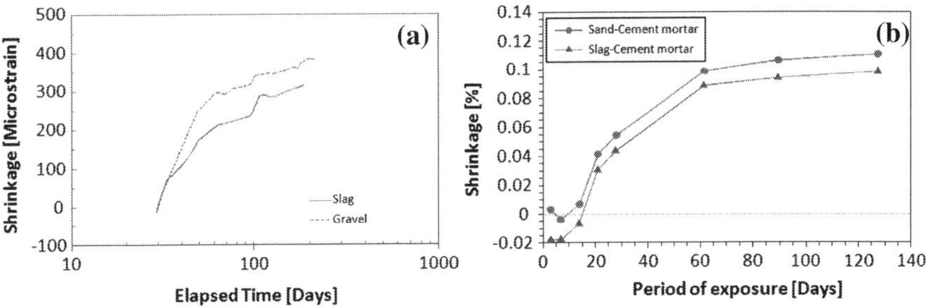

Fig. 4.19 Drying shrinkage of gravel and slag concrete. **a** Al-Negheimish et al. 1997; **b** Maslehuddin et al. 2003

4.3.3.1 Drying Shrinkage

Al-Negheimish et al. (1997) found lower drying shrinkage for concrete with steel slag aggregates than for natural gravel concrete due to their angular particle shape and honeycomb surface texture in comparison to irregular shaped and smooth surface textured gravel aggregates (Fig. 4.19a). The higher modulus of elasticity of slag aggregates by comparison with gravel aggregates was also responsible for the low shrinkage of concrete with slag aggregates. Masleduddin et al. (2003) also observed lower shrinkage for cement mortar with slag aggregates than for normal cement mortar (Fig. 4.19b). After 120 days of curing at 25 °C and 50 % room humidity, the shrinkage of slag and normal cement mortars was 0.097 and 0.11 %, respectively. The incorporation of steel mill scale into mortar as fine aggregates also lowers its drying shrinkage (Al-Otaibi 2008).

4.3.3.2 Expansion of Concrete with Steel Slag Aggregates

Normally, steel slag aggregates contain potentially expansive oxides like free lime and magnesium oxide (periclase); therefore, several tests are performed at normal moist as well as accelerated curing conditions to evaluate the expansion behaviour of concrete with steel slag aggregates. That information will be discussed here.

Lee and Lee (2009) reported the formation of pop-outs of concrete with EAF-slag as fine aggregates. Combined thermogravimetric and EDX analyses of the materials in the pop-out portion revealed the formation of expansive $Ca(OH)_2$ and $Mg(OH)_2$ due to the hydration of free CaO and MgO, present in EAF-slag aggregates.

Length Change Due to Moist Curing

Etxeberria et al. (2010) observed similar length change behaviour for concrete with various percentages of slag and conventional aggregates after submersing

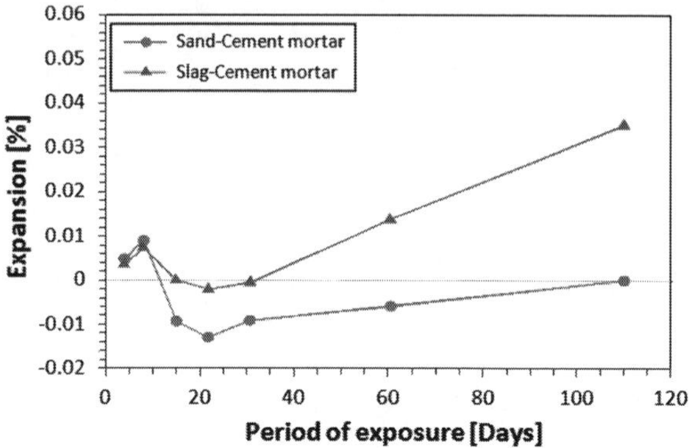

Fig. 4.20 Expansion of cement mortar exposed to moist environment (Maslehuddin et al. 2003)

them in water from 12 to 56 weeks. The specimens suffer a slight length change after 12 weeks of curing and then it almost stabilizes up to 56 weeks of curing. On the other hand, Maslehuddin et al. (2003) observed slightly higher length change for cement mortar with steel slag aggregates than for mortar with NA when both types of mortars were exposed to moist environment for 4 months. The length change observed for concrete with steel slag aggregates was 0.034 %, which was lower than the ASTM C33 prescribed limit of 0.05 %. Their results are presented in Fig. 4.20. The observed expansion of concrete due to the incorporation of steel slag aggregates was due to calcium carbonate, present in steel slag, which expanded on absorption of water.

Effect on Accelerated Ageing

Pellegrino and Gaddo (2009) used three kinds of accelerated ageing conditions to evaluate the effect of deleterious free lime and periclase, present in steel slag aggregates, on the expansion behaviour of the resulting concrete. They initially used the accelerated ageing method described in the ASTM D 4792 standard and found a reduction in strength of about 5 % for concrete with EAF-slag aggregates and an increase of 9 % for conventional concrete. The surface of the concrete with EAF-slag aggregates also exhibits higher efflorescence than conventional concrete due to the formation of white powder of calcium and magnesium hydroxides. However, ageing of concrete specimens obtained after the test by 3-month moist curing at room temperature improves their strength. Their results are presented in Table 4.6. Manso et al. (2006), on the other hand, did not observe any significant differences in compressive strength of conventional concrete as well as of concrete with steel slag aggregates after using the method described in ASTM D 4792 standard followed by 90 days of moist curing (Table 4.6).

Manso et al. (2006) also reported the results of a vigorous accelerating test where conventional concrete and concrete with EAF-slag replacing various percentages of fine and coarse NA were initially subjected to autoclave test followed by 90 days of weathering. Their results indicated that concrete with limestone

Table 4.6 Compressive strength behaviour of concrete after accelerated ageing

Concrete type	Compressive strength (MPa)		
	Before ageing	After 32 days curing at 70 °C	After curing at 70 °C for 32 days plus moist curing for 90 day
Pellegrino and Gaddo (2009)			
Control	30.4	33.1	32.9
EAF-slag	44.4	41.9	43.4
Manso et al. (2006)			
Control	38.5		39.6
EAF-slag-1	33.7		35.9
EAF-slag-2	35.3		39.4
EAF-slag-3	30.2		33.5
EAF-slag-4	30.7		34.1

Table 4.7 Compressive strength before and after autoclave ageing followed by weathering (Manso et al. 2006)

Concrete type	Compressive strength (MPa)		
	Before ageing	After ageing	Appearance
Control	38.5	18.4	Superficial cracking
EAF-slag-1	33.7	20.9	Slight superficial cracking
EAF-slag-2	35.3	23.8	Slight superficial cracking

aggregates exhibited poorer compressive strength than concrete with EAF-slag aggregates due to the difference in shape of these two aggregates. Their results are presented in Table 4.7.

Manso et al. (2004) evaluated the soundness of cement mortar with EAF-slag as partial substitution of fine aggregates in concrete according to ASTM C1012, in which cement mortar were subjected to ten cycles of repeated immersion in a saturated Na_2SO_4 solution followed by drying in an oven. They observed larger deterioration in mortar with EAF-slag than in conventional mortar.

4.3.3.3 Freeze–Thaw Resistance

Manso et al. (2006) reported that concrete with steel slag aggregates exhibited poorer performance than conventional concrete after 25 cycles of freezing and thawing. Their results are presented in Table 4.8. Out of four concrete mixes with steel slag aggregates, the authors observed better performance for concrete with EAF-slag-2, which exhibited slightly higher compressive strength as well as lower porosity. The addition of an air-entraining admixture improved the freeze–thaw resistance of concrete with steel slag aggregates. Pellegrino and Gaddo (2009) observed about 7 % reduction in compressive strength for concrete with EAF-slag,

Table 4.8 Compressive strength behaviour of concrete after freeze–thaw cycles

Concrete type[a]	Compressive strength (MPa)		Strength change (%) (strength gain: +) (strength loss: −)	Appearance
	Before ageing	After freezing and thawing		
Manso et al. (2006)				
Control (13)	38.5	32.7	−15	Good
EAF-slag-1 (16.2)	33.7	20.6	−39	Significant damage
EAF-slag-2 (16.0)	35.3	27.2	−23	Slight damage
EAF-slag-3 (17.6)	30.2	16.9	−44	One sample cracked
EAF-slag-4 (19.6)	30.7	16.0	−48	Significant damage
Pellegrino and Gaddo (2009)				
Control	30.4	33.9	+11.5	
EAF-slag	44.4	41.2	−7.3	

[a] Data in parenthesis indicates the porosity of concrete in percentage

when the concrete specimens were subjected to repeated freeze–thaw cycles for 25 days. Their results are also presented in Table 4.8. The lesser reduction in compressive strength in comparison to Manso et al.s' study (2006) was due to the incorporation of an air-entrainment agent, which caused the formation of closed pores in the specimens and therefore concrete specimens gained resistance against freezing and the thermal/expansive stress decreased.

4.3.3.4 Resistance Against Wet–Dry Cycles

Pellegrino and Gaddo (2009) observed a reduction of about 26.5 % in compressive strength of concrete with EAF-slag aggregates in comparison to a reduction of 7.7 % for conventional concrete when both types of concrete specimens, after 28 days normal curing, were subjected to 30 cycles of repeated 16-h moist curing followed by 8-h oven drying at 110 °C. The presence of free calcium and magnesium oxides in EAF-slag favours the more serious degradation of the resulting concrete than that observed in the conventional concrete. Manso et al. (2006) also observed significantly higher strength reduction in concrete with EAF-slag aggregates (except for one slag concrete composition) than in the control concrete when the four types of concrete mixes along with a control mix were subjected to a similar type of wet–dry cycles to that performed by Pellegrino and Gaddo (2009). These results are presented in Table 4.9.

Maslehuddin et al. (2003) observed a reduction of 3–7 % in compressive strength of concrete with limestone and EAF-slag as coarse aggregates, when the 28-day hardened concrete specimens were exposed to 120 cycles of mild thermal cycles. The concrete specimens were exposed 8 h at 70 °C followed by 16 h at

Table 4.9 Compressive strength behaviour of concrete after wetting and drying cycle

Concrete type	Compressive strength (MPa)		Strength loss (%)	Appearance
	Before ageing	After freezing- thawing		
Manso et al. (2006)				
Control	38.5	27.3	29	Good
EAF-slag-1	33.7	19.9	41	Slight damage
EAF-slag-2	35.3	24.7	30	Good
EAF-slag-3	30.2	16.6	45	Slight damage
EAF-slag-4	30.7	15.6	49	One sample cracked
Pellegrino and Gaddo (2009)				
Control	30.4	28.7	5.60	
EAF-slag	44.4	32.7	26.52	

25 °C to complete one thermal cycle. However, a reduction in pulse velocity and an increase in water absorption after the completion of thermal cycles indicated better performance of concrete with steel slag aggregates than of the limestone aggregates concrete due to the denser microstructure of the former concrete compared with the latter.

4.3.3.5 Other Durability Behaviour

Manso et al. (2006) investigated the alkali-aggregate reaction of EAF-slag to be used as aggregates in concrete by using the ASTM C1260 method. According to the authors, slag contains a significant amount of glassy phase, which can react with alkalis present in cement. The average value of expansion was 0.14 % after 16 days and 0.15 % after 28 days, both well below the specified value of 0.2 %. However, the presence of free CaO and MgO in EAF-slag overestimate the expansion value.

Maslehuddin et al. (2003) detected a better performance in concrete with steel slag aggregates than in limestone concrete when both were subjected to chloride induced corrosion. The time to initiation of reinforcement corrosion and time to cracking of concrete specimens were, respectively, 190 and 517 h for conventional concrete and 198–367 h and 509–774 h for steel slag aggregates concrete. This was mainly due to a denser microstructure in steel slag aggregates concrete than in NA concrete. Lower water absorption capacity and higher ultrasonic pulse velocity in concrete with steel slag aggregates than in NA concrete were also observed.

4.4 Blast Furnace Slag

The use of BFS as aggregates in concrete is not as common as its use as a component in cement. However, some recent reports indicate that this material (particularly air-cooled BFS) can be used as an aggregate in concrete preparation.

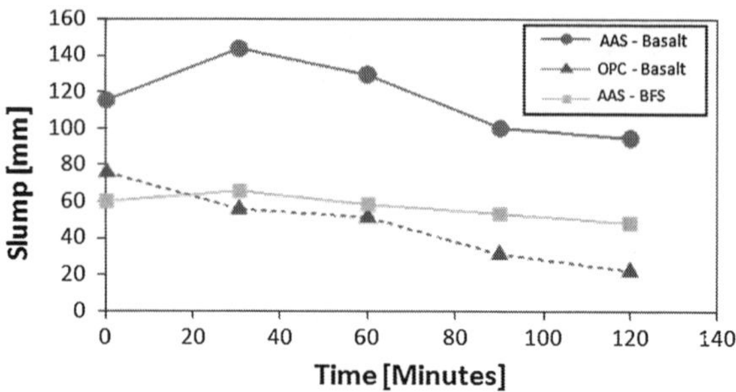

Fig. 4.21 Comparison of slump of AAS concrete with BFS and basalt as coarse aggregates and conventional concrete with basalt coarse aggregates (Collins and Sanjayan 1999)

4.4.1 Fresh Concrete Properties

There are vast differences in results presented in various references on the slump behaviour of concrete mix due to the addition of BFS aggregates. Etxeberria et al. (2010) observed a slight increase in slump when 25 % by volume of coarse NA were replaced by BFS aggregates. The mix was workable just like conventional concrete. On the other hand, replacing 50 % by volume of NA by BFS aggregates considerably reduced the slump of the resulting concrete. However, slump increased again when natural coarse aggregates were completely replaced by BFS aggregates. Collins and Sanjayan (1999) observed a slump of 65 mm for alkali-activated slag (AAS) concrete with BFS coarse aggregates and a slump of 115 mm for a similar concrete with basalt coarse aggregates. In this study, the BFS aggregates were presaturated with water before being used as aggregates due to their higher water absorption capacity (4.4 %) in comparison to that of basalt aggregates (1.2 %) (Fig. 4.21). The observed low slump of BFS aggregates was due to the differences in surface texture, shape and porosity from basalt aggregates. However, in some studies, no significant difference was found in terms of slump or workability between conventional concrete and concrete with BFS as coarse aggregates (Demirboga and Gul 2006; Haque et al. 1995). The incorporation of BFS as fine aggregates replacement also decreases the slump of the resulting concrete and increasing their content further decreases it (Yüksel et al. 2011).

The density of concrete with BFS aggregates depends on the bulk density of BFS aggregates. Demirboga and Gul (2006) observed an increase in fresh density of about 7.9–8.5 % in HSC with different w/c values due to BFS-aggregates incorporation. The bulk density of BFS aggregates was equal to 2.78 g/cm^3 and higher than that of natural coarse aggregates. Etxeberria et al. (2010) reported lower dry density for concrete with BFS aggregates than for conventional concrete. The dry density was further decreased as the content of BFS aggregates increased due to the lower bulk density of BFS aggregates (2.36 g/cm^3) than that NA (2.56 g/cm^3). The air content of alkali-activated concrete with BFS coarse aggregates and basalt coarse aggregates were 1.6 and 1.2 %, respectively (Collins and Sanjayan 1999).

4.4.2 Hardened Concrete Properties

Etxeberria et al. (2010) observed contrasting compressive strength behaviour, when concrete mixes were prepared by replacing 0, 25, 50 and 100 % by volume of natural coarse aggregates by BFS aggregates at low and high w/c ratios. At high w/c, the compressive strength for concrete with BFS aggregates used to replace 25 and 100 % by volume of coarse aggregates was lower than that of the conventional concrete. But the compressive strength of concrete with BFS aggregates used to replace 50 % by volume of coarse aggregates was higher than that of the conventional concrete. On the other hand, at low w/c, the compressive strength of concrete with BFS aggregates was higher than that of the conventional concrete and strength increased with the content of slag aggregates in concrete. In both cases, the w/c value of conventional concrete was lower than those of the BFS-aggregates concrete mixes.

Demirboga and Gul (2006) found higher compressive strength for HSC with BFS as coarse aggregates than for conventional concrete for various w/c values. However, the difference in strength between BFS-aggregates concrete and conventional concrete decreased as the w/c value increased (Fig. 4.22). Haque et al. (1995) reported that compressive strength of concrete with BFS aggregates could be as high as 107 MPa. Yüksel and Bilir (2007) observed similar compressive strength for concrete pavement blocks where 60–80 % by volume of fine aggregates were replaced by equal-size BFS aggregates and for a conventional pavement block. Yüksel et al. (2011) observed a decreasing trend of compressive strength with increasing substitution level of fine NA by BFS aggregates.

Collins and Sanjayan (1999) found higher 1-day compressive strength for AAS concrete with BFS as coarse aggregate than for coarse basalt aggregates (NA) AAS concrete when both were cured by immersion. However, the compressive strength of NA AAS at the later stages of immersion curing and sealed curing at all

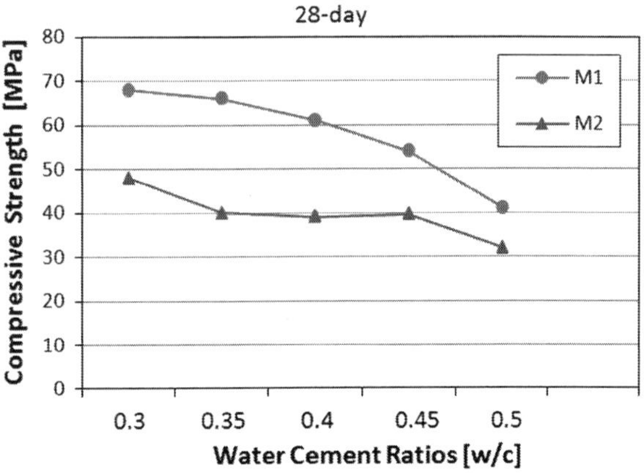

Fig. 4.22 Compressive strength of concrete with natural and BFS coarse aggregates (Demirboga and Gul 2006)

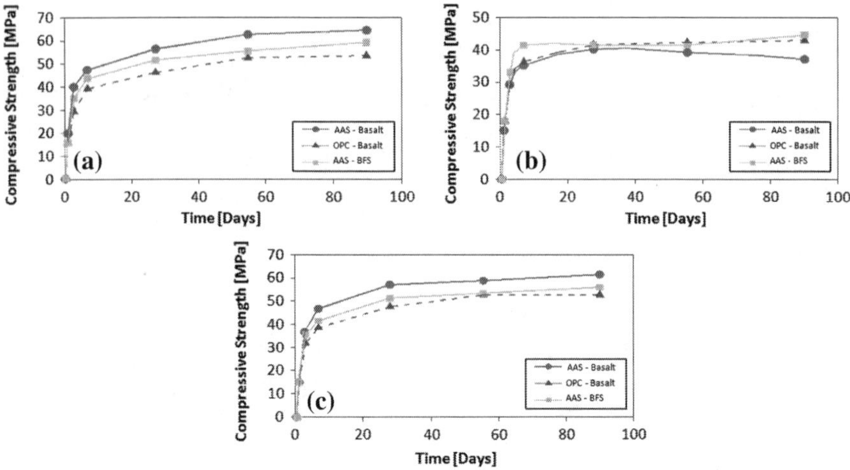

Fig. 4.23 Compressive strength of various types of concrete at three different curing conditions: **a** at 23 °C in immersion; **b** at 23 °C in 50 % humidity; **c** at 23 °C with sealed specimens (Collins and Sanjayan 1999)

curing period was higher than that of AAS with BFS aggregates. On the other hand, the compressive strength of AAS with BFS aggregates was higher than that of NA AAS at the whole curing period when both were cured at 23 °C with 50 % room humidity due to internal curing effect, where the moisture present in BFS aggregates came out at low humidity conditions (Fig. 4.23).

Etxeberria et al. (2010) observed lower modulus of elasticity and splitting tensile strength for concrete with BFS aggregates than for conventional concrete at two different ranges of w/c values. These values further decreased as the content of BFS aggregate in concrete increased. Lower modulus of elasticity was reported for BFS-aggregates concrete in comparison to conventional concrete in Haque et al.s' (1995) study too. On the other hand, in comparison to conventional concrete, about 8–10 % higher splitting tensile strength and higher elastic modulus were recorded for concrete with BFS coarse aggregates prepared at various w/c ratios.

Ashby (1996) observed similar elastic modulus of elasticity but marginally higher Poisson's ratio for concrete with air-cooled slag aggregates than for natural gravel aggregates concrete. The specific creep for grade 20 concrete with air-cooled aggregates was lesser but for grade 40 concrete it was similar to that of the natural gravel concrete.

Yüksel et al. (2007) observed higher abrasion resistance of concrete with BFS as partial replacement of fine aggregates. In this study, the fine aggregates were replaced by BFS aggregates up to a replacement level of 50 % (by weight). Maximum abrasion resistance was observed when 10 % of fine aggregates were replaced by BFS aggregates.

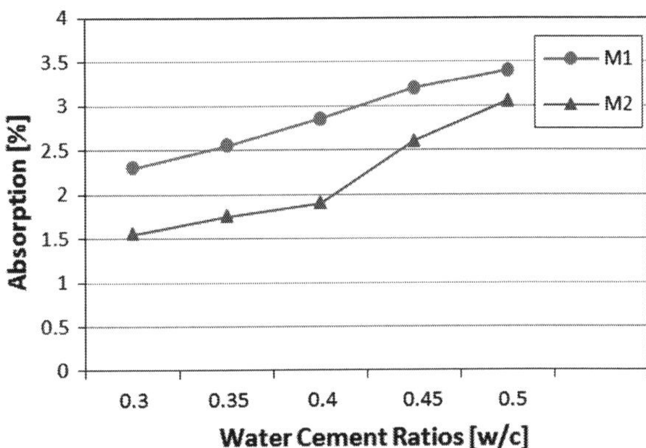

Fig. 4.24 Water absorption capacity of conventional concrete (M1) and concrete with BFS aggregates (M2) (Demirboga and Gul 2006)

4.4.3 Durability Performance

Etxeberria et al. (2010) observed almost identical length change behaviour for concrete with BFS coarse aggregates and conventional concrete when both were immersed in water for 12 and 56 weeks: all types of concrete specimens suffered a slight length change after 12 weeks of immersion and then it stabilized up to 56 weeks of immersion. However, in comparison to conventional concrete, BFS-aggregates concrete suffered lesser length change in the early 12-week period.

Etxeberria et al. (2010) observed lower capillary water absorption for AAS concrete with BFS aggregates than for conventional concrete. The lowest and highest absorptions were observed for mixes with 50 and 100 % by volume replacement of coarse aggregates by BFS aggregates, respectively. Demirboga and Gul (2006) also observed lower water absorption capacity for HSC with BFS as coarse aggregates than for conventional concrete at different w/c values (Fig. 4.24). On the other hand, Yüksel et al. (2007) reported higher capillary water absorption for concrete with BFS aggregates replacing 20–50 % by weight of fine aggregates than for conventional concrete; however, at 10 % replacement level the capillary water absorption of BFS concrete was lower than that of the conventional concrete due to pozzolanic reactions of some constituents of BFS aggregates with free lime. At high substitution level, concrete became porous due to porous aggregates addition and therefore increased capillary water absorption. Haque et al. (1995) observed lesser water absorption and water penetration for a high-performance concrete with air-cooled BFS aggregates than for conventional concrete.

Etxeberria et al. (2010) observed significantly higher residual strength for concrete with BFS aggregates than for conventional concrete when both were exposed to 800 °C for 4 h. The residual compressive strength for mixes where 25, 50 and 100 % by volume of NA were replaced by BFS aggregates was, respectively, 48, 53 and 51 % in comparison to 33 % for conventional concrete. The concrete with BFS as partial replacement of fine aggregates exhibited similar or slightly better compressive strength and dynamic elastic modulus behaviour than the conventional concrete when both were exposed to 800 °C (Yüksel et al. 2011).

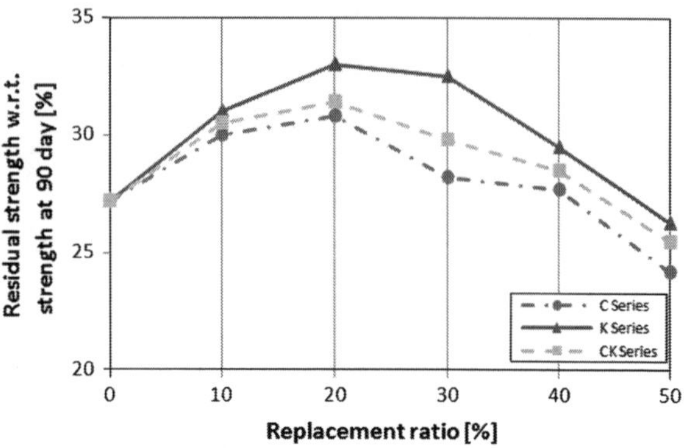

Fig. 4.25 Residual compressive strength of concrete after high temperature exposure of BFS-aggregates concrete (C-series) along with CBA aggregates concrete (K-series) (Yüksel et al. 2007)

Increasing addition of aggregates slightly improved these properties, which was more noticeable in the dynamic elasticity modulus results. In another work, Yüksel et al. (2007) observed higher residual strength for concrete where up to 40 % by weight of fine aggregates were replaced by BFS aggregates than for conventional concrete. In this study, the maximum strength was observed at 20 % replacement level. The residual strength was lower than that of the conventional concrete for 50 % replacement of fine aggregates by BFS aggregates, possibly due to changes in concrete microstructure and the generation of more porosity because of the substitution of fine aggregates by porous BFS aggregates. However, the residual strength of concrete with CBA as partial replacement of fine aggregates was better than that of the BFS-aggregates concrete (Fig. 4.25).

Yüksel et al. (2007) found higher freeze–thaw resistance for concrete with BFS as partial replacement of fine aggregates than for conventional concrete as well as CBA aggregates with concrete. The specimens were subjected to 50 cycles of repeated freezing and thawing. This resistance increased with the replacement ratio of fine aggregates by BFS aggregates and at 20 % replacement level it reached a maximum; further increasing the replacement level up to 50 % by weight decreased the resistance even though the performance was still better than that of conventional concrete. Their results are presented in Fig. 4.26.

Yüksel et al. (2007) observed an insignificant effect of wet–dry cycles on the strength loss of concrete with BFS fine aggregates as 0–50 % by weight replacement of NA, when the specimens were subjected to 25 cycles. Collins and Sanjayan (1999) observed lower drying shrinkage for AAS concrete with BFS as coarse aggregates than for concrete with basalt aggregates. The experiment was undertaken at 23 °C with 50 % room humidity. However, similar autogenous shrinkage in concrete with basalt and BFS aggregates indicated that the improvement of shrinkage because of BFS-aggregates addition was attributed to the internal curing caused by the moisture present in these aggregates. Ashby (1996) also observed lower drying shrinkages for concrete with air-cooled BFS aggregates than for conventional concrete with river gravel up to a period of 56 days. Haque et al. (1995) reported lower drying shrinkage for high-performance concrete with air-cooled BFS aggregates than for conventional concrete.

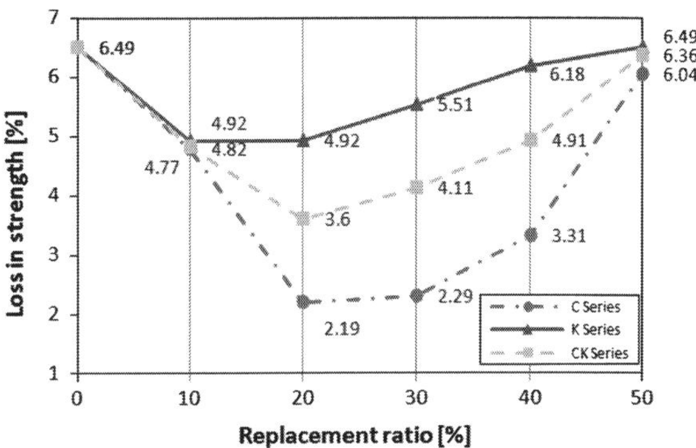

Fig. 4.26 Loss in strength after freeze–thaw testing: C-Series—BFS-aggregates concrete; K-series—CBA aggregates concrete (Yüksel et al. 2007)

4.5 Non-Ferrous Slag

4.5.1 Copper Slag as Aggregate in Concrete

Several reports are available on the use of copper slag as fine and coarse aggregates in concrete. Both normal and HSC are prepared with copper slag aggregates. In this section, normal and high-performance concrete properties will be discussed in the same section.

4.5.1.1 Fresh concrete properties

The incorporation of copper slag as a partial or full substitution of fine aggregate in concrete increases the slump value (Al-Jabri et al. 2011; Wu et al. 2010; Pezhani and Jeyaraj 2010). Al-Jabri et al. (2011) observed increased slump of concrete as the incorporation of copper slag as replacement of fine aggregates rose. The slump of conventional concrete and concrete with copper slag at 100 % fine aggregate level were, respectively, 65.5 and 200 mm. The improvement in slump was due to the presence of a greater amount of free water in slag concrete than in conventional concrete as the water absorption capacity of copper slag was lower than that of natural fine aggregates. However, segregation and bleeding was observed in fresh concrete mixes with high amount of copper slag, i.e. in this case concrete mixes prepared with 80 and 100 % of fine NA replaced by copper slag. Al-Jabri et al. (2009a, b) also observed a significant reduction in w/c ratio in HSC due to the replacement of fine aggregates by equal-size copper aggregates. The w/c value of conventional concrete and concrete with copper slag at 100 % fine aggregates replacement level were 0.35 and 0.27, respectively. The addition of copper slag as coarse aggregates also increased the slump of concrete (Khanjadi and Behnood 2009).

Increased bleeding was also reported in the Ishimaru study (2005) when the content of copper slag used to replace natural fine aggregates increased. Bleeding increased with copper slag incorporation due to the high bulk density, glass-like surface properties and irregular grain shape of copper slag (Shoya et al. 1997). Bleeding depends on several factors such as w/c ratio, air content and slag content in concrete. According to Shoya et al. (1997) 40 % copper slag can be used as partial replacement of aggregates to control the amount of bleeding to less than 5 l/m^2. The addition of an admixture with cellulose ether and powder-like material, such as limestone powder, is highly effective to improve the bleeding performance of concrete with copper slag as aggregates (Shoya et al. 1997). Hwang and Laiw (1989) obtained a concrete mix with satisfactory workability and minimal bleeding at the optimum fineness modulus of a mixture of copper slag and natural fine aggregates, which was roughly equal to 2.6.

The addition of copper slag as fine or coarse aggregates in concrete increases its fresh density due to the higher bulk density of copper slag than that of NA (Al-Jabri et al. 2011; Khanjadi and Behnood 2009). Al-Jabri et al. (2011) observed an increase in the density of fresh concrete of about 5 % when fine aggregates were totally replaced by copper slag. Khanjadi and Behnood (2009) reported a density of 2310 and 2668 kg/m^3 for HSC mixes with limestone and copper slag as coarse aggregates, respectively. Khanjadi and Behnood (2009) also observed air-content values of 2.5 and 2.4 % for natural and copper slag concrete, respectively. The difference in air content between control concrete and copper slag concrete increases with addition of silica fume with cement.

4.5.1.2 Hardened Concrete Properties

Al-Jabri et al. (2011) observed an increasing trend of 28-day compressive strength of concrete as the content of copper slag rose up to a 40 % replacement level (mix No. 4 in Fig. 4.29) of fine aggregates. Their results are presented in Fig. 4.29. Further increment of the content of copper slag decreased the compressive strength of resulting concrete and at a 60 % replacement level (mix No. 6) it became slightly higher than the compressive strength of conventional concrete. The mix with 40 % copper slag content yielded the highest 28-day compressive strength of 47.1 N/mm^2 compared with 45 N/mm^2 for the control mix, whereas the lowest compressive strength of 34.8 N/mm^2 was obtained for the mix with 80 % copper slag (mix No. 7). This reduction in compressive strength for concrete mixes with high copper slag contents was due to the increase in free water content that resulted from the low water absorption characteristics of copper slag in comparison with sand, which caused a considerable increase in the workability of concrete and thus reduced the compressive strength as shown in Fig. 4.27.

In the Birindha and Nagam (2011) study, the 28-day compressive strength of concrete prepared by replacing 40 % of fine aggregates by copper slag aggregates was 46.7 MPa in comparison to the equivalent compressive strength of 35.1 MPa for the control concrete. However, at 60 % substitution level of fine aggregates by copper slag the compressive strength decreased to 39.7 MPa. Similar observations were reported in some other studies, where copper slag was used as fine aggregates

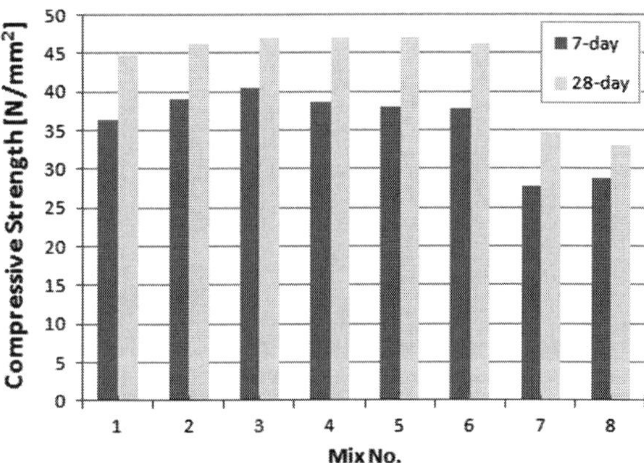

Fig. 4.27 Compressive strength of concrete with copper slag fine aggregates (Al-Jabri et al. 2011)

(Ayano and Sakata 2000; Caliskan and Behnood 2004; Hwang and Laiw 1989; Li 1999; Shoya et al. 1997; Zong 2003). Wu et al. (2010) observed slightly higher dynamic compressive strength of concrete with copper slag replacing 20 % of fine aggregates, which at 40 % replacement level became similar to that of the control concrete. The dynamic compressive strength continuously decreased when the replacement amount exceeded 40 %. According to the authors, the observed improvement at lower substitution level was due to the presence of angular sharp edged particles in copper slag as well as the improvement of cohesion between the cement paste and the aggregates. However, at higher copper slag contents the amount of free water increased due to the low water absorption capacity of copper slag, which increased bleeding, internal voids and capillary pores in concrete.

The 7- and 28-day compressive strength of HSC with copper slag up to 50 % replacement ratio (of natural sand) as fine aggregates were similar (or slightly better) than those of HSC with natural sand. The compressive strength was reduced significantly beyond that level of substitution, due to the separation of particles of the constituents and the formation of pores in concrete by excess free water present in concrete with copper slag aggregates (Fig. 4.28) (Al-Jabri et al. 2009a, b).

The compressive and splitting tensile strengths of HSC with copper slag as coarse aggregates prepared at constant w/c ratio are higher than those of HSC with limestone aggregates, due to the higher strength of copper slag aggregates compared to limestone aggregates and also the porous and rough surface texture of copper slag. This surface texture may produce a superior bond and transition zone in comparison with that of the limestone aggregates (Khanzadi and Behnood 2009). The incorporation of silica fume with cement can produce a stronger transition zone between the copper slag aggregates and the cement paste due to its pozzolanic reaction, and therefore increase the compressive strength (Khanzadi and Behnood 2009).

The tensile splitting strength and flexural strengths of HSC with copper slag behave similarly to compressive strength. However, by comparison with compressive strength a higher rate of development of splitting tensile strength was observed for HSC with copper slag coarse aggregates than for limestone

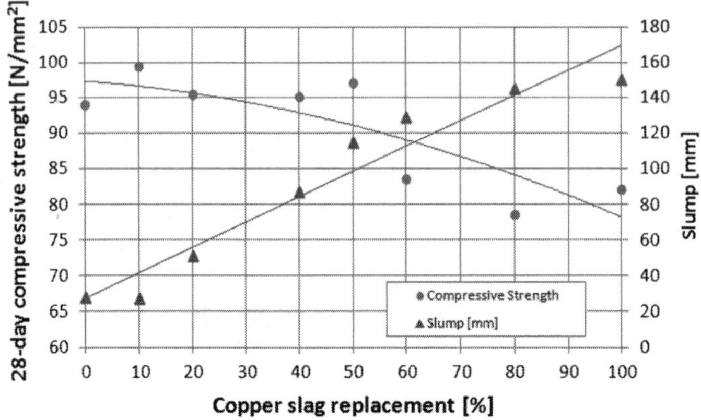

Fig. 4.28 Effect of fine copper slag aggregates addition on the compressive strength and slump of high-strength concrete (Al-Jabri et al. 2009b)

aggregates concrete (Fig. 4.29) (Khanjadi and Behnood 2009). Birindha and Nagam (2011) observed an increase in tensile strength of about 36.5 % when 40 % of natural fine aggregates were replaced by copper slag. However, the increase in percentage became 17 % at 60 % replacement level.

The abrasion resistance for cement mortar with copper slag aggregates is better than for conventional concrete (Tang et al. 2000). Khanjadi and Behnood (2009) also reported a rebound hammer value about 2.6 % higher for HSC with copper slag as coarse aggregates than for natural limestone aggregates concrete. The improvement of properties due to the addition of copper slag as aggregates is due to the higher hardness of copper slag aggregates in comparison with NA.

4.5.1.3 Durability Properties

Birindha et al. (2010) observed increasing ultrasonic pulse velocity of conventional concrete as the replacement of fine aggregates by copper slag aggregates rose. In this study, the copper slag aggregates replaced fine aggregates up to a 60 % level. The optimum value was observed for concrete where 40 % of fine

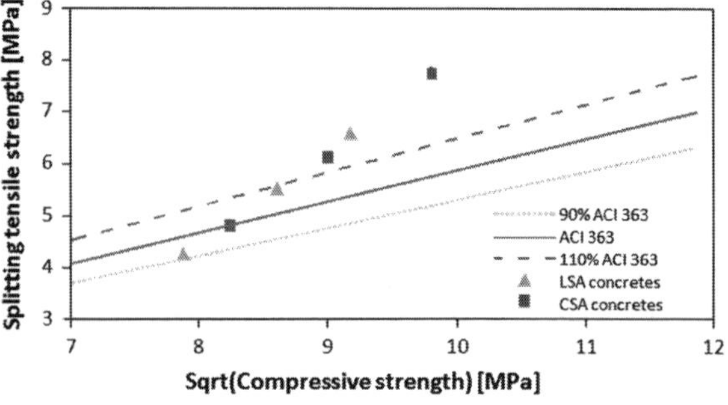

Fig. 4.29 Splitting tensile strength and compressive strength relationship according to ACI 363 (Khanzadi and Behnood 2009)

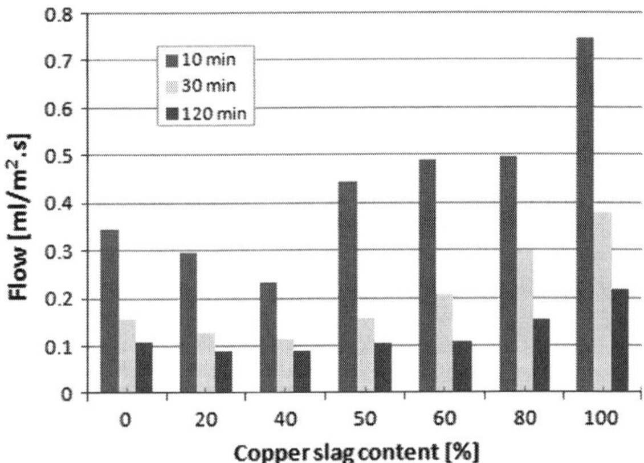

Fig. 4.30 Surface water absorption of HSC due to addition of copper slag as sand replacement (Al-Jabri et al. 2011)

aggregates were replaced, which indicates that the densest microstructure was observed at this replacement level. The drying shrinkage of concrete with copper slag as fine aggregates is similar or even less than that of specimens without copper slag (Ayano and Sakata 2000). Al-Jabri et al. (2011) observed a decreasing trend of water permeable voids in HSC with increasing replacement of sand by copper slag fine aggregates up to a 50 % replacement level; the voids increased again as the content of copper slag continued to rise. The same authors observed a decreasing trend of surface water absorption by HSC with increasing replacement of sand by copper slag aggregates up to a 40 % replacement level; however, after this substitution level, the surface absorption increased abruptly due to the presence of pores created by excessive free water (Fig. 4.30).

The freeze–thaw resistance of concrete with copper slag aggregates is lower than that of conventional concrete due to the internal defects originated by the upflow of bleeding water. However, the addition of an admixture and limestone powder improves this property (Shoya et al. 1997). Birindha et al. (2010) observed higher chloride corrosion rate of uncoated rebar in concrete with copper slag as partial replacement of fine aggregates than in control concrete. The corrosion rate increased with the slag content. But when the rebar was coated with zinc phosphate paint, no corrosion was observed in the corrosion period. The authors also observed higher penetration rate of chloride ions at 40 and 60 % replacement rates of fine aggregates by copper slag aggregates even though the amounts for all types of concrete were very low according to the ASTM C1202 specification. The sulphuric acid resistance capacity of concrete with copper slag aggregates was also observed to be low in comparison to control concrete and decreased as the content of copper slag in concrete increased (Fig. 4.31) (Birindha et al. 2010). The resistance to sulphate attack and the rate of carbonation of concrete with copper slag aggregates are similar to (or even better than) the ones of concrete with conventional aggregates (Ayano and Sakata 2000, Hwang and Laiw 1989).

Fig. 4.31 Compressive strength resistance behaviour of concrete with copper slag due to 5% sulphuric acid attack (Brindha et al. 2010)

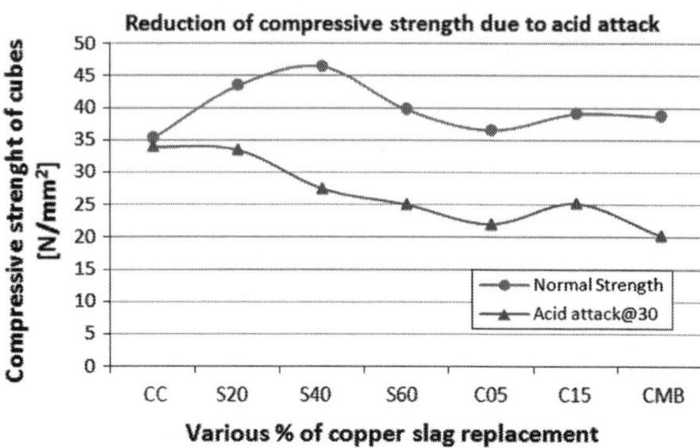

4.5.2 Other Non-Ferrous Slag

The use of several other non-ferrous industrial slags as aggregate in concrete is also reported. Here, some results will be highlighted.

Atzeni et al. (1996) observed no significant differences between the 30-day compressive strength of conventional concrete and concrete with two types of lead and zinc slag, used as partial replacement of sand. However, leaching of significant amounts of lead from concrete with slag is a big problem when using this material as aggregates in concrete, and therefore it needs to be addressed at disposal of demolished slag added concrete. Penpolcharoen (2005) reported the use of a slag, produced during processing of lead batteries, as partial and full replacement of coarse limestone aggregates and partial replacement of Portland cement. The partial and full replacement of limestone aggregates by lead slag increased compressive strength and rising slag addition further increased the strength due to the superior mechanical properties and better packing of slag over limestone aggregates, and the magnetic nature of slag (Fig. 4.32). On the other hand, the water absorption capacity of concrete with slag aggregates was higher than that of NA

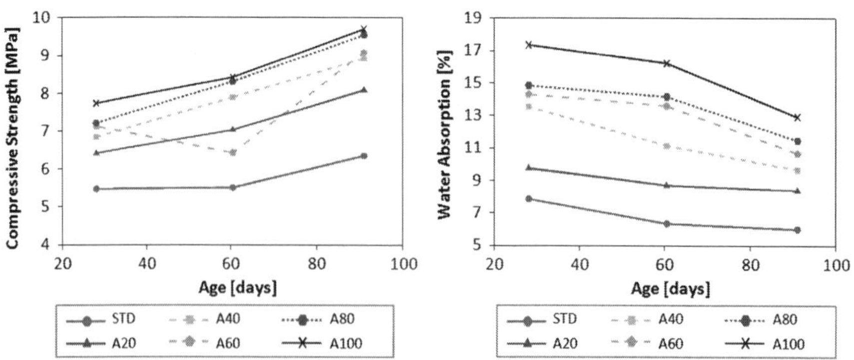

Fig. 4.32 Compressive strength and water absorption of concrete with lead slag (Penpolcharoen 2005)

concrete and rising slag addition further increased the water absorption due to the magnetic nature of slag aggregates (Fig. 4.32).

Metwally et al. (2005) reported the use of a slag produced during smelting of lead batteries as partial replacement of fine aggregates and full replacement of coarse aggregates. The slump of concrete with lead slag was lower than that of conventional concrete and increasing slag addition as replacement of fine aggregates further reduced slump. The highest compressive and tensile strengths of concrete with lead slag as fine aggregates were obtained at 20 % replacement level, when the slag was used as coarse aggregates. The same occurred at 40 % replacement level of sand by fine slag aggregates. An increase in cement content could increase the strength of slag aggregates concrete. An increasing addition of slag as aggregates in cement mortar lowers the abrasion resistance. Concrete with slag as 20 % replacement of fine aggregates showed almost equal absorption of a ray of α and β particles but better absorption of a γ-ray than conventional concrete with 100 % sand. The slag concrete absorbed 88 % of a ray of α and β particles and 90 % of a ray of γ particles in comparison to 86 % for α and β particles ray and 10 % for γ-ray absorption by conventional concrete.

Sorlini et al. (2004) reported the use of two slags (raw as well as 6 months weathered) generated during processing of EAF dust produced in steel production, which contains very high amounts of Zn. Their results suggest that slag addition in concrete does not change the concrete's mechanical performance (compressive, tensile and flexural strength) or in some cases improves these properties. However, slag addition lowers the modulus of elasticity of concrete. Saikia et al. (2008, 2012) also observed higher compressive strength and bending strength and lower water absorption of cement mortar with a slag obtained from lead and zinc smelting as partial replacement of aggregates than those of normal cement mortar.

The ferrosilicate slag, generated during the production of zinc in Imperial Smelting Furnace (ISF-slag) is successfully used as aggregate in concrete road construction (Morrison et al. 2003; Morrison and Richardson 2004). Monosi et al. (2001) observed a negligible reduction in strength, when a ground or unground slag obtained from zinc smelting was used to replace 20 % of sand and 15 % of Portland cement by weight. The reduction in compressive strength due to slag addition as aggregates was negligible especially at the later curing periods (7-day and onwards). However, the compressive strength significantly decreased in the whole curing period, when 15 % of Portland cement was replaced by an equal amount of ground slag in the concrete mix with 20 % slag aggregates. There was practically zero strength at the early periods (1–3 days) of curing, which indicates that the reduction in cement hydration due to deleterious component of slag was the major cause of strength reduction.

The use of other slags, e.g. ferronickel slag, ferrochromium slag, ferromolybdenum slag, aluminium with salt slag as fine or coarse aggregates, is also reported (Boheme and Van Den Hende 2011; Pereira et al. 2000; Shoya et al. 1997; Zelic 2005).

The concentrations of toxic elements present in the leachate generated from lead slag with cement mortar and concrete are generally higher than their concentrations in the leachate generated from conventional concrete; however, the concentrations generally meet the standard specifications (Atzeni et al. 1996; Penpolcharoen 2005; Saikia et al. 2008, 2012; Monosi et al. 2001). However, the use of these slags as aggregates in concrete can be considered only after thorough analysis of their environmental as well as economic suitability. Information on the long-term mechanical and durability performances of concrete with slag is also necessary for the effective application of these materials as aggregates in concrete, which can solve problem related to their disposal.

4.6 Plastic Waste

Significant work has been done on the use of various plastic wastes as aggregates or fibres or fillers in concrete. In this section, the properties of concrete with plastic waste as aggregates from existing literature data will be presented. The concrete properties will be discussed in three main sections: fresh concrete properties, mechanical and durability of hardened concrete properties. Some special properties of concrete will be highlighted in another section. Details about the properties of plastic as aggregates, the generation of plastic aggregates and other related issues are presented in Chap. 2.

4.6.1 Fresh Concrete Properties

The incorporation of plastic aggregates in concrete strongly affects the various fresh concrete properties due to their organic nature as well as their shape, size, porosity and lightweight nature. In this section, some fresh concrete properties available in various references will be highlighted.

4.6.1.1 Slump

Slump is used to measure the workability or consistency of fresh concrete mix. Being an important property, the slump of concrete and cement mortar mixes with plastic aggregates was studied extensively.

There are two parallel views on the workability behaviour of concrete with plastic aggregates. In the majority of the studies, a lower slump value of fresh concrete due to the incorporation of several types of plastic aggregates than that of conventional concrete was observed and increasing the incorporation level of plastic aggregates further lowers the slump (Albano et al. 2009; Batayneh et al. 2007; Frigione 2010; Ismail and Al-Hashmi 2008a; Kou et al. 2009). The reason for this is the sharp edge and angular particle size of plastic aggregates.

On the other hand, in a few studies, an increase in slump value due to the incorporation of plastic aggregates is reported (Al-Manaseer and Dalal 1997; Choi et al. 2005, 2009). According to Al-Manaseer and Dalal (1997) the increased slump of concrete mixes due to the incorporation of plastic aggregates is due to the presence of more free water in the mixes with plastic waste than in that with NA, since unlike NA, plastic aggregates cannot absorb water during mixing. Choi et al. (2005, 2009) reported an increase in slump of concrete with increasing content of two types of treated PET-bottle aggregates, due to the spherical shape of the PET-aggregates as well as the slippery surface texture, which decreases the inner friction between the mortar and the PET-aggregates and therefore increases the flowability.

Saikia and de Brito (2010) reported that the slump of concrete with cylindrical PET-aggregates with very smooth surface texture is slightly higher that of concrete with NA. The authors also found decreasing slump values in concrete due to the addition of fine and coarse sized flaky plastic aggregates, attributed to the fact that these PET-aggregates have sharper edges compared to NA. Moreover in comparison to NA, these flaky aggregates are angular and non-uniform by nature. The slump further decreased as the size of flaky aggregates increased.

The addition of some types of plastic aggregates such as rigid polyurethane (PUR) foam waste or heat treated expanded polystyrene foam (MEPS) decrease the slump of the resulting concrete mix due to the presence of large amounts of surface pores in these aggregates (Fraj et al. 2010; Mounanga et al. 2008; Kan and Demiboga 2009).

4.6.1.2 Density

Irrespective of the type and size of substitutions, the incorporation of plastic as aggregates generally decreases the fresh density of the resulting concrete due to the lightweight nature of these aggregates (Al-Manaseer and Dalal 1997; Ismail and Al-Hashmi 2008a; Hannawi et al. 2010; Marzouk et al. 2007; Kou et al. 2009; Choi et al. 2005, 2009; Saikia and de Brito 2010).

Ismail and Al-Hashmi (2008a) reported that the fresh density of concrete with 10, 15, and 20 % plastic aggregates as replacement of fine aggregates tends to decrease by 5, 7, and 8.7 %, respectively, by comparison with the reference concrete. Al-Manaseer and Dalal (1997) also found 2.5, 6 and 13 % lower densities of concrete with 10, 30, and 50 % plastic aggregates, respectively. Saikia and de Brito (2010) observed a reduction of the density of fresh concrete with increasing volume of PET-aggregates incorporated. The authors found a trend of this density reduction for the three different types of PET-aggregates they used: pellet-size aggregates > fine fraction of flaky aggregates > coarse fraction of flaky aggregates.

According to Fraj et al. (2010) the fresh density of different concrete mixes with dry and water-saturated PUR-foam aggregates classifies them as lightweight concrete and these values were 27–33 % lower than the control concrete's density. The density values decreased as foam incorporation increased.

Hannawi et al. (2010) reported that there was a decrease in fresh and dry densities as the plastic aggregates content increased. Dry density decreased from 2173 kg/m^3 for mixes with 0 % plastic aggregates to 1755 and 1643 kg/m^3, respectively, for mixes with 50 % PET and polycarbonate (PC) plastic aggregates, mainly due to the lower bulk density of plastic. These values were below 2000 kg/m^3, the minimum dry density required for structural lightweight concrete according to RILEM LC2 classification.

4.6.1.3 Air Content

No report is available on the evaluation of air content of cement mortar or concrete mixes with untreated plastic waste as aggregates. Choi et al. (2009) reported the air content of concrete with sand stone coated PET as partial replacement of fine aggregates (Table 4.10). An air-entrainment agent was used during preparation of concrete. The air content of concrete mixes with PET-aggregates was slightly lower than that of the control concrete for the same w/c value and a reducing trend was observed with increasing PET-content in concrete.

4.6.2 Mechanical Properties

The addition of plastic drastically changes the various hardened concrete properties, which will be highlighted in this section. The properties presented are: compressive, splitting tensile and flexural strengths, Young's modulus of elasticity, toughness behaviour: stress–strain curve, failure characteristics and abrasion resistance.

4.6.2.1 Compressive Strength

The compressive strength of concrete and cement mortar is a fundamental property, thoroughly studied in almost all studies related to plastic aggregates. The

Table 4.10 Air content of fresh concrete (Choi et al. 2009)

Amount of sand replaced by PET aggregate (%)	Air content		
	w/c = 0.53	w/c = 0.49	w/c = 0.45
0	4.5	5.0	5.0
25	4.2	4.5	4.8
50	4.1	4.3	4.0
75	4.1	4.2	–

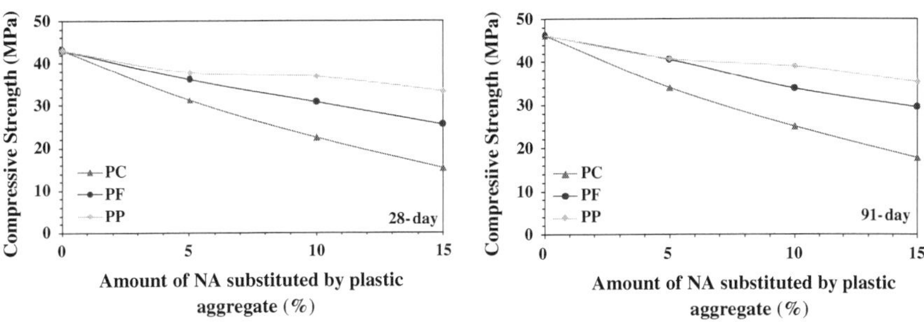

Fig. 4.33 Compressive strength of concrete with PET aggregates (Saikia and de Brito 2010)

incorporation of plastic as aggregates decreased the compressive strength of resulting concrete and mortar (Albano et al. 2009; Akcaozoglu et al. 2010; Batayneh et al. 2007; Choi et al. 2005, 2009; Fraj et al. 2010; Frigione 2010; Ismail and Al-Hashmi 2008a; Hannawi et al. 2010; Kan and Demirboga 2009; Kou et al. 2009; Marzouk et al. 2007; Panyakapo and Panyakapo 2008; Remadnia et al. 2009; Saikia and de Brito 2010). The compressive strength behaviour of concrete and mortar with three types of PET aggregate as partial substitution of fine and coarse NA are presented in Fig. 4.33.

The very low bond strength between the surface of plastic waste and cement paste as well as the hydrophobic nature of plastic waste, which can inhibit cement hydration reaction by restricting water movement, are the causes for low compressive strength of concrete with plastic aggregates. Another factor is the mismatch of particle size and shape between natural and plastic waste aggregates.

However, several authors reported that concrete with partial replacement of NA up to a certain level meet the standard strength values for various types of concrete such as concrete with moderate strength (Albano et al. 2009), minimum compressive strength requirement for structural concrete (Ismail and Al-Hashmi 2008a). Fraj et al. (2010) observed that concrete with dry PUR-foam aggregates almost satisfied the criteria for structural lightweight aggregates concrete as defined in ACI 318 and ASTM C 330. Panyakapo and Panyakapo (2008) reported that concrete with melamine waste aggregates as partial replacement of natural fine aggregates and FA as partial replacement of normal Portland cement (NPC) met most of the requirements for non-load-bearing lightweight concrete according to the ASTM C129-05 Type II standard. The percentage reduction of compressive strength of mortar and concrete due to partial replacement of natural fine aggregates by plastic aggregates at various substitution levels is presented in Table 4.11.

Akcaozoglu et al. (2010) investigated the use of shredded waste polyethylene terephthalate (PET) bottle granules as lightweight aggregates in mortar preparation using two types of binders: NPC and a 50:50 mixture of BFS and NPC. The authors found that the compressive strength of mortar with PET aggregate is higher for the NPC–BFS binder than for NPC only.

Table 4.11 Reduction of compressive strength of cement mortar and concrete (28-day) due to the substitution of natural aggregates by plastic aggregates

Reference	Types of substitution	Reduction in compressive strength for substitution level (%) of									
		3	5	10	15	20	30	45	50	75	100
Batayneh et al. (2007)	Fine/PET		23			72					
Frigione (2010)	Fine/PET		<2								
Hannawi et al. (2010)	Fine/PET	9.8		30.5		47.1			69		
	Fine/PC	6.8		27.2		46.1			63.9		
Kou et al. (2009)	Fine/PVC		9.1		18.6		21.8	47.3			
Saikia and de Brito (2010)	Fine flakes/PET		13.8	28.5	41.8						
	Coarse flakes/ PET		28.3	47.9	64.4						
	Fine pellet/PET		12.2	14.6	22.4						

Fraj et al. (2010) observed a 57–78 % lower 28-day compressive strength of concrete with 8–20 mm rigid PUR foam as aggregate compared to a control concrete, due to the lightweight nature of the modified concrete as well as the low mechanical properties and the high porosity of PUR-foam aggregates. Prewetting the PUR-foam aggregates further lowers the compressive strength due to an increase in the mortar's porosity. Using a superplasticizer along with increasing cement content, on the other hand, increases compressive strength. The use of superplasticizer made it possible to decrease cement content by 15 % and to increase PUR-foam content by 33 % compared, with an acceptable reduction (15 %) of compressive strength.

Mounanga et al. (2008) reported that water curing concrete with PUR-foam aggregates and NA slightly improved the compressive strength compared to dry curing. For conventional lightweight concrete, the increase in strength was about 69 % and this improvement for concrete with 13.1, 21.2 and 32.7 % by volume of PUR-foam aggregates was 39, 34 and 5 %, respectively.

Kan and Demirboga (2009) reported that lightweight concrete with heat-treated expanded polystyrene (MEPS) waste aggregates exhibited a compressive strength 40 % higher than that of concrete with vermiculite or perlite aggregates at equal concrete density. However, the compressive strength of concrete with MEPS aggregates decreased with increasing addition of aggregates. The development of compressive strength of concrete with 100 % MEPS aggregates at 90 days with respect to that at 7 days was about 83 % whereas it was 69 % for concrete with 25 % MEPS aggregates, which might be due to the high heat of hydration of the former type of concrete because of low specific thermal capacity of the MEPS aggregates. The compressive strength of concrete with coarse MEPS aggregates was lower than that of concrete with fine MEPS aggregates as the coarse MEPS aggregates had higher porosity, and therefore were more brittle and weaker than the fine MEPS aggregates.

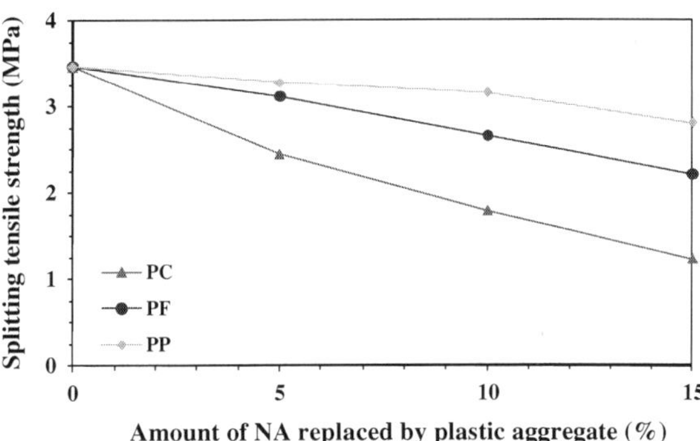

Fig. 4.34 Splitting tensile strength of concrete with plastic waste aggregates (Saikia and de Brito 2010)

4.6.2.2 Splitting Tensile Strength

Similarly to compressive strength, the incorporation of any type of plastic aggregates lowers the splitting tensile strength of concrete. The causes for the reduction observed in splitting tensile strength reported in various references were similar to those used to explain the decrease in compressive strength due to the addition of plastic aggregates. Some results on the tensile strength behaviour of concrete and mortar with various percentages of different types of plastic aggregates are presented in Fig. 4.34.

Kou et al. (2009) reported that splitting tensile strength was reduced with an increase in PVC content in a manner similar to that observed for compressive strength. According to them, the tensile splitting strength of concrete is influenced by the properties of the interfacial transition zone (ITZ); therefore, the smooth surface of the PVC particles and the free water accumulated at the surface of PVC granules could cause a weaker bonding between the PVC particles and the cement paste. According to Albano et al. (2009), the decrease in splitting tensile strength was due to the higher porosity of concrete caused by the increasing addition of PET-aggregates as well as the higher w/c value. Kan and Demirboga (2009) also reported that splitting tensile strength of concrete with heat treated expanded polystyrene (MEPS) aggregates decreases with their content in concrete, due to the generation of more porosity. Batayneh et al. (2007) reported a decreasing trend of splitting tensile strength but not as prominent as for compressive strength. Saikia and de Brito (2010) also reported lower 28-day tensile strength of concrete with three differently shaped PET-aggregates. The authors reported that the concrete cylinders with flakier PET-aggregate did not split into two fractions after the determination of tensile strength, which was generally observed for cylinders with natural and pellet-shaped plastic aggregates as the flaky shaped plastic aggregates could act as a bridge between the two split pieces (Fig. 4.35).

Kou et al. (2009) found an excellent correlation between 28-day splitting tensile strength and 28-day compressive strength of concrete with PVC aggregates as replacement of fine aggregates, which follows a linear relationship. Choi et al. (2009) also found an expression, $f_{st} = 0.23 \times f_c^{(1/3)}$, for the relationship between

Fig. 4.35 Concrete specimens after the determination of tensile splitting strength, from left to right: concrete with natural, pellet-shaped PET, fine and coarse PET flakes aggregates (Saikia and de Brito 2010)

28-day compressive strength and splitting tensile strength of concrete with PET aggregates and an expression, $f_{st} = 1.40 \times (f_c/10)^{(1/3)}$, for a similar relationship for conventional concrete.

4.6.2.3 Modulus of Elasticity

According to ASTM C 469, the modulus of elasticity is defined as a stress–strain ratio value for hardened concrete. The type of aggregates influences the modulus, since the deformation produced in the concrete is partially related to the elastic deformation of the aggregates.

From their study on the use of three different size fractions of PET waste aggregates in concrete production, where concrete was prepared at two different w/c values and at two different natural fine aggregates replacement levels, Albano et al. (2009) observed higher modulus at lower substitution rate of NA by PET-aggregates than at higher substitution rate and at low w/c value. They did not observe any effect of particle size. The modulus of elasticity of concrete with PET-aggregates met the requirement as described in "American Manual of Reinforced Concrete" (1952) except the concrete composition with 20 % large PET-aggregate at w/c of 0.60.

Frigione (2010) plotted the stress–strain curves (σ–ε curve) during the determination of the compressive strength of a reference concrete and a concrete with PET with w/c = 0.45 and cement content of 400 kg/m^3 to determine the modulus of elasticity. The calculated modulus was 48.1 and 41.8 GPa for the reference concrete and the concrete with PET, respectively.

Hannawi et al. (2010) found that increasing the plastic content in concrete decreased the resulting modulus of elasticity, probably due to the low stiffness of PET and PC plastics as well as the poor bond between the matrix and plastic aggregates (Fig. 4.36a). Saikia and de Brito (2010) also found lower modulus of elasticity for concrete with three differently shaped PET waste aggregates than for

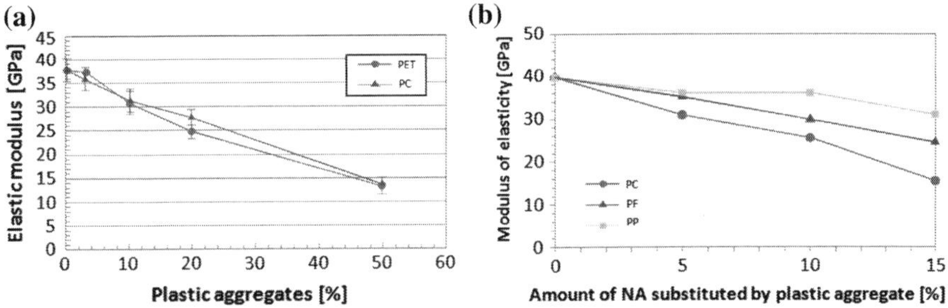

Fig. 4.36 Modulus of elasticity of concrete and cement mortar with plastic aggregates. **a** Hannawi et al. 2010; **b** Saikia and de Brito 2010

concrete with NA (Fig. 4.36b). According to them, the lower modulus of elasticity of concrete because of the incorporation of PET-aggregates is due to the lower stiffness of PET-aggregates than that of NA as well as to the higher porosity due to the high w/c value.

Compared to compressive strength, Fraj et al. (2010) observed a less significant effect on the modulus of elasticity due to the addition of fine expanded PUR-foam aggregates in lightweight concrete. The same authors found an increasing linear correlation between air-dry density and dynamic modulus of elasticity. As the PUR foam had a low stiffness due to its high porosity, increasing the content of PUR foam in concrete reduced its modulus of elasticity. Prewetting the PUR-foam aggregates, improving the cementitious matrix properties by using superplasticizer and decreasing the w/c value did not have an influence on the modulus of elasticity.

Increasing the replacement ratio of fine NA by PVC granules in concrete also reduced the resulting modulus of elasticity (Kou et al. 2009). The replacement of 5, 15, 30 and 45 % of fine NA by PVC granules reduced the modulus of elasticity by 6.1, 13.8, 18.9 and 60.2 %, respectively, when compared to that of the control concrete. According to the authors, the major causes of this reduction were (1) lower stiffness of PVC granules than of the cement paste; (2) lower compressive strength of the concrete with PVC than of the conventional concrete. They also reported that the prediction of the modulus of elasticity of concrete with PVC granules by using an equation suggested by ACI 318-83 overestimated the property value.

Choi et al. (2005) reported that the increasing addition of granulated BFS coated PET aggregates in concrete decreased the resulting modulus of elasticity. In another study, Choi et al. (2009) compared the relationship between the 28-day compressive strength and 28-day modulus of elasticity of concrete with different proportions of sand coated PET-aggregates as replacement of fine NA with CEB-FIP model code (CEB Bulletin Information No. 213/214: Committee Euro-international du Béton, Thomas Telford; 1993) and ACI code (ACI 318 M-05: Building code requirements for structural concrete and commentary. ACI Manual of concrete practice, ACI; 2005). The relationship between compressive strength and modulus of elasticity of concrete with plastic aggregates was in close agreement with the one suggested in ACI 318-05, in which the concrete's density was taken into consideration.

4.6.2.4 Flexural Strength

Flexural strength is defined as a material's ability to resist deformation under load and is measured in terms of stress. The flexural strength represents the highest stress experienced within the material in the moment of rupture. The transverse bending test is the most frequently employed, in which a rod specimen having either a circular or rectangular cross-section is bent until fracture using a three- or four-point flexural test technique.

Batayneh et al. (2007) reported a decreasing trend of flexural strength with increasing plastic waste aggregates content in concrete. However, this reduction was not as significant as for compressive strength. Ismail and Al-Hashmi (2008a) reported that the flexural strength of plastic waste concrete mixes at each curing age was prone to decrease with the increase of the plastic waste content in these mixes. Saikia and de Brito (2010) also found low flexural strength values for concrete with PET-aggregates than that for concrete with NA (Fig. 4.37).

Hannawi et al. (2010) did not find significant changes in the flexural strength of mortar specimens with up to 10 % PET-aggregates and up to 20 % PC-aggregates with similar composition. However, decreases of 9.5 and 17.9 % for mixes with 20 and 50 % PET-aggregates, respectively, were observed. For mixes with 50 % PC-aggregates a decrease of 32.8 % was measured. According to the authors, the elastic nature and the non-brittle characteristic under loading of the plastic aggregates might have an effect on the observed flexural strength. The bending strength of cement composites prepared by Laukaitis et al. (2005), using three different types of waste polystyrene granules followed a proportional relationship with its density.

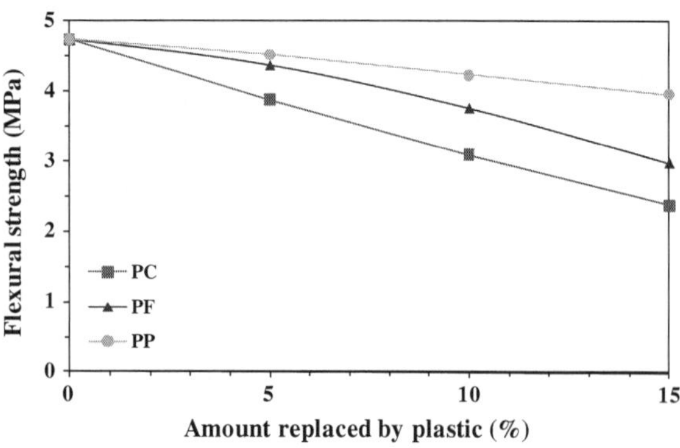

Fig. 4.37 Flexural strength behaviour of concrete with PET-aggregates (Saikia and de Brito 2010)

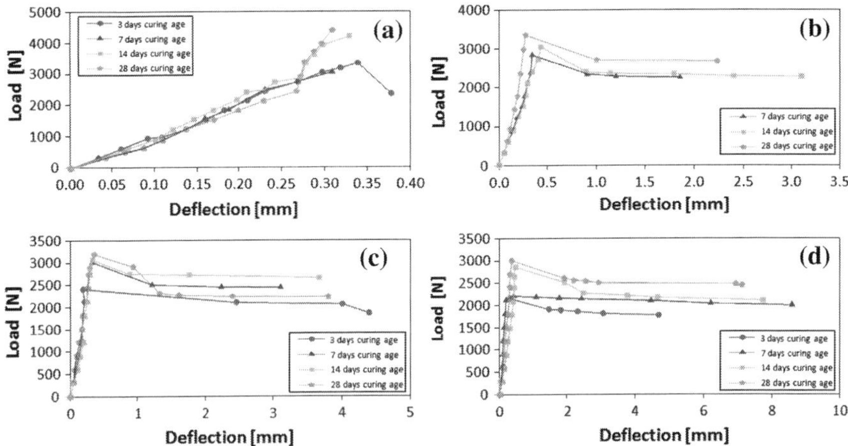

Fig. 4.38 Load–deflection curves of concrete with 0, 10, 15 and 15 % of fine aggregates by plastic aggregates (Ismail and Al-Hashmi 2008a). **a** 0 %, **b** 10 %, **c** 15 %, **d** 15 %

4.6.2.5 Toughness/Poisson's Ratio

Ismail and Al-Hashimi (2008a) plotted the load–deflection curves of a reference concrete and concrete mixes with 10, 15, and 20 % plastic waste as fine aggregate replacement at the curing ages of 3, 7, 14 and 28 days. The results are illustrated in Fig. 4.38. They show the propagation of microcracks is arrested by the introduction of plastic waste particles in concrete. The authors also determined the toughness indices for the concrete compositions with plastic waste aggregate at the curing ages of 3, 7, 14 and 28 days (Table 4.12).

The toughness indices of concrete mixes with plastic waste aggregates for all replacement levels after 14- and 28-day curing reached a plastic behaviour according to ASTM C1018, desirable for many applications that require high toughness.

Frigione (2010) plotted the stress–strain curves (σ–ε curve) during the determination of compressive strength of a reference concrete and a concrete with PET. Compared to the reference concrete, a higher strain value corresponding to the maximum stress was registered for the concrete with PET waste aggregates. The

Table 4.12 Toughness indices for concrete with plastics aggregates (Ismail and Al-Hashmi 2008a)

Percentages of plastic in concrete mixes (%)	Toughness indices at curing age											
	3-day			7-day			14-day			28-day		
	I_5	I_{10}	I_{10}:I_5	I_5	I_{10}	I_{10}:I_5	I_5	I_{10}	I_{10}:I_5	I_5	I_{10}	I_{10}:I_5
10	–	–	–	8.3	11.6	1.4	4.3	8.6	2.0	2.5	7.5	3.0
15	3.0	11.0	3.7	4.5	9.5	2.1	4.2	8.4	2.0	8.0	16.1	2.0
20	6.8	13.7	2.0	7.3	14.8	2.0	5.2	11.5	2.1	5.7	11.6	2.0

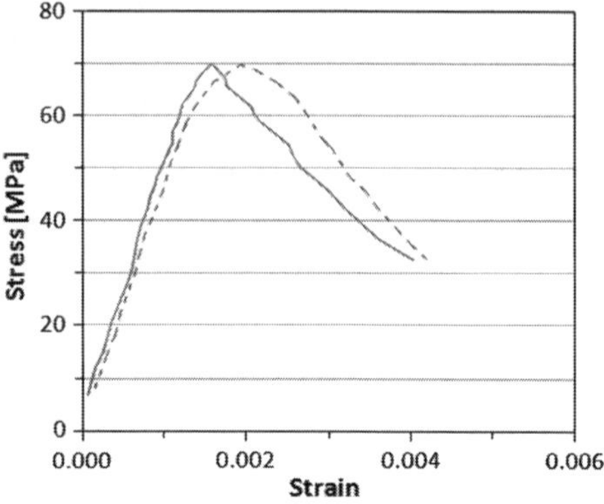

Fig. 4.39 Stress–strain curves for a reference concrete (*plain line*) and a concrete with PET waste aggregates (*dotted line*) (Frigione 2010)

peak shapes of the two curves also suggested that the concrete with PET waste aggregates is less brittle than the reference concrete and this type of concrete could withhold a larger deformation still keeping its integrity (Fig. 4.39). Kou et al. (2009) observed increasing Poisson's ratio values with increasing contents of PVC waste aggregates in concrete. Since the higher Poisson's ratios meant higher ductility, the addition of PVC improved the ductility of the resulting lightweight aggregates concrete, due to the elastic nature of PVC.

4.6.2.6 Failure Characteristics

After failure during the determination of compressive strength, specimens with plastic aggregates do not exhibit the typical brittle type of failure, obtained for conventional cement mortar and concrete. As the plastic aggregates content increased, the failure became more ductile. The specimens with plastic aggregates can carry load for a few minutes after failure without full disintegration, as was observed by various researchers (Hannawi et al. 2010; Marzouk et al. 2007, Saikia and de Brito 2010). The recycled PET-aggregates can delay crack initiation and prolong the crack propagation interval thereby increasing structural strength.

Albano et al. (2009) found various types of failure including normal cone type for concrete specimens with PET-aggregates, where 20 % of fine aggregates were replaced. As the smooth surface of the PVC particles and the free water accumulated at the surface of PVC granules may have caused weaker bonding between PVC particles and cement paste, most of the PVC granules in the concrete matrix did not fail but were debonded from the cement paste after reaching the ultimate strength of concrete (Kou et al. 2009). Fraj et al. (2010) reported that the rupture mechanism of concrete with PUR-foam aggregates was different from that of the normal weight control concrete: in the first case, the rupture occurred on the mortar matrix/PUR-foam aggregates interfaces as well as in the middle of the PUR-foam aggregates. In normal weight concrete, the rupture mainly took place in the ITZ because of the poor properties of this zone compared to the other concrete components. By observing the splitting behaviour of concrete blocks after tensile strength and flexural strength tests, Saikia and de Brito (2010) concluded that the flaky PET-aggregates can act as bridge between the two separated pieces of concrete block after failure, which was not observed for concrete blocks with natural as well as pellet-shaped PET-aggregates.

4.6.2.7 Abrasion Resistance

Compared to other properties, very few data are available on the abrasion resistance of concrete (or mortar with any type of plastic waste aggregates). Soroushian et al. (2003) reported the abrasion resistance of concrete with plastic waste fibres. The authors found a reduction of the abrasion resistance of concrete due to the addition of plastic waste fibres in concrete. However, the incorporation of commercial plastic aggregates in concrete improved its abrasion resistance (Nasvik 1991).

Recently, Saikia and de Brito (2010) reported that the incorporation of PET-aggregates can improve the abrasion resistance of concrete (Fig. 4.40a). The authors found that the abrasion resistance of concrete with pellet-shaped PET-aggregates increased with their content. On the other hand, for concrete with two types of flaky aggregates the best results were obtained for a 10 % substitution level. From the relationship between compressive strength and depth of wear for concrete with different types of plastic aggregates, the authors found a certain compressive strength level for concrete with PET-aggregates over which the abrasion resistance deteriorates (Fig. 4.40b).

4.6.3 Durability Performance

Several durability factors like permeability properties, shrinkage, carbonation resistance and resistance against freeze–thaw cycles are evaluated for concrete or mortar with plastic as aggregates. However, compared to the available information

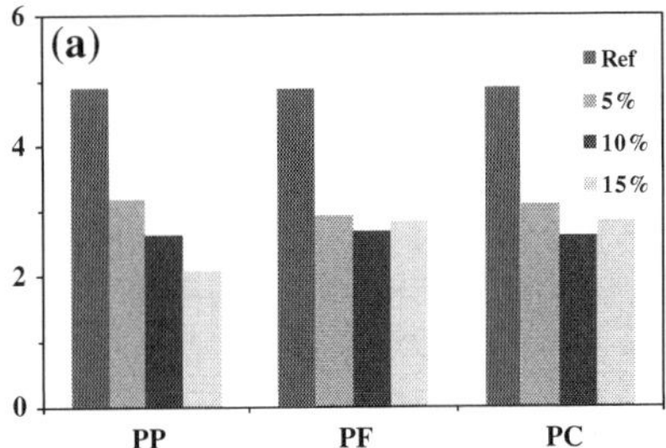

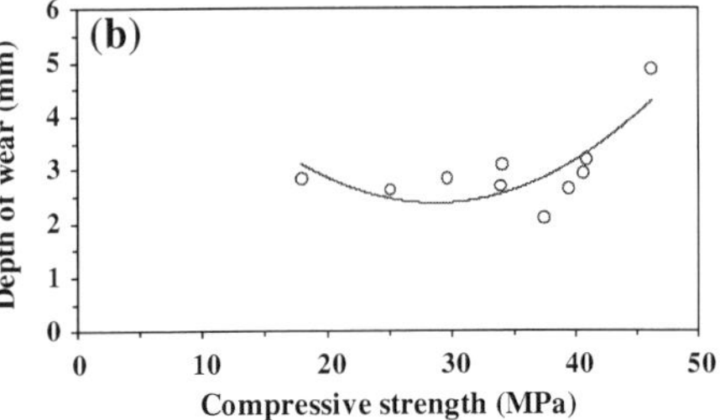

Fig. 4.40 **a** Depth of wear and **b** cubic compressive strength versus depth of wear of concrete with PET-aggregates after abrasion resistance test (Saikia and de Brito 2010)

on the mechanical performance of concrete with plastic aggregates, relatively less data are available on the durability behaviour.

4.6.3.1 Permeability Behaviour

Generally permeability of aggressive chemical species through the pores of concrete is the major factor, which controls several durability properties. Tests like water absorption, gas permeability and chloride permeability measurement can give information on the vulnerability of concrete for ingress of deleterious chemical species.

Water Absorption and Water Accessible Porosity

Albano et al. (2009) observed a higher water absorption value for concrete with PET-aggregates than that for concrete with NA. The water absorption value was further increased with increasing content of PET-aggregates in concrete, increasing size of PET-aggregates and increasing w/c value. According to the authors, the differences in size grading as well as in shape of plastic aggregates from the natural fine aggregates were responsible for this behaviour.

Choi et al. (2009) measured the sorptivity coefficient of 28-day cured cement mortars prepared by replacing 0, 25, 50 and 75 % of fine NA by sand powdered coated PET-aggregates. Their results indicated that the sorptivity of cement mortar with PET-aggregates at 25 % replacement level was lower than that of the control mortar and at 50 and 75 % replacement level it was higher than that of the control mortar. According to the authors, at 50 and 75 % replacement level the change in grading size of the fine aggregates mixture increased the inside porosity of mortar and thus increased the sorptivity.

Akcaozoglu et al. (2010) found higher water absorption and porosity values for a cement mortar with 100 % PET-aggregates than for a mortar with equal percentage in volume of PET-aggregates and sand. The authors found a similar trend for cement mortar with a mixture of equal weight of BFS and NPC though the BFS addition with NPC increased the water absorption and porosity of the resulting cement mortar. However, according to the authors, all the values for all types of mortar meet the range generally observed for lightweight concrete.

Fraj et al. (2010) recorded a higher value of the water accessible porosity of cement mortar with PUR-foam aggregates than that of the mortar with no plastic aggregates. The authors also reported that prewetting the PUR-foam aggregates further increased the porosity. However, the addition of superplasticizer in cement mortar with prewetted PUR-foam aggregates can decrease the porosity value.

Marzouk et al. (2007) reported that the volumetric substitution of plastic aggregates by less than 100 % decreased the rate of water adsorption with respect to the reference mortar that contained no waste. The authors found lower sorptivity for cement mortars with PET-aggregates than for mortars with no plastic waste.

The sorptivity further decreased with increasing volumetric substitution up to 50 %. Thus, their results suggest better durability performance of cement mortar with PET-aggregates than that of mortar with NA when in contact with aggressive solutions. Hannawi et al. (2010) measured the water absorption and apparent porosity values of the different mortar specimens with various amounts of PET and PC waste aggregates. Their results revealed that replacing 3 % by volume of sand by an equal volume of PET or PC do not exert influence either on water absorption or on the apparent porosity of the composites in comparison with the control mortar. However, apparent porosity and water absorption increased with increasing plastic content.

Gas Permeability

Fraj et al. (2010) reported higher gas permeability (2.2 times) of concrete with dry and prewetted PUR-foam aggregates than that of conventional concrete. Prewetting the PUR-foam aggregates can further increase the value considerably. Decreasing the w/c value and increasing superplasticizer content can reduce this value for concrete with prewetted PUR-foam aggregates.

Hannawi et al. (2010) found an increase of helium gas permeability coefficient with increasing plastic aggregates content in mortar, which indicated an increase of the percolated porosity of mortar due to the incorporation of plastic aggregates, because of weak bonding between the cement paste and plastic aggregates. They also reported greater helium gas permeability coefficient of mortar with PET aggregates than that of mortar with PC aggregates at the replacement level of 10, 20 and 50 % by volume of sand by plastic aggregates.

Chloride Migration

Kou et al. (2009) investigated the resistance to chloride-ion penetration of 28 and 91 days hardened concrete prepared by partially replacing natural fine aggregates by PVC waste granules. The chloride-ion penetration resistance of concrete was represented by the total charge passed in Coulomb during a test period of 6 h. Their results (presented in Fig. 4.41) indicated that this property improved with an increase in PVC content as well as with curing time. They found reduction of about 36 % in the total charge passed through the 28-day cured concrete, with 45 % replacement of NA by PVC granules in comparison to same-age concrete with no PVC granules. According to them, the increase in the resistance to chloride-ion penetration of concrete is attributed to the impervious PVC granules blocking the passage of the chloride ions.

Fraj et al. (2010) evaluated the chloride diffusion coefficient of concrete with rigid PUR foam as partial replacement of coarse NA. Their results are presented in Table 4.13. The authors observed lower chloride diffusion coefficient for concrete with dry PUR-foam aggregates than that of concrete with NA only. However, the

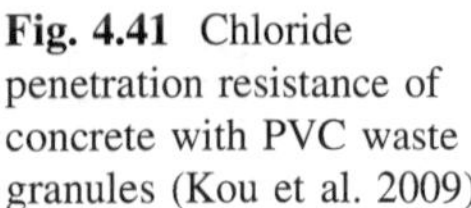

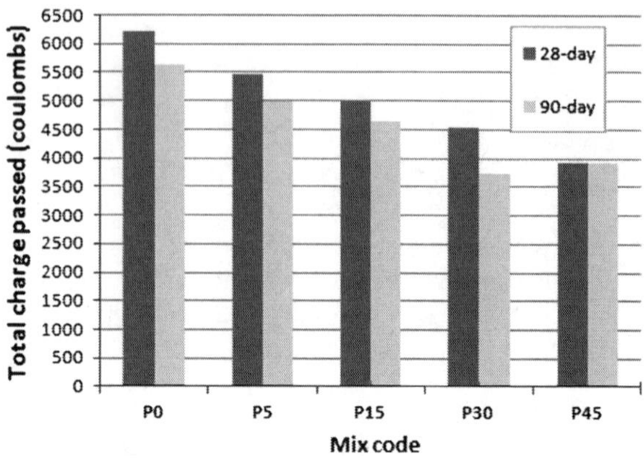

Fig. 4.41 Chloride penetration resistance of concrete with PVC waste granules (Kou et al. 2009)

Table 4.13 Chloride-ion penetration coefficient of concrete with PUR-foam aggregates (Fraj et al. 2010)

Volume of PUR-foam aggregate	w/c ratio	Cement content (kg/m^3)	Volume content of PUR foam (%)	Amount of superplasticizer (kg/m^3)	Effective chloride diffusivity coefficient (10^{-12} m^2/s)
Control	0.55	397	0	0	1.87
Dry PUR aggregate	0.55	397	34	0	1.62
Prewetted PUR aggregate	0.55	397	34	0	5.30
	0.44	415	35	1.405	2.70
	0.44	353	45	1.196	5.98

presaturation of PUR-foam aggregates in water resulted in a significant increase of the chloride diffusion coefficient, due to the increase in porosity of concrete with increasing incorporation of PUR-foam aggregates. They also reported that the reduction in w/c value and increase in cement content could significantly improve the chloride resistance performance of concrete with prewetted PUR-foam aggregates.

4.6.3.2 Carbonation

Akcaozoglu et al. (2010) measured the carbonation resistance of various types of cement mortars by measuring carbonation depth. A phenolphthalein solution was applied on the broken surfaces of the half pieces obtained after flexural tensile strength test. The compositions of various mixes along with the carbonation depth at various time periods are presented in Table 4.14. Irrespective of binder types, the carbonation depth of mortar with PET-aggregates only after 28 days of curing is lower than that of the mortar with an aggregate mixture of PET and sand. The authors also found a higher porosity for the mortar with sand and PET mixture than

Table 4.14 Carbonation depth of cement mortar specimens (Akcaozoglu et al. 2010)

Amount in mortar (%)					Depth of carbonation (mm) in			
Cement	Slag	PET-aggregate	Normal aggregate	Water	7 days	28 days	90 days	180 days
51.28	0	25.64	0	23.08	0.3	1.2	4.3	5.0
25.64	25.64	25.64	0	23.08	0.3	1.7	5.5	7.6
33.90	0	16.95	33.90	15.25	0.0	1.4	4.8	5.9
16.95	16.95	16.95	33.90	15.25	0.6	2.5	6.8	8.5

for the mortar with PET aggregates only. According to the authors, PET and sand aggregates used together did not combine sufficiently and the resulting mortar became porous. On the other hand, the depth of carbonation for concrete with slag is significantly higher than for the mortar prepared by using cement as the only binder.

4.6.3.3 Shrinkage

Frigione (2010) found an increase in drying shrinkage due to the incorporation of 5 % PET-aggregates in concrete at different experimental conditions due to the lower modulus of elasticity of concrete with plastic aggregates than that of conventional concrete. However, the shrinkage of concrete with PET-aggregates was acceptable for various uses as structural concrete.

From their experiments on the use of PVC waste granules as a partial volumetric replacement of natural sand in concrete, Kou et al. (2009) found decreasing drying shrinkage with increasing content of plastic aggregates (Fig. 4.42). According to the authors, PVC granules are impermeable and do not absorb water when compared to sand and do not shrink either, and hence are able to reduce the overall shrinkage of concrete.

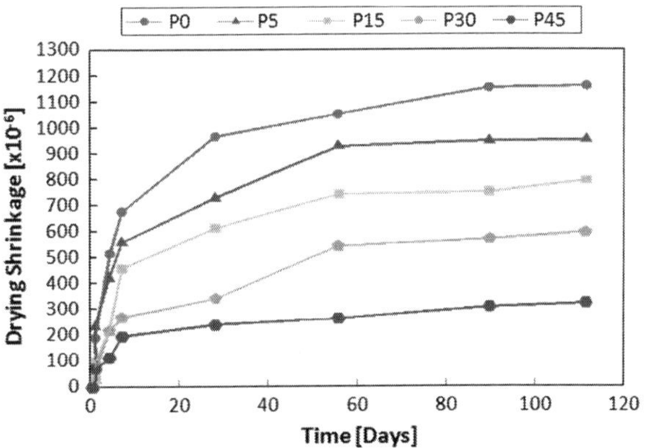

Fig. 4.42 Drying shrinkage of concrete with fine PVC aggregates (Kou et al. 2009)

Fraj et al. (2010) found higher drying shrinkage values for lightweight concrete with dry and prewetted polyurethane foam (PUR foam) as partial replacement of fine aggregates. Concrete with dry PUR-foam aggregate has 8.1 % higher 28-day drying shrinkage than control concrete. On the other hand, concrete mixes with prewetted PUR-foam aggregates at 34 and 45 % by volume replacement levels exhibited, respectively, 72.5 and 149.5 % higher 28-day drying shrinkage than control concrete. Lowering the w/c value or increasing superplasticizer, sand and cement content can reduce drying shrinkage of concrete with prewetted PUR-foam aggregates. In these conditions, the 28-day drying shrinkage of concrete with prewetted PUR-foam aggregates at 35 % by volume replacement level is 49.7 % higher than that of control concrete. According to the authors, the lower modulus of elasticity of PUR-foam aggregates and the higher amount of prewetting water in the case of concrete with prewetted aggregates are the causes of the higher drying shrinkage of concrete with PUR-foam aggregates.

Mounanga et al. (2008) reported higher drying shrinkage of lightweight concrete in which various fractions of fine aggregates were replaced by PUR-foam aggregates than that of control concrete. According to the authors, this behaviour was mainly due to effect of PUR-foam aggregates on the stiffness of concrete. However, other factors such as w/c value, sand content and thermal dilation during hydration also had a significant effect.

Akcaozoglu et al. (2010) observed significantly higher drying shrinkage of mortars with PET aggregates only than that of a mortar with equal percentage by weight of sand and PET-aggregates at the experimental drying periods. Mixing BFS with cement can reduce mortar shrinkage for both types of aggregates (PET only and sand-PET mixture).

4.6.3.4 Freeze–Thaw Resistance

Kan and Demirboga (2009) reported the freeze–thaw resistance of concrete with modified expanded polystyrene foam (MEPS) as partial or full substitution of natural fine and coarse aggregates by using standard ASTM 666 procedure B. The following conclusions were taken from the results: 1. increasing the MEPS aggregate ratio in the mixes the concrete is expected to exhibit a higher frost resistance and guarantee a higher durability; 2. coarse lightweight MEPS aggregates are more susceptible to freeze–thaw cycles than fine lightweight MEPS aggregates.

4.6.4 Other Properties

There are other concrete properties that are reported to be altered due to the incorporation of plastic aggregates. In this section, the fire behaviour and thermophysical properties of concrete with plastic aggregates are highlighted from the literature results.

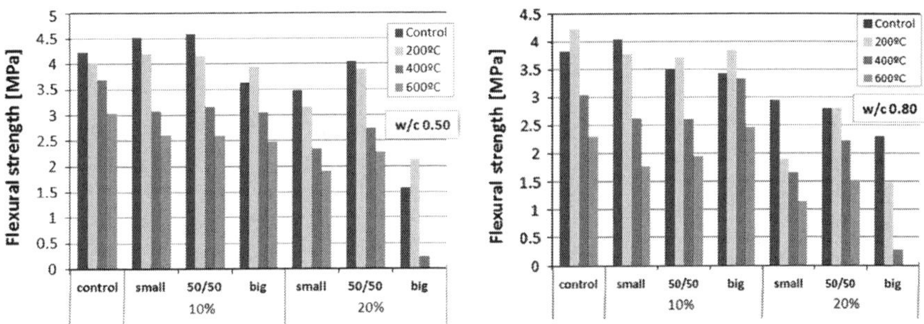

Fig. 4.43 Flexural strength behaviour of concrete with fine PET-aggregates before and after heat treatment (Albano et al. 2009)

4.6.4.1 Fire Behaviour

Albano et al. (2009) determined the fire behaviour of concrete with various percentages of shredded PET-aggregates as partial replacement of natural fine aggregates. The authors placed the cured slabs in a muffle furnace, the temperature inside the furnace was increased up to a given temperature, the slabs were kept at that temperature for 2 h and then the heating was stopped immediately. The temperatures chosen for this study were 200, 400 and 600 °C. After cooling the specimen to room temperature, the flexural strength was determined. In parallel, unheated specimens were tested. Their results are presented in Fig. 4.43.

As the temperature increased, the flexural strength decreased regardless of the substitution ratio and the PET particle size. However, the decrease in flexural strength was more significant when PET content was 20 % than 10 % due to the presence of more porosity (voids), which act as stress concentration spots. Moreover, PET-aggregates were more susceptible to temperature than natural fine aggregates. The volume change and the degradation of the PET particles produce less cohesion between concrete components and a greater number of voids. The decrease in flexural strength also increased with the w/c value. According to the authors, at high w/c value the thermal stability of PET-aggregates decreased due to the hydrolytic degradation of PET particles. The formation of carboxyl and hydroxyl end groups occurred due to the reaction of one water molecule with one PET molecule, which accelerated its decomposition. Besides, the water vapour was difficult to discharge at high temperatures, so the vapour pressure favours crack formation in concrete.

4.6.4.2 Thermophysical Properties

Mounanga et al. (2008) observed significantly low thermal conductivity for concrete with PUR-foam aggregates used to partially replace fine NA due to the porous nature of PUR-foam aggregates. These pores contain air, whose thermal conductivity is much lower than that of the other concrete constituents. The decrease in thermal conductivity was prominent for concrete with dry PUR-foam aggregates compared to concrete with saturated PUR-foam aggregates. Yesilata et al. (2009) observed an improvement of thermal insulation performance of plain concrete due to the incorporation of plastic aggregates, which was also dependent on the shape of the plastic aggregates.

The heat capacity of concrete with dry PUR-foam aggregates is lower than that of the reference concrete since the heat capacity of PUR-foam aggregates is also lower than that of the NA (Mounanga et al. 2008). On the other hand, the heat capacity of concrete with saturated PUR-foam aggregates is higher than that of the reference concrete due to the higher heat capacity of water present in the pores of prewetted PUR-foam aggregates.

4.7 Rubber Waste

Disposal of rubber tyre waste has become a serious problem due to the generation of huge amounts of tyres, which are non-biodegradable by nature. Tyre rubber in asphaltic concrete mixes, in incinerator to produce steam, to produce different plastic and rubber products, as a fuel for cement kiln, as feedstock for making carbon black, and as artificial reefs in marine environment are some attractive utilisation options (Siddique and Naik 2004). Extensive references including excellent reviews are available on the use of rubber tyre as coarse or fine aggregates or as a filler material for the preparation of various types of concrete (Kumaran et al. 2008; Siddique and Naik 2004). In this section, the properties of concrete with rubber tyre waste particles as aggregates will be discussed. The properties of these aggregates are presented in detail in Chap. 2.

4.7.1 Fresh Concrete Properties

The incorporation of rubber aggregates in concrete affects the various fresh concrete properties due to their organic nature as well as their shape, size and lightweight nature. In this section, changes in fresh concrete properties due to the addition of rubber aggregates available in the various references will be highlighted.

4.7.1.1 Slump

Just like for plastic aggregates, there are two parallel views on the workability behaviour of mortar and concrete mixes with rubber tyre aggregates. Sukontasukkul and Chaikaew (2006) observed lower slump for concrete with rubber aggregates and they added more water to obtain similar consistence to that of a conventional concrete. The water requirement increases with the rubber content and as the average particle size of the rubber aggregates decrease. Guneyisi et al. (2004) reported that the slump of concrete at two w/c values with and without silica fume gradually decreased with increasing rubber aggregates content. At a rubber content of 50 % by total aggregate volume the slump decreased near to zero and the mix was not workable so that an extra effort was required for the compaction of the concrete. The decrease in the slump was more remarkable for low w/c concrete mixes. Nayef et al. (2010) also found a near zero slump of a concrete mix with coarse rubber content of 20 % by total coarse aggregate volume and a very low slump value for a concrete mix with fine rubber aggregates (Fig. 4.44a). The slump of rubberized concrete mixes can be improved by adding 5 % microsilica. Taha et al. (2008) also observed heavy reduction in slump of concrete due to increasing substitution of NA by rubber aggregates (Fig. 4.44b).

Li et al. (2004) did not found any significant change in slump due to the replacement of 15 % coarse aggregates by rubber tyre chips or fibre. Khaloo et al. (2008) found contrasting slump behaviours of concrete mixes due to the incorporation of fine and coarse rubber tyre aggregates as a partial replacement of NA (Fig. 4.44c). The slump increased with the replacement ratio of sand by fine rubber aggregates up to 15 %, beyond which slump decreased. On the other hand, slump of concrete mixes with coarse rubber aggregates decreases to a minimum with tyre aggregates contents of 15 % and then it fluctuates slightly over the minimum value for higher rubber aggregate contents.

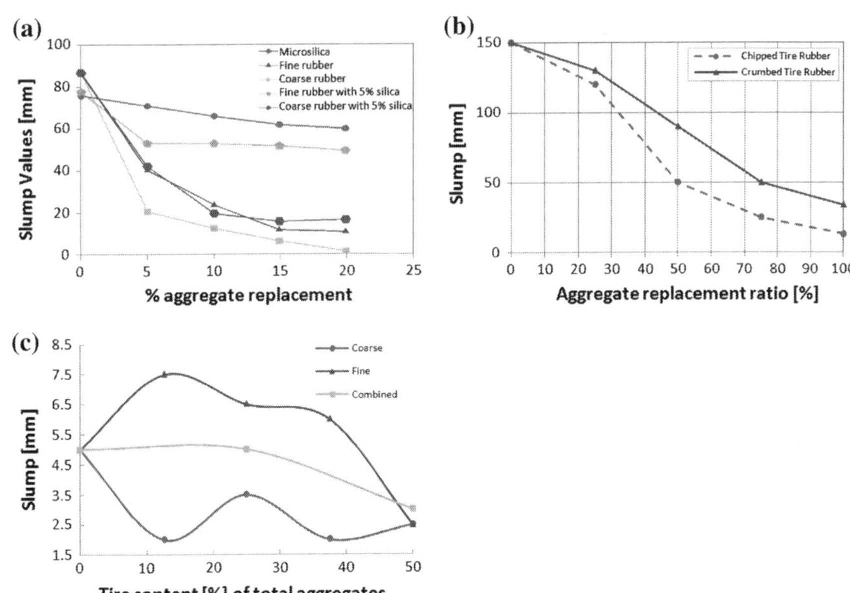

Fig. 4.44 Slump of concrete with rubber aggregates

Turki et al. (2009) reported a decreasing w/c value (for same slump) and therefore increasing slump (for same w/c value) of mortar mixes with increasing rubber aggregates replacement of sand up to 30 %. However, increasing the replacement ratio to 40 % does not change the w/c value observed for 30 %. On the other hand, the w/c value increased slightly at 50 % replacement of sand. Aiello and Leuzzi (2010) also found a slightly improved workability of fresh concrete with partial substitution of coarse or fine aggregates by rubber shreds. Raghvan et al. (1998) achieved comparable or better workability for rubber tyre aggregates cement mortar mixes than for control mortar.

In several investigations, it was reported that the rubberized concrete specimens have acceptable workability in terms of ease of handling, placement, and finishing (Khalloo et al. 2008; Li et al. 2004, Raghvan et al. 1998; Aiello and Leuzzi 2010). According to Fattuhi and Clark (1996), the process of mixing concrete with rubber by hand (i.e. manually) was easy and less strenuous than mixing concrete with natural stone aggregates. They did not encounter any problems with placing and compacting concrete with rubber aggregates. However, the workability of mixes with rubber crumbs was slightly better than that of mixes with low-grade rubber, possibly due to the small surface of rubber crumbs as well as the presence of a lesser amount of textile fibres.

4.7.1.2 Density

In general, using rubber aggregates in concrete and mortar mixes decreases their density due to the replacement of much heavier NA by lighter rubber tyre aggregates. Increasing the rubber content further reduces the density of concrete. The lightweight nature of concrete with rubber aggregates can be used for several purposes like in structures to reduce earthquake damage, architectural applications such as false facades and interior construction.

Some typical experimental results are presented in Fig. 4.45. Although there is a global consensus that the addition of rubber aggregates reduces the density of resulting concrete, large variations in the scale of the reduction in density are observed in various studies. In some investigations, a heavy reduction in density due to incorporation of rubber aggregates was reported. For example, Guneyisi et al. (2004) found that the unit weight of concrete ranged from 2427 to 1805 kg/m^3 depending on the silica fume and rubber contents. At 50 % rubber content, the unit weight was as low as about 75 % of that of the conventional concrete, irrespective of the silica fume content. Nayef et al. (2010) reported similar reduction for a concrete mix with 20 % by volume replacement of NA. In the Khaloo et al. (2008) study, the unit weight of concrete mixes with coarse, fine and coarse–fine aggregates mix at 50 % were reduced by 45, 34 and 33 %, respectively, compared to reference concrete. Fattuhi and Clark (1996) observed unit weights of concrete mix in the range of 2380–1880 kg/m^3 due to the addition of rubber aggregates in the range 0–13 % of total concrete mix.

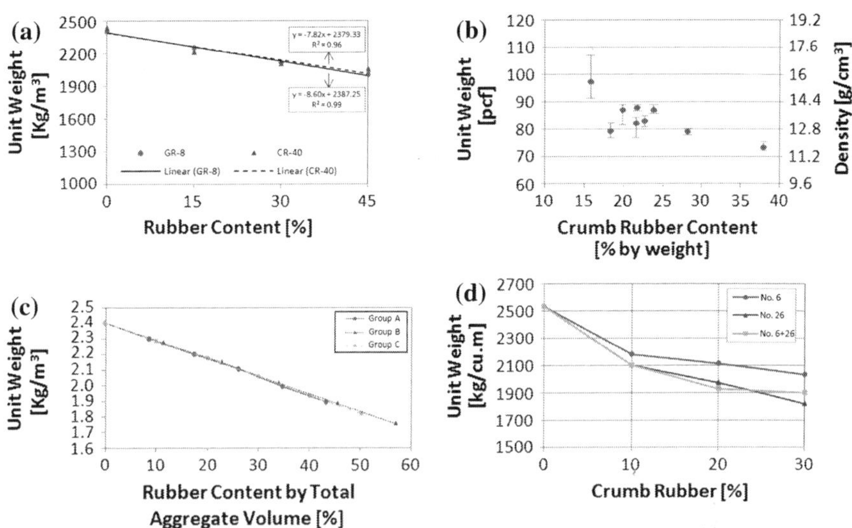

Fig. 4.45 Density and unit weight of concrete with rubber aggregates. **a** Zheng et al. 2008a; **b** Pierce and Blackwell 2003; **c** Khatib and Bayomy 1999; **d** Sukontasukkul and Chaikaew 2006

Aiello and Leuzzi (2010) observed a reduction in density of 5.8 and 6.0 % for concrete with 50 % by volume replacement of NA by coarse and fine rubber aggregates. They also observed an 8.8 and 8.3 % unit weight decrease for concrete with 75 % by volume replacement of NA by two-size fractions of rubber aggregates. Topcu (1995) also reported a 12.6 % reduction in unit weight for mixes with 45 % fine and coarse rubber chips by volume of total aggregate by comparison with the reference concrete. Ling (2011) found a density range of 2200–2000 kg/m^3 for concrete mixes with rubber aggregates content of 0–50 % by fine aggregates volume.

The size and quality of rubber aggregates also has some influence on the unit weight of rubberized concrete. In the Khaloo et al. (2008) study, a higher reduction in unit weight was observed for concrete with coarse rubber aggregates than that with fine rubber aggregates and a mixture of fine and coarse rubber aggregates. Fattuhi and Clark (1996) also reported that concrete with low-grade rubber aggregates had lower density than that with rubber crumb for similar rubber content and the difference in density increased with the rubber content. For example, at a rubber to cement ratio of 0.4 (by mass), the density of concrete with low-grade rubber was about 2 % lower than that of a similar concrete with rubber crumb. This difference may be due to the higher content of textile fibres in the low-grade rubber, and hence lower mass. On the other hand, Aiello and Leuzzi (2010) observed little difference in the density of concrete mixes prepared by partially replacing NA by fine and coarse rubber aggregates.

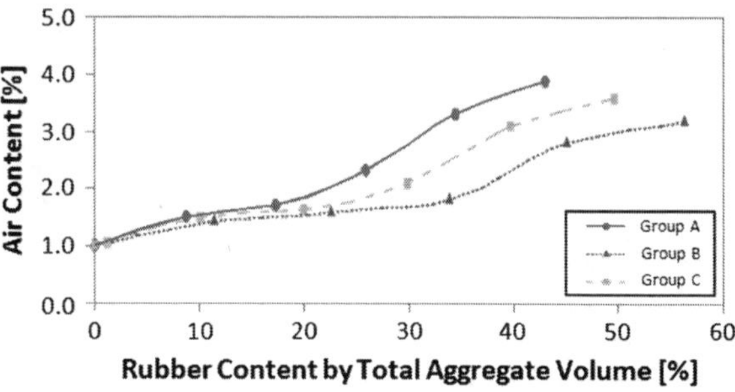

Fig. 4.46 Change in air content due to the addition of rubber aggregates in concrete: *Group A* fine aggregates replaced by crumb rubber, *Group B* coarse aggregates replaced by rubber chips, *Group C* fine and coarse aggregates replaced by crumb and chip rubber, respectively (Khatib and Bayomy 1999)

4.7.1.3 Air Content

The air content of concrete mixes with rubber particles is generally higher than that of conventional concrete with NA and it increases with the amount of rubber particles (Khatib and Bayomy 1999). Li et al. (2004) observed an increasing trend of air content of a concrete mix where 15 % by volume of natural coarse aggregates were replaced by rubber chips or fibres. Benazzouk et al. (2006) also reported a sharp increase in the air content of cement paste due to the incorporation of rubber particles. On the other hand, Figueiredo and Mavroulidou (2007) observed a reduction in air content of concrete due to the incorporation of crumb and fine rubber aggregates used to replace 10 % of coarse and fine aggregates, respectively. A typical air content behaviour of concrete due to the addition of rubber aggregate is presented in Fig. 4.46.

4.7.2 Hardened Concrete Properties

4.7.2.1 Dry Density

As the dry density of rubber aggregates is considerably lower than that of NA, similarly to fresh density, the dry density of hardened concrete decreases with increasing rubber aggregates content. Typical data are presented in Fig. 4.47.

Topcu (1995) observed a systematic decrease in the density of concrete with increasing contents of tyre chips. The average density of control concrete was 2300 kg/m^3. On the other hand, the values for concrete with 15, 30 and 40 % by volume replacement of fine and NA were 2220, 2140 and 2010 kg/m^3, respectively. Benazzouk et al. (2006) found 22–35 % reductions in 28-day dry density of concrete specimens with two types of rubber aggregates. The decrease in density was higher for concrete with small-size rubber aggregates. The effect was also more significant with expanded rubber type.

An increase in rubber aggregates content in concrete increases the air content which in turn reduces the dry density of the specimens (Yilmaz and Degirmenci

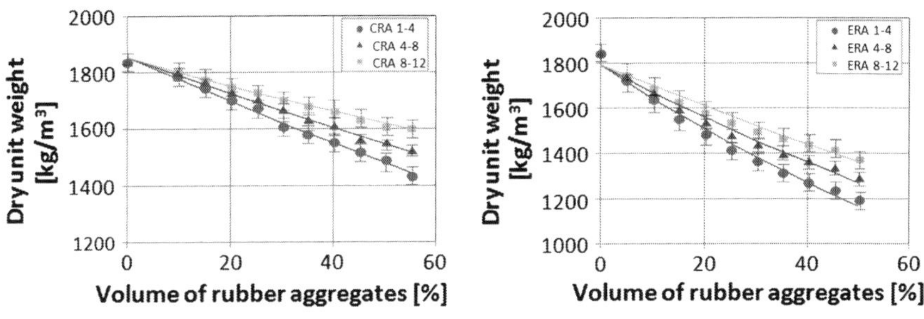

Fig. 4.47 Dry unit weight of rubberized concrete (Benazzouk et al. 2003)

2009). Sukontasukkul and Chaikaew (2006) speculated that the flocculation of the rubber particles during mixing of concrete with higher rubber contents may have some effect on the lower density of concrete specimens. Flocculation can create large voids inside the block, leading to a higher porosity which ultimately lowers the density. Due to the high water absorption of tyre particles, the ratio between the fresh density and the hardened density in rubber tyre concrete is greater than in conventional concrete. Therefore, rubber tyre concrete is expected to be more porous than conventional concrete.

4.7.2.2 Compressive Strength

It is universally accepted that the addition of rubber aggregates reduces the compressive strength of the resulting concrete and the increase of rubber content further deteriorates the compressive strength. In some of the studies, about 80–90 % reduction in compressive strength was reported depending on the size and type of rubber aggregates. Khatib and Bayomy (1999) observed about 90 % lower compressive strength in concrete with 100 % gravel replaced by chipped rubber than in conventional concrete. However, in a few studies like that of Ganjian et al. (2009) a slight improvement of 28-day compressive strength of concrete with rubber chips used to replace coarse NA in 5 % by volume was also reported, possibly due to the improvement of aggregates grading curve due to the incorporation of rubber particles.

Possible reasons for this strength reduction are (Ganjian et al. 2009): (i) reduction of the quantity of solid load-carrying material with increasing rubber content; (ii) the soft and smooth surfaces of rubber particles may significantly degrade the adhesion between the boundaries of rubber particles and cement paste, and thus increase the volume of the weakest phase and ITZ; (iii) non-uniform distribution of rubber particles at the concrete top surface tends to produce non-homogeneous samples and leads to a reduction in concrete strength at those parts, resulting in failure at lower stresses.

The compressive strength behaviour of concrete due to the incorporation of rubber aggregates reported in the literature is presented in Table 4.15. These

Table 4.15 Compressive strength of concrete with rubber aggregates

Reference	Size of aggregates/type of replacement	Type of concrete	Amount of replacement (%)	Compressive strength (MPa)
Aiello and Leuzzi (2010)	12.5–20 mm/volume	Normal	0	45.80
			25	23.90
			50	20.87
			75	17.42
	10–12.5 mm/volume	Normal	0	27.11
			15	23.97
			30	20.41
			50	19.45
			75	17.06
Bignossi and Sandrolini (2006)	Sand/volume	SCC	0	33.0
			22.2	24.7
			33.3	20.2
Emiroglu et al. (2007)	0–4 mm/volume	Normal	0	45.69
			5	41.71
			10	33.69
			15	24.75
			20	22.14
	4–8 mm/volume	Normal	0	45.69
			5	42.49
			10	37.30
			15	26.96
			20	23.91
Futtuhi and Clark (1996)[a]	Low grade rubber/mass	Normal	0	37.45
			~9.9	12.66
	Rubber crumb	Normal	~11.2	11.69

[a] Amount of replacement with respect to total solid content

results indicate that the size, proportions and surface textures of rubber particles noticeably affect the compressive strength of rubberized concrete mixes (Eldin and Senouci 1993; Topcu 1995). Gesoglu and Guneyisi (2007) observed a relatively higher strength development between 3 and 7 days of curing and the rate gradually decreased with curing age. However, the strength development pattern was almost similar in conventional concrete.

Benazzouk et al. (2003) found a sharp reduction in 28-day compressive strength of concrete due to the addition of different size fractions of two types of rubber aggregates. Some results of their investigation are presented in Table 4.16. They also found a high dependency of strength on several parameters such as substitution ratio, size and properties of rubber aggregates. The compressive strength of concrete specimens prepared by using compacted rubber aggregates was considerably higher than that using expanded rubber aggregates. Similarly, compressive strength decreased drastically when the content of rubber aggregates increased.

Table 4.16 28-day compressive strength of concrete with two types of rubber aggregates (Benazzouk et al. 2003)

Volume of rubber (%)	Size of rubber aggregates	Compressive strength (MPa)	
		CRA	ERA
0	–	82.5	82.5
5	1–4	68.0	59.0
	4–8	63.0	54.0
	8–12	60.5	51.0
10	1–4	55.0	42.0
	4–8	48.0	36.0
	8–12	43.0	32.0
25	1–4	26.0	15.0
	4–8	20.0	13.0
	8–12	15.0	11.0
50	1–4	6.5	3.4
	4–8	5.0	2.6
	8–12	3.5	2.0

CRA compacted rubber aggregates; ERA: expanded rubber aggregates

The rubber particles are less stiff than the surrounding cement paste which lowers the compressive strength of concrete. The cracks are initiated around the rubber particles, which accelerates the failure in the matrix. Larger incorporation of rubber particles in the concrete mix creates difficulty in the packing of lightweight rubber particles, and therefore voids are introduced in the matrix. The trend is slightly influenced by aggregate size; e.g. for a given amount of rubber, finer aggregates lead to lower losses in compressive strength than coarse aggregates.

Similar effect of particle size on the strength behaviours of rubberized concrete was also reported in other studies (Topcu 1995; Son et al. 2011; Khatib and Bayomy 1999; Ali and Goulias 1998; Ali et al. 1993). Khatib and Bayomy (1999) showed that rubberized concrete made with coarse chipped rubber replacing coarse aggregates has less strength than concrete made with fine crumb rubber. Ali and Goulias (1998) and Ali et al. (1993) also observed higher reduction in compressive strength due to the addition of coarse sized rubber aggregates than of fine rubber particles. This is due to the high compressibility of rubber particles, which generates localised stresses and bonding problems between them and the cement matrix. According to Topcu (1995), the interfacial bond in a coarse tyre rubber chips cement paste is weaker than in a fine tyre rubber chips cement paste, which ultimately affects the compressive strength. However, in some studies such as that of Emiroglu et al. (2007) the exact opposite effect of particle size is reported (Table 4.15).

Li et al. (2004) reported that the compressive strength of concrete with rubber chips replacing 15 % by volume of coarse NA was lower than that of concrete with an equal volume percentage of elongated or fibre type coarse rubber aggregates (Table 4.15). According to the authors, this is possibly due to the difference between their load transfer capabilities. Once debonded from the concrete matrix, chips do not have enough length to transfer the applied load through interfacial frictional force, while fibres have longer length to transfer the applied load, resulting in higher strength.

In a few studies, the effect of chemical or physical treatments of rubber aggregates on the compressive strength behaviour of the resulting concrete was also reported. This type of technique is generally adopted to improve the weak ITZ between rubber aggregates and cement paste. Li et al. (2004) reported that the surface treatment of fine rubber aggregates by NaOH solution increased the mechanical performance including compressive strength. However, this technique could not improve the properties for coarse rubber aggregates. The same author also tried to improve the mechanical performance by making holes in the rubber aggregates but it did not improve the studied properties. Naik and Singh (1991) also reported that the surface treatments of rubber particles could enhance the hydrophilicity of the rubber surface and therefore could improve mechanical performance including compressive strength.

4.7.2.3 Tensile Strength

Just like for compressive strength, the addition of rubber aggregates decreases the splitting tensile strength of the resulting concrete. The development of microcracks due to weak interfacial binding of rubber aggregates and cement paste as well as a surface segregation between rubber aggregates and cement paste due to the exerted stress are the major causes that lower the tensile strength of concrete due to the incorporation of rubber aggregates (Ganjian et al. 2009). However, for a given substitution ratio the reduction in splitting tensile strength of concrete with rubber aggregates is less prominent than that observed in compressive strength (Eldin and Senoucci 1993; Mavroulidou and Figueiredo 2010). The reduction in splitting tensile strength of concrete with fine rubber aggregates is also smaller than that with coarse rubber aggregates.

Ganjian et al. (2009) reported that the percentage reduction of tensile strength in concrete using chipped rubber as a partial replacement of NA was about twice than that in concrete using ground rubber particles for the same replacement level. The reduction in tensile strength with 7.5 % replacement was 44 % for concrete with chipped rubber and 24 % for concrete with ground rubber as compared to the control mix. In the Topcu (1995) study, the splitting tensile strength of C 20 type conventional concrete was 3.21 MPa, while it was 2.17, 1.53 and 1.13 MPa for concrete with fine rubber chips and, 1.50, 1.06 and 0.82 MPa for concrete with coarse rubber chips at the replacement ratios of 15, 30 and 45 %, respectively.

Instead of the brittle failure usually exhibited by conventional concrete specimens under compression, specimens with rubber aggregates generally show ductile failure due to the plastic behaviour of the rubber aggregates. Topcu (1995) found that the failed specimens withstood measurable post-failure loads during tensile strength test and underwent significant displacement, which was partially recoverable. Therefore, concrete specimens with rubber aggregates showed high capacity of absorbing plastic energy during the splitting tensile strength test. A similar type of tensile behaviour was reported by Eldin and Senoucci (1993). Kang et al. (2009) reported that the incorporation of rubber particles in RCC increased the tensile strength, as well as the ultimate tension elongation if the compressive strength was kept at the level of about 40 MPa. This was due to the higher deformation capability and lower modulus of elasticity of rubber particles than those of NA.

Kang et al. (2009) reported that the splitting tensile strength of concrete specimens is about one-tenth to one-fifteenth of cubic compressive strength.

4.7.2.4 Flexural Strength

The incorporation of rubber aggregate decreases the flexural strength of the resulting concrete and rising the rubber content further deteriorates the flexural strength due to the weak bond between cement paste and rubber particles. However, Benazzouk et al. (2003) observed higher flexural strength values for concrete prepared by replacing 20 % by volume of coarse and fine aggregates by two types of rubber aggregates with three size ranges than for conventional concrete. However, after substitution of 35 % by volume of NA by any type of rubber aggregates and any size range, the flexural strength decreased drastically due to the rupture of the rubber/cement matrix connection. Concrete with expanded rubber aggregate showed better flexural strength behaviour than concrete with compacted rubber aggregates, which was exactly the opposite trend of compressive strength of concrete with these aggregates.

In several studies, it was reported that the reduction in flexural strength was not as significant as that observed in the reduction of compressive strength of concrete due to the incorporation of rubber aggregates (Mavroulidou and Figueiredo 2010; Toutanji 1996). Toutanji (1996) found a significantly smaller reduction in flexural strength in comparison to compressive strength as the tyre chip content increased. Khatib and Bayomy (1999) observed a steeper initial rate of flexural strength reduction than that of compressive strength.

From the load–deflection curves during flexural strength measurement of concrete beam specimens with various amounts of rubber, several authors reported that the failure of specimens with rubber tyre chips exhibited a ductile mode of failure as compared to control specimens (Toutanji 1996; Sukontasukkul and Chaikaew 2006). The specimens with rubber could also withstand measurable post-failure loads due to the ability of the rubber aggregates to undergo large elastic deformation before the failure of the specimen took place.

Aiello and Leuzzi (2010) observed a larger reduction in flexural strength for concrete when coarse aggregates rather than fine aggregates were replaced by rubber particles. The rubberized concrete mixes prepared with 50 and 75 % by volume of coarse NA replacement both exhibited a decrease in flexural strength, referred to the control mix, of about 28 %. Whereas, mixes with substitution by volume of fine aggregates of 50 and 75 % showed a decay of about 5.8 and 7.3 %,

Table 4.17 Dynamic and static elastic moduli of concrete with ground and crushed rubber as coarse aggregates replacement (Zheng et al. 2008a)

Properties	Conventional	Concrete prepared by replacing coarse aggregates with					
		Ground rubber in volume (%)			Crushed rubber in volume (%)		
		15	30	45	15	30	45
E_d (GPa)	43.7	41.2	35.2	31.2	35.4	36.5	32.8
E_s (GPa)	31.8	27.1	24.1	22.3	23.1	24.3	22.1

respectively, referred to the control mix. A decrease in flexural strength with the increase in particle size of rubber aggregates was reported in other studies too (Benazzouk et al. 2003; Mavroulidou and Figueiredo 2010).

4.7.2.5 Modulus of Elasticity

Just like strength properties, the incorporation of crumb or chip rubber as aggregates in concrete considerably reduces both the static and dynamic moduli of elasticity. Aggregates characteristics affect the modulus of elasticity: concrete with aggregates with higher stiffness normally has high modulus of elasticity. Since the rubber aggregates have very low stiffness as compared to NA, the addition of rubber aggregates lowers the modulus of elasticity of the resulting concrete.

The type of rubber (i.e. chips or ground rubber) may have some effect on the modulus of elasticity. Zheng et al. (2008a) reported higher values of both static and dynamic moduli for concrete with 15 % by volume of coarse aggregate replaced by ground rubber than for concrete with crushed rubber at similar replacement level (Table 4.17). However, at higher replacement level the elasticity behaviour of concrete with ground and crushed rubber aggregates becomes similar. Skripkiunas et al. (2007) observed a reduction of about 11 % in the modulus of elasticity of concrete due to the addition of rubber aggregates that replaced fine aggregate by about 3 % by weight. Mavroulidou and Figueiredo (2010) observed a higher static modulus of elasticity for concrete with coarse rubber aggregates (19–10 mm) than for concrete incorporating finer rubber aggregates (10–4.75 mm). Both types of aggregates were used to replace 10 % by weight of natural coarse aggregates.

Azmi et al. (2008) found reductions in the modulus of elasticity with increasing rubber aggregates content in concrete as well as with increasing w/c value. The authors found a reduction of about 30 % in modulus of elasticity when the replacement ratio of fine aggregates by crumb rubber increased from 0 to 30 % by volume. According to the authors, the inclusion of crumb rubber implies defects in the internal structure of the composite material, producing a reduction of strength and stiffness. Benazzouk et al. (2003) reported that the decrease in dynamic modulus of elasticity was greater with expanded type rubber aggregates compared with compacted rubber aggregates for the same size and same amount of rubber content. Ganjian et al. (2009) reported lower modulus of elasticity for concrete with rubber aggregates than for conventional concrete. Kang et al. (2009) also observed a reduction in modulus of elasticity with increasing rubber content; however, the modulus of elasticity increased with the curing time. The reduction amount with respect to the modulus of elasticity of conventional concrete was slightly low at all substitution levels.

Guenisiyi et al. (2004) reported that the static modulus of elasticity of concrete decreased with increasing rubber content in a similar fashion to that observed in compressive and splitting tensile strengths. By increasing the rubber content to 50 % of the total aggregate volume, the modulus of elasticity dropped to about 6.5 and 8.0 GPa for w/c ratios of 0.60 and 0.40, respectively. These were respectively about 20 and 17 % of the modulus of elasticity of a similar type of conventional concrete. The use of silica fume slightly improved the modulus of elasticity of concrete even though the improvement was smaller than that observed for compressive and splitting tensile strengths. The results are presented in Fig. 4.48. Peisller et al. (2011) observed lower modulus of elasticity for concrete with various percentages of rubber aggregate used to replace fine NA in concrete. The reduction in the modulus of elasticity was smaller than that in compressive strength. The decrease in modulus of elasticity was 49 % on an average for the concrete with rubber by comparison with the reference concrete. The addition of 15 % silica fume with cement increased the modulus of elasticity but it was still lower than that observed for conventional concrete.

4.7.2.6 Stress–Strain Curve: Toughness Behaviour

It is consensual that the addition of rubber aggregates can substantially improve the post-cracking behaviour of concrete by absorbing a significant amount of energy. Thus special types of concrete can be prepared by incorporating rubber aggregate that can be used for applications where impact or blast resistance is

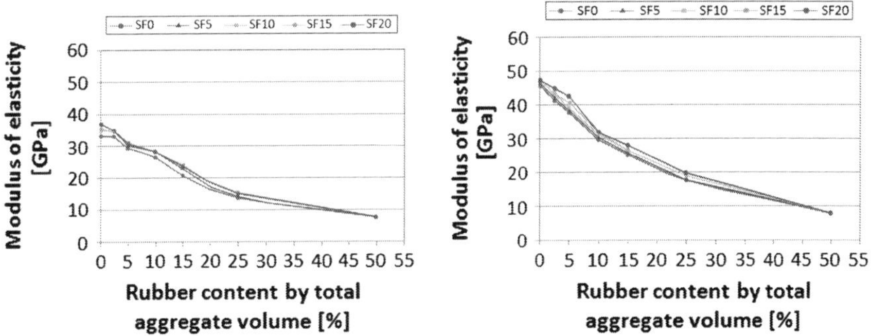

Fig. 4.48 Static modulus of elasticity of concrete with various percentages of rubber aggregates incorporation (Guneyisi et al. 2004)

Table 4.18 Toughness indices and some other parameters of rubberized concrete (Aiello and Leuzzi 2010)

Amount of rubber in concrete (%)	Toughness indices[a]			Residual strength factor[a]		Toughness (kN/mm^3)
	I_5	I_{10}	I_{20}	$R_{5,10}$	$R_{10,20}$	
50	4.06	8.72	14.4	93.2	56.8	113
75	4.96	9.92	17.8	99.2	78.8	196

[a] For details, see ASTM C1018-97

needed, such as bunkers and jersey barriers, or where vibration damping is required such as foundation pads in railway stations. Due to the positive influence of rubber aggregate, substantial work has been done to evaluate the stress–strain curve and the toughness behaviour of concrete with rubber aggregates. This behaviour is generally evaluated during the determination of various strength properties.

Aiello and Leuzzi (2010) observed substantial improvement of post-cracking behaviour of concrete due to the addition of coarse rubber aggregates. In Table 4.18, the toughness indices and energy absorption capacities (toughness) measured during the determination of the flexural strength of concrete with rubber aggregates used to replace 50 and 75 % by volume of natural coarse aggregates are presented. The toughness indices determined from the curves were in the specified limit of the standard range defined in ASTM C1018-97 and these increased with rubber content. However, in the same investigation insignificant enhancement of toughness behaviour due to the incorporation of fine rubber aggregates used to replace 25 and 50 % by volume of natural fine aggregates in concrete was also reported.

Batayneh et al. (2008) also found two distinct behaviours in the stress–strain curves of concrete depending on rubber content (Fig. 4.49). 0.075–4.75 mm rubber aggregates were used to replace fine NA in concrete. The stress–strain behaviour of specimens with rubber content up to 40 % follows a trend similar to that of the control specimen. In this case, concrete behaved like a brittle material i.e. there was a linear increase of stress until it reached its peak value before specimen's fracture. However, the curves became nonlinear for concrete mixes with 60 and 80 % rubber, which indicated that concrete behaves like a ductile material. Kang et al. (2009) observed a similar type of ductility behaviour for concrete with shredded rubber aggregates. Concrete with rubber aggregates did not disintegrate and some cracks closed after unloading.

Benazzouk et al. (2003) also observed increasing ductility in the stress–strain curve of concrete due to increasing addition of rubber aggregate as well as due to increasing particle size of rubber aggregates. The brittleness index (BI) was also measured to estimate the ductility of different concrete specimens. These values for different mixes as a function of rubber aggregates volume are presented in Fig. 4.50. The peak was obtained at a rubber addition level of 10 % for all aggregate sizes and characterised the transition from brittle to ductile material after

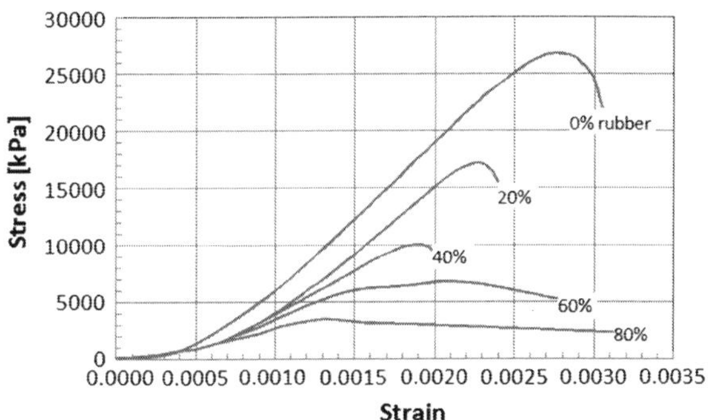

Fig. 4.49 Stress-strain behaviour of concrete with various percentage of rubber aggregates (Batayneh et al. 2008)

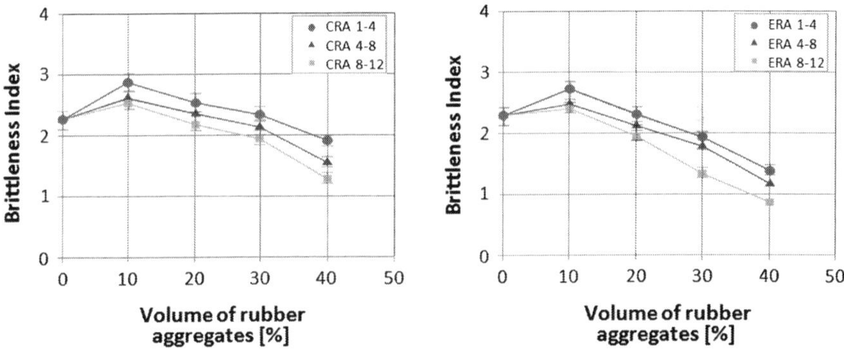

Fig. 4.50 Brittleness index of various types of concrete (Benazzouk et al. 2003)

this rubber content. The decrease in BI values with rubber content over 10 % reflected an increase in plastic deformation energy. This increase became even greater as the rubber size increased. For the same rubber content, the BI was lower for expanding type rubber aggregates than for compacted rubber aggregates. The alveolar character of rubber, therefore, helped to increase the deformability of cement–rubber composites.

Khaloo et al. (2008) observed increasing nonlinearity of stress–strain curves due to the incorporation of rubber aggregates in concrete. To compare the nonlinearity between the control concrete and the rubber tyre concrete, a nonlinearity index was defined as the ratio between the slope of the line connecting the origin to 40 % of the ultimate stress and the slope of the line connecting the origin to the ultimate stress. A higher nonlinearity index implies a more nonlinear stress–strain curve. The nonlinearity index increases as the rubber content increases for all mixes. The substitution of rubber for mineral aggregates appears to allow more uniform crack development and provide gentler crack propagation, compared to conventional concrete. The authors also determined the toughness indices of the concrete mixes. Rubber tyre concrete exhibited greater toughness as compared to conventional concrete. Toughness indices maximise as rubber concentration approaches 25 % of the total aggregates volume. Beyond rubber concentrations of 25 %, toughness indices decrease due to the systematic reduction in strength.

4.7.2.7 Impact Resistance

Futtuhi and Clark (1996) evaluated the impact resistance of concrete with low-grade rubber, which contains textile fibres and dust as impurities. Two slabs were made and tested simultaneously. One slab was made with ordinary concrete (without rubber), while the other contained about 11 % of low-grade rubber relative to the total solids content by weight. The rubber to cement ratio was maintained at about 0.44. After impact by a hammer, examination of the slabs showed that both suffered cracking in all directions. However, the slab with rubber had a larger spread of cracks over the tension face. After the second hit, the maximum crack width in the ordinary concrete slab was 0.16 mm, while that for the slab with rubber was 0.50 mm. After the third hit, the maximum crack widths (at the same locations) increased to 0.3 and 2.0 mm for the plain concrete and rubberized concrete slabs, respectively. These results show that both slabs sustained the impact of the drop hammer; despite the compressive strength of the rubberized concrete slab being about 30 % of the strength of the ordinary concrete slab.

Ling et al. (2009) reported that using rubber aggregates as partial substitution of fine aggregates in concrete pavement blocks improved the impact resistance. Their results are presented in Table 4.19. They reported that the energy absorption and toughness of rubberized blocks were much larger than those of the control block. Extra forces were needed to fully open the blocks with high amounts of rubber aggregates because they maintained the integrity of the broken pieces even after a number of falling weight hits. The rubberized blocks exhibited higher displacement and did not show any clean split into two halves at failure mode as the rubber content increased.

Table 4.19 Numbers of hits that cause damage in concrete pavement blocks

Type of concrete	Type of damage							
	Small crack		Transverse crack		All directions crack		Completely broken	
	Block 1	Block 2	Block 1	Block 2	Block 1	Block 2	Block 1	Block 2
Type I	1	2	2	3	–	–	3	4
Type II	3	3	6	5	–	–	10	8
Type III	7	7	–	–	13	12	13	12
Type IV	9	10	–	–	15	16	15	16

Concrete slab prepared by replacing fine aggregates fraction by rubber aggregates that is used to replace 0, 10, 20 and 30 % by volume of total aggregates (type I, type II, type III and type IV, respectively)

Drop height 100 cm (Ling et al. 2009)

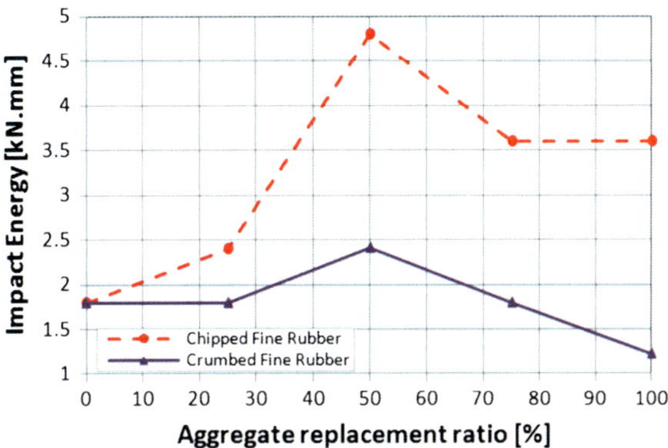

Fig. 4.51 Effect of rubber aggregates incorporation on the impact energy of concrete

Taha et al. (2008) evaluated the impact energy (the energy required to failure) of concrete prepared by replacing various percentages of coarse and fine aggregates, measured through an impact test. Their results suggested that increasing the replacement level of coarse aggregates by chipped tyre rubber particles up to 100 % significantly improved the impact energy of the concrete and that a maximum was reached for a replacement level of 50 % (Fig. 4.51). The impact resistance also peaked at 50 % replacement level for mixes with crumbed tyre rubber particles. However, it was lower than for the control mix at 75 and 100 % replacement levels. At low to medium replacement levels, the low stiffness of the tyre particles allowed the rubber-cement composite to have a relatively high flexibility, and thus absorb higher amount of energy than the conventional concrete.

4.7.2.8 Skid Resistance

Sukontasukol and Chaikaew (2006) evaluated the skid resistance of concrete using standard ASTM E303-93 and a pendulum type apparatus. 10 and 20 % by weight of coarse and fine aggregates of concrete were replaced by two sizes of crumb rubber (passing ASTM No. 6 and 20 sieves) and by a mixture of these two sizes. The results of the seven mixes are presented in Fig. 4.52a. Results show that crumb rubber concrete blocks (except those made with sieve No. 20 crumb rubber) exhibited better skid resistance than the control block. The highly elastic properties of rubber allowed the block surface to deform more and create more friction as the pendulum passed across it. Mixes with large rubber particles performed better than those with small particles.

On the other hand, a systematic reduction in skid resistance was observed for concrete slab where the fine aggregates were replaced by rubber aggregates at 0, 10, 20 and 30 % by volume of total aggregates (types I, II, III and IV respectively in Fig. 4.52b) (Ling et al. 2009). However, all the values met the minimum requirement in accordance with ASTM standard specification. No damage was

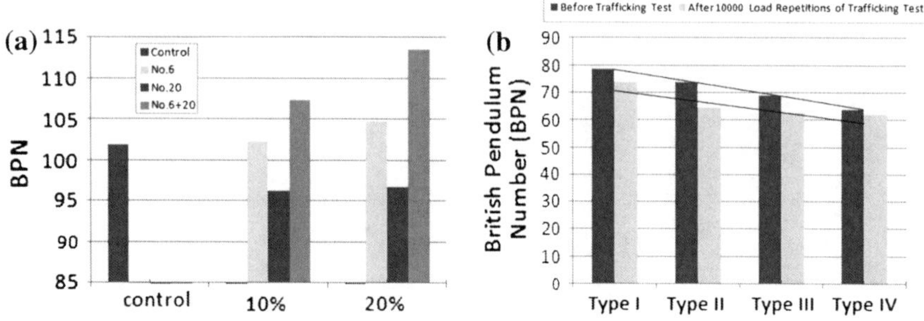

Fig. 4.52 Skid resistance of **a** concrete with rubber aggregates (Sukontasukkul and Chaikaew 2006); **b** concrete pavement block (Ling et al. 2009)

caused to any of the pavement blocks. The slightly higher skid resistance for low percentages of crumb rubber in pavement block was partly due to the rough surface texture of the paving blocks.

4.7.2.9 Abrasion Resistance

Sukontasukol and Chaikaew (2006) reported the abrasion resistance of concrete with various contents of rubber aggregates. The results in terms of percentile weight loss are shown in Fig. 4.53a. They found lower abrasion resistance for mixes with rubber crumb aggregates than for the control mix, as indicated by increasing weight loss with increasing crumb rubber content. Out of three types of rubber aggregates, abrasion resistance was lowest for the mixes of two-size fractions rubber and highest for the coarse rubber aggregates. Segre and Joekes (2000) reported that the NaOH treatment of rubber aggregates considerably improved the abrasion behaviour of rubberized concrete (Fig. 4.53b).

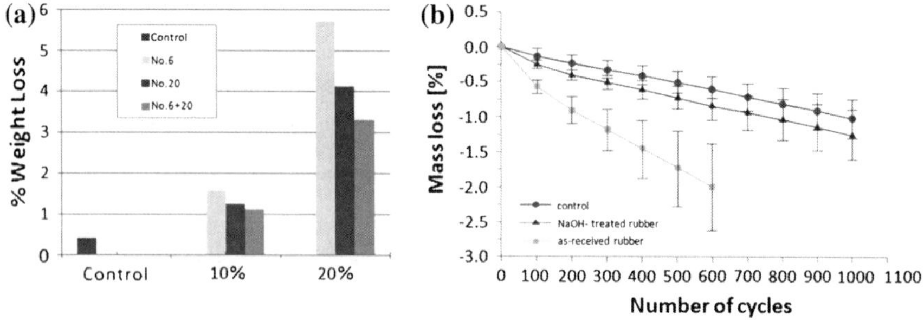

Fig. 4.53 Abrasion resistance of normal and rubberized concrete mixes

Table 4.20 Plastic shrinkage cracking behaviour of mortar (Raghvan et al. 1998)

Type of addition	Amount of cement replaced (%)	Number of cracks	Crack length (mm)			Average crack with (mm)			Time of first crack (min)
			1 h	2 h	3 h	1 h	2 h	3 h	
None	0	1	158	212	246	0.3	0.6	0.9	2
RS4.75	5	2	174	212	212	0.2	0.4	0.6	30
	10	2	156	203	203	0.2	0.4	0.4	60
	15	4	103	142	178	0.2	0.3	0.4	60
RS2.36	15	4	163	181	203	0.2	0.3	0.3	35
GR	15	3	107	204	219	0.2	0.2	0.4	45

4.7.3 Durability Parameters

Several durability parameters of concrete with rubber aggregates were reported in the literature. These include shrinkage, water absorption and water sorptivity, water and chloride permeability, and freeze–thaw resistance.

4.7.3.1 Drying Shrinkage

The incorporation of rubber aggregates decreases the drying shrinkage of concrete and increasing the amount of rubber aggregates further decreases it.

Raghvan et al. (1998) evaluated the plastic shrinkage of mortar with mass fractions of 0, 5, 10 and 15 % rubber shred with a size range of 4.75–2.36 mm. At 15 % rubber content, a fine fraction of rubber shred with size range of 2.36–1.18 mm and a fraction of granular rubber with about 2 mm diameter were also used to evaluate the properties. The width of the cracks for all the mixes was measured at 1, 2 and 3 h in the drying chamber. The results are summarised in Table 4.20. All the specimens cracked within the first 3 h of exposure and the cracks always occurred over the central stress raiser. After 3 h, the control mortar specimen developed a crack with an average width of about 0.9 mm, while the average crack width for the specimens with 5–15 % rubber shreds was 0.4–0.6 mm. The number of crack also increased due to the addition of rubber in cement mortar. The onset of cracking was delayed by the addition of rubber shreds. The content of rubber shreds in the mortar affected the onset time of cracking, the crack length, and the crack width.

Kang et al. (2009) reported that RCC with different contents of rubber tyre aggregates exhibited a shrinkage pattern similar to that of conventional concrete and the drying shrinkage developed at a higher rate in the first month than later on (Fig. 4.54). They found almost similar shrinkage for RCC with 50 kg/m^3 rubber aggregates and NA. However, higher shrinkage was recorded for rubber contents of 100 and 120 kg/m^3. Uygunoglu and Topcu (2010) reported lower drying

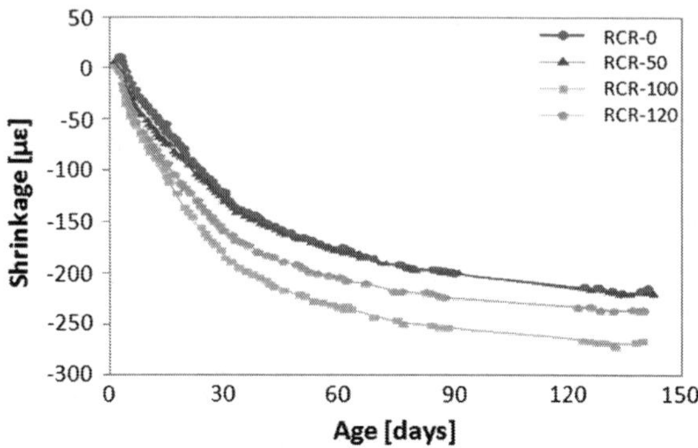

Fig. 4.54 Drying shrinkage of concrete with various amounts of rubber aggregates (Kang et al. 2009)

shrinkage of self-consolidating mortar prepared by replacing 10 and 20 % of sand by rubber aggregates. However, at higher substitution levels it increases sharply.

Ho et al. (2009), from a test according to ASTM standard C 1581-04, reported that the incorporation of rubber aggregate in concrete reduced the sensitivity of concrete to cracking due to shrinkage-related length change. This was due to the enhanced strain capacity of rubberized concrete. The compressive strain developed in the steel ring caused by the restrained shrinkage of the concrete specimen measured from the time of casting show that in comparison with the control concrete, the development of compressive strain in the steel ring slowed down for rubberized concretes, which confirms the stress relaxation resulting from the presence of rubber particles. The incorporation of rubber into concrete delayed the time of crack initiation and increasing rubber amounts further delayed it. These results are presented in Table 4.21.

4.7.3.2 Water Absorption

The amount of water absorbed is related to the porosity of the test specimens and gives an insight of the internal microstructure. Several reports are available on the water absorption behaviour of concrete due to incorporation of rubber aggregates.

The water absorption behaviour of concrete with rubber aggregates depends on their particle size. In general, the presence of large size rubber aggregates

Table 4.21 Effect of rubber aggregate on the cracking potential of concrete (Ho et al. 2009)

Mix	Time to cracking (day)	Average stress rate (MPa/day)	Potential for cracking[a]
COR	9.25	0.39	High
C20R	15.50	0.16	Moderate-low
C40R	33.25	0.05	Low

[a] According to ASTM C1581-04
COR conventional concrete
C20R and *C40R* concrete prepared by replacing 20 and 40 % of sand by an equal volume of rubber aggregates

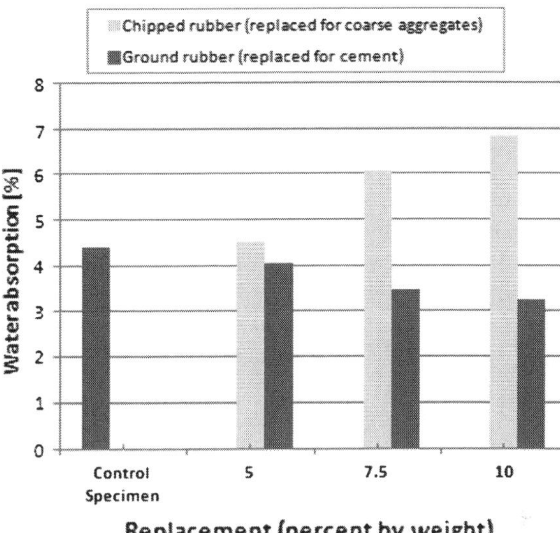

Fig. 4.55 Water absorption of concrete with rubber aggregates (Ganjian et al. 2009)

increases the water absorption of the resulting concrete due to the weak cement paste-rubber aggregates interactions, which is presented in Fig. 4.55 (Ganjian et al. 2009). Bignozzi and Sandrolini (2006) observed a slight increase in the water absorption of self-compacting concrete due to the incorporation of fine rubber aggregates, possibly due to deviations of rubber particles size from sand grain size distribution and an increase in air amount trapped during mixing procedures. Uygunoglu and Topcu (2010) also observed higher water absorption of self-consolidating cement mortar due to partial substitution of fine aggregates by rubber aggregates and these values further increased with the water to powder ratio. On the other hand, fine rubber particles can reduce the water absorbed. According to some authors, rubber particles do not absorb water, which ultimately lowers the amount of water absorbed (Yilmaz and Degirmenci 2009; Segre and Joekes 2000). Some authors argued that water reduction was due to a reduction in porosity of concrete as fine rubber particles filled the voids (Ganjian et al. 2009). Gesoglu and Guneyisi (2011) reported that the addition of FA can also reduce the amount of water absorbed in rubberized concrete due to the filling effect of FA at early ages and its pozzolanic reaction at later ages.

Segere et al. (2003) reported that using rubber aggregates at partial substitution of sand reduced the capillary water absorption (Fig. 4.56a). Rubber aggregate treated with NaOH can further improve the capillary water absorption behaviour (i.e. lower the absorption) of the resulting concrete due to an improvement in binding between rubber aggregates and cement paste. The authors reported a sorptivity coefficient of 0.29 mm/min$^{1/2}$ for conventional mortar and 0.06 mm/min$^{1/2}$ for the mortar with 10 % rubber particles. Benazzouk et al. (2007) also reported a decrease in capillary water absorption and in water absorption rate of cement composites with an increase in rubber content, which may be due to the

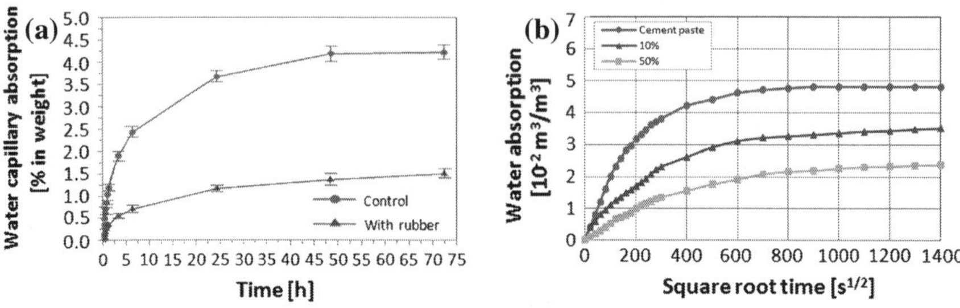

Fig. 4.56 Capillary water absorption of rubber-cement composites

capability of rubber to repel water (non-sorptive nature) and to an increase of air-entrainment, as manifested by closed empty pores that are not accessible to water (Fig. 4.56b). The decrease in water absorption is also attributed to a reduction in the porosity near particle/matrix interfacial zone, due to the high bonding between rubber aggregates and cement paste.

Bennazzouk et al. (2007) determined the sorptivity of cement composites with rubber particles. Value decreased from 0.193×10^{-3} m/s$^{1/2}$ for cement paste to 0.037×10^{-3} m/s$^{1/2}$ for specimens with 50 % of shredded rubber particles. The water sorptivity of concrete with crumb rubber aggregates was higher than that of composites with fine rubber particles; however, these values are lower than for conventional concrete. On the other hand, Gesoglu and Guneyisi (2011) found higher sorptivity coefficient of self-compacting concrete due to the addition of crumb rubber aggregates. The addition of FA can decrease the sorptivity of rubberized aggregate, which is particularly significant for concrete cured for 90 days.

In general, the depth of water penetration into concrete increased due to the incorporation of rubber aggregates. The increasing size of rubber particles further increases this parameter. The reasons for water absorption to increase are also accountable for depth of penetration. Ganjian et al. (2009) reported that concrete with replacements of 5 and 7.5 % of tyre rubber is classified as low permeability according to DIN 1048 standard but the mix with 10 % tyre rubber incorporation is classified as medium (Fig. 4.57).

4.7.3.3 Chloride Permeability

Very few data are available on chloride-ion permeability behaviour of concrete with rubber aggregates. Gesoglu and Guneyisi (2007) found a systematic increase in the depth of chloride penetration with increasing rubber content for concrete with and without silica fume, especially at high w/c ratio (Fig. 4.58). As the rubber content increased from 0 to 25 % by total aggregate volume, the chloride permeability of the rubberized concrete with and without silica fume was about 6–40 % and about 27–59 % greater than that of the controlled concrete at w/c ratio of 0.6 and 0.4, respectively, depending on the moist curing period. They also reported that increasing the moist curing period as well as adding silica fume

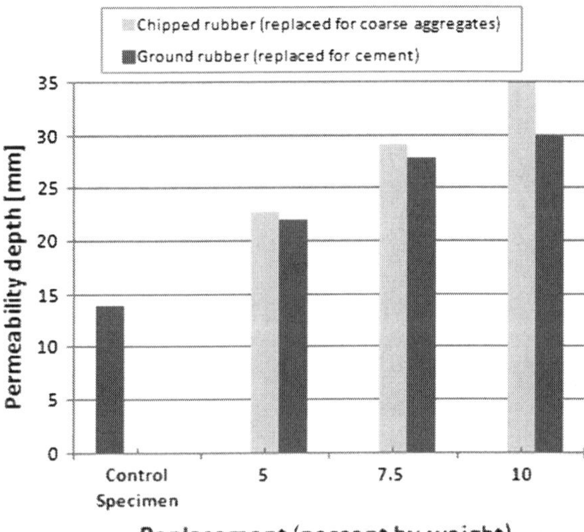

Fig. 4.57 Depth of water penetration (Ganjian et al. 2009)

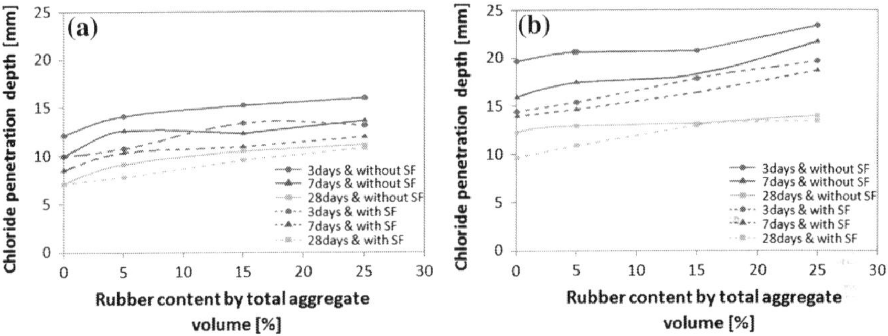

Fig. 4.58 Chloride permeability of concrete with rubber aggregates: **a** w/c: 0.4; **b** w/c: 0.6 (Gesoglu and Guneyisi 2007)

decreased the effect of rubber aggregate on the chloride-ion permeability of concrete.

The same authors (Gesoglu and Guneyisi 2011) reported the effect of FA on the chloride-ion permeability behaviour of self-compacting concrete with rubber aggregates. There was a progressive increase in the chloride-ion penetration as the rubber content rose. Extending the curing period from 28 to 90 days slightly improved the chloride-ion penetration behaviour. Incorporating FA slightly improved the chloride-ion permeability behaviour of the rubberized concretes at 28 days. However, when the curing period was prolonged to 90 days, incorporating the FA into the self-compacting rubberized concrete mixes significantly enhanced the resistance of the mixes against chloride-ion ingress. This finding was attributed to the long-term reaction of FA, which refined the pore structure of concrete and reduced the ingress of chloride ions.

4.7.3.4 Resistance to Chemical Attack

Topcu and Demir (2007) reported that the effect on the decrease of compressive strength of mortars with various amounts of rubber aggregates replacing natural sand was stronger in sea-water curing than in normal curing. The authors therefore recommended using sulphate resistant cement or high-strength cement in rubberized mortars to be used in sea-water environments.

4.7.3.5 Freeze–Thaw Resistance

Topcu and Demir (2007) also reported the effect of freeze–thaw cycles on the performance of rubberized concrete. In this study, concrete specimens had a cement content of 300 kg/m^3, a w/c ratio of 0.5, and 0, 10, 20 and 30 % replacement of fine aggregates by equal volume of rubber aggregates with size 1–4 mm. The results revealed that the concrete's compressive strength decreased with the increment of rubber incorporation after the freeze–thaw test. However, this reduction was slightly lower than the one observed for a similar concrete due to increasing addition of rubber aggregates before the freeze–thaw test. These reductions for all cylindrical specimens with 10, 20 and 30 % rubber incorporation compared to cylindrical control specimens not exposed to freeze–thaw cycles were, respectively, 16, 19 and 21 %. The reductions in cylindrical specimens with 10, 20 and 30 % rubber incorporation compared to control specimens, both exposed to freeze–thaw cycles, were respectively 15, 16 and 16. A similar behaviour was observed for cubic specimens. The authors also evaluated the freeze–thaw durability according to weight loss where they found that concrete prepared with 10 % replacement of fine aggregates by rubber aggregates exhibited better performance than conventional concrete.

4.7.3.6 High Temperature Behaviour

Topcu and Demir (2007) reported that rubber incorporation in cement mortar did not have significant effect on the compressive strength reduction due to increase in temperature. The highest decrease was observed for rubberized mortar after treatment at 400 °C. Nayaf et al. (2010) reported that the addition of 5 % microsilica to cement and the use of fine rubber aggregates with a maximum size of 0.07 mm could improve the rubberized concrete's compressive strength behaviour at high temperature. On the other hand, microsilica does not have any effect on concrete with coarse rubber aggregates with maximum size of 20 mm. Both aggregate sizes were used to replace coarse NA by 5–30 % in volume. Their results are presented in Fig. 4.59.

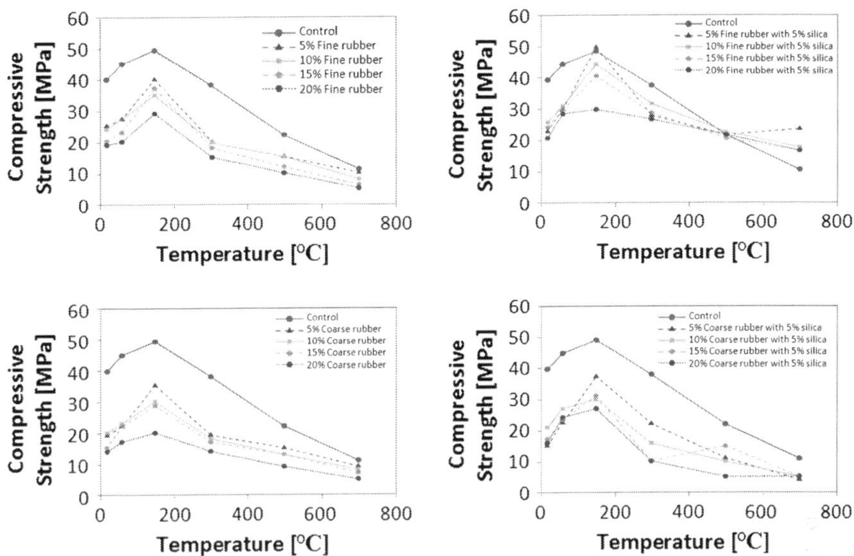

Fig. 4.59 Effect of temperature on the compressive behaviour of rubberized concrete (Nayef et al. 2010)

4.7.3.7 Damping Ratio and Base Isolation Property

Vibration damping is valuable for concrete structures because it mitigates hazards that may arouse from various factors like accidental loading, wind, ocean waves, or earthquakes. It can also increase the comfort of a person who uses the structures and enhances their reliability.

Zheng et al. (2008b) measured the damping ratio of conventional as well as rubberized concrete. These results suggest that the use of coarse and fine rubber aggregates as partial substitution of coarse NA increased the damping ratio and this effect further increased with the particles size and the rubber aggregates content. They also observed that the damping ratio increased with an increase in the maximum response amplitude. These results are presented in Fig. 4.60. From the analysis of the dynamic modulus of elasticity and the damping behaviour, the authors concluded that the rubber aggregates content in concrete should be below 30 % since higher contents dramatically reduced the modulus of elasticity of rubberized concrete.

Owing to the excellent flexibility and energy absorbency of rubberized concrete, Li et al. (1998) evaluated the base isolation capability of conventional and rubberized concrete. They determined the top dynamic response of the structure due to base excitation. Their results showed that the fundamental frequency of the structure shifted from 8 to 5 Hz when part of the base structure is replaced by rubberized concrete (Fig. 4.61). Moreover, the maximum acceleration frequency

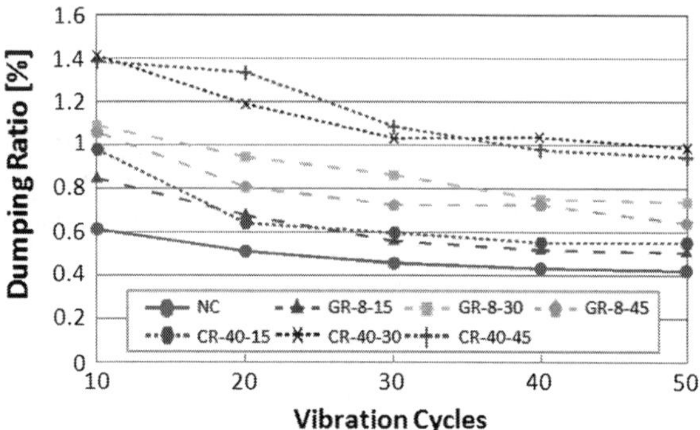

Fig. 4.60 Effect of rubber aggregates incorporation on the dumping ratio of concrete (Zheng et al. 2008b)

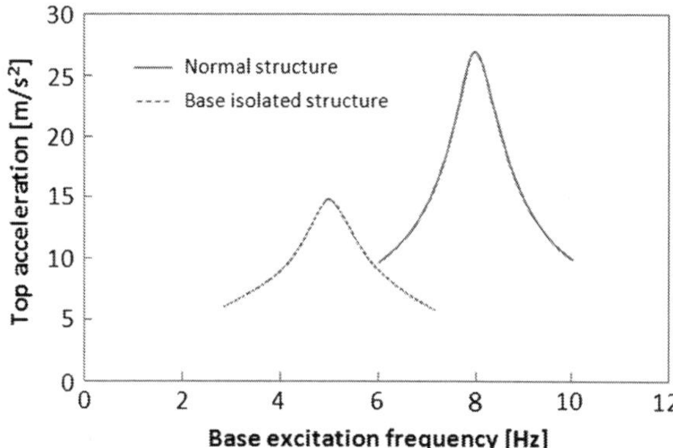

Fig. 4.61 Dynamic response of conventional and rubberized concrete (Li et al. 1998)

for rubberized concrete was significantly lower than that for conventional concrete, which indicates that rubber incorporation in concrete reduced the resonant response.

4.7.3.8 Thermal Insulation Properties

Yesilata et al. (2009) observed improvement of the thermal insulation performance of concrete due to the incorporation of rubber elements with thickness of 2 mm. This improvement was 18.52 % by adding of square rubber matrix in ordinary concrete.

4.8 Ceramic Industry Waste

In Chap. 2, the properties of some ceramic waste as aggregate were presented in detail. Here, the fresh and hardened concrete properties with different types of ceramic waste will be presented.

4.8.1 Fresh Concrete Properties

The slump of a concrete mix with ceramic waste aggregates depends on the nature of the aggregates. The majority of ceramic aggregates reported in the literature have higher porosities than normal aggregates, and therefore the incorporation of these aggregates in the concrete mix decreases slump due to their high porosity, rough surface texture and angular nature. Topcu and Canbaz (2007) observed workability problems due to the use of tile waste as partial and full replacement of coarse aggregate because of rough surface texture of the tile aggregates. Lopez et al. (2007) and Guerra et al. (2009) observed a similar workability of concrete with ceramic aggregates and natural concretes, when the latter was replaced by various amounts of fine and coarse ceramic aggregates. On the other hand, ceramic aggregates have some properties like lower water absorption than NA and smooth surface texture that can increase the slump of the resulting concrete mix (Senthamarai and Devadas 2005). Debeib and Kenai (2008) observed some segregation of concrete mix when brick waste was used as fine and coarse aggregates. The variations of concrete's slump due to the incorporation of ceramic aggregates given in various references are presented in Table 4.22.

The density of concrete with waste ceramic aggregates is generally lower than (or similar to) that of conventional concrete. Binici (2007) reported similar density and air contents for conventional concrete and concrete mixes prepared by replacing 40, 50 and 60 % by weight of sand by ceramic aggregate. Torkittikul and Chaipanich (2010) observed a decreasing trend in the density of fresh concrete and cement mortar mixes due to use of ceramic aggregates as replacement of sand (Fig. 4.62).

Brito et al. (2005) reported that the density of concrete decreased as the replacement ratio of coarse limestone aggregates by similar size ceramic aggregates increased, due to the lower density of ceramic aggregates compared to the limestone aggregates. The bulk density of fresh concrete mixes with ceramic waste with specific gravity of 2.45 as coarse aggregates with different w/c ratios in the Senthamarai et al. (2011) study was in the 2215–2281 kg/m^3 range in comparison to the equivalent range of 2383–2480 kg/m^3 for conventional concrete with granite coarse aggregates with specific gravity of 2.68. Cachim (2009) observed about a 5 and 6 % decrease in fresh density at w/c ratios of 0.45 and 0.5, respectively, when coarse NA were replaced by brick waste aggregates. Debeib and Kenai (2008) observed a reduction of up to 17 % in the fresh density of concrete with brick waste aggregates by comparison with NA concrete. The air content in concrete also increased as the content of ceramic waste aggregates rose.

Table 4.22 Slump of concrete with various types of ceramic aggregates

Torkittikul and Chaipanich (2010)	Fine aggregates	*W* (%)		0, *w*	50, *w*	100, *w*				WA: 1.25 (*C*);
		L (mm)		120	110	5				Texture: rough (*C*)
										Shape: angular (*C*)
Senthamarai and Devadas (2005)	Coarse aggregates	w/c		0.35	0.40	0.45	0.50	0.55	0.60	WA: 0.72 (*C*); 1.2 (*N*)
		L (mm)	*N*	10	18	35	48	80	148	Texture: smooth (*C*)
			C	13	24	45	64	99	155	Shape: angular (*C*)
Suzuki et al. (2009)	Coarse aggregates	*W* (%)		0, *v*	10, *v*	30, *v*				WA: 9.31 (*C*); 0.88 (*N*)
		L (mm)		600	550	530				
Binici (2007)	Fine aggregates	*W* (%)		0, *w*	40, *w*	50, *w*	60, *w*			WA: 2.44 (*C*); 2.65 (*N*)
		L (mm)		110	90	85	80			Texture: smooth (*C*)
										Shape: angular (*C*)
Lopez et al. (2007)	Fine aggregates	*W* (%)		0, *w*	10, *w*	20, *w*	30, *w*	40, *w*	50, *w*	–
		L (mm)		30	49	30	42	36	34	
Guerra et al. (2009)	Coarse aggregates	*W* (%)		0, *w*	3, *w*	5, *w*	7, *w*	9, *w*		
		L (mm)		40	42	39	41	43		

W replacement amount, *L* slump/consistency, *w* and *v* substituted by weight and by volume respectively, *WA* water absorption, *C* and *N* ceramic and natural aggregates

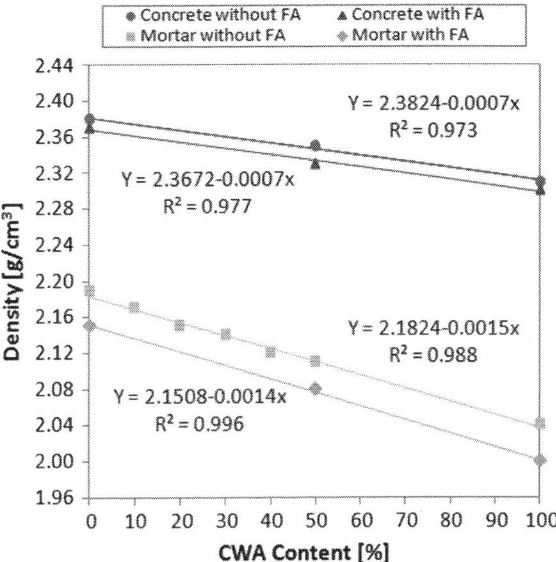

Fig. 4.62 Effect of ceramic waste aggregates incorporation in the density of fresh concrete and mortar mixes (Torkittikul and Chaipanich 2010)

4.8.2 Mechanical Properties

The compressive strength behaviour of concrete with ceramic aggregates depends on the properties of these aggregates. In several studies, it was observed that the incorporation of ceramic aggregates in concrete increased the compressive strength. This is particularly true for ceramic aggregates with low water absorption capacity like aggregates made of glazed ceramic waste. On the other hand, concrete with ceramic aggregates with very high water absorption capacity like aggregates generated from brick type ceramics exhibited lower compressive strength than conventional concrete.

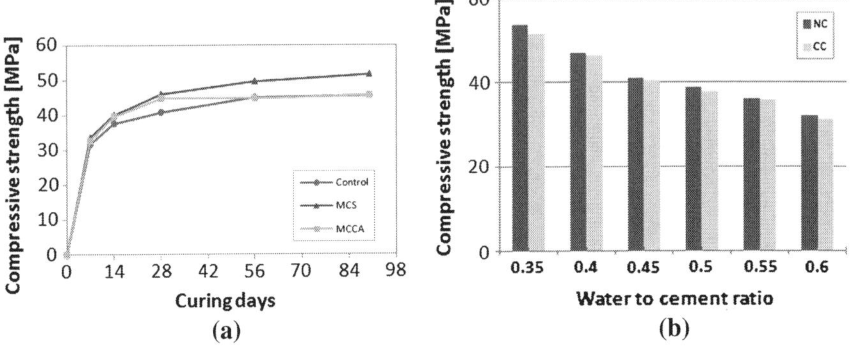

Fig. 4.63 Compressive strength of concrete with various types of ceramic aggregates: **a** Pacheco-Torgal and Jalali 2010; **b** Senthamarai and Devadas 2005

Binici (2007) observed higher compressive strength in concrete with 40, 50 and 60 % by weight of natural sand replaced by fine ceramic aggregates than in conventional concrete after 1 year of curing. In the Torkittikul and Chaipanich (2010) study, the 28-day compressive strength of concrete with ceramic earthenware waste aggregates as 50 and 100 % by weight replacement of natural fine aggregates was respectively 40.0 and 38.5 MPa that compares with 37.0 MPa for conventional concrete. This increase was attributed to improved interfacial zone due to the rough surface texture of ceramic aggregates and the presence of hard crystalline material like mullite in sintered ceramics. However, a slight drop at 100 % replacement level was observed, as the angular nature of ceramic aggregates deteriorated the workability of fresh concrete.

Pacheco-Torgal and Jalali (2010) observed higher compressive strength for two types of concrete with water-saturated white ceramic waste as complete replacement of fine and coarse NA than for conventional concrete (Fig. 4.63a). Ceramic aggregates replacing sand (MCS) were more effective than coarse ceramic aggregates (MCCA) in increasing the compressive strength of concrete after 28-day of curing. Lopez et al. (2007) also observed higher early compressive strength (up to 28 days) of concrete with ceramic aggregates content in the 10–50 % by weight range as replacement of natural sand.

Guerra et al. (2009) observed similar 28-day compressive strength for concrete with aggregate from sanitary porcelain waste replacing 3 % by weight of natural coarse aggregates; however, compressive strength increased with the content of ceramic aggregates at 5 and 7 % replacement levels but slightly decreased at 9 % replacement level even though still higher than that of the reference concrete. Compressive strength also increased with curing time. Senthamarai and Devadas (2005) observed a maximum 3.8 % reduction in 28-day compressive strength of concrete with coarse white ceramic waste aggregates with various w/c ratios when compared to concrete with NA (Fig. 4.63b).

Brito et al. (2005) observed lower compressive strength in concrete pavement blocks prepared by replacing 33, 66 and 100 % by volume of coarse limestone aggregates by aggregates from ceramic hollow bricks waste due to the lower density and lower crushing strength of ceramic aggregates than those of the limestone aggregates. The strength decreased as the content of ceramic aggregates increased. Topcu and Canbaz (2007) reported that using tile waste as replacement of coarse aggregates could decrease up to 43 % the compressive strength exhibited by the reference concrete due to the lower crushing strength of tile aggregates than that of crushed stone as well as the higher pores content in tile aggregates concrete. Debeib and Kenai (2008) observed up to 35 and 30 % reduction in compressive strength of concrete when coarse and fine NA were, respectively, replaced by coarse and fine recycled brick aggregates. Compressive strength was further decreased up to 40 % when both fine and coarse aggregates were replaced by brick aggregates. Cachim (2009) reported the compressive strength of two types of ceramic brick waste with different physical properties as partial (15 and 30 %) replacement of natural coarse aggregates in concrete. The author observed that the incorporation of brick aggregates with higher crushing strength than the natural

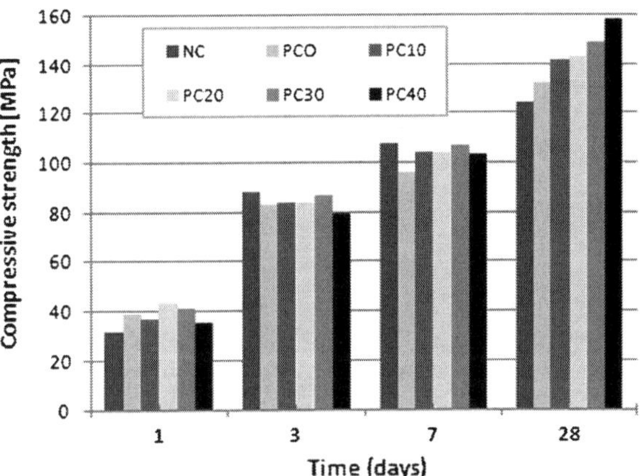

Fig. 4.64 Effect of the incorporation of porous ceramic aggregates on the compressive strength behaviour of internally cured concrete samples (Suzuki et al. 2009)

ones as well as of other type of brick aggregates having low crushing strength but with similar shape index as that of NA gave slightly higher 90-day compressive strength than for conventional concrete when 15 % by volume of natural coarse aggregates were replaced by this type of brick aggregates. On the other hand, the compressive strength of concrete with other type of brick aggregates was lower than that of conventional concrete at all curing ages. According to the author, the observed increase in strength was due to internal curing of concrete as the water absorbed by brick aggregates was used for hydration at later stages of curing.

Suzuki et al. (2009) reported the effect of internal curing on various properties including compressive strength of concrete with porous red ceramic aggregates as a 0–40 % by volume replacement of natural coarse aggregates. Incorporating ceramic aggregates did not affect the 3- and 7-day compressive strength but the 28-day strength increased with the content of ceramic aggregates. The 28-day compressive strength of concrete with 40 % by volume replacement of NA by ceramic aggregates was 20 % higher than that of NA concrete. These results are presented in Fig. 4.64.

There are few reports available on the evaluation of other strength properties (splitting tensile and flexural strength) and modulus of elasticity of concrete with ceramic aggregates. Lopez et al. (2007) and Guerra et al. (2009) did not observe any significant differences in the indirect tensile and fracture strengths of concrete where 10–50 % by weight of sand was replaced by white ceramic aggregates even though this incorporation significantly increased its compressive strength. Senthamarai and Devadas (2005) observed that the 28-day splitting tensile and flexural strengths of concrete with white ceramic waste aggregates as complete replacement of coarse NA were in the ranges of 3.2–4.5 MPa and 4.7–6.9 MPa at various w/c ratios (0.35–0.60). In this study, the corresponding ranges of splitting tensile and flexural strengths for conventional concrete were 3.9–5.5 MPa and 5–7 MPa, respectively. The authors observed lower tensile and flexural strengths to compressive strength ratios for ceramic aggregates concrete than for conventional concrete. The modulus of elasticity of conventional as well as of ceramic waste aggregates concrete varied in the ranges of 16.5–25.1 GPa and 16.1–22.2 GPa, respectively, at various w/c ratios.

Topcu and Canbaz (2007) observed a reduction of up to 40 % in the splitting tensile strength of concrete when coarse NA were replaced by tile waste aggregates. Brito et al. (2005) observed a linear decreasing trend in flexural strength of concrete pavement blocks with increasing content of aggregates from ceramic hollow brick as partial replacement of coarse limestone aggregates. However, the reduction in flexural strength of concrete was 26 % in comparison to 45 % in compressive strength when coarse NA were completely replaced by ceramic aggregates (Fig. 4.65). A considerable reduction in flexural strength and modulus of elasticity was also observed when crushed ceramic brick was used as partial and full replacement of natural fine and coarse aggregates in concrete (Debieb and Kenai 2008). On the other hand, Cachim (2009) observed slightly higher or similar flexural strength and modulus of elasticity of concrete with coarse brick aggregates at 15 and 30 % replacement level of coarse aggregates than that observed for conventional concrete.

Suzuki et al. (2009) observed a decreasing trend in 28-day splitting tensile strength of high-performance concrete with increasing content of porous ceramic aggregates (after observing a slight increase at the of 10 % by volume replacement level of coarse aggregates), due to a weaker interfacial bonding of ceramic aggregates–cement paste than for NA–cement paste. The Young's modulus of elasticity also decreased as the content of ceramic aggregates rose.

Cachim (2009) drew similar stress–strain curves for conventional concrete and concrete with two types of coarse brick aggregates replacing 15 and 30 % by volume of natural coarse aggregates. Topcu and Canbaz (2007) observed lower toughness of concrete with tile waste as partial and full replacements of coarse aggregates than that of conventional concrete.

Contrasting results are available on the effect of ceramic aggregates incorporation on the abrasion resistance of concrete due to changes in the properties of the aggregates. In some studies, the incorporation of fine or coarse ceramic aggregates improved the abrasion resistance of the resulting concrete due to good adhesion of porous ceramic aggregates to the cement paste (Brito et al. 2005; Binici 2007).

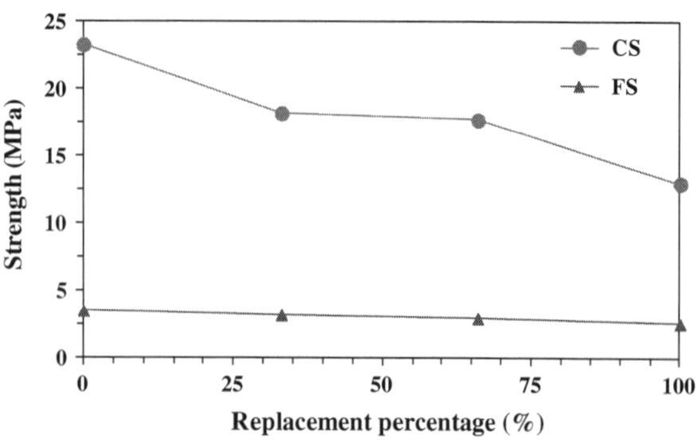

Fig. 4.65 Compressive strength (CS) and flexural strength (FS) of concrete with waste ceramic aggregate (Brito et al. 2005)

However, Topcu and Canbaz (2007) observed a significant reduction in the abrasion resistance of concrete due to the addition of ceramic tile waste as partial or full replacement of fine NA.

4.8.3 Durability Behaviour

Durability properties such as water absorption, chloride migration, gas permeation, freeze–thaw resistance and shrinkage were reported in various references even though the numbers of references for each type of ceramic aggregates is not substantial.

Debeib and Kenai (2008) observed higher drying shrinkage for concrete with crushed brick as partial or full replacement of fine and coarse aggregates than for NA concrete. Shrinkage increased with the content of both types of aggregates. The increase in shrinkage was more prominent for concrete with fine ceramic aggregates than for coarse ceramic aggregates, possibly due to the movement of water present in fine brick aggregates as progressive drying changed the moisture conditions (Fig. 4.66). The presence of porous red ceramic waste as partial replacement of coarse NA can significantly reduce the autogenous shrinkage of high-performance concrete when it is subjected to internal curing (Suzuki et al. 2009). The shrinkage reduction increased with the content of ceramic aggregates in concrete (Fig. 4.67).

The use of crushed brick as partial and full substitution of fine and coarse aggregates significantly increased the water absorption by capillarity of concrete (Debeib and Kenai 2008). The absorption was more pronounced for concrete with coarse brick aggregates. The water permeability of concrete with crushed brick aggregates was found to be 2.0–2.5 times higher than that of conventional concrete.

Use of a superplasticizer can improve the water absorption behaviour of concrete with brick aggregates (Debeib and Kenai 2008). Correia et al. (2006) also

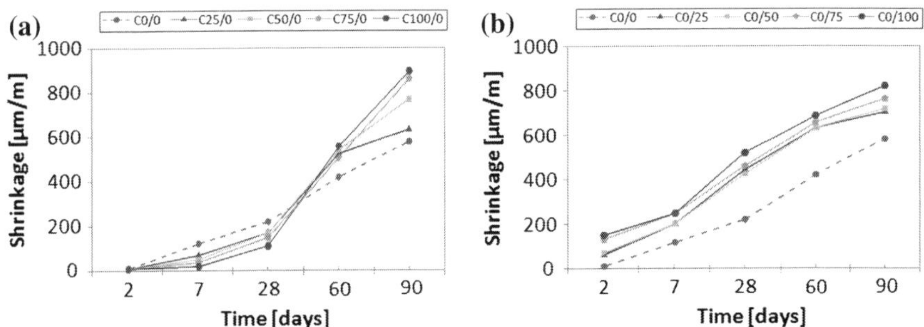

Fig. 4.66 Behaviour of drying shrinkage of concrete ceramic waste aggregates: **a** coarse; **b** fine (Debeib and Kenai 2008)

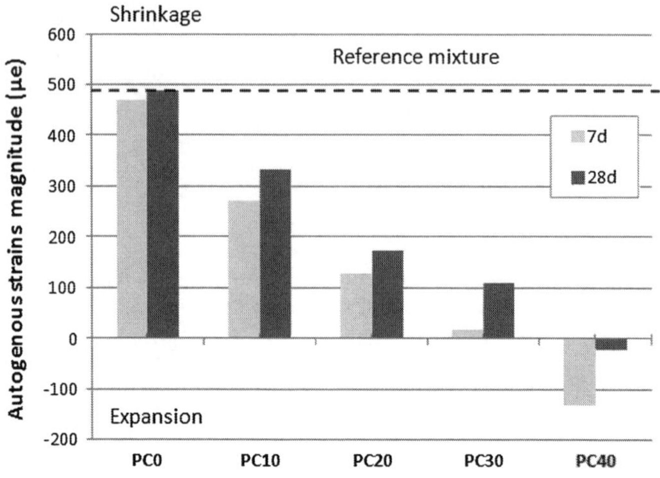

Fig. 4.67 Autogenous shrinkage behaviour of concrete with ceramic waste aggregates (Suzuki et al. 2009)

observed increasing water absorption of concrete pavement blocks as the replacement of natural coarse aggregates by ceramic hollow bricks aggregates increased. The water absorption capacity of conventional concrete blocks and blocks with 1/3, 2/3 and 3/3 (by mass) replacement of natural coarse aggregates by ceramic aggregates were 17.05, 21.11, 23.97 and 27.64 %, respectively. Senthamarai et al. (2011) observed higher water absorption by capillarity and volume of voids for concrete with coarse white porcelain waste aggregates than for conventional concrete with granite coarse aggregate. The water absorption by capillarity of concrete with waste aggregates and conventional concrete were in the ranges of 3.74–7.21 % and 3.10–6.52 %, respectively. Pacheco-Torgal and Jalali (2010) found lower water and oxygen permeability for concrete with ceramic waste as fine and coarse aggregates than for conventional concrete (Fig. 4.68a); however, the vacuum water permeability of conventional concrete waste was negligibly lower than that of ceramic aggregates concrete (Fig. 4.68b).

Binici (2007) observed lower depth of chloride permeation for concrete with ceramic waste aggregates replacing 40, 50 and 60 % by volume of fine NA than for conventional concrete. The depth of penetration for ceramic waste aggregates concrete was 10–15 mm in comparison to about 45 mm in conventional concrete. In the same study, the compressive strength of concrete with ceramic aggregates was also higher than that of conventional concrete and strength increased with the content of ceramic aggregates. Pacheco-Torgal and Jalali (2010) also observed lower chloride diffusion through concrete with fine and coarse ceramic waste aggregates than that through conventional concrete and best performance was observed for concrete with fine ceramic aggregates (Fig. 4.69a). Like in the Binici (2007) study the results can be related with the compressive strength of concrete. Senthamarai et al. (2011) observed higher electrical charge for concrete with

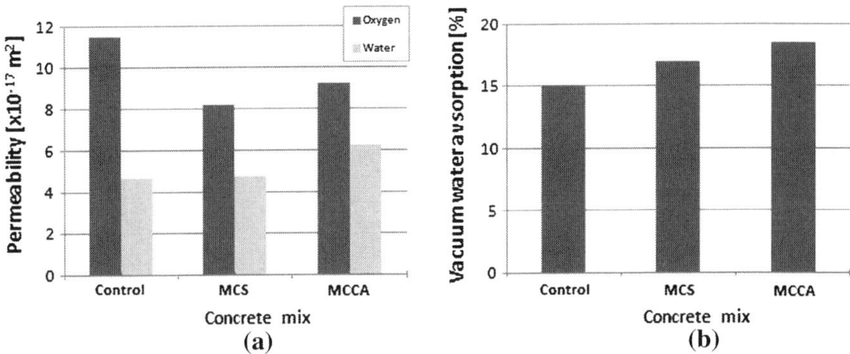

Fig. 4.68 Permeability of concrete with conventional and ceramic waste aggregates (Pacheco-Torgal and Jalali 2010)

coarse porcelain waste aggregates than for conventional concrete when both were subjected to ASTM C1292-10 specified rapid chloride penetration test (Fig. 4.69b). The increase in porosity in the ITZ of ceramic aggregates and cement paste due to their smoother surface texture than that of NA increased chloride diffusion.

Pacheco-Torgal and Jalali (2010) performed an accelerated ageing test to evaluate the effect of very harsh environmental conditions on the behaviour of concrete with ceramic waste aggregates as well as conventional concrete. The adopted procedures and results are presented in Fig. 4.70. The compressive strength after ageing of concrete with ceramic waste aggregates was higher than that of conventional concrete; however, the reduction in compressive strength of conventional concrete (14 %) was lower than that of the concrete with ceramic sand aggregate (18 %) and ceramic coarse aggregate (19 %). The incorporation of waste ceramic tile as 50 and 100 % replacement of coarse NA increased the weight loss during the freeze–thaw test due to lower hardness of tile aggregates and weaker binding of cement paste-tile aggregates than those of NA (Topcu and Canbaz 2007).

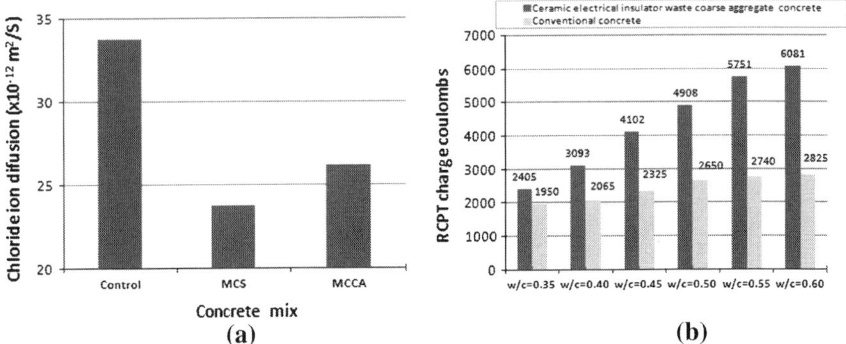

Fig. 4.69 Results of the chloride penetration test

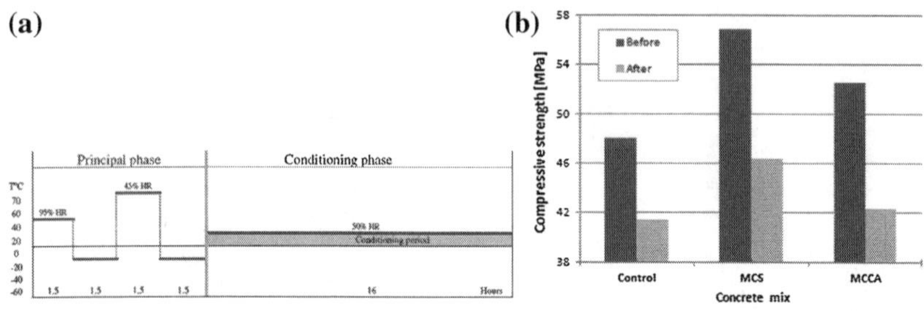

Fig. 4.70 Sequence of accelerated ageing tests and compressive strength results before and after ageing (Pacheco-Torgal and Jalali 2010)

4.9 Other Waste Materials

Other waste materials have also been used as aggregates in concrete and cement mortar production. The behaviour of these waste materials, which can be considered as industrial byproducts, will be discussed here.

4.9.1 Oil Palm Shell

The use of oil palm shell (OPS), an agricultural waste, created in palm oil producing countries as coarse aggregates in structural lightweight concrete was reported in detail (Basri et al. 1999; Jumaat et al. 2009; Mannan et al. 2006; Mannan and Ganapathy 2001a, b, 2002, 2004; Teo et al. 2006, 2007, 2010). According to Mannan and Ganapathy (2004) concrete with OPS as coarse aggregates can be used for several purposes such as road pavement, kerbs, concrete drains and flooring of buildings.

The workability of concrete with OPS as coarse aggregates was better than that of conventional concrete, due to the smooth surface texture of OPS aggregates (Basri et al. 1999; Mannan and Ganapathy 2004). The air content of concrete with OPS aggregates was also higher than that of conventional concrete due to the lower compaction of concrete with OPS aggregates. According to these authors, the obstructions in compaction due to variations in shape of OPS aggregates as well as the porous nature of OPS aggregates increased the air content of concrete. The fresh and dry densities of OPS concrete were about 20 % lower than those of conventional concrete and this mixes can be considered as structural lightweight concrete (1450–1900 kg/m^3) (Basri et al. 1999; Mannan and Ganapathy 2004).

Basri et al. (1999) observed that the compressive strength of OPS concrete was about 40–55 % lower than that of conventional concrete, when both were cured in three curing conditions. The highest compressive strength was seen in concrete cured in standard moist curing conditions. The reduction in strength was mainly due to the weaker crushing strength of OPS aggregates than that of NA. They observed a drop in compressive strength due to use of FA in OPS concrete. Mannan and Ganapathy (2001a) reported that a lightweight concrete with OPS as coarse aggregates with 28-day compressive strength of 24.2 N/mm^2 and a density of about 1900 kg/m^3 can be produced from a mixture of cement, normal sand and OPS in a proportion of 1:1.71:0.77 with free w/c ratio of 0.41. Adding 1 % $CaCl_2$ can be considered to increase concrete strength up to 29 N/mm^2. In another study, Mannan and Ganapathy (2002) reported that concrete with OPS aggregates has considerably lower compressive, flexural and splitting tensile strengths and dynamic elastic modulus than conventional concrete. However, the strength properties of concrete with coarse OPS aggregates meet the standard requirement for structural lightweight concrete. Jumaat et al. (2009) observed better shear capacity without shear reinforcement in the concrete with OPS aggregates beam than in that of conventional concrete. The concrete beam with OPS aggregates also showed better ductility behaviour with more shear and flexural cracks than the conventional concrete beam (Jumaat et al. 2009; Teo et al. 2006).

Concrete with OPS aggregates also had higher drying shrinkage than conventional concrete; however, the increase was in the normal range generally observed for lightweight concrete. The water absorption capacity of OPS concrete was also higher than that of conventional concrete due to the higher porosity of the OPS aggregates concrete. The results presented by Teo et al. (2010) indicated similar permeability properties (water and chloride) of concrete with OPS aggregates like in other lightweight aggregates concrete. However, proper curing is essential for OPS concrete to achieve better performance at later ages. Generally, normal immersion curing was best for better durability performance of OPS aggregates lightweight concrete.

Treatment of OPS by PVA (20 % solution was best) can improve mechanical and durability performance of lightweight concrete with OPS aggregates because it strengthens cement paste–OPS aggregates binding by forming a thin layer on the surfaces of OPS aggregates (Mannan et al. 2006). PVA can form a thin layer on the surface of OPS aggregates and therefore increase the aggregates–cement interaction.

4.9.2 *Crushed Oyster Shell*

Yang et al. (2005) studied the effect of crushed oyster shell (OS) as partial replacement of sand (10 and 20 % by weight) in concrete. From the experimental results they concluded that:

1. The slump of concrete with OS decreased with as the fineness modulus of OS (though it is insignificant) and the substitution rate of sand by OS increased, due to the dry and flaky nature of OS aggregates; however, setting time and air content were not affected by the partial substitution of sand by OS;

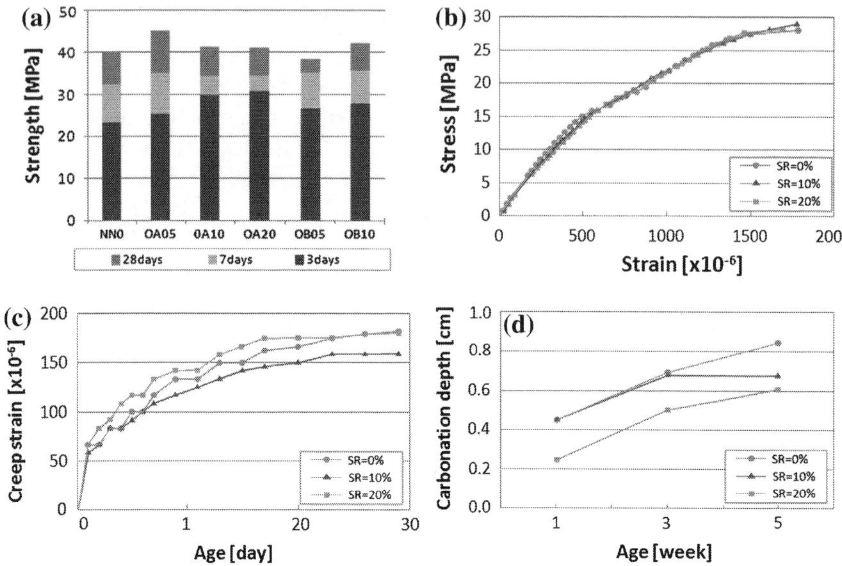

Fig. 4.71 Properties of concrete with crushed oyster shell as partial substitution of fine natural aggregates (Yang et al. 2010)

2. The early age compressive strength (up to 28 days) of concrete with OS at 10 and 20 % level was higher than that of conventional concrete due to the lower free water content in this type of concrete than in conventional concrete (Fig. 4.71a); however, the 1-year compressive strength of concrete with OS at 10 and 20 % level was already lower than that of conventional concrete; the modulus of elasticity of concrete with OS at all replacement levels was lower than that of conventional concrete at all curing period of 1 year of curing due to the lower stiffness of OS aggregates than that of fine NA; the addition of OS as aggregates in concrete did not have any effect on the stress–strain curve (Fig. 4.71b);
3. The drying shrinkage of concrete with OS at all replacement levels was higher than that of conventional concrete due to the lower stiffness and fineness modulus of OS aggregates than those of natural sand; the creep of concrete with OS aggregates at 10 and 20 % replacement levels of sand were, respectively, lower and higher or similar to that of conventional concrete (Fig. 4.71c); the permeability of concrete with OS was lower than that of conventional concrete; the use of OS as partial replacement of sand at 10 and 20 % levels improved concrete's freeze–thaw resistance as fine grains of OS filled the entrapped air voids scattered in concrete; the partial substitution of OS as fine aggregates decreased the carbonation depth of the resulting concrete and the decrease was higher at 20 % level than at 10 % level (Fig. 4.71d); the hydrochloric and sulphuric acid resistances of concrete with OS as partial substitution of sand were similar to those of conventional concrete.

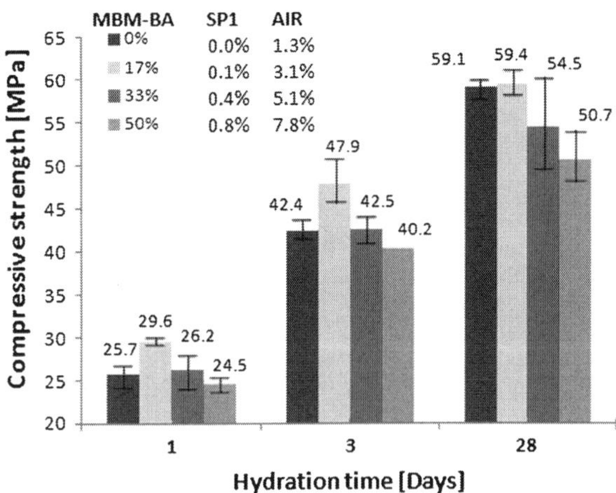

Fig. 4.72 Compressive strength at 1, 3 and 28 days of mortars with 17, 33 and 50 % of MBM-BA in replacement of sand and reference mortar (0 % MBM-BA)

4.9.3 Low-Risk Meat and Bone Meal Bottom Ash

Cyr and Ludmann (2006) reported the use of low-risk meat and bone meal bottom ash (MBM-BA) as fine aggregates in cement mortar production. This use up to a 17 % by weight level increases the early age compressive strength of cement mortar due to its high chloride content (Fig. 4.72). However, the presence of considerable amounts of chlorides (0.38 %) in MBM-BA, which restricts the use of this material as aggregates in reinforced concrete, must be considered before any cement based applications.

4.9.4 Tobacco Waste and Spent Mushroom Substrate

Ozturk and Bayrakli (2005) prepared a lightweight concrete using tobacco waste, collected from a cigarette factory, as fine aggregates. The concrete meets the specifications of lightweight concrete class that can be used as coating and dividing material in construction according to values of consistency, density (0.50–0.56 kg/dm^3), porosity, compactness, compressive strength (0.20–0.60 N/mm^2) and the thermal insulating behaviour (thermal conductivity 0.194–0.210 W/mK[13]).

Pang et al. (2007) reported that concrete with quicklime treated spent mushroom substrate (SMS) as partial substitution of sand can be used for sidewalks, concrete curbs, concrete barricades, sound walls, and other non-structural applications. However, the durability of this kind of concrete needs further investigation.

4.9.5 Pulp and Paper Mill Waste

The use of waste generated from pulp and paper mills as fine aggregates, fillers or fibres in concrete was also reported (Ahmadi and Al-Khaja 2001; Gallardo and Adajar 2006; Naik 2002; Naik et al. 2004). The dry paper mill sludge used as fine aggregates in concrete up to 10 % replacement of fine NA can improve the compressive and splitting tensile strengths of the resulting concrete (Gallardo and Adajar 2006). On the other hand, Ahmadi and Al-Khaja (2001) reported that the incorporation of paper sludge as aggregates decreased the compressive and splitting tensile strengths of the resulting concrete. Concrete masonry blocks made with sludge at 5 % replacement level exhibit compressive strength of 8 MPa, splitting strength of 1.3 MPa, water absorption of 11.9 %, and density of 20 kg/dm^3 (Ahmadi and Al-Khaja 2001).

Fibres and waste from pulp and paper mill production as well as deinking solids from paper-recycling plants were also considered for incorporation in concrete. However, these materials should be properly dispersed in water, preferably hot water, before using such sludges in making structural-grade Portland cement concrete. Chloride-ion penetration can be reduced by adding residual solids to concrete. This type of concrete also showed higher resistance to salt scaling and freeze–thaw damage than that of control concrete (Naik 2002; Naik et al. 2004).

4.9.6 Wastes from the Shoe Industry

Baffa and Akasaki (2005) investigated the performance of lightweight concrete prepared with leather waste, mostly residue from cattle, as aggregates. Results showed that hardened mortar specimens that contain leather pieces with dimensions larger than 10 mm can be permanently deformed if the mortar is cured in humid condition. The compressive strength, splitting tensile strength and drying shrinkage are inside the standards limits and may be different with different types of leather. The expansion of concrete with leather is lower than that of control concrete and it decreases as the content of leather pieces in concrete increases (Baffa and Akasaki 2005). Santiago et al. (2009) reported the use of ethylene vinyl acetate (EVA), a waste from the shoe industry as aggregates in concrete. The EVA aggregates were obtained by cutting off the waste of EVA expanded sheets used to produce insoles and innersoles of the shoes. Two concrete mixes were prepared by replacing 50 % by volume of natural coarse aggregates by EVA aggregates, and by a 1:1 mixture of EVA and construction and demolition waste (CDW) aggregates. Their results suggests that EVA waste can be used as aggregates in the production of structural lightweight concrete by mixing it with CDW aggregate with a 28-day compressive strength of about 18 MPa. The toughness value of concrete was also increased due to the addition of EVA aggregates.

4.9.7 Different Sludges

Rao et al. (2009) reported that partial replacement of river sand by bone char sludge, a waste material generated during purification of fluoride contaminated water, did not degrade the strength or environmental integrity of dense concrete specimens and met the British standard specification (BIS 2185) for load-bearing blocks (Fig. 4.73). The concrete specimens with bone char sludge at a replacement level of 3 % of sand yielded higher compressive strength than conventional concrete specimens due to the improvement of packing of small NA and larger bone char sludge particles within concrete. The leachability of fluoride from concrete blocks was much lower than the specified limit for disposal of treated leachate into inland surface water (2 mg/l).

Kuo et al. (2007) studied the use of organic-modified reservoir sludge (OMRS) as fine aggregates in cement mortars. The sludge contained high amounts of hydrophilic smectite clay, and therefore the sludge was treated with organic surfactant to convert it into a hydrophobic material.

The results indicated that increasing OMRS use to replace sand gradually decreased the compressive strength of mortar; however, the 28-day compressive strength of mortar made replacing 1–30 % of fine NA by OMRS aggregates was more than 30 MPa, and therefore up to a 30 % replacement of fine aggregates by OMRS aggregates can be accepted for production of normal-strength cement mortars with improved water permeability (Fig. 4.74). On the other hand, the OMRS particles can be used in controlled low-strength materials if the replacement ratio for fine aggregates exceeds 80 %.

Sales and De Souza (2009) reported the possibility of recycling water treatment sludge to produce medium strength structural concrete, underlayment concrete, and block laying mortar. Their results suggest that sludge may be applied as a regulator of consistency and plasticity and, in suitable quantities (e.g. 2 % of sand) can even increase the compressive strength of medium strength concrete with NA. The axial compressive strength, modulus of elasticity and stress–strain curve of

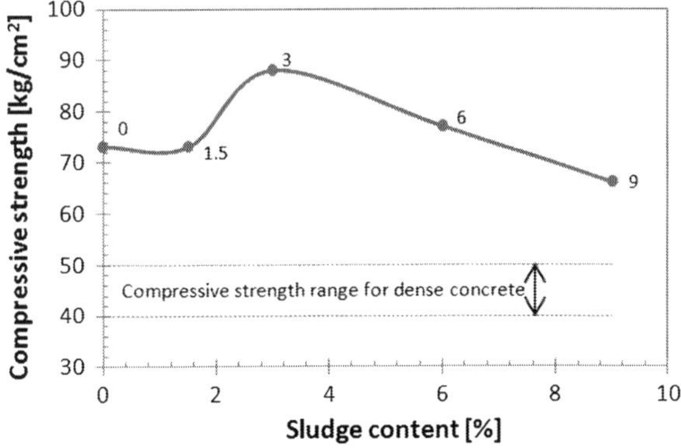

Fig. 4.73 Change in compressive strength of dense concrete blocks due to bone char sludge incorporation (Rao et al. 2009)

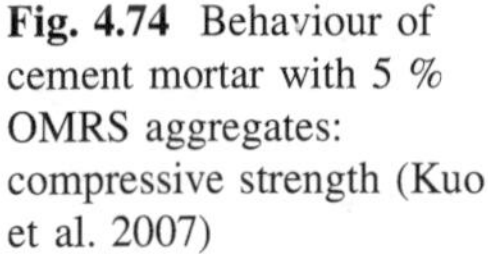

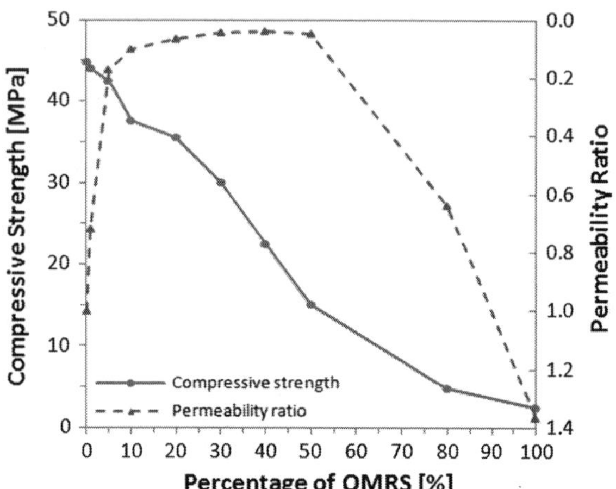

Fig. 4.74 Behaviour of cement mortar with 5 % OMRS aggregates: compressive strength (Kuo et al. 2007)

underlayment conventional concrete and concrete with 3 % replacement of sand by sludge were similar.

Aspiras and Manalo (1995) produced a concrete type composite material by using a mixture of textile waste cuttings with an average length of 2 cm (short fibre) and 6 cm (long fibre), taken from disposed trimmings of a garments producer and a textile manufacturer, cement and water. The results indicated that textile waste cuttings can be used to prepare lightweight cement composite with maximum 28-day compressive, tensile and flexural strengths of 8.48, 9.24 and 16.14 MPa, respectively. This composite has various potential uses like in ceilings, walls, or as a wooden board substitute.

Agostini et al. (2007) investigated the use of dredged polluted sediment with high concentrations of toxic elements such as Cd, Cu, Cr, Pb and Zn after treatment as partial and complete replacement of fine aggregates in cement mortar. In this work, the dredged sediment was initially treated by a well-established technique to stabilize toxic elements by forming phosphate minerals followed by a heat treatment at 650 °C to remove organic matters. The results suggest that the incorporation of sediment in mortar considerably increases its drying shrinkage. However, the incorporation of sediment in mortar does not affect water permeability and it significantly increases strength for low-to-moderate substitution levels, while high levels of incorporation of sediment lead to strength on the same order of that of the reference mortar.

4.9.8 Waste Generated from Mining Industry

Yellishetty et al. (2008) reported the use of iron ore mineral waste of 12.5 and 20 mm size ranges as coarse aggregates in concrete. Their results indicate that the 28-day uniaxial compressive strength of concrete with iron-ore mineral waste as aggregates was 21.93 MPa while the equivalent compressive strength of concrete with conventional granite aggregates was 19.91 MPa. The higher strength of concrete with mine waste aggregates than that of conventional aggregate was due to an internal curing effect whereby water from the aggregate was gradually released into the concrete to further hydrate the paste. Their results are presented in Fig. 4.75. Leaching results indicated very low amount of dissolution of toxic element into the leaching solution.

Nataraja and Nalanda (2008) reported that quarry dust generated from granite mining could be used as complete replacement of sand in the preparation of controlled low strength materials (CLSM). Bouzalakos et al. (2008) reported the use of a waste generated from bioleaching of pyrite ore (OMW) and a jarosite residue (JR) generated from zinc production as 10 % by weight substitution of silica sand in the preparation of CLSM. Their results indicated that OMW can be used as CLSM and met the strength requirements. Excessive leaching of arsenic, copper and chromium in a follow-up study, however, negate the use of this material in such purposes and therefore further study is necessary before proper application (Chan et al. 2009). On the other hand, CLSM with required strength from JR can only be prepared by mixing JR with a binder with an equal amount of Portland cement and lime mixture. The acidic nature of JR as well as its high content of Pb and Zn in JR was possibly the reason for this behaviour. Their results are presented in Fig. 4.76.

Kinuthia et al. (2009) prepared a well-graded aggregate from the colliery spoil (minestone), a byproduct of coal mining, by mixing its coarse and fine fraction. This material has several drawbacks such as excessive wear, expansive behaviour, leaching of toxic elements and radioactivity. The use of this material as fine and coarse aggregates in concrete significantly increased the w/c value and therefore reduced the resulting compressive strength. Adding an admixture could reduce water requirement, and therefore improve the compressive strength of the resulting concrete. Despite the several drawbacks, the compressive strength of the resulting concrete indicated its possibility to be used in the production of low and medium category concrete usable for blinding concrete and for use in bound granular fill or foundations.

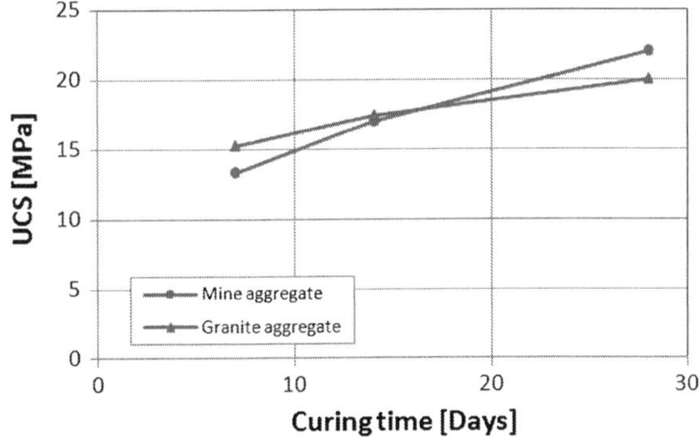

Fig. 4.75 Compressive strength of concrete with coarse aggregates of iron-ore mining waste and granite (Yellishetty et al. 2008)

Negm and Abouzeid (2008) reported that coarse solid phosphate mill tailings could be used as coarse aggregates to prepare concrete with 240 kg/cm^2, to be used in construction of small buildings.

Madany et al. (1991) reported the use of sand blasting grit waste (copper slag) as replacement of sand in the preparation of concrete blocks. The compressive strength of the concrete blocks with grit waste was 12 N/mm^2 and higher than the Bahrain specification for precast concrete blocks.

4.9.9 Waste from Metal Processing

Shinzato and Hypolito (2005) reported the use of a waste, called non-metallic product (NMP) generated during recovery of Al from aluminium black dross and salt cake by tertiary aluminium industry as aggregate in non-structural cement mortar block production. The NMP mainly contains silica, alumina and spinel along with alkali chlorides. The produced block met the requirement of size, water absorption and humidity prescribed in Brazilian standard, NBR 7173/1982. However, strength was slightly lower than the standard limit (2 MPa). However, according to the authors, the mortars were prepared using cement to aggregate ratio of 1:6 instead of the prescribed ratio of 1:3. According to the authors, NMP could be used in some construction applications after removing undesirable constituents such as chlorides and alkalis, which can reduce the usable amount.

Ismail and Al-Hashmi (2008b) reported the use of iron waste generated from an industrial workshop due to ironsmith processes. The material with maximum size of 4.75 mm and fineness modulus of 2.65 was used to replace 10, 15 and 20 % of sand in the preparation of concrete. The slump of concrete decreased 3.3, 4 and 8 % with respect to the reference concrete when 10, 15 and 20 % of sand, respectively, was replaced by iron waste aggregates, due to the heterogeneity and

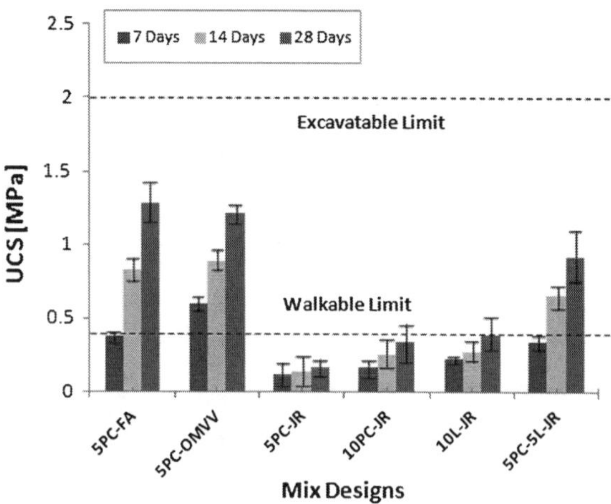

Fig. 4.76 Compressive strength of CLSM (Bouzalakos et al. 2008)

angular shape of the waste iron particles. On the other hand, fresh and dry densities of concrete increased with the incorporation of iron waste aggregates, due to their higher specific gravity by comparison with that of natural sand. The compressive and flexural strengths of concrete with various contents of iron waste aggregate were also higher than those of conventional concrete at all curing period up to 28 days (except the 14-day compressive strength) due to the higher specific gravity and strength of iron aggregate than those of natural sand (Fig. 4.77). However, after 14 days, the compressive strength of all compositions with waste iron aggregates were lower than that of the control mix, which may be due to the formation of a layer of iron compounds on the hydrating cement particles.

4.9.10 Marble Waste

Several reports are available on the use of waste generated from marble waste stored in marble quarries or sizing industries. This waste can be used either as filler material in cement or fine aggregates during preparation of cement mortar and concrete (Rai et al. 2011; Cornaldesi et al. 2010; Topcu et al. 2009; Alyamaç and Ince 2009).

Rai et al. (2011) uses marble granules with fineness modulus of 2.72 as 0–20 % by weight replacement of fine aggregates in cement mortar and concrete. The compressive strength of cement mortar with marble granule up to a 15 % replacement level was higher than that of the reference mortar; however, at 20 % level the compressive strength was lower than that of the reference mortar. The slump of concrete slightly increased with the replacement ratio of fine NA by marble aggregates. The compressive strength and flexural strengths of concrete with marble waste as partial replacement of fine NA increased with the content of marble aggregates and peaked at the 15 % replacement level. The mean compressive strength of concrete with marble aggregates was about 5–10 % higher than that of the reference concrete. Cornaldesi et al. (2010) observed a compressive strength of cement mortar with very fine marble powder as 10 % by weight replacement of sand about 10 % lower than that of the reference mortar. However,

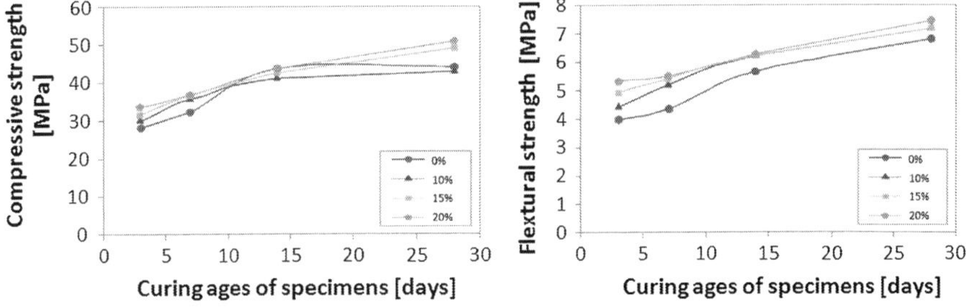

Fig. 4.77 Strength behaviour of concrete with iron waste fine aggregates

they observed similar compressive strength for concrete with marble aggregates and reference concrete when the former mix was prepared with a superplasticizer.

4.9.11 Waste from Wood Sawing

The waste generated from sawing of wood can be used as lightweight aggregates in the preparation of special types of concrete composites. Sales et al. (2010) prepared lightweight composite aggregates by mixing sawdust and water treatment sludge. The composite concrete prepared by using this composite aggregate had a lower thermal conductivity by about 23 % than conventional concrete. The authors proposed to use this concrete as non-structural sealing elements. Al Rim et al. (1999) produced a clay–cement–wood composite using wood pieces with maximum size of 22 mm as aggregates. The incorporation of wood aggregate into clayey concrete improved the thermal insulation and deformability behaviour but reduced compressive strength. Several treatments methods were proposed by Ledhem et al. (2000) to treat wood aggregate to improve the compressive strength of wood with clayey concrete.

Becchio et al. (2009) reported that the use of waste generated from wood-sawing industry as aggregates in concrete reduced the density and compressive strength but significantly improved the thermal insulation of mineralised wood concrete; furthermore, the compressive strength of this type of concrete was significantly higher than that of commercial wood concrete. Turgut (2007) prepared a composite brick using Portland cement, limestone powder waste and wood sawdust waste. The wood sawdust waste with different size fractions was used as aggregates in this composite. The compressive strength, flexural strength, density, ultrasonic pulse velocity and water absorption of the composite satisfied the relevant international standards. The incorporation of high amounts of wood aggregates with limestone powder increased the ductility of the resulting composite brick. According to the author, the product could be used for several purposes such as walls, wooden board substitute, an alternative to concrete blocks, ceiling panels, sound barrier panels, and absorption materials.

4.10 Concluding Remarks

Several types of waste material generated from various industries are currently used as aggregates in the preparation of concrete and cement mortar. The use of waste materials in concrete as aggregates significantly changes the properties of concrete.

The incorporation of ground plastics and shredded rubber can enhance some specific properties of concrete such as toughness and post-cracking behaviour, damping behaviour, thermophysical behaviour, which have several technical applications. The incorporation of rubber and plastic wastes and CBA, waste from wood sawing and some ceramic waste as aggregates can reduce the density of concrete and therefore these wastes can be used to produce lightweight concrete. On the other hand, several industrial wastes such as the majority of non-ferrous slags and some ferrous slags that have significantly higher bulk density than that of NA can be used as aggregates in the production of heavyweight concrete. CBA and some ceramic wastes can absorb high amounts of water and therefore they are used as aggregates for internal curing of concrete. FA generated from coal fired powered plant can beneficially be used as fine aggregates in concrete and cement mortar preparation, which can increase the consumption volume of these materials in the construction sector.

The incorporation of rubber and plastic wastes as well as some ceramic aggregates deteriorates the mechanical properties of the resulting concrete. On the other hand, the addition of some ferrous slags causes the expansion of concrete due to the presence of deleterious constituents in these slags, like free lime and magnesia. Therefore, treatments of these wastes before application or some innovation during the production of concrete and cement mortar are occasionally necessary to overcome these problems. The presence of toxic elements in some industrial wastes such as coal ash, some mining industries wastes and the majority of slags will increase the concentrations of these elements in cement mortar and concrete with these waste aggregates. Therefore, the fate and leachability of toxic elements from waste with cement mortar and concrete are important aspects that need to be considered for proper use of these wastes in constructions.

Some other solid waste materials generated from various industries such as mining, shoe, tobacco production, food and metal processing, pulp and paper mills, marble processing and waste sludge generated from processes such as contaminated water treatment, petroleum exploration are also reported as aggregates in various types of concrete production. The use of industrial wastes as aggregates in concrete has the potential to consume a vast amount of waste materials since aggregates are the major constituent of concrete and cement mortar. The use of waste materials in construction can solve most of the problems associated with their disposal as well as save natural resources related to aggregates mining. Therefore, the production of cheaper and more durable concrete using these waste aggregates can solve to some extent the ecological and environmental problems. However, the lack of widespread reliable data on the use of the majority of waste materials as aggregates in cement mortar and concrete can hinder their use in construction industry. Therefore, more research is required to design consistent and durable concrete with such waste aggregates.

References

Aggarwal P, Aggarwal Y, Gupta SM (2007) Effect of bottom ash as replacement of fine aggregates in concrete. Asian J Civil Eng (Build Hous) 8(1):49–62

Agostini F, Skoczylas F, Lafhaj Z (2007) About a possible valorisation in cementitious materials of polluted sediments after treatment. Cement Concr Compos 29(4):270–278

Ahmadi B, Al-Khaja W (2001) Utilization of paper waste sludge in the building construction industry. Resour Conserv Recycl 32(2):105–113

Aiello MA, Leuzzi F (2010) Waste tyre rubberized concrete: properties at fresh and hardened state. Waste Manage (Oxf) 30(8–9):1696–1704

Akcaozoglu S, Atis CD, Akcaozoglu K (2010) An investigation on the use of shredded waste PET bottles as aggregate in lightweight concrete. Waste Manage (Oxf) 32(2):285–290

Al Rim K, Ledhem A, Douzane O, Dheilly RM, Queneudec M (1999) Influence of the proportion of wood on the thermal and mechanical performances of clay-cement-wood composites. Cement Concr Compos 21(4):269–276

Albano C, Camacho N, Hernandez M, Matheus A, Gutierrez A (2009) Influence of content and particle size of waste pet bottles on concrete behaviour at different w/c ratios. Waste Manage (Oxf) 29(10):2707–2716

Ali AH, Goulias DG (1998) Evaluation of rubber-filled concrete and correlation between destructive and nondestructive testing results. Cement Concr Aggreg 20(1):140–144

Ali NA, Amos AD, Roberts M (1993) Use of ground rubber tires in portland cement concrete. Concrete 2000, University of Dundee, Scotland, UK, pp 379–390

Al-Jabri KS, Hisada M, Al-Saidy AH, Al-Oraimi SK (2009a) Performance of high strength concrete made with copper slag as a fine aggregate. Constr Build Mater 23(6):2132–2140

Al-Jabri KS, Hisada M, Al-Oraimi SK, Al-Saidy AH (2009b) Copper slag as sand replacement for high performance concrete. Cement Concr Compos 31(7):483–488

Al-Jabri KS, Al-Saidy AH, Taha R (2011) Effect of copper slag as a fine aggregate on the properties of cement mortars and concrete. Constr Build Mater 25(2):933–938

Al-Manaseer AA, Dalal TR (1997) Concrete with plastic aggregates. Concr Int 19(8):47–52

Almusallam AA, Beshr H, Maslehuddin M, Al-Amoudi OSB (2004) Effect of silica fume on the mechanical properties of low quality coarse aggregate concrete. Cement Concr Compos 26(7):891–900

Al-Negheimish AI, Al-Sugair FH, Al-Zaid RZ (1997) Utilization of local steel making slag in concrete. J King Saud Univ Eng Sci 9(1):39–55

Al-Otaibi S (2008) Recycling steel mill scale as fine aggregate in cement mortars. Eur J Sci Res 24(3):332–338

Alyamaç KE, Ince R (2009) A preliminary concrete mix design for SCC with marble powders. Constr Build Mater 23(3):1201–1210

Andrade LB, Rocha JC, Cheriaf M (2007) Evaluation of concrete incorporating bottom ash as a natural aggregates replacement. Waste Manage (Oxf) 27(9):1190–1199

Andrade LB, Rocha JC, Cheriaf M (2009) Influence of coal bottom ash as fine aggregate on fresh properties of concrete. Constr Build Mater 23(2):609–614

Ashby J (1996) Air cooled blast furnace slag as a concrete aggregate for engineering construction. National symposium on the use of recycled materials in engineering construction, Sydney, Australia, pp 110–115

Aspiras FF, Manalo JRI (1995) Utilization of textile waste cuttings as building material. J Mater Process Technol 48(1):379–384

Atzeni C, Massidda L, Sanna U (1996) Use of granulated slag from lead and zinc processing in concrete technology. Cem Concr Res 26(9):1381–1388

Ayano T, Sakata K (2000) Durability of concrete with copper slag fine aggregate. Fifth CANMET/ACI international conference on durability of concrete, SP-192, American Concrete Institute, Farmington Hills, USA, pp 141–158

Azmi NJ, Mohammed BS, Al-Mattarneh HMA (2008) Engineering properties of concrete with recycled tire rubber. International conference on construction and building technology, ICCBT 2008, Kuala Lumpur, Malaysia, vol B(34), pp 373–382

Baffa IPD, Akasaki J (2005) Light-concrete with leather: durability aspects. http://www.ppgec.feis.unesp.br/producao2005/Lightconcrete%20with%20leather%20durability%20aspects.pdf. Accessed 25 Oct 2011

Bai Y, Darcy F, Basheer PAM (2005) Strength and drying shrinkage properties of concrete with furnace bottom ash as fine aggregate. Constr Build Mater 19(9):691–697

Basri HB, Mannan MA, Zain MFM (1999) Concrete using waste oil palm shells as aggregate. Cem Concr Res 29(4):619–622

Batayneh M, Marie I, Ibrahim A (2007) Use of selected waste materials in concrete mixes. Waste Manage (Oxf) 27(12):1870–1876

Batayneh MK, Marie I, Asi I (2008) Promoting the use of crumb rubber concrete in developing countries. Waste Manage (Oxf) 28(11):2171–2176

Becchio C, Corgnati SP, Kindinis A, Pagliolico S (2009) Improving environmental sustainability of concrete products: Investigation on MWC thermal and mechanical properties. Energy Build 41(11):1127–1134

Benazzouk A, Mezreb K, Doyen G, Goullieux A, Queneudec M (2003) Effect of rubber aggregates on the physico-mechanical behaviour of cement–rubber composites-influence of the alveolar texture of rubber aggregates. Cement Concr Compos 25(7):711–720

Benazzouk A, Douzane O, Mezreb K, Queneudec M (2006) Physico-mechanical properties of aerated cement composites with shredded rubber waste. Cement Concr Compos 28(7):650–657

Benazzouk A, Douzane O, Langlet T, Mezreb K, Roucoult JM, Queneudec M (2007) Physico-mechanical properties and water absorption of cement composite with shredded rubber waste. Cement Concr Compos 29(10):732–740

Beshr H, Almusallam AA, Maslehuddin M (2003) Effect of coarse aggregate quality on the mechanical properties of high strength concrete. Constr Build Mater 17(2):97–103

Bignossi MC, Sandrolini F (2006) Tyre rubber waste recycling in self-compacting concrete. Cem Concr Res 36(4):735–739

Binici H (2007) Effect of crushed ceramic and basaltic pumice as fine aggregates on concrete mortars properties. Constr Build Mater 21(6):1191–1197

Boheme L, Van Den Hende D (2011) Ferromolybdenum slag as valuable resource material for the production of concrete blocks. 2nd international slag valorization symposium. The transition to sustainable material management, Leuven, Belgium, pp 129–143

Bouzalakos S, Dudeney AWL, Cheeseman CR (2008) Controlled low-strength materials with waste precipitates from mineral processing. Miner Eng 21(4):252–263

Brindha D, Nagan S (2011) Durability studies on copper slag admixed concrete. Asian J Civil Eng (Build Hous) 12(5):563–578

Brindha D, Baskaran T, Nagan S (2010) Assessment of corrosion and durability characteristics of copper slag admixed concrete. Int J Civil Eng Struct Eng 1(2):192–211

Brito J, Pereira AS, Correia JR (2005) Mechanical behaviour of non-structural concrete made with recycled ceramic aggregates. Cement Concr Compos 27(4):429–433

Cachim PB (2009) Mechanical properties of brick aggregate concrete. Constr Build Mater 23(3):1292–1297

Caliskan S, Behnood A (2004) Recycling copper slag as coarse aggregate: hardened properties of concrete. 7th international conference on concrete technology in developing countries, Kuala Lumpur, Malaysia, pp 91–98

Chan BKC, Bouzalakos S, Dudeney AWL (2009) Cemented products with waste from mineral processing and bioleaching. Miner Eng 22(15):1326–1333

Choi YW, Moon DJ, Chung JS, Cho SK (2005) Effects of waste PET bottles aggregate on the properties of concrete. Cem Concr Res 35(4):776–781

Choi YW, Moon DJ, Kim YJ, Lachemi M (2009) Characteristics of mortar and concrete with fine aggregate manufactured from recycled waste polyethylene terephthalate bottles. Constr Build Mater 23(8):2829–2835

Collins F, Sanjayan JG (1999) Strength and shrinkage properties of alkali-activated slag concrete with porous coarse aggregate. Cem Concr Res 29(4):607–610

Cornaldesi V, Moriconi G, Naik TR (2010) Characterization of marble powder for its use in mortar and concrete. Constr Build Mater 24(1):113–117

Correia J, de Brito J, Pereira AS (2006) Effects on concrete durability of using recycled ceramic aggregates. Mater Struct 39(2):151–158

Cyr M, Ludmann C (2006) Low risk meat and bone meal (MBM) bottom ash in mortars as sand replacement. Cem Concr Res 36(3):469–480

Debeib F, Kenai S (2008) The use of coarse and fine crushed bricks as aggregate in concrete. Constr Build Mater 22(5):886–893

Demirboga R, Gul R (2006) Production of high strength concrete by use of industrial by-products. Build Environ 41(8):1124–1127

Dhir RK, McCarthy MJ, Tittle PAJ (2000) Use of conditioned PFA as a fine aggregate component in concrete. Mater Struct 33(1):38–42

Eldin NN, Senouci AB (1993) Rubber-tire particles as concrete aggregate. J Mater Civ Eng 5(4):478–496

Emiroglu M, Kelestemur MH, Yildiz S (2007) An investigation on ITZ microstructure of the concrete with waste vehicle tire. 8th international fracture conference, Istanbul, Turkey, pp 453–459

Etxeberria M, Pacheco C, Meneses JM, Berridi I (2010) Properties of concrete using metallurgical industrial by-products as aggregates. Constr Build Mater 24(9):1594–1600

Fattuhi NI, Clark LA (1996) Cement-based materials containing shredded scrap truck tyre rubber. Const Build Mater 10(4):229–236

Figueiredo J, Mavroulidou M (2007) Reducing tyre waste by using discarded tyre rubber as concrete aggregate. 10th international conference on environment science and technology, Kos Island, Greece, vol A, pp 379–380

Fraj AB, Kismi M, Mounanga P (2010) Valorization of coarse rigid polyurethane foam waste in lightweight aggregate concrete. Constr Build Mater 24(6):1069–1077

Frigione M (2010) Recycling of PET bottles as fine aggregate in concrete. Waste Manage (Oxf) 30(6):1101–1106

Gallardo RS, Adajar MAQ (2006) Structural performance of concrete with paper sludge as fine aggregates partial replacement enhanced with admixtures. Symposium on infrastructure development and the environment 2006, SEAMEO-INNOTECH, University of the Philippines, Diliman, Quezon City, Philippines, pp 1–10

Ganjian E, Khorami M, Maghsoudi AA (2009) Scrap-tyre-rubber replacement for aggregate and filler in concrete. Constr Build Mater 23(5):1828–1836

Gesoglu M, Guneyisi E (2007) Strength development and chloride penetration in rubberized concrete with and without rubberized silica fume. Mater Struct 40(9):953–964

Gesoglu M, Guneyisi E (2011) Permeability properties of self-compacting rubberized concrete. Constr Build Mater 25(8):3319–3326

Ghafoori N, Bucholc J (1996) Investigation of lignite-based bottom ash for structural concrete. J Mater Civ Eng 8(3):128–137

Guerra I, Vivar I, Llamas B, Juan A, Moran J (2009) Eco-efficient concretes: the effects of using recycled ceramic material from sanitary installations on the mechanical properties of concrete. Waste Manage (Oxf) 29(2):643–646

Guneyisi E, Gesoglu M, Ozturan T (2004) Properties of rubberized concretes with silica fume. Cem Concr Res 34(12):2309–2317

Hannawi K, Kamali-Bernard S, Prince W (2010) Physical and mechanical properties of mortars with PET and PC waste aggregates. Waste Manage (Oxf) 30(11):2312–2320

Haque MN, Kayyali OA, Joynes BM (1995) Blast furnace slag aggregate in the production of high-performance concrete, vol 153. ACI special publication SP 153-48, Farmington Hills, pp 911–930

Ho AC, Turatsinze A, Vu DC (2009) On the potential of rubber aggregates obtained by grinding end-of-life tyres to improve the strain capacity of concrete. Concrete repair, rehabilitation and retrofitting II. Taylor & Francis Group, London, pp 123–129. ISBN 978-0-415-46850-3

Hwang CL, Laiw JC (1989) Properties of concrete using copper slag as a substitute for fine aggregate. 3rd international conference on fly ash, silica fume, slag, and natural pozzolans in concrete, ACI special publication SP-114-82, Farmington Hills, USA, pp 1677–1695

Hwang KR, Noguchi T, Tomosawa F (1998) Effects of fine aggregate replacement on the rheology, compressive strength and carbonation properties of fly ash and mortar, vol 178. ACI special publication SP 178–22, Farmington Hills, pp 401–410

Ismail ZZ, Al-Hashmi EA (2008a) Use of plastic waste in concrete mixture as aggregate replacement. Waste Manage (Oxf) 28(11):2041–2047

Ismail ZZ, Al-Hashmi EA (2008b) Reuse of waste iron as a partial replacement of sand in concrete. Waste Manage (Oxf) 28(11):2048–2053

Jumaat MZ, Alengaram UJ, Mahmud H (2009) Shear strength of oil palm shell foamed concrete beams. Mater Des 30(6):2227–2236

Kan A, Demirboga R (2009) A novel material for lightweight concrete production. Cement Concr Compos 31(7):489–495

Kang J, Han C, Zhang Z (2009) Strength and shrinkage behaviors of roller-compacted concrete with rubber additives. Mater Struct 42(8):1117–1124

Kasemsaisiri R, Tantermsirikul S (2008) Properties of self-compacting concrete incorporating bottom ash as a partial replacement of fine aggregate. Sci Asia 34(1):87–95

Khaloo AR, Dehestani M, Rahmatabadi P (2008) Mechanical properties of concrete with a high volume of tire–rubber particles. Waste Manage (Oxf) 28(12):2472–2482

Khanzadi M, Behnood A (2009) Mechanical properties of high-strength concrete incorporating copper slag as coarse aggregate. Constr Build Mater 23(6):2183–2188

Khatib ZK, Bayomy FM (1999) Rubberized Portland cement concrete. J Mater Civ Eng 11(3):206–213

Kim HK, Lee HK (2011) Use of power plant bottom ash as fine and coarse aggregates in high-strength concrete. Constr Build Mater 25(2):1115–1122

Kinuthia J, Snelson D, Gailius A (2009) Sustainable medium-strength concrete (CS-concrete) from colliery spoil in South Wales UK. J Civil Eng Manag 15(2):149–157

Kou SC, Poon CS (2009) Properties of concrete prepared with crushed fine stone, furnace bottom ash and fine recycled aggregate as fine aggregates. Constr Build Mater 23(8):2877–2886

Kou SC, Lee G, Poon CS, Lai WL (2009) Properties of lightweight aggregate concrete prepared with PVC granules derived from scraped PVC pipes. Waste Manage (Oxf) 29(2):621–628

Kumaran GS, Mushule N, Lakshmipathy M (2008) A review on construction technologies that enables environmental protection: rubberized concrete. Am J Eng Appl Sci 1(1):40–44

Kuo WY, Huang JS, Tan TE (2007) Organo-modified reservoir sludge as fine aggregates in cement mortars. Constr Build Mater 21(3):609–615

Kurama H, Topcu IB, Karakurt C (2009) Properties of the autoclaved aerated concrete produced from coal bottom ash. J Mater Process Technol 209(2):767–773

Laukaitis A, Zurauskas R, Keriene J (2005) The effect of foam polystyrene granules on cement composite properties. Cement Concr Compos 27(1):41–47

Ledhem A, Dheilly RM, Queneudec M (2000) Reuse of waste oils in the treatment of wood aggregates. Waste Manage (Oxf) 20(4):321–326

Lee MH, Lee JC (2009) Study on the cause of pop-out defects on the concrete wall and repair method. Constr Build Mater 23(1):482–490

Lee HK, Kim HK, Hwang EA (2010) Utilization of power plant bottom ash as aggregates in fibre-reinforced cellular concrete. Waste Manage (Oxf) 30(2):274–284

Li F (1999) Test research on copper slag concrete. J Fuzhou Univ 127(5):59–62

Li Z, Li F, Li JSL (1998) Properties of concrete incorporating rubber tyre particles. Mag Concr Res 50(4):297–304

Li G, Stubblefield MA, Garrick G, Eggers J, Abadie C, Huang B (2004) Development of waste tire modified concrete. Cem Concr Res 34(2):2283–2289

Ling TC (2011) Prediction of density and compressive strength for rubberized concrete blocks. Constr Build Mater 25(11):4303–4306

Ling TC, Hasanan MN, Hainin MR, Chik AA (2009) Laboratory performance of crumb rubber concrete block pavement. Int J Pavement Eng 10(5):361–374

Lopez V, Llamas B, Juan A, Moran JM, Guerra I (2007) Eco-efficient concretes: impact of the use of white ceramic powder on the mechanical properties of concrete. Biosyst Eng 96(4):559–564

Madany IM, Al-Sayed MH, Raveendran I (1991) Utilization of copper blasting grit waste as a construction material. Waste Manage (Oxf) 11(1):35–40

Mannan MA, Ganapathy C (2001a) Long-term strengths of concrete with oil palm shell as coarse aggregate. Cem Concr Res 31(9):1319–1321

Mannan MA, Ganapathy C (2001b) Mix design for oil palm shell concrete. Cem Concr Res 31(9):1323–1325

Mannan MA, Ganapathy C (2002) Engineering properties of concrete with oil palm shell as coarse aggregate. Constr Build Mater 16(1):29–34

Mannan MA, Ganapathy C (2004) Concrete from an agricultural waste-oil palm shell (OPS). Build Environ 39(4):441–448

Mannan MA, Alexander J, Ganapathy C, Teo DCL (2006) Quality improvement of oil palm shell (OPS) as coarse aggregate in lightweight concrete. Build Environ 41(9):1239–1242

Manso JM, Gonzalez JJ, Polanco JA (2004) Electric arc furnace slag in concrete. J Mater Civ Eng 16(6):639–645

Manso JM, Polanco JA, Losanez M, Gonzalez JJ (2006) Durability of concrete made with EAF slag as aggregate. Cement Concr Compos 28(6):528–534

Marzouk OY, Dheilly RM, Queneudec M (2007) Valorisation of post-consumer plastic waste in cementitious concrete composites. Waste Manage (Oxf) 27(2):310–318

Maslehuddin M, Al-Mana AI, Shamim M, Saricimen H (1989) Effect of sand replacement on the early-age strength gain and long-term corrosion-resisting characteristics of fly ash concrete. ACI Mater J 86(1):56–62

Maslehuddin M, Sharif AM, Shameem M, Ibrahim M, Barry MS (2003) Comparison of properties of steel slag and crushed limestone aggregate concretes. Constr Build Mater 17(2):105–112

Mavroulidou M, Figueiredo J (2010) Discarded tyre rubber as concrete aggregate: a possible outlet for used tyres. Global Nest J 12(4):359–367

Metwally MEA, Seleem MH, Balaha MM, H Abd El-Rahman (2005) Utilizing of slag produced from recycling of spent lead-batteries as concrete aggregate. Alex Eng J 44(6):883–892

Monosi S, Giretti P, Moriconi G, Favoni O, Collepardi M (2001) Non-ferrous slag as cementitious material and fine aggregate for concrete, vol 202. ACI special publication SP-202-03, Farmington Hills, pp 33–44

Morrison C, Richardson D (2004) Re-use of zinc smelting furnace slag in concrete. Eng Sustain 157(4):213–218

Morrison C, Hooper R, Lardner K (2003) The use of ferro-silicate slag from ISF zinc production as a sand replacement in concrete. Cem Concr Res 33(12):2085–2089

Mounanga P, Gbongbon W, Poullain P, Turcry P (2008) Proportioning and characterization of lightweight concrete mixtures made with rigid polyurethane foam wastes. Cement Concr Compos 30(9):806–814

Naik TR (2002) Greener concrete using recycled materials. Concr Int 24(8):49–53

Naik TR, Friberg TS, Chun YM (2004) Use of pulp and paper mill residual solids in production of cellucrete. Cem Concr Res 34(7):1229–1234

Naik TR, Singh SS (1991) Utilization of discarded tires as construction materials for transportation facilities. Report no. CBU-1991-02, UWM Center for by-products utilization, University of Wisconsin-Milwaukee, Milwaukee, USA, 16 p

Nasvik J (1991) Plastic aggregate offers colorful alternative to mineral aggregate: color flexibility is offered in a material made from recycled plastic in some colors. The Aberdeen Group, Publication #J910701a, Aberdeen

Nataraja MC, Nalanda Y (2008) Performance of industrial by-products in controlled low-strength materials (CLSM). Waste Manage (Oxf) 28(7):1168–1181

Nataraja MC, Nagaraj TS, Bhavanishankar S, Ramalinga Reddy BM (2007) Proportioning cement based composites with burnt coal cinder. Mater Struct 40(6):543–552

Nayef A, Fahad A, Ahmed B (2010) Effect of microsilica addition on compressive strength of rubberized concrete at elevated temperatures. J Mater Cycles Waste Manage 12(1):41–49

Negm AA, Abouzeid AZM (2008) Utilization of solid waste from phosphate processing plants. Physicochem Probl Min Process 42(1):5–16

Ozturk T, Bayraklı M (2005) The possibilities of using tobacco waste in producing lightweight concrete. Agricultural Engineering International: the CIGR Ejournal, vol VII, Manuscript BC 05 006

Pacheco-Torgal F, Jalali S (2010) Reusing ceramic wastes in concrete. Constr Build Mater 24(5):832–838

Pang X, Liu C, Suri R (2007) Recycling spent mushroom substrate as fine aggregates in concrete. TRB 86th annual meeting compendium of papers CD-ROM, Transportation Research Board 86th annual meeting, Transportation Research Board, Washington, USA, paper #07-1726

Panyakapo P, Panyakapo M (2008) Reuse of thermosetting plastic waste for lightweight concrete. Waste Manage (Oxf) 28(9):1581–1588

Papadakis VG (1999) Effect of fly ash on Portland cement systems. Part I: Low-calcium fly ash. Cement Concr Res 29(11):1727–1736

Papadakis VG (2000) Effect of fly ash on Portland cement systems. Part II: High-calcium fly ash. Cement Concr Res 30(10):1647–1654

Papayianni I, Anastasiou F (2010) Production of high-strength concrete using high volume of industrial by-products. Constr Build Mater 24(8):1412–1417

Park SB, Jang YI, Lee J, Lee BJ (2009) An experimental study on the hazard assessment and mechanical properties of porous concrete utilizing coal bottom ash coarse aggregate in Korea. J Hazard Mater 166(1):348–355

Pazhani K, Jeyaraj R (2010) Study on durability of high performance concrete with industrial wastes. Appl Technol Innov 2(2):19–28

Pelisser F, Zavarise N, Longo TA, Bernardin AM (2011) Concrete made with recycled tire rubber: effect of alkaline activation and silica fume addition. J Clean Prod 19(6–7):757–763

Pellegrino C, Gaddo V (2009) Mechanical and durability characteristics of concrete with EAF slag as aggregate. Cement Concr Compos 31(9):663–671

Penpolcharoen M (2005) Utilization of secondary lead slag as construction material. Cem Concr Res 35(6):1050–1055

Pereira DA, De Aguiar B, Castro F, Almeida MF, Labrincha JA (2000) Mechanical behaviour of Portland cement mortars with incorporation of Al-containing salt slags. Cem Concr Res 30(7):1131–1138

Pierce CE, Blackwell MC (2003) Potential of scrap-tire rubber as lightweight aggregate in flowable fill. Waste Manage (Oxf) 23(3):197–208

Pofale AD, Deo SV (2010) Comparative long term study of concrete mix design procedure for fine aggregate replacement with fly ash by minimum voids method and maximum density method. KSCE J Civil Eng 14(5):759–764

Polanco JA, Manso JM, Setien J, Gonzalez JJ (2011) Strength and durability of concrete made with electric steelmaking slag. Mater J 108(2):196–203

Qasrawi H, Shalabi F, Asi I (2009) Use of low CaO unprocessed steel slag in concrete as fine aggregate. Constr Build Mater 23(2):1118–1125

Raghvan D, Huynh H, Ferraris CF (1998) Workability, mechanical properties and chemical stability of a recycled tire rubber-filled cementitious composite. J Mater Sci 33(7):1745–1752

Rai B, Khan NH, Abhishek K, Tabin RS, Duggal SK (2011) Influence of marble powder/granules in concrete mix. Int J Civil Struct Eng 1(4):827–834

Rao SM, Venkatarama Reddy BV, Lakshmikanth S, Ambika NS (2009) Re-use of fluoride contaminated bone char sludge in concrete. J Hazard Mater 166(2–3):751–756

Ravina D (1997) Properties of fresh concrete incorporating a high volume of fly ash as partial fine sand replacement. Mater Struct 30(8):473–479

Remadnia A, Dheilly RM, Laidoudi B, Quéneudec M (2009) Use of animal proteins as foaming agent in cementitious concrete composites manufactured with recycled PET aggregates. Constr Build Mater 23(10):3118–3123

Saikia N, Brito J (2010) Mechanical performance of structural concrete containing recycled polyethylene terephthalate (PET) as a partial substitution of natural aggregate. Technical Report, Department of Civil Engineering and Architecture, ICIST, Instituto Superior Técnico (IST), Technical University of Lisbon, ISSN:0871-7869, DTC No. 09/2010

Saikia N, Cornelis G, Mertens G, Elsen J, Van Balen K, Van Gerven T, Vandecasteele C (2008) Assessment of Pb-slag, MSWI bottom ash and boiler and fly ash for using as a fine aggregate in cement mortar. J Hazard Mater 144(1–3):766–777

Saikia N, Cornelis G, Ozlem C, Elsen J, Van Gemert D, Van Balen K, Vandecasteele C, Van Gerven T (2012) Utilization of Pb-slag as a partial substitution of fine aggregates in cement mortar. J Mater Cycles Waste Manag 14(2):102–112

Sales A, De Souza FR (2009) Concretes and mortars recycled with water treatment sludge and construction and demolition rubble. Constr Build Mater 23(6):2362–2370

Sales A, De Souza FR, Santos WN, Zimer AM, Almeida FCR (2010) Lightweight composite concrete produced with water treatment sludge and sawdust: thermal properties and potential application. Constr Build Mater 24(12):2446–2453

Santiago EQR, Lima PRL, Leite MB, Filho RDT (2009) Mechanical behavior of recycled lightweight concrete using EVA waste and CDW under moderate temperature. Ibracon Struct Mater J 2(3):211–221

Segre N, Joekes I (2000) Use of tire rubber particles as addition to cement paste. Cem Concr Res 30(9):1421–1425

Segre N, Galves AD, Rodrigues JA, Monteiro PJM, Joekes I (2003) Use of tyre rubber particles in slag-modified cement mortars. 11th International Congress on the Chemistry of Cement (ICCC): cements contribution to the development in the 21st century, Durban, South Africa, pp 1546–1554

Senthamarai RM, Devadas MP (2005) Concrete with ceramic waste aggregate. Cement Concr Compos 27(9–10):910–913

Senthamarai RM, Devadas MP, Gobinath D (2011) Concrete with ceramic waste aggregate. Constr Build Mater 25(5):2413–2419

Seo T, Lee M, Choi C, Ohno Y (2010) Properties of drying shrinkage cracking of concrete with fly ash as partial replacement of fine aggregate. Mag Concr Res 62(6):427–433

Shinzato MC, Hypolito R (2005) Solid waste from aluminum recycling process: characterization and reuse of its economically valuable constituents. Waste Manage (Oxf) 25(1):37–46

Shoya M, Nagataki S, Tomosawa F, Sugita S, Tsukinaga Y (1997) Freezing and thawing resistance of concrete with excessive bleeding and its improvement. Fourth CANMET/ACI international conference on durability of concrete, ACI special publication SP-170-45, Farmington Hills, MI, pp 879–898

Siddique R (2003a) Effect of fine aggregate replacement with Class F fly ash on the mechanical properties of concrete. Cem Concr Res 33(4):539–547

Siddique R (2003b) Effect of fine aggregate replacement with Class F fly ash on the abrasion resistance of concrete. Cem Concr Res 33(11):1877–1881

Siddique R, Naik TR (2004) Properties of concrete with scrap-tire rubber—an overview. Waste Manage (Oxf) 24(6):563–569

Skripkiunas G, Grinys A, Černius B (2007) Deformation properties of concrete with rubber waste additives. Mater Sci 13(3):219–223

Son KS, Hajirasouliha I, Pilakoutas K (2011) Strength and deformability of waste tyre rubber-filled reinforced concrete columns. Constr Build Mater 25(1):218–226

Sorlini S, Collivignarelli C, Plizzari G, Foglie MD (2004) Reuse of Waelz slag as recycled aggregate for structural concrete. International RILEM conference on the use of recycled materials in building and structures, Barcelona, Spain, pp 1086–1094

Soroushian P, Plasencia JS, Ravanbakhsh S (2003) Assessment of reinforcing effects of recycled plastic and paper in concrete. ACI Mater J 100(3):203–207

Sukontasukkul P, Chaikaew C (2006) Properties of concrete pedestrian block mixed with crumb rubber. Constr Build Mater 20(7):450–457

Suzuki M, Meddah MS, Sato R (2009) Use of porous ceramic waste aggregates for internal curing of high-performance concrete. Cem Concr Res 39(5):373–381

Taha MMR, El-Dieb AS, Wahab MAA, Hameed MEA (2008) Mechanical, fracture, and microstructural investigations of rubber concrete. J Mater Civ Eng 20(10):640–649

Tang M, Wang B, Chen Y (2000) The research on super high strength, high wearability cement mortar with the incorporation of copper slag as aggregates. Concrete 4:30–32

Teo DCL, Mannan MA, Kurian JV (2006) Flexural behaviour of reinforced lightweight concrete beams made with oil palm shell (OPS). J Adv Concr Technol 4(3):459–468

Teo DCL, Mannan MA, Kurian VJ, Ganapathy C (2007) Lightweight concrete made from oil palm shell (OPS): structural bond and durability properties. Build Environ 42(7):2614–2621

Teo DCL, Mannan MA, Kurian JV (2010) Durability of lightweight OPS concrete under different curing conditions. Mater Struct 43(1):1–13

Topcu IB (1995) The properties of rubberized concrete. Cem Concr Res 25(2):304–310

Topcu IB, Canbaz M (2007) Utilization of crushed tile as aggregate in concrete. Iran J Sci Technol Trans B Eng 31(5):561–565

Topçu IB, Demir A (2007) Durability of rubberized mortar and concrete. J Mater Civ Eng 19(2):173–178

Topçu IB, Bilir T, Uygunoglu T (2009) Effect of waste marble dust content as filler on properties of self-compacting concrete. Constr Build Mater 23(5):1947–1953

Torkittikul P, Chaipanich A (2010) Utilization of ceramic waste as fine aggregate within Portland cement and fly ash concretes. Cement Concr Compos 32(6):440–449

Toutanji HA (1996) The use of rubber tyre particles in concrete to replace mineral aggregates. Cement Concr Compos 18(2):135–139

Triches G, Pinto RCA, Silva AJ (2007) A feasibility study of incorporating bottom ash roller compacted concrete pavements. AFCM concrete road symposium 2007, Selangor, Malaysia. www.cbtu.gov.br/estudos/pesquisa/anpet/PDF/3_152_AC.pdf

Turgut P (2007) Cement composites with limestone dust and different grades of wood sawdust. Build Environ 42(11):3801–3807

Turki M, Bretagne E, Rouis MJ, Quéneudec M (2009) Microstructure, physical and mechanical properties of mortar–rubber aggregates mixtures. Constr Build Mater 23(7):2715–2722

Uygunoğlu T, Topçu JB (2010) The role of scrap rubber particles on the drying shrinkage and mechanical properties of self-consolidating mortars. Constr Build Mater 24(7):1141–1150

Wu W, Zhang W, Ma G (2010) Mechanical properties of copper slag reinforced concrete under dynamic compression. Constr Build Mater 24(6):910–917

Yang EI, Yi ST, Leem YM (2005) Effect of oyster shell substituted for fine aggregate on concrete characteristics. Part I: Fundamental properties. Cement Concr Res 35(11):2175–2182

Yang EI, Kim MY, Park HG, Yi ST (2010) Effect of partial replacement of sand with dry oyster shell on the long-term performance of concrete. Constr Build Mater 24(5):758–765

Yellishetty M, Karpe V, Reddy EH, Subhash KN, Ranjith PG (2008) Reuse of iron ore mineral waste in civil engineering constructions: a case study. Resour Conserv Recycl 52(11):1283–1289

Yesilata B, Isıker Y, Turgut P (2009) Thermal insulation enhancement in concretes by adding waste PET and rubber pieces. Constr Build Mater 23(5):1878–1882

Yilmaz A, Degirmenci N (2009) Possibility of using waste tire rubber and fly ash with Portland cement as construction materials. Waste Manage (Oxf) 29(5):1541–1546

Yüksel I, Bilir T (2007) Usage of industrial by-products to produce plain concrete elements. Constr Build Mater 21(3):686–694

Yüksel I, Bilir T, Ozkan O (2007) Durability of concrete incorporating non-granulated blast furnace slag and bottom ash as fine aggregate. Build Environ 42(7):2651–2659

Yüksel I, Siddique R, Özkan O (2011) Influence of high temperature on the properties of concretes made with industrial by-products as fine aggregate replacement. Constr Build Mater 25(2):967–972

Zelic J (2005) Properties of concrete pavements prepared with ferrochromium slag as concrete aggregate. Cem Concr Res 35(12):2340–2349

Zheng L, Huo SH, Yuan Y (2008a) Strength, elasticity, and brittleness index of rubberized concrete. J Mater Civ Eng 22(11):692–699

Zheng L, Huo SH, Yuan Y (2008b) Experimental investigation on dynamic properties of rubberized concrete. Constr Build Mater 22(5):939–947

Zong L (2003) The replacement of granulated copper slag for sand concrete. J Qingdao Inst Archit Eng 24(2):20–22

Chapter 5
Use of Construction and Demolition Waste as Aggregate: Properties of Concrete

5.1 Introduction

Studies on recycling of construction and demolition waste (CDW) as aggregate have been conducted for a long time. Nowadays, this material is considered as aggregate to produce several types of concrete. To use CDW as aggregate in concrete, it is necessary to process the waste to remove impurities and to comply with size grading requirements. The production process and properties of various CDW aggregates have been presented in Chap. 3. In this chapter, the fresh and hardened properties of concrete containing various types of CDW aggregates are discussed from existing literature data.

"CDW" is used here to indicate waste generated during construction and demolition activities. In several references RA is meant to represent recycled aggregate, which contains impurities like brick, wood, ceramics, asphalt and other materials, and RCA for aggregate represents waste generated from crushing concrete or that contains very small amounts of impurities (<1 %). A similar nomenclature will also be used in this chapter.

5.2 Fresh Concrete Properties

As the properties of CDW aggregate are different from those of conventional aggregate, the use of CDW aggregate in concrete substantially changes various properties of the concrete mix. In this section, fresh properties such as workability, density and air-content will be presented from various references.

J. de Brito and N. Saikia, *Recycled Aggregate in Concrete*,
Green Energy and Technology, DOI: 10.1007/978-1-4471-4540-0_5,

5.2.1 Workability

The workability of fresh concrete is a very important property, which controls various other fresh and hardened-state properties of concrete such as density, air-content and strength. The workability of concrete depends on various properties of its constituents. The workability performance of concrete containing CDW aggregate was studied extensively since various properties of CDW aggregate, which controls the workability of concrete, do not match those of natural aggregate (NA). The workability of concrete is determined by various methods; the most versatile one of which is slump.

The slump of concrete containing any type of CDW aggregate should be lower than that of conventional concrete due to the higher water absorption capacity of CDW aggregate than that of natural aggregate. The surface texture and angularity of CDW aggregate have also considerable influence on the workability performance of concrete (Buyle-Buddin and Zaharieva 2002).

Topçu (1997) observed 75 mm of slump of concrete containing coarse RCA generated from crushed concrete, while that of the equivalent conventional concrete was 100 mm. A lower workability of concrete containing CDW aggregate than of concrete containing natural aggregate was reported in several earlier publications and therefore additional amount of water was added to control the workability (Topçu 1997; Rasheeduzzafar and Khan 1984; Mukai and Kikuchi 1978; Buck 1977; Frondistou-Yannas 1977; Malhotra 1978; Hansen and Narud 1983).

Due to the higher water absorption capacity and consequent porosity of CDW aggregate by comparison with natural aggregate, in several studies, a water pre-saturation of the CDW aggregate was adopted before adding it to the mix. However, depending on the pre-saturation technique, the slump varies greatly. Table 5.1 shows some typical literature data on the slump performance of concrete containing pre-saturated CDW aggregate along with the water absorption capacity.

Poon et al. (2009) studied the slump performance of concrete mixes containing coarse granite aggregate as well as various amounts of coarse recycled concrete aggregate (RCA) at three different moisture states: air-dried (AD), oven dried at 105 °C for 24 h and then cooled down to room temperature prior to mixing (OD), and saturated surface dried (SSD). The water and aggregate contents were adjusted to keep the design proportions similar in all concrete mixes.

Due to the higher water absorption capacity of RCA at OD and AD states, higher amounts of water need to be added to the mix containing this aggregate. The initial slump (0 min) was measured immediately after mixing and then slump values were measured in 15 min intervals. The changes in slump of the mixes containing various types of aggregates are presented in Fig. 5.1. Their results can be summarised as follows: (1) the initial slump of the mix containing OD aggregate was higher than that of the other two types of aggregate; (2) the initial slump of mix containing OD and AD aggregates increased with their contents in the mix; (3) the higher slump of the mix containing OD aggregate or with higher

Table 5.1 Slump of concrete containing pre-saturated recycled aggregate

Reference	Type of aggregate (replacement amount)	w/c value	Slump (mm)	Water absorption capacity (%)	Pre-saturation method
Sagoe-Crentsil et al. 2001	Coarse basalt	0.76	90	1.0	Pre-saturated for 10 min before mixing
	Coarse recycled concrete (100 %)	0.73	75	5.6	
		0.74	95		
		0.70[a]	80		
Yang et al. 2011	Natural aggregate		33	1.4	Pre-soaked before mixing
	Coarse recycled concrete (100 %)		24	4.2 (RCA)	
	Coarse recycled concrete (80 %) + coarse crushed clay brick (20 %)		20	10.2 (CCB)	
	Coarse recycled concrete (50 %) + coarse crushed clay brick (50 %)		10		
Vieira et al. 2011	Natural aggregate	0.43	89 ± 2.8	1.0	10 min pre-saturation in water before mixing
	Recycled coarse aggregate (20 %, v/v)	0.44	91 ± 6	6.7	
	Recycled coarse aggregate (50 %, v/v)	0.46	88 ± 7.4		
	Recycled coarse aggregate (100 %, v/v)	0.49	82 ± 4		
Etxeberria et al. 2007a	Natural aggregate	0.50	80–100	0.88	Moistened by sprinkler system and covered by plastic for 1 day
	Recycled aggregate	0.40–0.50	80–100	4.44	
Gonzalez-Fonteboa et al. 2011	Natural aggregate	0.65	120	2.25 (average)	Pre-soaked in water for 10 min before use
	Coarse recycled aggregate (20 %, v/v)	0.66[b]	110	5.01	
	Coarse recycled aggregate (20 %, v/v)	0.68[b]	110		
	Coarse recycled aggregate (20 %, v/v)	0.68[b]	110		

[a] Higher cement content than the other mixes; [b] Estimated w/c

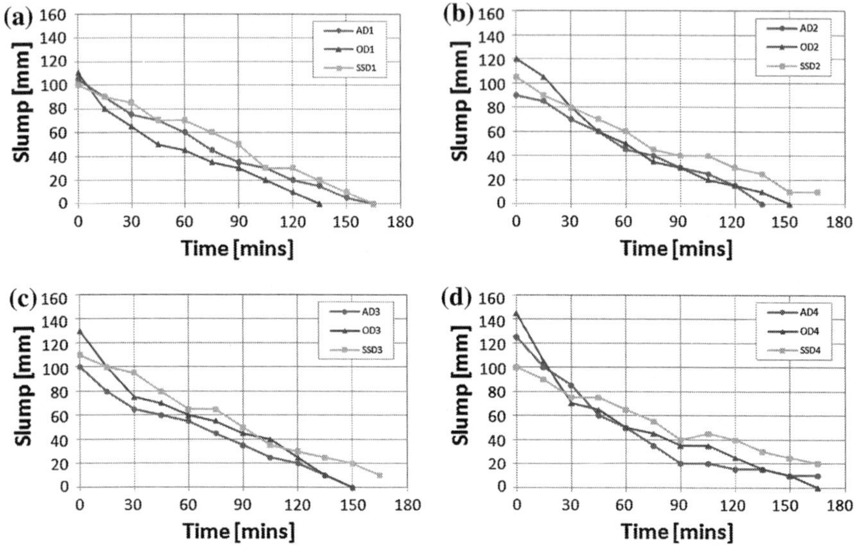

Fig. 5.1 Changes in slump of concrete containing various types of aggregate (Poon et al. 2009) **a** 100 % crushed granite, **b** 80 % crushed granite + 20 % recycled aggregate, **c** 50 % crushed granite + 50 % recycled aggregate, **d** 100 % recycled aggregate

content of OD and AD aggregates was due to the increase in free water in the mix; (4) the initial slump of the mix containing RCA in SSD state was almost constant for all replacement levels of natural aggregate; (5) the slump loss was faster and slower for mixes containing OD and SSD aggregates, respectively.

Poon et al. (2009, 2004a) also reported that the replacement of natural coarse aggregate by RCA in concrete prolonged the slump loss time. The slump value and the rate of slump loss with time of mixes containing RCA respectively increased and decreased with the addition of fly ash (Poon et al. 2004a). On the other hand, Thangchirapat et al. (2008) observed faster water loss in mixes containing coarse RCA aggregate than in the mix containing natural aggregate, when water was added to reach the required slump range of 50–100 mm. Sagoe-Crentsil et al. (2001) did not observe any difficulty in achieving the desired workability and subsequent compaction of concrete mixes containing RCA even though the amount of water absorbed by RCA was higher than for basalt aggregate.

Several mixing methodologies were developed to control the workability-related problems of concrete containing CDW aggregate. They can be summarised as: (1) increasing the amount of added water according to the water demand of concrete mixes containing dry CDW aggregate; (2) pre-soaking CDW aggregate in water for 10–20 min or for 24 h before use; (3) increasing the moisture content of CDW aggregate up to 70–80 % of total water absorption capacity for 24 h before followed by covering with plastic to control water loss due to evaporation; (4) increasing the super-plasticizer amount; (5) increasing cement content in the concrete composition.

Evangelista and de Brito (2007) adopted two different techniques to mix concrete, where fine natural aggregate was replaced by fine RCA. In the first technique, fine aggregates (recycled as well as natural) were mixed with water (2/3 of the required mixing water, plus the estimated absorbed water by fine aggregate) for 10 min before adding the other constituents. They increased the mixing time to 20 min without changing the remaining procedure in the second technique. However, they suggested the use of a superplasticizer to overcome the huge water demand of the mixes containing high amounts of RCA. Their results are presented in Table 5.2. Gonzalez-Fonteboa et al. (2011) pre-soaked the RA for 10 min before their use in concrete preparation to reach around 70 % of the RA's water absorption capacity. Padmini et al. (2009) also reported that the 10 min' water absorption value of CDW aggregate satisfied the desirable workability performance of concrete containing this aggregate. Etxeberria et al. (2007a) used partially saturated CDW aggregate to prepare concrete. The authors recommended 80 % humidity of the total water absorption capacity and therefore the aggregate was moistened by a sprinkler system 1 day before the preparation of concrete. The wet aggregate was then covered with a plastic sheet in order to maintain a high humidity level.

On the other hand, Gonzalez-Fonteboa and Martinez-Abella (2008) increased the cement content by 6.2 % in concrete prepared by replacing 50 % (by volume) of coarse natural aggregate with coarse recycled concrete aggregate keeping the slump similar to that of the control concrete. The average slump of 10 different concrete mixes containing recycled aggregate was 76 mm versus the average slump of 73 mm of control mixes. The water to cement ratio of both types of concrete remained constant at 0.55 (including the moisture content of both aggregates) and in both types of concrete around 1.2 % of superplasticizer was used. Limbachiya et al. (2000) changed the water to cement ratio by changing the water content, the cement content or both to obtain a mixes containing various amounts of coarse CDW aggregate with adequate fresh concrete properties. Etxeberria et al. (2007a) added higher amounts of superplasticizer to mixes containing CDW aggregate than to the control mix to guarantee constant slump and water to cement ratio.

The water to cement ratio and therefore the slump of fresh concrete is also dependent on the composition of CDW aggregate used to replace natural aggregate. Normally, the mortar and crushed brick contained in CDW aggregate absorb

Table 5.2 Effective and actual water to cement ratio of concrete containing recycled aggregate (Evangelista and de Brito 2007)

Constituents	Conventional concrete	Amount of recycled fine aggregate (%)				
		10	20	30	50	100
Cement (kg)	380	380	380	380	380	380
Water (kg)	155.8	160.6	165.4	170.2	175.6	180.9
Actual w/c	0.41	0.42	0.44	0.45	0.46	0.48
Effective w/c	0.41	0.42	0.43	0.44	0.45	0.45
Superplasticizer (g)	4.9	4.9	4.9	4.9	4.9	4.9

high amounts of water and therefore their content in the CDW aggregate control the slump of the resulting mix. Table 5.3 shows a typical water to cement ratio for concrete mixes with 80 ± 10 mm slump and containing coarse CDW aggregate from crushed concrete as well as a mixture of masonry brick and cement mortar (Gomes and de Brito 2009). The water absorption capacity of aggregate generated from crushed brick and mortar was considerably higher than that of the aggregate generated from crushed concrete and therefore the water to cement ratios of concrete mixes containing the former type of aggregate was also higher than that of the concrete mixes contacting the latter type of aggregate.

Yang et al. (2011) observed a reduction of about 27 % in slump due to the addition of pre-saturated recycled coarse concrete aggregate (RCA) as complete replacement of natural coarse aggregate (NCA). This reduction increased to about 40 % when the NCA was replaced by a mixture of 20 % recycled coarse crushed brick (CCB) and 80 % RCA. The percentage reduction in slump further rose to 70 %, when the NCA was replaced by a mixture of 50 % CCB and 50 % RCA. The reduction in slump was due to: (1) the presence of porous adhered mortar in RCA; (2) the generation of fine particles due to the relatively weak particles in RCA and particularly in CCB; (3) the decrease in density in the mix when 100 % of high-density natural coarse aggregate was replaced by low-density RCA and CCB particles. The fine RCA contains higher amounts of attached mortar than the coarse RCA and therefore a higher amount of water is needed to reach workability similar to that of the concrete mix containing coarse RCA (Buyle-Buddin and Zaharieva 2002).

Khatib (2005) reported an increase in slump due to the addition of crushed concrete (CC) as fine aggregate without using any type of admixture at free water to cement ratio of 0.5. In the case of fine aggregate generated from crushed brick (CB), the slump of the mixes increased up to a replacement ratio of 25 % by

Table 5.3 Variation of water to cement ratios of concrete mix prepared by different types of recycled aggregate considering water absorption capacity of recycled aggregate (Gomes and de Brito 2009)

Type of aggregate	Replacement ratio (%, v/v)	Water absorption capacity (%)	w/c	Effective w/c
Natural aggregate	0	2.29	0.430	0.43
Coarse recycled aggregate from concrete (1)	12.5	8.49	0.436	
	25		0.442	
	50		0.453	
	100		0.476	
Coarse recycled aggregate from mortar and bricks (2)	6.25	16.34	0.442	
	12.5		0.454	
	25		0.477	
	50		0.524	
Mixtures of (1 + 2)	12.5 (1) + 6.25 (2)		0.448	
	25 (1) + 12.5 (2)		0.465	
	50 (1) + 25 (2)		0.500	

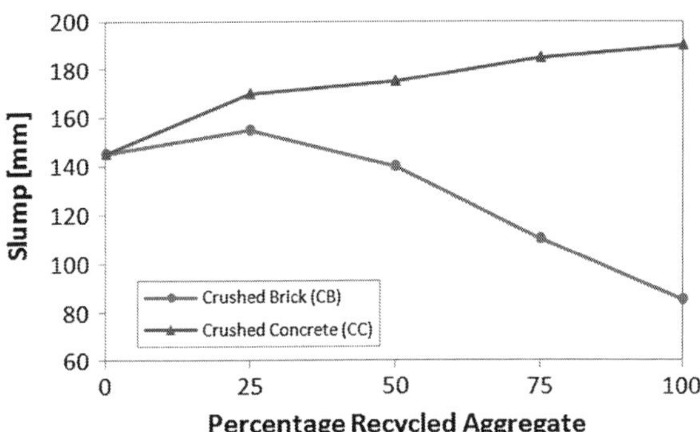

Fig. 5.2 Slump of concrete containing two types of CDW aggregate (Khatib 2005)

weight of sand and then gradually decreased and, at 100 % replacement level, the slump of concrete mix was about 80 mm in comparison to 145 mm of the control mix. These results are presented in Fig. 5.2.

Li et al. (2009) observed higher slump in mixes containing coarse RCA coated with pozzolanic powder than in mixes containing coarse RCA prepared by conventional methods. The poor workability of the mix containing RCA due to the presence of attached mortar was significantly improved due to the surface coating of RCA by pozzolanic powder, and therefore reduced the amount of water absorbed by RCA.

Katz (2003) observed similar slump and comparable contents of free water in conventional concrete and mixes containing coarse recycled aggregate (RCA) from crushed concrete cured for 1, 7 and 28 days. However, due to insufficient content of fine aggregate in RCA, an extra amount of fine natural aggregate was added to the mix containing RCA to maintain workability and cohesivity. Kou et al. (2011a) reported that the addition of silica fume (SF) and metakaolin (MK) can reduce the slump of concrete mix containing RCA, which otherwise increases due to the addition of extra water to compensate the increased water requirement of RCA present in the mix. On the other hand, fly ash and granulated blast furnace slag addition to similar concrete containing RCA can further increase the slump. Tu et al. (2006) reported that the workability loss of high-performance concrete (HPC) containing RCA was much higher than that of conventional HPC. The RCA with higher water absorption capacity had a slight influence on the workability at the initial stages of mixing of HPC; however, the workability of the HPC mix deteriorated after 1 h of mixing as the added water amount was insufficient to satisfy the water demand of the mix.

5.2.2 Density

Fresh concrete density is the mass of fresh, normally compacted concrete, including its remaining voids per unit volume. This property depends on several others such as aggregate and cement types, water content and void content. The density of fresh concrete also controls various hardened-state properties of concrete. For example, given the same quantity of cement and aggregate, lower fresh concrete density indicates lower strength because the density decreases as the water and voids content increases.

The density of fresh concrete containing CDW aggregate is slightly lower than that of the mix containing natural aggregate since the density of CDW aggregate is lower than that of natural aggregate. The presence of lower density residual cement mortar particles attached to the aggregate is the main factor for lowering density of concrete due to the addition of CDW aggregate (Hansen and Narud 1983; Gonzalez-Fonteboa et al. 2011). Table 5.4 shows some typical values of the density of various concrete mixes containing natural as well as CDW aggregate along with the density of the aggregate.

Katz (2003) observed no significant difference in densities of concrete mixes containing RCA prepared from old concrete of three different ages (1, 7 and 28 days), which suggested that the amount of adhered mortar content in the various concrete aggregates was similar regardless of their crushing age (Table 5.4). Soutsos et al. (2011) observed marginally lower wet density in concrete containing coarse RCA than in concrete containing coarse limestone aggregate due to volumetric rather than weight-based substitution of coarse aggregate. Lopez-Gayarre et al. (2009) reported that the variations of properties of RCA have little effect on the resulting density of concrete. They observed a reduction of about 5 % in the density of concrete mix when all natural coarse aggregate was replaced by RCA. They used the analysis of variance (ANOVA) method to study the effects of various parameters of aggregate on the density behaviour, which is presented in Table 5.5.

5.2.3 Air-Content

The presence of a certain amount of air bubbles trapped during concrete mixing has several beneficial effects for fresh and hardened concrete properties. In fresh concrete, an air-content of around 3 % can reduce the water demand of concrete and make the mix stickier, which helps to reduce segregation and bleeding. However, if the air-content is higher than the specified amount, the increased stickiness makes concrete finishing more difficult. Thus the air-content of a concrete mix is an important property and a few studies are available on the behaviour of air-content of fresh concrete due to the incorporation of CDW aggregate. Table 5.6 shows the air-content of different concrete mixes containing natural as well as CDW aggregates of various types.

In some of the studies, it was reported that the addition of CDW aggregate to concrete increases the air-content (Katz 2003; Lopez-Gayarre et al. 2009). Lopez-Gayarre et al. (2009) observed an increase in the air-content of concrete with the RCA aggregate content, which was visible above 50 % replacement of aggregate

Table 5.4 Density of concrete mix containing pre-saturated recycled aggregate

Reference	Type of aggregate (replacement amount)	Density (kg/m^3)	Density/specific gravity[a] of aggregate (kg/m^3)
Sagoe-Crentsil et al. 2001	Coarse basalt	2,466	2,890 (bulk)
	Coarse recycled (100 %)	2,335 2,321[b] 2,335[c]	2,394 (bulk)
Gomez-Soberon (2002)	Natural (coarse + fine)	2,130	2593.3 (average)
	Recycled concrete (coarse + fine)	2,090	2236.7 (average)
Vieira et al. 2011	Natural	2413.5	2,600 (coarse)
	Recycled concrete (coarse, 20 %, v/v)	2392.3	2,400
	Recycled concrete (coarse, 50 %, v/v)	2355.0	
	Recycled concrete (coarse, 100 %, v/v)	2299.8	
Etxeberria et al. 2007a	Natural	2,420	2,670
	Recycled concrete (coarse, 25 %, v/v)	2,400	2,430
	Recycled concrete (coarse, 50 %, v/v)	2,390	
	Recycled concrete (coarse, 100 %, v/v)	2,340	
Gonzalez-Fonteboa et al. (2011[d] (MatStruc)	Natural	2,340 and 2,360	2,725
	Recycled concrete (coarse, 25 %, v/v)	2,320 and 2,330	2,400
	Recycled concrete (coarse, 50 %, v/v)	2,300 and 2,310	
	Recycled concrete (coarse, 100 %, v/v)	2,270 and 2,270	
Katz 2003	Natural	2,463	–
	Recycled concrete (coarse + fine)	2,175 (1-day[e]) 2,145 (7-day[e]) 2,156 (28-day[e])	2.23–2.59[a] 2.25–2.60[a] 2.23–2.55[a]
Buyle-Budin and Zaharieva 2002	Natural	2,360–2,410	2,600 (fine); 2,680 (coarse
	Recycled concrete (coarse + fine)	2,195–2,220	2,160 (fine); 2,250 (coarse)
Topçu and Sengel 2004	Natural	~2386	2,700
	Recycled concrete (coarse, 50 %, v/v)	2,301 (50 %)	2,470
	Recycled concrete (coarse, 100 %, v/v)	2,251	

[a] Specific gravity; [b] Binder is slag cement; [c] 5 % more cement was used; [d] w/c ratios: 0.65 and 0.50; [e] Crushing age of old concrete

Table 5.5 Influence of various parameters of RCA on concrete density (Lopez-Gayarre et al. 2009)

Parameters	Levels	Density Influence	Density Variation (%)	Parameters	Levels	Density Influence	Density Variation (%)
Type of aggregate	OV	XXX		Declassified content (%)	0		
	MA		1		5		
Replacement ratio (%)	0	XXX			10		
	20		−1.2	Base concrete (MPa)	35		
	50		−2.8		25		
	100		−5.2	Targeted slump (cm)	8	X	<1
Type of sieve curves	CF	X	−1		13		
	CC			Replacement criteria	SR		
	D				CR		<0.5

XXX Very influential; *X* Slightly influential; *OV* and *NA* RCA with different properties (*OV* has higher adhered mortar and declassified contents, los Angeles coefficient and water absorption capacity than *NA*); *CF* Fine sieve continuous curve; *CC* Coarse sieve continuous curve; *D* Discontinuous curve; *SR* Coarse natural aggregate replaced by identical volume of RCA; *CR* Fine natural aggregate replaced by identical volume of declassified content in RCA and natural coarse aggregate replaced by coarse fraction present in RCA

Table 5.6 Air-content of concrete mix containing pre-saturated recycled aggregate

Reference	Type of aggregate (replacement ratio)	Air-content (%)	Comment
Sagoe-Crentsil et al. (2001)	Coarse basalt	2.4	
	Coarse recycled (100 %)	2.4	
		1.8	slag cement as binder
		2.3	5 % more cement
Katz (2003)	Normal	1.3	White Portland cement as binder
	Recycled concrete (coarse + fine)	5.4 (1 day) 4.1 (7 day) 5.0 (28 day)	
	Recycled concrete (coarse + fine)	4.8 (1 day) 5.4 (7 day) 5.6 (28 day)	Ordinary Portland cement as binder
Rustom et al. (2007)	Normal	1.0–2.0	Ordinary Portland cement as binder
	Recycled concrete	1.8–3.3	

(Fig. 5.3). Their results also indicated lower air-content in the mix containing better quality RCA aggregate (MA) than that of inferior one (OV). The quality of the two aggregates used in this investigation is presented in Table 5.7.

Katz (2003) observed higher air-content in concrete containing RCA aggregate than in concrete containing natural aggregate, when the air-content was measured by a gravimetric method. The air-content of the aggregate was also measured in the method. However, the air-content of concrete containing RCA was not affected by the crushing age of the original hardened concrete. On the other hand,

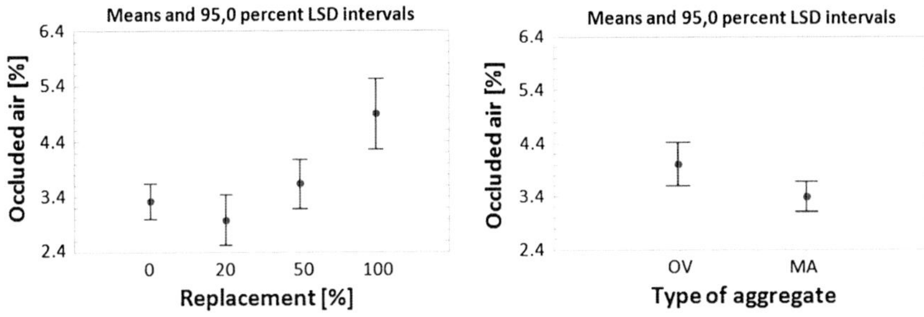

Fig. 5.3 Effect of replacement ratio (%) and type of aggregate on the air-content of concrete containing CDW aggregate

Table 5.7 Properties of two types of RCA aggregate (Lopez-Gayarre et al. 2009)

Aggregate name	Size (mm)	Dry density (kg/m^3)	24 h water absorption (%)	Los Angeles coefficient (%)	Attached mortar content (%)	Declassified content (%)
OV	4–20	2,200	5.0	37.2	34.2	2.6
MA	4–20	2,360	3.8	33.1	23.0	1.5

Sagoe-Crentsil et al. (2001) observed no difference in air-content between concrete containing normal aggregate and concrete containing RCA aggregate.

5.2.4 Bleeding

Bleeding of concrete is the upward movement of water during settling of concrete mix. It is a particular form of segregation, in which some of the water within concrete comes out of its surface. Sometimes, with this water along comes a certain quantity of cement. A higher bleeding to the surface increases the water to cement ratio and therefore decreases the strength of the concrete surface. The increase in capillary porosity of hardened concrete due to bleeding can also affect the durability performance. However, in some cases, the bleeding water may not come up to the surface and is trapped by flat or flaky pieces of aggregates and also by reinforcement and accumulates below such obstacles. This is known as internal bleeding, which can affect several properties of hardened concrete such as cement paste-aggregate bonding and enhance micro cracks. High bleeding of concrete occurs due to various factors such as high water to cement ratio, low cement content, coarse cement particles and poorly graded aggregates. Few references are available on the bleeding behaviour of concrete containing CDW aggregate.

The old cement paste on the surface of the recycled aggregate can absorb some mixing water. Thus, the total amount of bleeding of concrete decreases as the replacement level of recycled coarse aggregate increases (Kim et al. 1993). Yang et al. (2008) also found a reduction in total bleeding due to the replacement of 100 % coarse NA by coarse RCA and 100 % fine NA by fine RCA. Bleeding of concrete containing fine RCA was lower than that of concrete containing coarse RCA. Total bleeding and rate of bleeding of RCA concrete decreased as the relative water absorption by the aggregate increased.

Poon et al. (2007) reported the bleeding performance of concrete containing coarse air-dried RCA, where 0, 20, 50, 80 and 100 % by mass of natural aggregate was replaced by RCA. The bleeding of concrete was measured in terms of bleeding rate (defined as the amount of water in ml collected per cm^2 surface area of concrete per second during the first 60 min of the test) and bleeding capacity (defined as the total volume of bleeding water collected during the entire course of the experiment and expressed as a fraction of the initial volume of concrete).

Their results showed that adding air-dried RCA to concrete increases its bleeding rate and bleeding capacity and this trend rises as the content of RCA in concrete increases. The use of 100 % RCA increased the bleeding rate and bleeding capacity by 26 and 22 %, respectively. On the other hand, delaying the starting time of the experiment or adding fly ash to cement can reduce the bleeding rate and bleeding capacity of concrete with natural or RCA aggregates. These results are presented in Table 5.8. Poon et al. (2004a) also reported that over-wetting of RCA should be avoided to reduce concrete bleeding, which also had some effect on the hardened concrete properties.

Table 5.8 Bleeding capacity and bleeding rate of concrete due to the addition of RCA (Poon et al. 2007)

Concrete mix	Immediately after mixing		30 min after mixing		120 min after mixing	
	Bleeding capacity, 10^{-3}, ml/ml	Bleeding rate, 10^{-6}, ml/cm^2.s	Bleeding capacity, 10^{-3}, ml/ml	Bleeding rate, 10^{-6}, ml/cm^2.s	Bleeding capacity, 10^{-3}, ml/ml	Bleeding rate, 10^{-6}, ml/cm^2.s
R0	18.8	47.9	13.2	19.6	5.2	9.6
R0F25	16.2	43.5	9.6	17.4	4.6	6.5
R20	19.9	50.9	14.2	20.5	5.4	10.0
R20F25	16.9	45.7	10.4	18.3	4.8	6.7
R50	21.2	53.6	15.2	20.9	5.6	10.4
R50F25	18.1	49.2	10.8	20.5	5.0	7.1
R80	22.6	56.6	16.2	21.3	5.8	10.5
R80F25	19.5	52.3	11.4	21.8	5.1	7.3
R100	23.0	60.1	17.1	22.2	6.1	10.5
R100F25	20.0	54.5	12.0	22.7	5.4	7.8

R0, R20, R50, R80 and R100: concrete mixes containing ordinary Portland cement (OPC) and prepared by replacing 0, 20, 50, 80 and 100 % by mass of natural aggregate by RCA

R0F25, R20F25, R50F25, R80F25 and R100F25: concrete mixes containing a blended cement with 75 % OPC plus 25 % fly ash (FA) and prepared by replacing 0, 20, 50, 80 and 100 % by mass of natural aggregate by RCA

5.3 Hardened Concrete Properties

As the properties of CDW aggregate are significantly different from those of NA, the various hardened properties of conventional concrete change with the addition of CDW aggregate. In this section, the hardened properties of concrete containing various types of CDW aggregate are discussed from information available in various references. As indicated in the introductory section, a similar terminology (RA and RCA) will be used to indicate recycled aggregate with different origin.

5.3.1 Compressive Strength

In this section, the compressive strength (CS) of concrete containing CDW aggregate (both RA and RCA) is highlighted. Results will be analysed in terms of size and composition of CDW aggregate. The relationship with fresh properties is also presented.

Normally, the CS of concrete decreases with the addition of CDW aggregate (Oliveira and Vazquez 1996; Dhir et al. 1999; Topçu and Sengal 2004) and the reduction in strength can reach 40 % (Katz 2003; Chen et al. 2003). In some studies, it was pointed out that the reduction in CS was between 12 and 25 %, when 25–30 % (Corinaldesi 2011; Etxeberria et al. 2007a) or 100 % NA was replaced by CDW aggregate (Li et al. 2009; Rahal 2007; Safiuddin et al. 2011). However, a negligible influence is observed when the coarse or fine recycled aggregate is used to replace up to 30 % of coarse NA (Gomez-Soberon 2002; Li et al. 2009; Limbachiya et al. 2004, 2012; Rao et al. 2011; Yang et al. 2011) or 20 % of fine NA (Dhir et al. 1999), respectively. Some typical results are presented in Table 5.9. An increase in concrete porosity due to the addition of CDW aggregate (due to old mortar content) and weak aggregate-matrix interface bond are the major reasons for the reduction in CS of CDW aggregate concrete (Kwan et al. 2012).

The reduction in CS due to the addition of CDW aggregate can be controlled by changing various factors of the concrete mix such as adjusting the water to cement ratio, changing the mixing procedure, treating the aggregate and using a mineral addition. The information gathered so far on the use of CDW aggregate in concrete shows that modifications in the concrete mixing procedure are the key step to develop a good quality concrete containing any type of CDW aggregate.

Etxeberria et al. (2007a) observed that the 28-day CS of concrete made with 100 % coarse RA (RAC) was 20–25 % lower than that of conventional concrete. A similar trend was reported by other researchers (Gonzalez-Fonteboa and Martinez-Abella 2005; Sani et al. 2005). Sani et al. (2005) observed about 40 % lower 90-day CS in RAC than in conventional concrete due to the incorporation of RA as complete replacement of coarse and fine aggregates. The mix was prepared with the same cement content and water to cement ratio without using any water-reducing

Table 5.9 Reduction in CS (%) due to the addition of CDW aggregate

References	Type of aggregate	Reduction in strength, (substitution level %)	Comment
Oliveira and Vasquez (1996)	Coarse/RCA	10	Varied depending on moisture content in RCA aggregate
Topçu and Sengel (2004)	Coarse/RCA	23.5 33	C16 type concrete C20 type concrete
Sani et al. (2005)	Coarse/fine/RA	40	At constant water to cement ratio
Gomez-Soberon (2002)	Coarse/RA/10 min pre-soaked in water	~2, (15) ~ 5, (30) ~11.5, (100)	Substitution level (v/v): 15, 30 and 100 %
Kou and Poon (2008)	Coarse/RCA/SSD[a]	4–6, (20) 13–17, (50) 16–22, (100)	Substitution level (v/v): 20, 50 and 100 %; strength decreasing with increasing mortar content;
Khatib (2005)	Fine/RCA	~24.5–25, (25–75) ~36, (100)	Substitution level (w/w): 25, 50, 70 and 100 %; prepared at free water to cement ratio of 0.5
Yang et al. (2011)	Coarse/RCA	5.7, (100)	Prepared at w/c of 0.47 with a slump of 24 mm

[a] *SSD* Saturated surface dry

admixture. They also observed an inverse relationship between CS of RAC and open porosity of RA. The addition of fly ash to replace a part of fine RA can recover part of the CS and therefore in this case a reduction of about 30 % in CS was observed. Dapena et al. (2011) did not observe any change in 28-day CS of resulting concrete due to 20, 50 and 100 % (by volume) replacement of coarse NA by coarse RA owing to the small amount of impurities in RA (<2 %) and an initial reduction of water to cement ratio of RAC due to higher water absorption capacity of RA. The 5 and 10 % (by volume) of fine (<4 mm) content in RCA did not affect the strength behaviour of concrete containing 20 and 50 % RCA; however, a drop of around 3.6 MPa in CS was observed in concrete containing 100 % RCA. Poon et al. (2009) observed a reduction in compressive strength of RAC containing RA as the only coarse aggregate as the replacement of fine natural aggregate by fine RA increased.

Etxeberria et al. (2007a) reported that concrete prepared by replacing 25 % (v/v) coarse NA by coarse RA can be used as medium strength (30–45 MPa) concrete having similar CS to conventional concrete, when both mixes are prepared with the same cement content and effective water to cement ratio of 0.55. However, a reduction of effective water to cement ratio of 0.52 or an increase of 6 % in cement content was necessary for RAC containing 50 % RA to achieve a CS similar to that of conventional concrete. These values became 0.50 and 8.3 %, respectively, for RCAC containing 100 % RCA. Gonzalez-Fonteboa and Martinez-Abella (2005) also increased the cement content by about 6.2 % without changing the w/c (including moisture content in aggregate before mixing) in RAC where 50 % (by volume) of natural coarse aggregates were replaced to obtain CS and consistency similar to those observed in a conventional concrete. They observed CS about 2.5 and 0.4 % lower and 2 % higher for RAC than for conventional concrete after 7, 28 and 115 days of curing, respectively.

The detrimental factors that control the CS of RAC are the crushing strength and mortar content in RA (Etxeberria et al. 2007a). Gonzalez-Fonteboa and Martinez-Abella (2005) specified three reasons, which controlled the CS performance of RAC. The higher absorption of water by RA than NA reduced the amount of free water in the mix. These decrease up to a certain level could increase the CS; however, an excessive decrease in free water can deteriorate the CS of RAC due to lesser hydration of cement particles and poor workability of the mix than in conventional concrete. The weak bond of cement paste and RA also lowers the CS of RAC. The addition of an extra amount of cement to the mix boosts the CS by improving bond between the cement paste and the RA and improves porosity and consistency.

The strength development of RAC was faster than for conventional concrete after 28 days of curing (Gonzalez-Fonteboa and Martinez-Abella 2005). Etxeberria et al. (2007a) observed gain of about 12–15 % in CS between 7 and 28 days of curing of RAC prepared by replacing 25, 50 and 100 % (by volume) of NA by RA in comparison to around 20 % in conventional concrete.

Compared to the studies on the use of RA in concrete, more information is available on the use of RCA in preparation of concrete. Similarly to RA, the CS of concrete containing RCA (RCAC) is normally lower than that of the corresponding control concrete and increasing the addition of RCA to concrete further lowers it (Table 5.8). However, RCA addition to concrete does not have an adverse effect on its strength development trend.

Frondistou-Yannas (1977) reported that the substitution of natural gravel by RCA and recycled aggregate that contained mortar only led to a lower CS than that of conventional concrete. The failure observed in both types of concrete was by fracture in the aggregate. Eguchi et al. (2007) observed higher reduction in CS of concrete due to the addition of RCA originated from a concrete which consisted of low strength aggregate than that observed for RCA originated from a concrete consisted of high-strength aggregate. Lopez-Gayarre et al. (2009) also observed a strong dependence of CS of RCAC on the quality of RCA, mainly concerning the adhered mortar content. However, they did not observe any effect of increased addition of coarse RCA on the mean CS of RCAC if the water to cement ratio was kept constant and the loss of workability due to addition of RCA was compensated by using chemical admixtures.

Santos et al. (2002) observed a reduction of about 20 % in 7- and 28-day CS of two types of RCAC from the corresponding strength of conventional concrete, when both types of RCAC were prepared with coarse RCA originated from two concrete mixes with different strengths. Gomez-Soberon (2002) observed that the CS of concrete decreased with increasing content of coarse RCA. The 28-day CS of concrete prepared by replacing coarse NA with 15, 30, 60 and 100 % coarse RCA were respectively about 98, 95, 92 and 88 % that of conventional concrete. After 90 days of curing, these values became 99, 94, 91, 89 %, indicating a similar development of the CS of RCA concrete to that of conventional concrete as curing time increased. Topçu and Sengel (2004) observed a systematic reduction in cubic and cylindrical CS of 16 and 20 MPa RCAC with increasing content of coarse RCA. Compared to a control concrete, the reduction in 28-day CS was 33 and 23.5 % for 16 and 20 MPa RCAC when 100 % NA by volume was replaced by RCA.

Rao et al. (2011) observed higher early strength gain (0–7 days) in RCAC than in conventional concrete due to the high water absorption capacity of old mortar present in RCA and the rough texture of RCA that provides improved bond and interlocking characteristics between mortar and RCA. However, they also observed 8 % gain in CS between 28 and 90 days of curing for mixes prepared by replacing 25 % (by volume) of natural coarse aggregate by coarse RCA compared with 12 % gain for conventional concrete and did not observe any gain in CS when the replacement ratio increased to 50 and 100 %. Just like Xiao et al. (2006a), they observed a linear relationship between RCAC's density and CS but with a different slope. Fonseca et al. (2011) also observed about 80 and 95 % of 56-day CS for normal aggregate and RCA mixes after 7 and 28 days of curing, respectively. Safiuddin et al. (2011) observed more than 80 % of 28-day CS after 7-day curing of conventional concrete and concrete containing various ratios of coarse RCA due to the higher cement content than in conventional concrete as well as the higher workability of the concrete mixes.

In contrast with the majority of the observations, some results are available where the addition of coarse RCA to concrete does not have adverse effect on the CS performance. On the contrary, this type of addition can increase the CS of concrete if some modification is done by improving concrete mixing methodology. For example, Sagoe-Crentsil et al. (2001) observed no difference in 28-day and 1-year CS of mixes containing pre-saturated RCA and NA as coarse aggregate (Fig. 5.4). However a 5 % increase in cement content or the use of slag cement in RCA concrete considerably increased the CS, which was particularly more significant when using slag cement in RCA concrete. Domingo-Cabo et al. (2010) observed an increase in CS as the replacement of NA by RCA increased, possibly due to the reduction in effective water to cement ratio on account of the higher

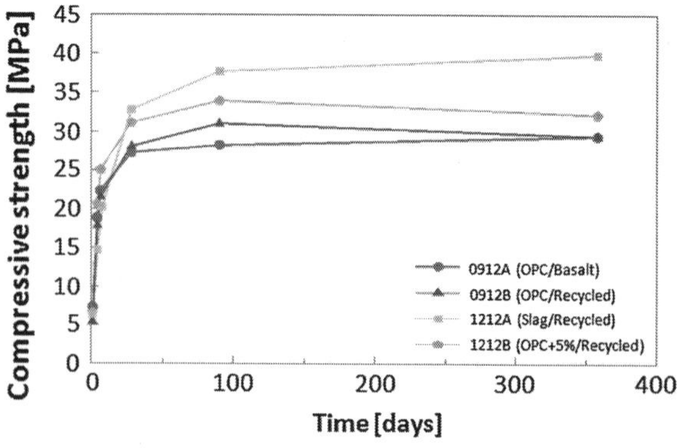

Fig. 5.4 CS of concrete containing RCA (Sagoe-Crentsil et al. 2001)

water absorption capacity of RCA. Evangelista and de Brito (2007) also observed higher CS of concrete incorporating fine RCA, which will be discussed later.

Some studies indicated a relationship between aggregate to cement ratio and the CS of RCAC (Fig. 5.5): decreasing the aggregate to cement ratio is beneficial for the CS of RCAC (Poon and Chan 2007; Poon and Lam 2008). The authors pointed out that the low crushing strength of the aggregate as well as the weak cement paste-aggregate bond are the causes of these results. In fact, the cement content in concrete plays a vital role in the CS (and other mechanical) behaviour of RCAC. A typical example is presented in Fig. 5.6 (Courad et al. 2010).

Similarly to conventional concrete, a linear relationship between water to binder ratio and CS also exists in RCAC (Corinaldesi 2010; Nagataki and Iida 2001). Nagataki and Iida (2001) observed this relationship for RCAC up to a binder to water ratio of 2.5–3.3 depending on the strength of the source concrete from which RCA originates (Fig. 5.7). Kou and Poon (2006) observed a decrease in CS as the content of coarse RCA in concrete increases at water to cement ratios of 0.45 and 0.55. However, the CS of the control concrete and RCAC increased with curing time and decreased with water to cement ratio. However, in a few studies, it was pointed out that the strength characteristics of RCAC were not significantly affected by the quality of RAC at high water to cement ratio, since it was affected only when the water to cement ratio was low (Ryu 2002; Padmini et al. 2002).

Chen et al. (2003) observed much lower differences between the CS of conventional concrete and that of RAC, when the w/c ratio increased to a certain value and they became lowest for w/c values between 0.58 and 0.80 (Fig. 5.8). Katz (2003) observed that the CS of RCAC was comparable to that of reference concrete up to the replacement level of 75 % at a w/c ratio between 0.6 and 0.75.

Padmini et al. (2009) specified several factors that can affect the CS performance of RCAC: to achieve the design CS RCAC required lower w/c ratio and higher cement content than concrete containing granite aggregate; for a targeted mean CS, the actual strength of RCAC increased with the maximum size of RCA used as the maximum size of RCA in concrete decreased and the strength of the source concrete increased.

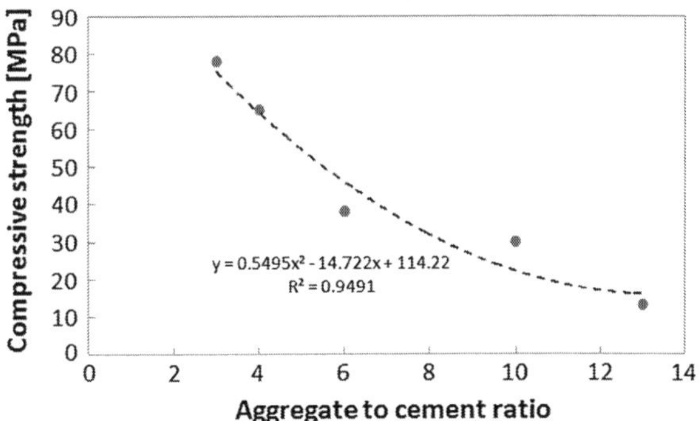

Fig. 5.5 Relationship between aggregate to cement ratio and CS of RCAC (Poon and Lam 2008)

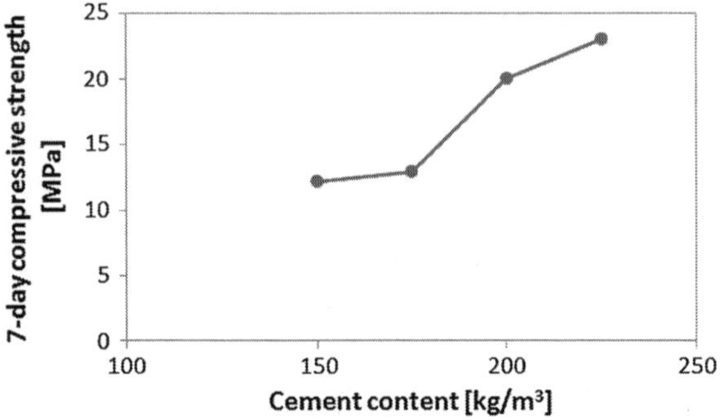

Fig. 5.6 Cement content and CS of roller compacted RCAC relationship (Courad et al. 2010)

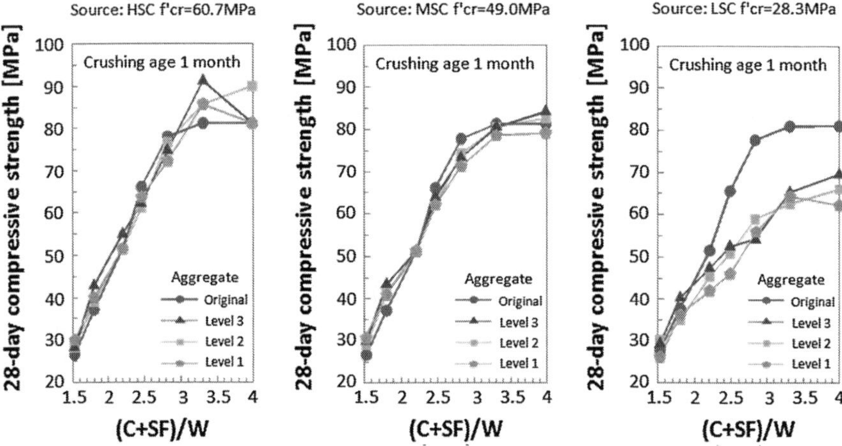

Fig. 5.7 Binder to water ratio *versus* CS (Nagataki and Iida 2001)

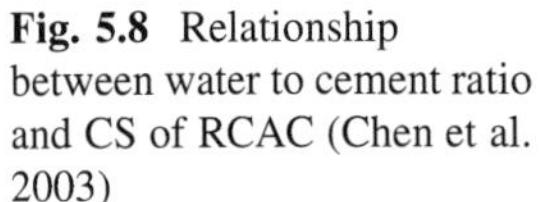
Fig. 5.8 Relationship between water to cement ratio and CS of RCAC (Chen et al. 2003)

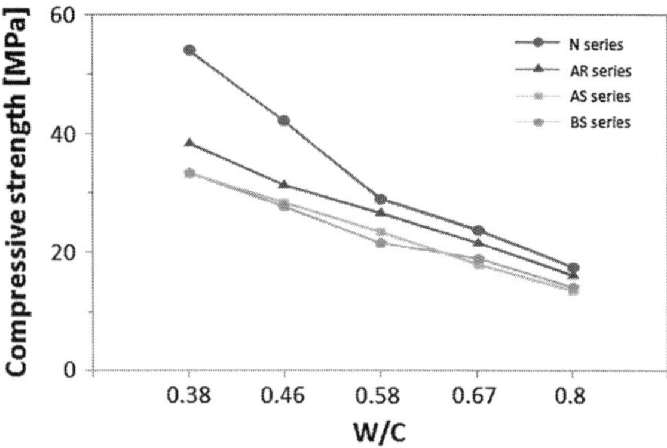

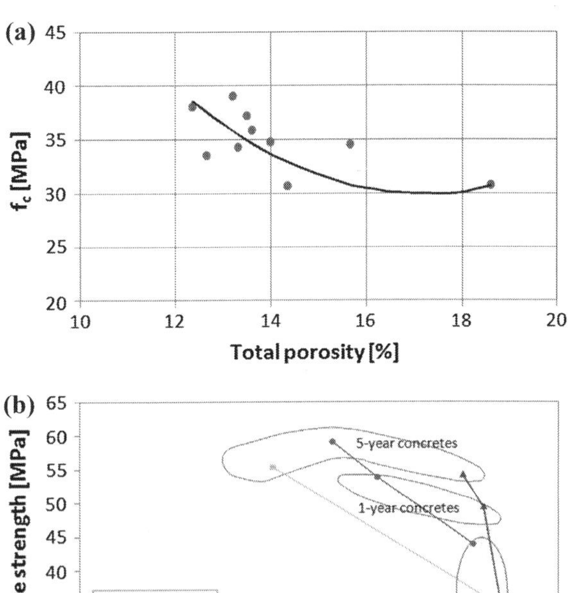

Fig. 5.9 Porosity versus CS of RCAC. **a** Gomez-Soberon (2002), **b** Kou et al. (2011a)

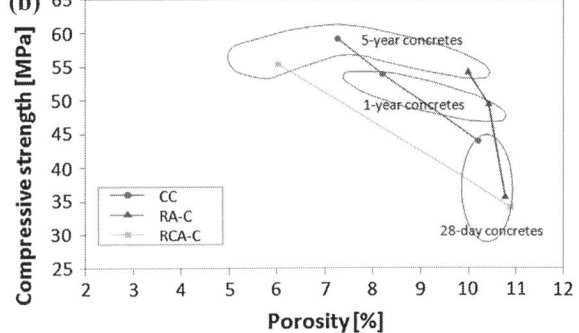

The CS of RCAC decreased as the porosity of concrete containing coarse RCA increased (Fig. 5.9a) (Gomez-Soberon 2002; Kou et al. 2011a). Kou et al. (2011a) observed significant differences in the relationship between the porosity and the CS of RCAC and RAC after 5 years of curing (Fig. 5.9b). The higher improvement of porosity (thus compressive strength) in RCAC than in RAC was due to the filling of pores and therefore improvement in interfacial zone due to continuous hydration of old cement mortar. Park et al. (2005) observed decreasing CS with increasing void ratio of porous concrete where various amounts of coarse NA were replaced by RCA. The CS decreased rapidly above the void ratio of 25 %. In contrast to the above results, Nagataki et al. (2004) observed comparable or even better CS (and splitting tensile strength) of concrete containing coarse high quality RCA than that of conventional concrete even though the RCAC had 20–52 % more permeable voids than conventional concrete.

The moisture content in RCA also controls the CS of RCAC (Oliveira and Vazquez 1996). The CS of RCAC with about 90 % saturated coarse RCA was marginally better than that for dry coarse RCA. The decrease of CS was especially felt when the RCAC contained 100 % saturated coarse RCA (Fig. 5.10). The compressive strength of RCAC containing RCA at three moisture levels as observed in Poon et al. (2004b) study can be rated as: air-dried (AD) > oven dried (OD) > saturated surface dry (SSD). The relatively high w/c ratio in the RCA-

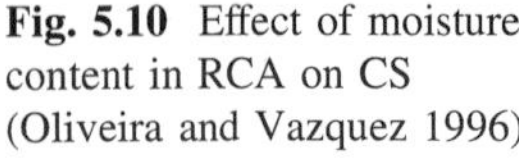

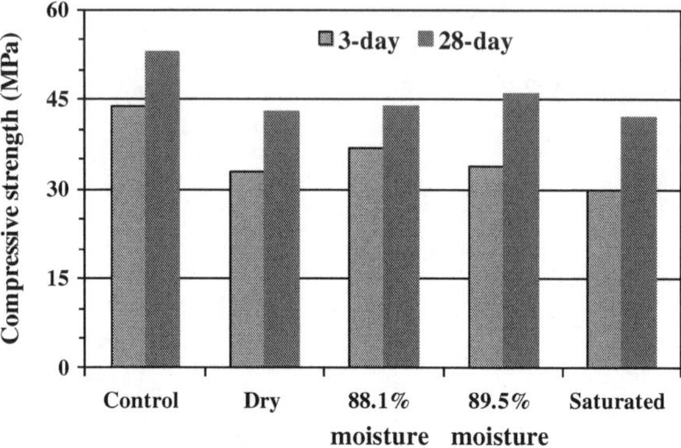

Fig. 5.10 Effect of moisture content in RCA on CS (Oliveira and Vazquez 1996)

cement matrix region due to movement of water from SSD aggregate weakens the cement paste/aggregate bond and lowers the compressive strength. On the other hand, the opposite movement of water from cement matrix to aggregate strengthens the bond and led to higher CS of concrete containing OD RCA.

Concrete containing larger RCA has higher CS than similar concrete containing smaller RCA as the mortar content in larger RCA is generally lower than in smaller RCA (Tavakoli and Soroushian 1996; Hansen and Narud 1983). Corinaldesi (2010) reported that the CS of RCAC containing larger coarse RCA is higher than that for smaller coarse RCA at the same water to cement ratio and at 30 % (by volume) substitution ratio of natural coarse aggregate (Fig. 5.11). Both types of aggregate were generated in same crushing plant at the same crushing period. According to the authors, the coarser RCA came from concrete with a higher CS and hence less friability than the other concrete, which generated the finer RCA. In this study, two classes of concrete (C30/37, C32/42) were prepared from coarse and fine RCA aggregate at the water to cement ratios of 0.5 and 0.4, respectively. Nagataki et al. (2004) observed the dependence of mechanical properties including CS on the size of coarse aggregate of the original concrete as well as on the amount of sand particles present in RCA. The stress concentration at the zone between RCA and mortar in RCAC is lower for smaller coarse aggregate. Low amounts of sand particles in RCA also enhance the CS of RCAC.

A strong relationship between the CS of RCAC and the properties of the source concrete from which RCA originated was reported in various studies. Hansen and Boegh (1985) observed that the 47-day CS of high and medium strength concrete containing RCA generated from high and medium strength concrete were higher than that of the original high and medium strength concrete; however, the CS of low strength concrete containing RCA originated from medium and low strength concrete was lower than that of conventional low strength concrete. Their results are presented in Fig. 5.12.

Tavakoli and Soroushian (1996) reported that the CS of concrete containing RCA depends on several factor such as the strength of the source concrete from which RCA is generated, the mixing procedure, the water to cement ratio and the

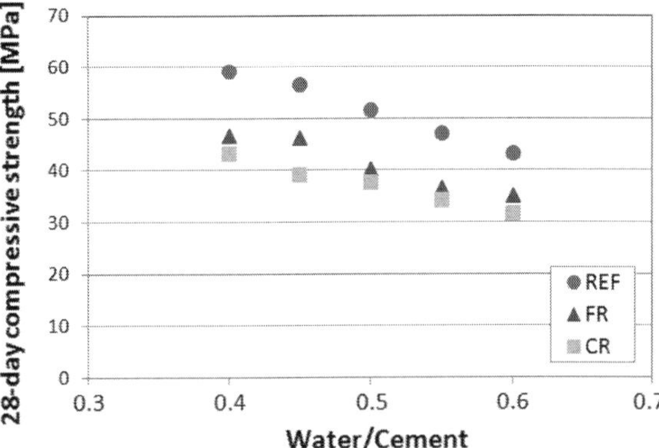

Fig. 5.11 CS of concrete containing coarse and fine coarse recycled aggregates and of reference concrete (Corinaldesi 2010)

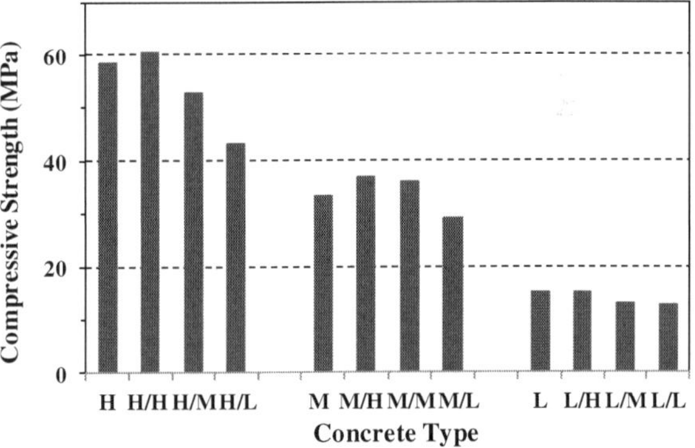

Fig. 5.12 CS of concrete (H, M, L: Original high, medium and low strength concrete; H/H, M/L: high and medium strength concrete containing RCA produced from H and L concrete, respectively, etc.) (Hansen and Boegh 1985)

aggregate size. They observed higher CS in concrete containing RCA than in concrete containing NA provided that the CS of concrete from which RCA was generated that was higher than that of the concrete containing NA used for comparison purposes. Hansen and Narud (1983) observed lower CS in concrete containing RCA than in control concrete, when the CS of control concrete exceeded the CS of concrete from which RCA was generated. The variations of CS of RCAC and control concrete containing NA were due to the bond between cement mortar and coarse aggregate; low bond gave a poor quality RCA and hence lower strength RCAC.

Nagataki and Iida (2001) also observed the dependence of CS on the strength of the source concrete; the CS is higher for RCAC produced from higher CS concrete than for that originated from lower CS concrete. Poon et al. (2004a) reported that the CS of RCAC originated from high CS concrete is higher than the RCAC originated from normal CS concrete due to formation of a denser microstructure in ITZ in the former one than in the latter one and also a higher aggregate strength of RCA originated from the high CS concrete than that of the RCA originated from normal CS concrete. The ITZ of concrete containing RCA originated from high CS concrete was denser than that of natural aggregate concrete too. Although the CS of both RCAC was lower than that of the conventional concrete after 7 and 28-days of curing, the CS of 90-day cured RCAC originated from high CS concrete was comparable to that of the conventional concrete. On the contrary, Santos et al. (2002) observed an insignificant effect of the source concrete's CS on the CS of RCAC. They observed a difference in CS of <1.5 MPa between two types of RCAC containing RCA originated from 56 and 45 MPa concrete mixes (Fig. 5.13).

Katz (2003) observed a moderate effect of the crushing age and strength of the source concrete from which RCA was generated on the CS of RCAC. The concrete made with aggregates crushed at 3 days exhibited higher CS than that made with aggregates generated after 1 and 28 days of curing, when white cement (WC) was used to make the source concrete. On the other hand, the concrete containing RCA made with the ordinary Portland cement (OPC)-based source concrete had slightly higher strength at the crushing age of 1-day and the lowest strength for RCA crushed at 3-day. The strength of the source concrete made with white cement was considerably higher than that of the OPC-based source concrete. The author indicated two factors that seemed to control the CS of the new RCAC: the strength of the source concrete and the presence of un-hydrated cement in the recycled aggregate. Both of the effects were active in the RCA generated from WC-based concrete after crushing age of 3 days and therefore led to the highest strength.

Nagataki and Ida (2001) observes no prominent effect of the crushing level during the preparation of RCA on the CS performance; however, the crushing age had some effect on the CS of RCAC: the crushing age of RCA produced from high (61 MPa) and medium (49 MPa) strength concrete had no effect on the CS of RCAC; but for low strength (28 MPa) RCAC the concrete containing RCA reclaimed after 1 or 2 years exhibited higher CS than the concrete containing RCA reclaimed after 28-days.

Poon and Chan (2007) reported that the presence of contaminants such as crushed brick, tiles, glass and wood in RCA deteriorated the CS of RCAC. However, concrete paving blocks made with crushed brick, tile and glass incorporated RCA at aggregate to cement ratio of 3 met the various standard specifications. Yang et al. (2011) observed 5.3 and 14.9 % drops in 28-day CS of RCAC

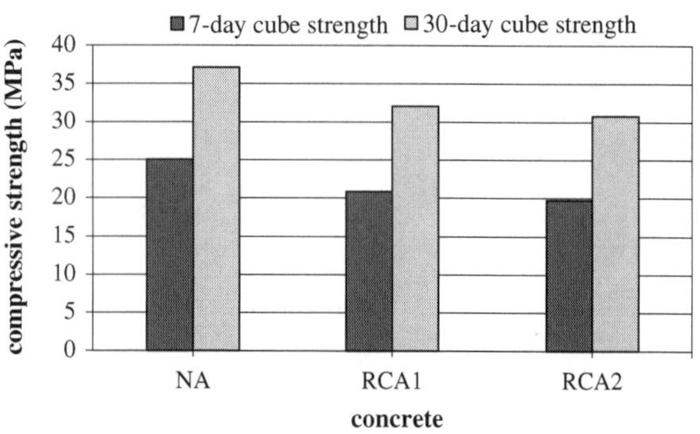

Fig. 5.13 CS of conventional and recycled aggregate concrete (Santos et al. 2002)

due to the replacement of 20 and 50 % by volume of coarse RCA by recycled crushed brick aggregate. Chen et al. (2003) observed little effect of brick and tile content in RCA on the concrete's CS if the ratio was <67 % (Fig. 5.14). The use of unwashed RCA had some effect on the CS of RCAC due to the presence of powdery impurities and other harmful materials; the effect of impurities on the CS of RCAC was observed prominently at lower w/c. Gomes and de Brito (2009) observed an insignificant CS loss due to the simultaneous incorporation of recycled brick and mortar, up to a maximum ratio of 25 % brick-mortar mix and 50 % RCA, or the incorporation of 100 % RCA, by comparison with conventional concrete with natural aggregates only.

A beneficial effect of RCA on the strength performance was observed when the RCAC was subjected to dry curing conditions due to the higher water absorption capacity of RCA. Buyle-Buddin and Zaharieva (2002) observed lower CS by around 9–12 % and 3–6 % for conventional and RCA concrete with 100 % replacement of fine and coarse natural aggregate by same size RCA respectively, when the curing conditions for both concrete were changed from water curing to air curing. The water absorbed by fine RCA gradually made its way to the cement paste and compensated the water loss of cement paste due to air-drying. On the other hand, Rao et al. (2011) observed higher CS for partially moist cured followed by air-cured RCAC than for moist cured similar RCAC (Fig. 5.15). The possible cause was the higher free water content in the old ITZ of moist cured concrete, which weakens the ITZ and therefore lowered the CS. Fonseca et al. (2011) did not observe any significant changes in CS due to variation of curing conditions. In this study, after casting, the concrete mixes were subjected to four curing conditions: (1) laboratory environment; (2) outer environment (atmospheric condition); (3) wet chamber; (4) water immersion.

Razaqpur et al. (2010) evaluated the CS of RCAC, prepared by using a new mixing method (equivalent mortar volume, EMV method). In this method, the amount of residual mortar present in RCA was included in the total amount of mortar present in RCAC. The total amount of mortar in RCAC was equivalent to the mortar amount present in the control concrete. The CS of concrete containing

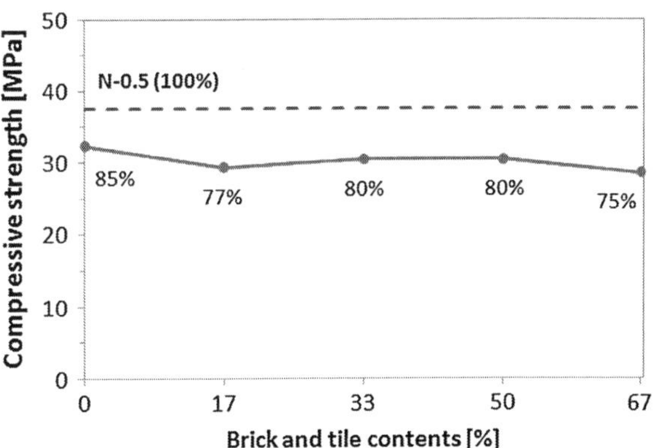

Fig. 5.14 Effect of brick and tile content in RCA on the CS of RCAC (Chen et al. 2003)

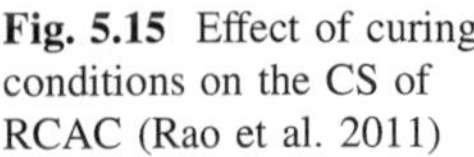

Fig. 5.15 Effect of curing conditions on the CS of RCAC (Rao et al. 2011)

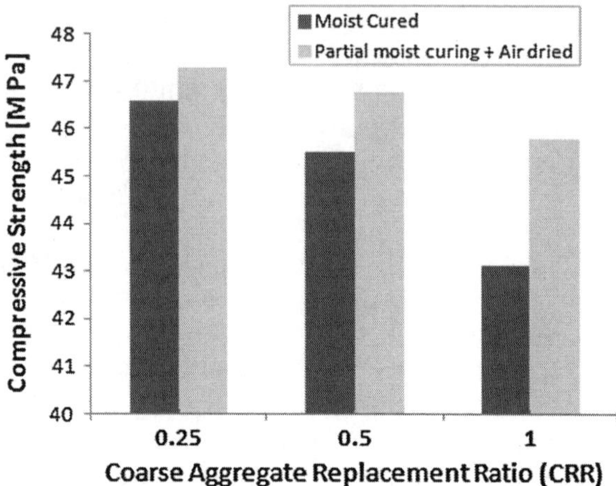

coarse RCA was 10–14 % higher than that of the control concrete when RCA was used to replace partially or fully coarse NA (limestone and gravel). This is due to the addition of a water reducing admixture to the RCA mix, which improved the quality of the mortar more than observed in conventional concrete. One advantage of this method was that the amount of cement necessary to make concrete was about 10 % less than for conventional concrete.

Ferreira et al. (2011) compared the CS of mixes containing coarse RCA using two mixing methods. In the pre-saturation method, the RCA was initially pre-soaked before mixing. In the water compensation method, extra water absorbed by RCA was added during mixing (Fig. 5.16).

The 7- and 28-day CS of RCAC using the pre-saturation method was lower than that using the other method, possibly due to a lower “nailing effect”, which results from the penetration of cement paste into the superficial pores of aggregate particles. Before mixing, pre-saturated RCCA exhibited not only a high level of humidity but also water on the surface and within surface pores. This might have impaired the penetration of the cement paste into the pores, leading to a decrease of the “nailing effect” and, consequently, to a weaker ITZ between cement paste and RCCA. However, the differences of CS between the mixes prepared by two methods became insignificant as the RCCA’s incorporation ratio increased, which might be a consequence of a higher number of weak zones in concrete.

Using a certain amount of mineral additions such as fly ash (FA), metakaolin (MK), silica fume (SF), rice husk ash or latent hydraulic materials such as ground blast furnace slag (gbfs) into cement have some beneficial effect on the CS performance of RCAC.

Kou and Poon (2006) observed an improvement in 90-day CS of RCAC due to 25 % replacement of OPC by FA; however, the addition of 35 % FA had a negative effect on the CS of RCAC. The gain in CS between 28 and 90 days of RCAC containing 25 and 35 % FA is higher than that observed in conventional concrete. The strength gain of concrete containing RCA between 28 days and

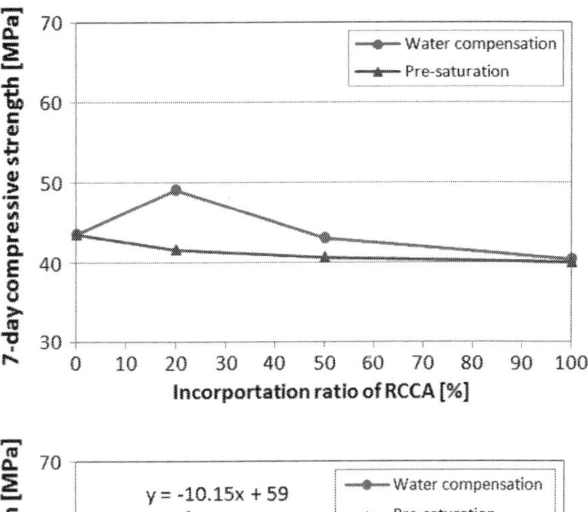

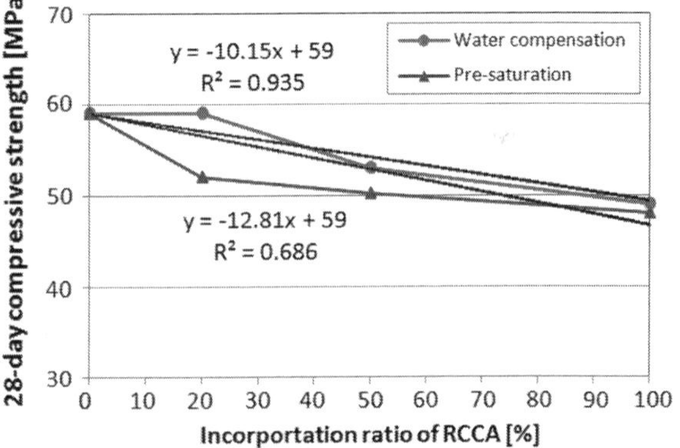

Fig. 5.16 Effect of the mixing procedure on the CS of RCAC (Ferreira et al. 2011)

10 years increased with the FA content (Poon and Kou 2010). Corinaldesi (2009) observed higher CS for RCAC prepared replacing Portland cement with 30 % FA or 15 % silica fume than the RCAC containing OPC when the mixes were moist cured up to 56-days. The strength of RCAC containing silica fume was significantly higher than that of the RCAC containing FA and even better than the conventional concrete at all curing periods. Gonzalez-Fonteboa and Martinez-Abella (2008) also observed a significant improvement of CS of RCAC due to the replacement of 8 % Portland cement by silica fume. Sagoe-Crentsil et al. (2001) observed much higher long-term CS in RCAC with incorporation of slag partially replacing cement than in conventional concrete owing to the higher reactivity of slag than Portland cement at the later stage of hydration. The gain in CS between 7 and 28 days was 12.4 MPa in comparison to 6 MPa for conventional concrete. Razaqpur et al. (2010) observed that the addition of 25 % (w/w) fly ash (FA) into cement reduced the CS of the resulting RCAC due to slow hydration of FA in the concrete matrix. On the other hand, the CS of RCAC containing 35 % gbfs was higher than that of conventional concrete.

Ann et al. (2008) compared the CS of two types of concrete containing RCA as coarse aggregate and made by replacing 30 and 65 % of Portland cement by

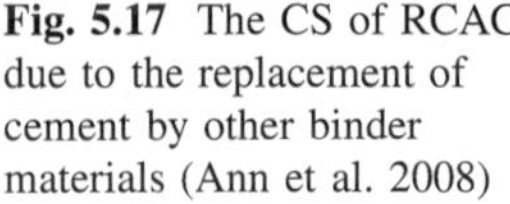
Fig. 5.17 The CS of RCAC due to the replacement of cement by other binder materials (Ann et al. 2008)

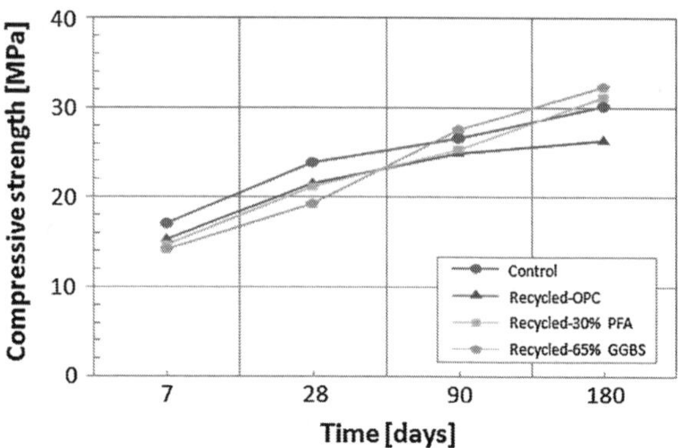

pulverized fuel ash (PFA) and gbfs with RCAC and conventional concrete mixes (Fig. 5.17). They observed lower CS for RCA concrete containing PFA and gbfs than for the control and RCAC mixes at the ages of 7 and 28 days, due to the lower hydraulic reactivity of these binders. However, the CS of concrete containing gbfs and PFA overtook the CS of the conventional concrete after 90 and 180 days of curing, respectively.

Kou et al. (2011b) observed a substantial reduction in CS due to the replacement of cement by 55 % gbfs or 35 % FA. However, they observed an improvement in the CS of RCAC with 50 and 100 % (by volume) replacement of coarse NA by RCA due to the replacement of cement by 10 % SF or 15 % MK. The CS of RCAC with SF and MK was similar to that of the conventional concrete for a 50 % replacement ratio of NA by RCA (Fig. 5.18). However, for a 100 % RCA replacement ratio, the 90-day CS of RCAC containing all type of mineral additions was substantially lower than that of the conventional concrete. The gain in CS of RCAC concrete with SF and MK was higher than that of the concrete containing FA and gbfs in the early ages (up to 28 days); on the other hand, later on (28–90 days), the gain in CS was higher for RCAC with FA and gbfs than for RCAC with SF and MK. The improvement of ITZ due to the filling of pores as well as cracks in RCA by the hydration products was the reason for the observed improvement. The contribution of mineral additions to the improvement of CS of RCAC was more than in conventional concrete. Ajdukiewicz and Kliszczzewicz (2002) observed a significant improvement in CS of RCAC due to the addition of SF and superplasticizer. The improvement was more prominent for RCA produced from high-strength concrete aggregates. Thangchirapat et al. (2008) reported that the use of rice husk-bark ash (RHBA) to replace 20 and 35 % by weight of Portland cement in RCAC yielded higher CS than in RCAC prepared without RHBA.

Several other methods were adopted to overcome the negative effect of CDW aggregate on the CS of concrete. Increasing cement content or lowering the water to cement ratio in the mix and improving the concrete mixing process have already been mentioned during the discussion of CS of concrete containing RA and RCA aggregate. Since the major reason of the reduction in CS of RAC and RCAC is the

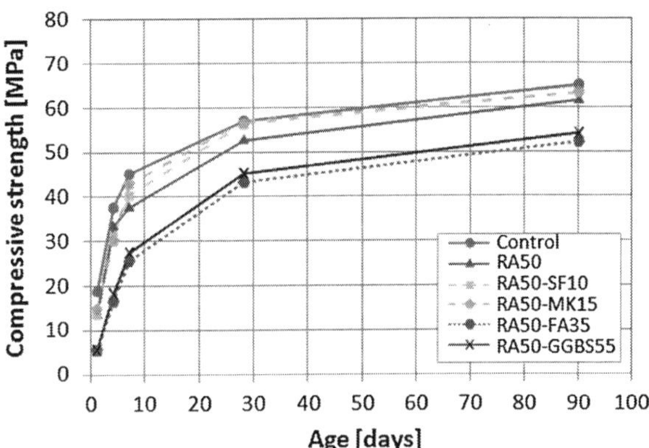

Fig. 5.18 CS of concrete due to the replacement of cement by mineral addition (Kou et al. 2011b)

weaker bond between the recycled aggregate and the cement paste, some studies are also directed at improving the bond strength by using some other materials. Out of these, coating the aggregate's surface with silica fume or nano-silica is most promising one.

Chen et al. (2003) observed a 10–15 % higher CS of concrete containing washed RCA than that of concrete containing unwashed RA at various w/c ratios. The RA was washed to remove the sand content and other impurities like bricks and tiles. Shayan and Xu (2003) observed that the surface treatment of coarse RCA by silica fume slurry can substantially improve the CS of the resulting concrete; however, sodium silicate treatment did not have any beneficial effect. They reported that structural concrete with 50 MPa strength grade could be produced by replacing all the coarse NA with silica fume treated coarse RCA and by replacing a maximum 50 % by weight of fine NA with untreated fine RCA. Katz (2004) also observed an increase of about 23–33 and 15 % in 7- and 28-day CS due to the impregnation of SF on the surface of coarse RCA. The improvement of the microstructure of the ITZ between the cement paste and the RCA aggregate surface and the mechanical performance of RCA can be the major cause for the observed increase in CS. Although it was not as prominent as SF impregnation, ultrasonic treatment to remove unbound particles of RCA also increased by about 7 % the 7- and 28-day CS of RCAC.

Akbarnezhad et al. (2011) observed significantly smaller reduction in the CS of concrete due to the incorporation of RCA obtained after microwave heating (MRCA) than that observed for normal RCA incorporation due to the removal of a part of the adhered mortar content as well as of weak RCA particles (Fig. 5.19). The RCA was heated using microwaves to remove adhered mortar. The differences in 28-day CS between conventional concrete and mixes containing MRCA were negligible up to 40 % replacement of coarse NA. These differences were respectively 10 % and around 30 % for the mixes containing RCA and MRCA as sole coarse aggregate.

Tam et al. (2007) observed up to 21 % improvement in 28-day CS of concrete with 20 % by volume replacement of coarse natural aggregate by RCA using a

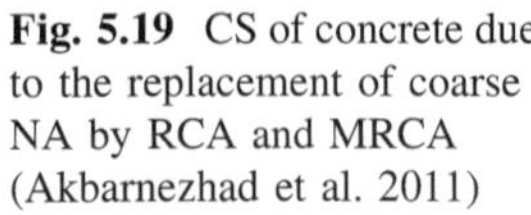
Fig. 5.19 CS of concrete due to the replacement of coarse NA by RCA and MRCA (Akbarnezhad et al. 2011)

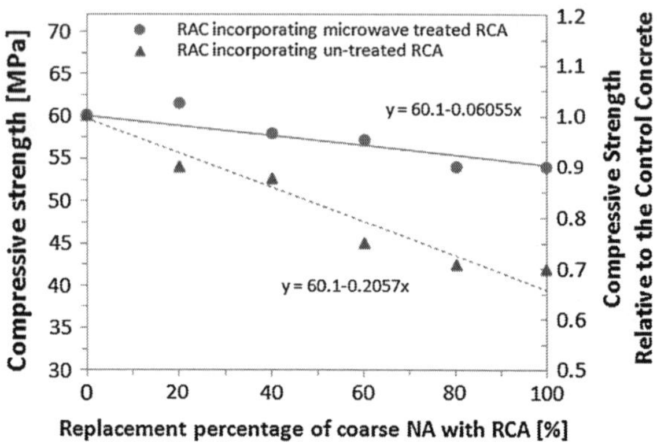

two-stage mixing approach. In the first stage, a cement layer is formed by mixing the cement with aggregates and half of the total water content. The cement layer can fill the pores, cracks and voids of the old mortars and improve the bond during the hardening stage. The addition of a small amount of silica fume to the RCA in the pre-mix procedure can improve by about 20 % the 28-day CS of concrete containing 25 % RCA (Tam and Tam 2008). Further improvement of CS could be achieved by adding silica fume and a fraction of cement with RCA during the pre-mix stage (Tam and Tam 2008). The filling of the pores and cracks of the old mortar with silica fume and the improvement of the aggregate—matrix bond associated with the formation of a less porous transition zone and a better interlock between the paste and the aggregate are the reasons for the observed improvement. Mixing the cement with silica fume provides relatively thick and soft coatings of the silica fume slurry and the necessary cement paste surrounding RA in the pre-mix process and therefore further enhances the strength of the ITZ.

Tsujino et al. (2007) observed an increase in the CS of RCA concrete, when the RCA is coated with mineral oil; however, silane coating of the surface of RCA decreased the CS of RCAC. Kou and Poon (2010) observed that the 90-day CS of RCAC prepared by using air-dried 10 % polyvinyl alcohol impregnated RCA was similar to that of the conventional concrete. The observed improvement was due to the various physic-chemical changes such as the filling of pores and cracks of RCA, the improvement of flocculation and coagulation of the cement particles, the improvement in the ITZ section, the reduction of w/c at the paste-aggregate interface and the reduction in bleeding. However, oven drying of PVA impregnated RCA has no effect on the CS performance of concrete.

Kutcharlapati et al. (2011) reported that the treatment of RCA by a colloidal solution of nano-silica can improve the mechanical properties of RCA and RCAC. The cubic CS of nano-silica containing RCAC and conventional concrete at the same age was 22 and 16 MPa, respectively. An improvement of the mix's cohesiveness, a reduction in segregation and bleeding, improvement in pore structure and a densification in ITZ of the RCA-cement paste due to the filler effect and hydration of the nano-silica particles were the major causes for the observed CS improvement due to the nano-silica treatment of RCA.

Topçu and Saridemir (2008) successfully predicted the 3- to 90-day CS of RCAC by applying artificial neural networks and fuzzy logic systems. Lin et al. (2004) used the optimal mix proportioning of RCAC by orthogonal array, ANOVA and significance test with F statistics to prepare a RCA concrete mix with suitable mechanical strength. The optimum RCAC mix observed in this investigation contains RCA and natural river sand as 100 % coarse and fine aggregates, respectively, with water/cement ratio of 0.5 and volume ratio of recycled coarse aggregate in RCAC 42.0 %. The mix should contain no crushed brick and unwashed RCA should be used. The slump and 28-day CS of this optimum mix was 180 mm and 30.17 MPa, respectively.

The use of fine fraction (<4 mm fraction) of CDW aggregate in preparation of concrete is not as thoroughly studied as the coarse fraction. It is believed that the greater water absorption capacity of the fine fraction of CDW waste can jeopardize the use of this fraction as fine aggregate in concrete (Evangelista and de Brito 2007).

Merlet and Pimienta (1993) observed 19–39 % lower CS of concrete made with fine and coarse recycled aggregate than that of a control concrete. Leite (2001), on the other hand, observed increasing and decreasing CS of concrete due to increasing incorporation of fine and coarse recycled aggregate (FRA and CRA, respectively), respectively (Fig. 5.20). The author justifies the results with the stronger bond created between FRA and the matrix, because of the precipitation of cement crystals inside the FRA.

Khatib (2005) observed a reduction in CS with increasing content of fine recycled concrete aggregate (FRCA) in concrete. The particle size of FRCA was <5 mm. The concrete was prepared at free water to cement ratio of 0.5. The reduction in 90-day CS was in the range of 15–27 % when 25–100 % by weight of fine natural aggregate (FNA) was replaced by FRCA. However, 28-day CS of concrete containing 25–75 % and 100 % FCRA were about 25 and 36 % lower than that of the control, respectively. The relative CS of FRCA concrete (FCRAC) in the 28–90-day period increased due to the hydration of un-hydrated cement

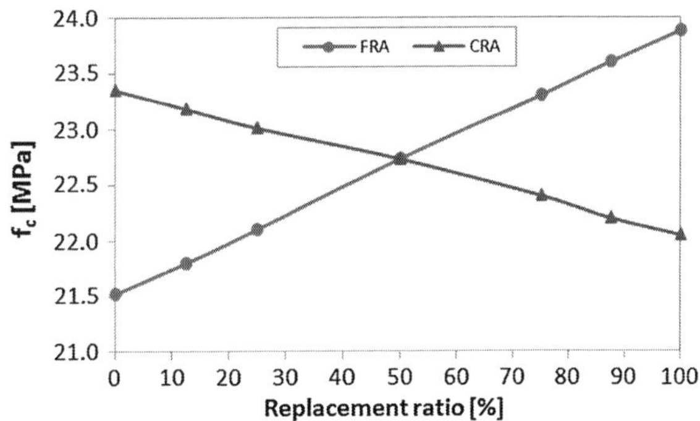

Fig. 5.20 CS of RCAC due to incorporation of fine (FRA) and coarse (CRA) recycled aggregates (Leite 2001)

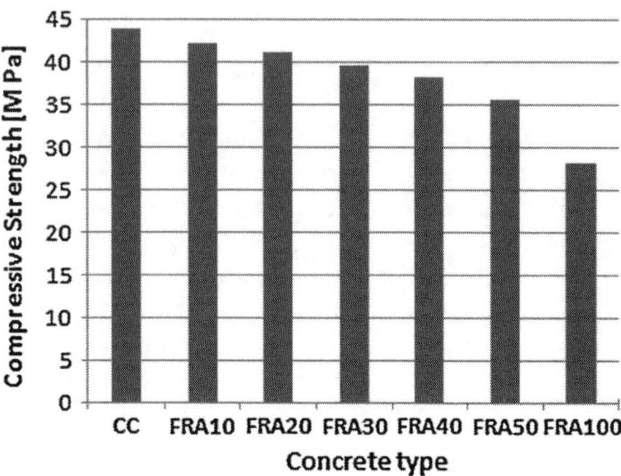

Fig. 5.21 Reduction in CS of concrete due to the incorporation of fine recycled aggregate (Yaprak et al. 2011)

particles of FCRA. Yaprak et al. (2011) also observed a gradual drop in CS as the replacement of FNA by FRCA increased (Fig. 5.21).

Evangelista and de Brito (2007) conducted a comprehensive study on the use of pre-saturated FRCA (<2.36 mm) as a partial or complete replacement of same size FNA in the preparation of structural concrete. The concrete mixing time was maintained as 10 and 20 min for series I and series II concrete, respectively. The results are presented in Table 5.10. The 28-day CS of series I FRCAC was marginally higher (2–5 %) than the conventional concrete, due to the pozzolanic reaction of un-hydrated cement present in FRCA. On the other hand, the CS decreased by about 0.6–7.6 % with respect to the conventional concrete in series II FRCAC since the increased soaking time of FRCA weakened the cement paste-aggregate bond by increasing the w/c in that particular region. The development of CS of conventional concrete almost stabilised after 28 days; however, the CS development of FRCAC continued after 28 days due to the hydration of cement present in FRCA. Zega and Di Miao (2011) also observed a slight decrease in 28- and 84-day CS of concrete due to a 20 and 30 % by volume replacement of fine NA by fine RCA.

Table 5.10 CS of concrete containing fine recycled concrete aggregate (Evangelista and de Brito 2007)

Concrete type	Amount of substitution of FNA by FRCA (%, v/v)	Compressive strength (MPa)	
		Series I	Series II
FNAC	0	59.4	59.3
FRCAC10	10	62.2	59.0
FRCAC20	20	58.4	57.3
FRCAC30	30	61.3	57.1
FRCAC50	50	60.8	58.8
FRCAC100	100	61.0	54.8

Pereira et al. (2012) reported that the compressive strength of concrete containing FRCAC could be improved by using superplasticizer and a lower w/c ratio. The 28-day compressive strengths of conventional concrete and FRCACs containing FRCA as the only fine aggregate with and without two types of superplasticizers were respectively 39.5, 38.6, 45.1 and 63 MPa indicating a significant increase in compressive strength due to the addition of superplasticizer in the mix of RCAC. They also observed much a lower influence of FRCA on CS performance in comparison to the change in w/c ratio. They proposed the following relationship between CS (f_c) and effective w/c ratio $\left(\left(\frac{\mathrm{w}}{\mathrm{c}}\right)\mathrm{ef}\right)$ (correlation coefficient of 0.96) from their experimental results:

$$f_c = \frac{230.3}{(25.9)^{\left(\frac{\mathrm{w}}{\mathrm{c}}\right)\mathrm{ef}}} \times (1 - (-0.077) \cdot W_{24} \cdot r) \tag{5.1}$$

where W_{24} is 24 h water absorption capacity of concrete, r is replacement ratio and numerical values are determined by regression analysis.

Kou and Poon (2009a) prepared two concrete series by replacing 25, 50, 75 and 100 % by weight of fine natural aggregate by fine recycled aggregate (FRA) with particle size below 5 mm. The first and second series of concrete were prepared using the same cement content at constant water to cement ratio (w/c) of 0.53 and a close slump range of 60–80 mm. At same w/c, the CS of concrete decreased with the content of FRA due to higher bleeding as well as poor aggregate-cement paste bond owing to the higher initial free water content. At constant slump, the CS of series II concrete also decreased with the FRA content; however, the deterioration of CS was marginally higher than that observed in series I. According to the authors, this was due to the weaker mechanical properties of FRA and FNA.

By replacing 100 % of fine natural aggregate by fine RCA, Kou and Poon (2009b) prepared a self-compacting concrete, which can yield CS values as high as 64 MPa. The authors concluded that the inclusion of FRCA up to a ratio of 25–50 % does not significantly change the CS of the resulting concrete. Dapena et al. (2011) observed a drop of around 7.3–9.4 % in the CS of concrete due to the replacement of 10 % coarse RCA by FRCA, where natural coarse aggregate was replaced by 20, 50 and 100 % (by volume) of coarse RCA.

Kou and Poon (2008) reported up to 5 years experience of CS of concrete prepared by replacing 0, 20, 50 and 100 % of coarse NA by an equal volume of RA and RCA. They observed lower CS of RCA concrete than that of RA concrete after 28-day but after 5 years the CS was highest for RCA for all substitution ratios. The CS of concrete containing RCA and RA was always lower than that of the conventional concrete; however, the reduction of CS decreases as curing time increases. The 28-day to 5-year gain in the CS of concrete containing RA and RCA was higher than that of the control concrete and it was highest for RCA concrete. The strengthening of the paste-aggregate bond, the healing of cracks in interfacial zone due to the deposition of new hydration products and the reduction in the preferred orientation of $Ca(OH)_2$ crystals were the main causes for the observed improvement of the CS of RCAC after prolonged curing. The gain in CS also increased with the RA and RCA contents in concrete.

5.3.2 Splitting Tensile Strength

Like for CS, the splitting tensile strength (STS) of concrete containing RA or RCA is normally lower than that of conventional concrete and increasing the addition of CDW aggregate into concrete further lowers it. Table 5.11 shows some typical data. The causes for CS reduction are also responsible for STS reduction. According to de Brito and Alves (2008) the lower mechanical properties of the recycled aggregates when compared to the natural ones lead to a fall in the STS of the concrete containing CDW aggregate as the substitution rate increases.

After analysing literature data de Brito and Robles (2010) reported that the replacement ratio of natural aggregate by CDW aggregate had slightly less effect on the reduction in STS than that observed in CS. Table 5.12 shows the differences in 28-day STS and CS of RAC and RCAC from conventional concrete determined from the data presented in various references. A great variation in the reduction in STS due to the incorporation of RCA or RA in concrete was observed, probably due to the variations in experimental parameters and properties of CDW aggregates used in the various studies.

From the results presented in Table 5.12 it can also be concluded that the inclusion of RA or RCA into concrete had little impact on STS or that the CDW aggregate mixes had higher STS than conventional concrete up to a given replacement level. This is probably due to the improvement in aggregate-cement paste bond strength, which induces a higher increase in STS than in CS (Kou and Poon 2008).

Corinaldesi (2009) observed similar STS for RCAC prepared with and without mixing FA into Portland cement. Etxeberria et al. (2007a) observed higher STS of

Table 5.11 28-day splitting tensile strength of concrete containing various amount of CDW aggregate

Reference	Type of aggregate	Tensile strength (MPa)/substitution level (%)
Gomez-Soberon (2002)	RA/fine + coarse	3.7/0; 3.7/15; 3.6/30; 3.4/60; 3.3/100 (v/v)
Gonzalez-Fonteboa and Martinez-Abella (2008)	RCA/coarse	3.15/0; 3.00/50 (v/v)
Kou and Poon (2008)	RCA/coarse	2.43/0; 2.40/20; 2.35/50; 2.26/100 (v/v)

Table 5.12 Reduction in 28-day CS and STS due to addition of CDW aggregate in conventional concrete

Reference	Type of aggregate	Reduction (%)		Replacement amount of similar size aggregate (%)
		Compressive strength	Tensile strength	
Corinaldesi and Moriconi (2009)	RCA/fine + coarse	6.3	6.2	100
Gomez-Soberon (2002)	RA/fine + coarse	10.8	11.5	100
		8.1	8.2	60
		2.7	5.1	30
Gonzalez-Fonteboa et al. (2011)	RCA/coarse	13.2	8.48[a]	20, 50 and 100 %; w/c = 0.50
		25.9	2.41[a]	
		20.4	3.53[a]	
Gonzalez-Fonteboa and Martinez-Abella {2008}	RA/coarse	0.38	4.77	50 (by volume)
		11.6[a]	6.78[a]	50 (by volume) and 8 % silica fume with cement
Katz {2003}	RCA/coarse + fine	23.1	3.03[a]	~100; crushing age: 1-day
		25.4	12.1	Crushing age: 3-day; replacement: ~100 %;
		22.5	6.06	~100; crushing age: 28-day;
Kou et al. (2011a)	RCA/coarse	21.7	9.05	100;
	RA/coarse	18.7	7.00	100;
Mas et al. (2012)	RA/coarse (8/40)	18	14	25; w/c = 0.65
		19	6	50; w/c = 0.65
		21	21	75; w/c = 0.65
	RA/coarse (8/40)	13	11[a]	20; w/c = 0.75
		13	10	40; w/c = 0.75
	RA/coarse (8/20)	26	25	20; w/c = 0.45
		39	34	40; w/c = 0.45
Etxeberria et al. (2007a)	RA/coarse	3.5	19.3[a]	25

[a] Increasing with respect to conventional concrete strength

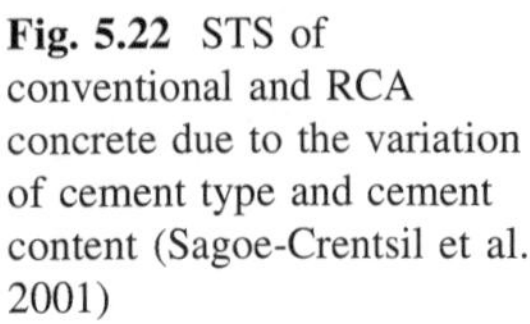

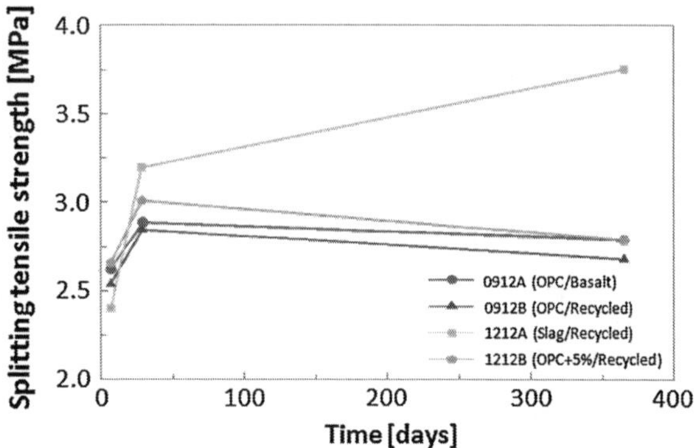

Fig. 5.22 STS of conventional and RCA concrete due to the variation of cement type and cement content (Sagoe-Crentsil et al. 2001)

RAC containing RA as 25 and 50 % replacement of coarse natural aggregate than in conventional concrete but not for the 100 % replacement. In this study, the CS of RAC with 100 % coarse RA was also comparable to that of conventional concrete. The authors point out that the absorption capacity of the adhered mortar present in the partly saturated (humid) recycled aggregate and the effectiveness of the new interfacial transition zone of the recycled aggregate concrete increased the STS.

Sagoe-Crentsil et al. (2001) observed slightly low and significantly high 28-day and 1-year STS, respectively, for RCAC containing OPC and slag cement than for conventional concrete containing OPC (Fig. 5.22). On the other hand, increasing the cement content led to higher STS than for conventional concrete at the early ages, but after 1-year the values became similar.

However, in some studies, a substantial reduction in STS in comparison to that of CS due to the addition of CDW aggregate was also reported. Some typical examples are presented in Table 5.13. Evangelista and de Brito (2007) reported that the un-hydrated cement content in fine recycled aggregate can affect the CS

Table 5.13 Reduction in 28-day CS and STS of concrete containing CDW aggregate in comparison to conventional concrete

Reference	Type of aggregate	Reduction (%)		Replacement ratio of similar size aggregate (%)
		Compressive strength	Tensile strength	
Yang et al. (2011)	RCA/coarse	5.7	13.8	100
Evangelista and de Brito (2007)	RCA/fine	2.7[a]	30.5	100
Rao et al. (2011)	RAC/coarse	7.48	13.9	25
		14.1	18.0	50
		17.5	23.2	100
Gonzales-Fonteboa et al. (2011)	RCA/coarse	10.7	17.2	20, 50 and 100 %; w/c = 0.65
		9.38	14.8	
		10.6	9.96	

[a] Increasing with respect to conventional concrete strength

Table 5.14 STS of concrete with various ratios of CDW aggregate at different curing ages

Reference	Type of aggregate	Substitution level (%, v/v)	Tensile strength (MPa)/day
Gomez-Soberon (2002)	RA/fine + coarse	0	3.6/7d; 3.7/28d; 3.9/90d;
		15	3.3/7d; 3.7/28d; 3.9/90d;
		30	3.3/7d; 3.6/28d; 3.9/90d;
		60	3.2/7d; 3.4/28d; 3.7/90d;
		100	3.5/7d; 3.3/28d; 3.6/90d;
Gonzalez-Fonteboa and Martinez-Abella (2008)	RCA/coarse	0	3.12/7d; 3.15/28d; 3.32/115d;
		50	3.17/7d; 3.00/28d; 3.37/115d;
Kou and Poon (2008)	RCA/coarse	0	2.43/28d; 2.68/90d; 2.83/180d; 2.94/1Y; 3.16/2Y; 3.32/5Y
		20	2.39/28d; 2.56/90d; 2.78/180d; 2.91/1Y; 3.21/2Y; 3.40/5Y
		50	2.34/28d; 2.52/90d; 2.74/180d; 3.04/1Y; 3.28/2Y; 3.52/5Y
		100	2.21/28d; 2.48/90d; 2.76/180d; 3.12/1Y; 3.36/2Y; 3.64/5Y

performance but it does not affect the STS performance and therefore in comparison to CS, a substantial reduction in STS was observed in RCA due to porous nature of the RCA. They observed a lowering in STS as the replacement ratio of fine RCA in concrete increased.

In several studies, it was reported that the STS of concrete containing RCA and RA improved substantially at the latter stages of curing and in some cases, the strength of concrete containing RCA or RA was even better than that of conventional concrete. Table 5.14 presents some typical results of STS of concrete containing natural and CDW aggregates with increasing curing time. Kou and Poon (2008) pointed out that the improvement in the microstructure of the interfacial transition zone (ITZ) and therefore an increase in the bond strength between the new cement paste and the old aggregates might be the factor for the observed improvement of STS.

Kou et al. (2011a) observed a 10 and 7 % lower 28-day STS of concrete containing coarse RA and RAC than that of conventional concrete. However, the STS of both recycled aggregate mixes was higher than that of conventional concrete after 1 and 5 years of curing (Fig. 5.23). The development of STS between 28 days and 5 years for RCAC and RAC were respectively 65 and 56 % compared to 37 % for conventional concrete. The percentage gain in STS from 28 days to 5 years of curing also increased with the RA or RCA contents. An improvement in the microstructure of the ITZ, increasing the bond strength between the new cement paste and the old aggregates after prolonged hydration, and the self-cementing ability of recycled aggregate were the causes for the observed increase in STS of the RCAC at the later stages of curing. The higher improvement in STS than CS observed in this study was due to the improvement in cement paste-aggregate bond strength. In another study, Kou and Poon (2008) also observed

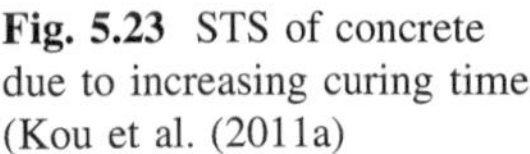

Fig. 5.23 STS of concrete due to increasing curing time (Kou et al. (2011a)

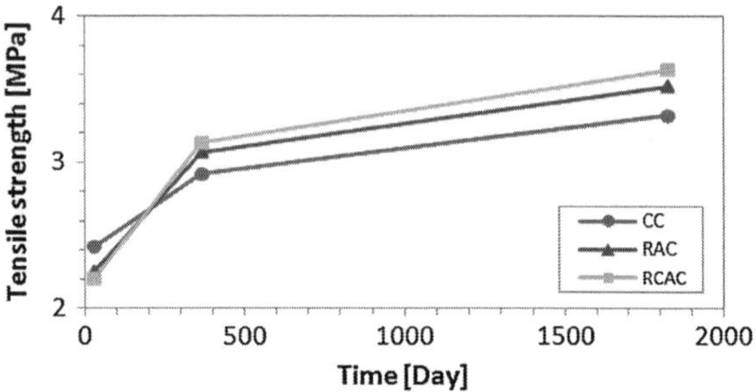

Fig. 5.24 STS of concrete after 28 days and 5 years of curing for various replacement of coarse NA by RCA (Kou and Poon 2008)

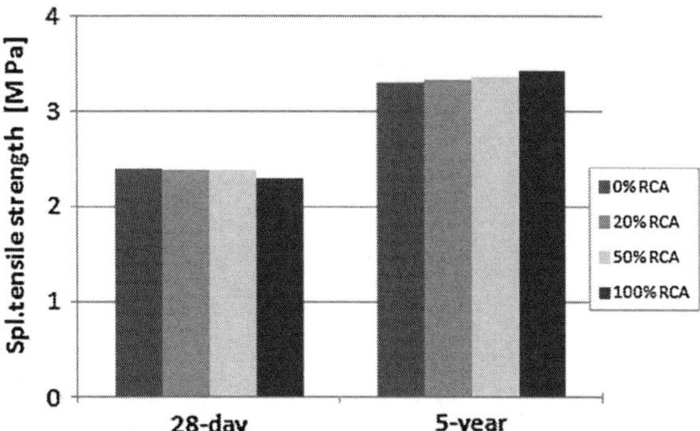

higher STS in RCAC than in conventional concrete after 5 years of curing, which also increased with the replacement ratio of coarse NA by RCA (Fig. 5.24). Conversely, Yong and Teo (2009) observed an improvement in the STS of concrete due to the replacement of 100 % of natural aggregate by coarse normal or saturated surface dry RCA at the early stage (up to 28 days) of curing; however, strength development of both types of RCAC slowed down in the 28–56 days curing period.

Gonzales-Fonteboa et al. (2011) observed an effect of w/c ratio on the STS performance of concrete containing RCA (Table 5.12 and 5.13). They observed higher STS in RCAC than in conventional concrete at the w/c ratios of 0.50; on the other hand, the STS of conventional concrete was higher than that of the RCAC at the w/c ratio of 0.65. However, Kou and Poon (2006) observed higher STS for concrete prepared by replacing various percentages of natural coarse aggregate by RCA at high w/c ratio than that prepared at low w/c ratio. They justified their results with the variability in surface texture of RCA, which gave the paradoxical results.

Yang et al. (2008) observed decreasing STS of RCAC with increasing water absorption capacity of incorporated RCA. Similarly to CS, Gomez-Soberon (2002) and Kou et al. (2011a) observed an inverse relationship between STS and porosity of RCAC. The relationship between open porosity and STS observed by Gomez-Soberon (2002) study is presented in Fig. 5.25.

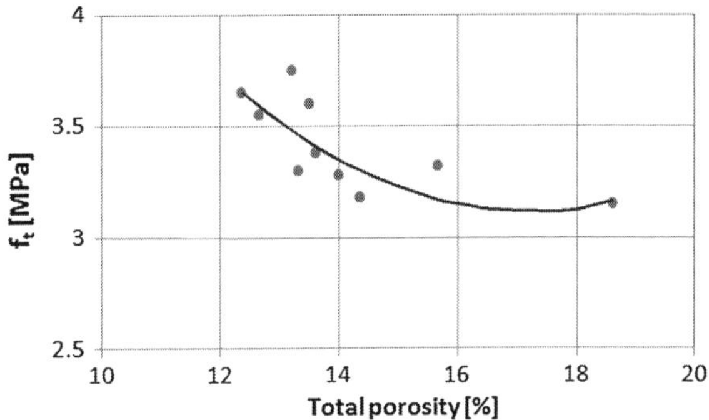

Fig. 5.25 STS versus total porosity curve of RCAC (Gomez-Soberon 2002)

The presence of impurities in RCA such as crushed clay brick and tiles has positive or negative effects on the STS of resulting RCAC depending on the substitution ratio. Yang et al. (2011) observed lower 7- and 28-day STS in RCAC prepared by replacing 20 and 50 % of RCA by crushed brick than in conventional RCAC (Fig. 5.26), owing to the higher porosity of crushed brick aggregate than of RCA. In crushed brick containing concrete, most of the tensile failures occurred within the crushed brick particles while for other mixes, failure seemed to occur between the aggregate and the mortar matrix interfaces. The STS of concrete paving blocks containing RCA decreased with increasing replacement of fine or coarse RCA by crushed clay brick in the Poon and Chan (2006) study too. Kou et al. (2011a) also observed lower STS for RAC than for RCAC especially at the later stages of hydration.

However, Poon and Chan (2007) observed higher STS in concrete paving blocks due to the 10 % replacement of RCA by crushed tiles or a 1:1 mixture of crushed tiles and bricks. Mixing crushed glass with tiles or tiles/bricks gave strength comparable to that of conventional RCA concrete. On the other hand, the addition of crushed wood drastically lowered the STS. Strengthening of the

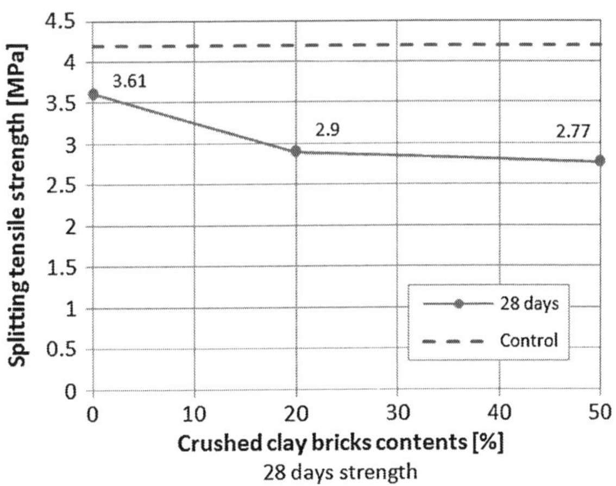

Fig. 5.26 Effect of crushed brick content in RCA on the STS performance (Yang et al. 2011)

cement paste/aggregate binding due to the penetration of cement paste into the porous tile and brick aggregate, filling of pores by fine tiles and brick aggregate as well as the presence of more tiles and bricks aggregate in concrete due to their lower density were the major causes for this improvement.

Nagataki et al. (2004) observed comparable STS for RCAC incorporating RCA with minimum adhered mortar content to that of conventional concrete made with the original aggregates. The RCA with minimum adhered mortar content even exhibited higher STS than conventional concrete. The smaller size, lower sand content as well as the elastic compatibility between RCA and cement paste were the causes of the good performance of RCAC. Padmini et al. (2009) observed lower STS in RCAC than in conventional concrete and the difference narrowed down as the CS decreased. In contrast to interfacial bond failure between cement mortar and aggregate observed in conventional concrete, RCAC exhibited both interfacial bond failure and aggregate failure in the STS tests (Padmini et al. 2009; Rao et al. 2011).

Tabsh and Abdelfatah (2009) reported that the STS for 50 and 30 MPa classes of conventional concrete as well as that of RCAC were similar when the RCA was generated from 50 MPa concrete. On the other hand, a drop by 25–30 % and 10–15 % in STS was observed for both concrete classes when RCA was generated from 30 MPa concrete. Tavakoli and Soroushian (1996) observed a negligible effect of aggregate size or dry mixing time on the STS of RCA concrete. The 28-day STS of RCAC with two types of RCA is either higher or statistically comparable to that of the control concrete for limited ranges of various experimental parameters such as size of coarse RCA, mixing time and w/c ratio.

The addition of several mineral admixtures such as silica fume, fly ash, rice husk ash does not have prominent beneficial effect on STS improvement as observed in CS (Gonzalez-Fonteboa and Martinez-Abella 2008; Thangchirapat et al. 2008). Gonzalez-Fonteboa and Martinez-Abella (2008) observed around 6.8 % higher STS in RCAC prepared at w/c of 0.55 than in conventional concrete due to the incorporation of silica fume as mineral admixture into cement but the improvement was not as significant as for CS (around 11.6 %). Ajdukiewicz and Kliszczzewicz (2002) observed improvement in STS of RCAC due to the addition of SF and superplasticizer, but the improvement was not as significant as in CS. Kou et al. 2007) also observed lower STS for RCAC using a blended cement prepared by replacing 25 % (by weight) OPC by FA than for RCAC using OPC. The increasing addition of FA into 35 % further lowered the strength. On the other hand, the same authors in another publication (2008) reported that the addition of fly ash as a replacement of 25 % of cement by weight can increase the STS. The major difference between these two studies was the larger amount of binder content in the mix containing FA in the later study than in the former one. The improvement was due to the pozzolanic activity of FA which densified the concrete matrix by improving porosity.

The replacements of 10 % OPC by SF or 15 % OPC by metakaolin (MK) gave higher STS to the resulting mixes prepared by replacing 50 and 100 % (by volume) of coarse natural aggregate by RCA than that of the control and of the RCAC

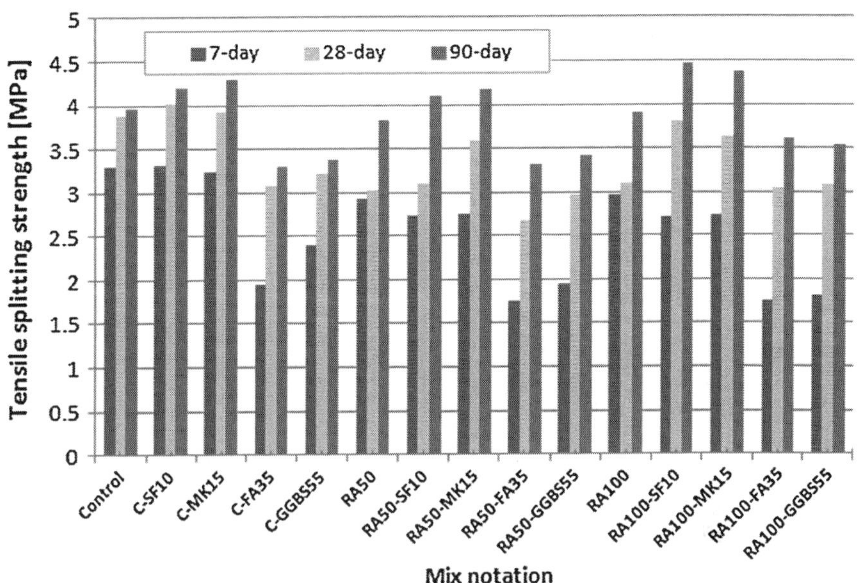

Fig. 5.27 STS of control concrete and RCAC containing various types of mineral admixtures (Kou et al. (2011b)

prepared with OPC at the curing ages of 7, 28 and 90 days (Kou et al. (2011b) (Fig. 5.27). On the other hand, the RCAC prepared by replacing 35 or 55 % of OPC by FA or ground granulated blast furnace slag (ggbfs) respectively had lower STS than the control concrete and the RCAC containing OPC at all the curing periods. The formation of more hydration products due to the SF and MK hydration and the consequent improvement of the microstructure of ITZ increased the binding of RCA and cement paste and hence improved the STS. The increase in STS between 7 and 90 days was higher for RCAC using blended cement than for the control concrete and the RCAC using OPC. The increase in tensile strength was higher for FA and ggbfs than for SF and MK too.

Ann et al. (2008) observed that the 28-day STS for RCAC was lower than for conventional concrete. The strength for RCAC using OPC-30 % pulverized fuel ash and OPC-65 % ground blast furnace slag as binder was similar but lower than that of the RCAC concrete. However, the ratios of STS to CS were comparable for all types of concrete (Fig. 5.28). In the Berndt's (2009) study, though the 28-day STS of RCAC using 50 and 70 % ggbfs as replacement of OPC and RCA as sole coarse aggregate was lower than that of concrete using slag cements and natural aggregate as sole coarse aggregate, the STS of mixes having former composition was higher than that of concrete using 100 % OPC and natural coarse aggregate as well as of concrete using 100 % OPC and 100 % coarse RCA.

The STS of concrete with coarse RCA obtained after treatment by polyvinyl alcohol followed by air-drying (PI-R(A) in Fig. 5.29) was higher than that of concrete with untreated RCA at the curing ages of 7–90 days. However, oven

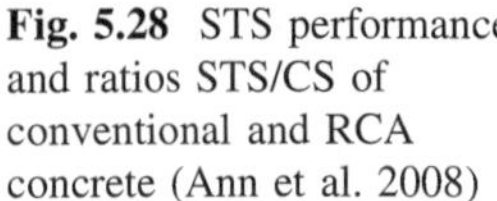

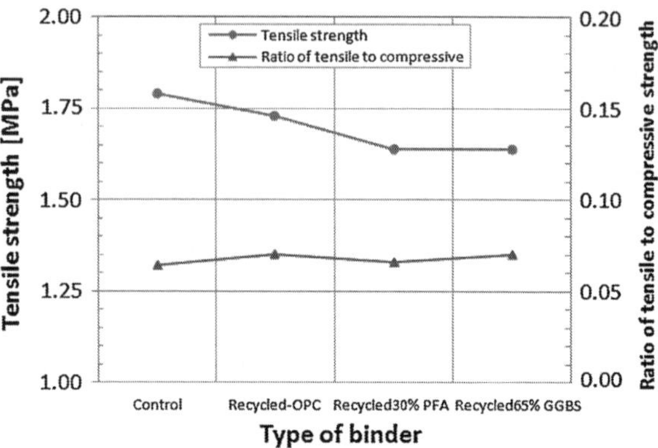

Fig. 5.28 STS performance and ratios STS/CS of conventional and RCA concrete (Ann et al. 2008)

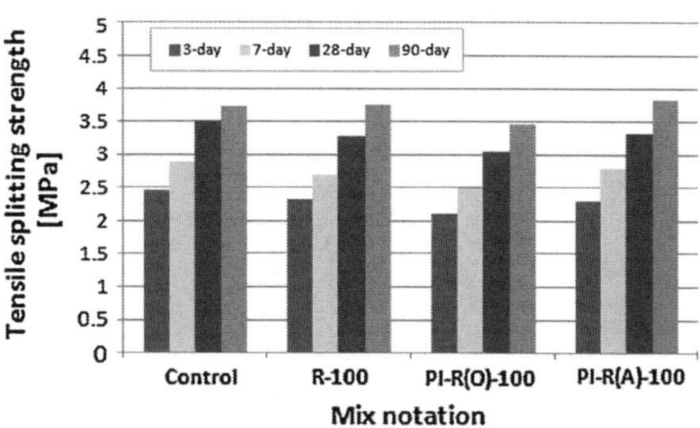

Fig. 5.29 STS of control concrete and RCAC containing normal or polyvinyl alcohol treated RCA (Kou and Poon 2010)

drying of polyvinyl alcohol treated RCA (PI-R(O) in Fig. 5.30) gave the resulting concrete lower strength than concrete with untreated RCA (Kou and Poon 2010). Tsujino et al. (2006) observed a beneficial effect of the surface treatment of coarse RCA by mineral oil in the STS performance of the resulting RCAC. On the other hand, silane treatments of RCA deteriorate the SRS of the resulting concrete.

The STS/CS ratio of RCAC was reported in some studies (Gonzalez-Fonteboa et al. 2011; Sagoe-Crentsil et al. 2001; Thangchirapat et al. 2008; Ravindrarajah and Tam 1985) to be similar to that of conventional concrete. Sagoe-Crentsil et al. (2001) observed ratios of STS to CS in the range of 0.89–1.21 depending on the type of cement used, cement content, curing age and RCA content and these values were similar to the range of 0.99–1.19 exhibited by conventional concrete at various curing ages. Thangchirapat et al. (2008) observed ratios in the range of 0.80–0.82 for RCAC in comparison to 0.95 for conventional concrete. According to Sagoe-Crentsil et al. (2001), the absence of detrimental effect of RCAC on the STS is partly indicative of good bond characteristics between the aggregate and the mortar matrix.

Paine et al. (2009), from their limited data, observed that the existing Eurocode 2 relationship between CS and STS could be applied to RCA concrete as to conventional concrete. Kou and Poon (2008) observed that the relationship between STS and CS presented in ACI 318-89 overestimates their STS data on

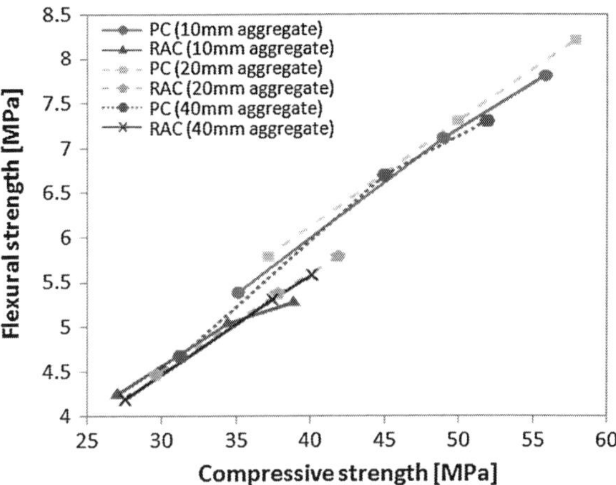

Fig. 5.30 Relationship between FS and CS of conventional concrete and RCAC (Padmini et al. 2009)

RCAC. They proposed the next relationship between STS (f_{sp}) and CS (f_{cu}) with a correlation coefficient of 0.87:

$$f_{\mathrm{sp}} = 0.0931\, f_{\mathrm{cu}}^{0.8842} \tag{5.2}$$

Xiao et al. (2006b) observed a significant overestimation of STS data using American ACI 318-02 and Chinese GB 50010-2002 codes. The authors proposed the following relationship between STS (f_{sp}) and cubic CS (f_{cu}):

$$f_{sp} = 0.24\, f_{cu}^{0.65} \tag{5.3}$$

5.3.3 *Flexural Strength*

Similarly to CS and the STS, the addition of CDW aggregate in concrete lowers the flexural strength (FS). However, in several studies, it is reported that this addition does not reduce FS as substantially as CS. The variation in FS between conventional concrete and concrete containing CDW aggregate was negligible in some studies and was lower than 30 % in others depending on the variations of different factors such as replacement amount, origin and quality of CDW aggregate, w/c ratio, design strength of concrete. Table 5.15 shows some of the results from various references.

Limbachiya et al. (2000, 2004) observed comparable 28-day FS for 50, 60 and 70 MPa concrete classes, prepared by replacing 0, 30, 50 and 100 % (by weight) of natural coarse aggregate by RCA. Safiuddin et al. (2011) did not observe any significant differences in 7- and 28-day FS of conventional and RCA concrete. The strength increased with the curing time like in conventional concrete. The improvement in interfacial bonding and mechanical interlocking due to the

Table 5.15 28-day flexural strength of concrete containing various amount of CDW aggregate

Reference	Type of aggregate	Tensile strength (MPa)/substitution level(%)/day	Comment
Limbachiya et al. (2004)	RCA/coarse	4.4/0/28; 4.3/30/28; 4.3/50/28; 4.5/100/28	–
Limbachiya et al. (2000)	RCA/coarse	5.2/0/28; 5.2/30/28; 4.9/50/28; 5.0/100/28[a] 6.0/0/28; 6.1/30/28; 6.1/50/28; 6.0/100/28[b] 7.0/0/28; 6.9/30/28; 7.0/50/28; 7.2/100/28[c]	[a, b, c] = 50, 60 and 70 MPa Concrete classes
Mas et al. (2012)	RCA/coarse	2.29/0/28; 2.73/20/28; 2.00/40/28[a] 3.83/0/28; 2.85/20/28; 2.70/40/28[b] 2.27/0/28; 2.09/25/28; 1.90/50/28; 1.81/75/28[c]	[a, b, c] = concrete prepared at w/c of 0.65, 0.72 and 0.45
Casuccio et al. (2008)	RCA/coarse 5-21 %	3.9/0/28; 3.7/100/28[a]; 3.2/100/28[b]; C18* 5.2/0/28; 5.3/100/28 [a]; 4.7/100/28[b]; C37 7.3/0/28; 6.0/100/28 [a]; 5.8/100/28[b]; C48	C18, C37 and C48 concrete classes; [a, b]: RCA from C55 and C30 concrete classes
Gupta et al. (2010)	RCA/coarse	6.0/0/28; 5.64/100/28; 6.2/0/28[a]; 6.18/100/28[a]; 4.78/100/28[b]	[a] and [b] = 10 and 20 % FA containing concrete
Gull (2011)	RCA/coarse + fine	6.67/0/3; 4.43/100/3; 6.38/100/3[a] 9.7/0/7; 6.0/100/7; 9.5/100/7[a] 13.0/0/28; 8.2/100/28; 13/100/28[a]	[a] = concrete contains water reducing admixture
Yang et al. (2011)	RCA/coarse	3.18/0/7; 4.19/0/28; 2.94/100/7; 3.61/100/28	–
Ahmed (2011)	RCA/fine	4.87/0/28; 4.77/25/28; 4.86/50/28; 4.03/75/28; 3.97/100/28	–
Heeralal et al. (2009)	RCA/coarse	4.19/0/28; 4.00/50/28; 3.46/100/28	–

*C18 indicates a concrete class with design 28-day strength of 28 MPa; the others are similar

angularity and surface roughness of RCA aggregate as well as the effectiveness of interfacial bonding due to the orientation of larger coarse RCA along the specimen's length compensated the negative impact of the weakness of RCA and therefore maintained a FS similar to that of conventional concrete.

Chen et al. (2010) observed a slight increase in FS due to the replacement of up to 40 % of coarse NA by RCA and similar values to that of conventional concrete above this replacement level. In this study, the ratios of FS to CS were in the range of 0.11–0.13 when 10–100 % coarse NA was replaced by RCA. Ahmed (2011) observed similar 28 and 56 days FS of concrete due to the replacement of 25 and 50 % of natural fine aggregate by fine RCA. However, at 75 and 100 % replacement level, the FS was lower than that of conventional concrete. The 28-day FS of RCAC prepared by replacing 50 and 100 % of coarse NA by RCA in the Malesev et al. (2010) study were respectively 5.7 and 5.2 MPa in comparison to the 5.4 MPa of conventional concrete.

Yang et al. (2011) observed a 7.5–13.8 % reduction in FS due to replacement of 100 % coarse NA by RCA at various ages. Gull (2011) observed a reduction of around 37 % in 28-day FS of concrete due to the replacement of fine and coarse natural aggregates by RCA when both mixes were prepared at w/c ratio of 0.5. However, the 28 day FS of RCAC prepared at the same w/c ratio but by using a water reducing agent was similar to that of conventional concrete. Casuccio et al. (2008) observed a 5–21 % reduction in 28-day FS of concrete due to the replacement of 100 % coarse NA by RCA. Mas et al. (2012) observed 20, 13 and 30 % reductions in FS due to the replacement of up to 75 % (by volume) of coarse natural aggregate by low quality RCA in three types of concrete prepared at w/c of 0.65, 0.72 and 0.45 respectively.

Singh and Sharma (2007) observed a 4–15 % reduction in 1- to 28-day FS of 20 and 25 MPa concrete mixes due to replacement of coarse natural aggregate by RCA aggregate. James et al. (2011) observed a 28-day FS about 2.5 % lower due to the replacement of 25 % by mass of NA by RCA at a w/c ratio of 0.55. Like in conventional concrete, the FS increased with curing time. The differences in FS between conventional concrete and RCAC are lower at higher w/c ratios than at lower ones. The authors did not observe any effect of the w/c ratio on the CS of RCAC either.

Yong and Teo (2009) observed higher 3-day FS for RCAC than for conventional concrete up to a 100 % substitution level of coarse NA by RCA. However, the FS of conventional concrete was higher than that of RCAC when the curing age increased to 28 days. They also reported that the FS performance of RCAC was not as good as that observed for CS and STS due to the lower modulus of elasticity of RCA than NA's; therefore RCA tended to deform more than NA. In comparison to CS, Akbarnezhad et al. (2011) observed a lower reduction in the modulus of rupture as the replacement of coarse NA by RCA increased. At 100 % replacement, the reduction in modulus of rupture and CS was 15 and 30 % respectively. The higher water absorption capacity of RCA might enhance the bond strength between the new mortar and aggregate, which can partially

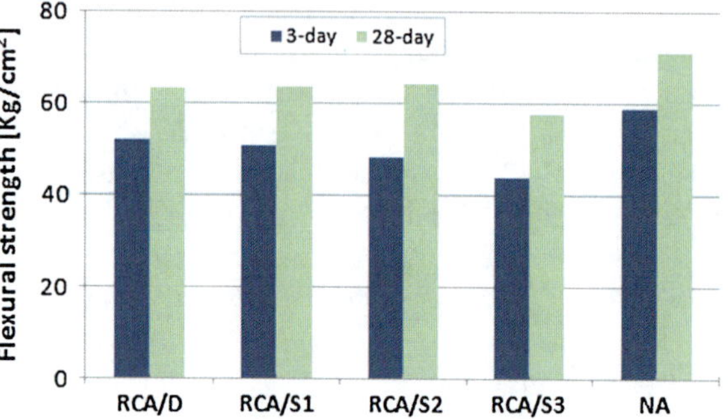

Fig. 5.31 FS of concrete containing dry NA and RCA with various moisture content (D: dry; S1, S2 and S3: 89.5, 88.1 and 100 % water saturated RCA) (Oliveira and Vazquez 1996)

compensate the negative effect related to the weakness of the old ITZ in RCA as the FS is largely dependent on the bond strength between aggregate and mortar matrix.

The concrete containing RCA from high-strength concrete exhibited higher FS than the one containing RCA from low strength concrete Limbachiya et al. (2000). Topçu and Sengel (2004) observed a systematic decrease in the FS of 16 and 20 MPa (cylindrical) concrete classes as the content of coarse RCA in concrete increased. The reduction was 13 and 27 % for the 16 and 20 MPa mixes respectively at a 100 % substitution level. Padmini et al. (2009) observed lower FS for RCAC than for conventional concrete and the differences in terms of CS and FS decreased with a reduction of the design CS (Fig. 5.30).

Takavoli and Soroushian (1996) observed higher FS for RCAC using smaller coarse RCA than for bigger coarse RCA and in some cases the FS was higher than for the control concrete. Oliveira and Vazquez (1996) observed a reduction from the control concrete of about 10 % in the 3- and 28-day FS of mixes made with coarse RCA with different moisture levels (Fig. 5.31). The FS of concrete containing dry RCA or around 90 % saturated RCA were comparable; however, the FS of concrete containing saturated surface dried RCA was significantly lower than that of the others.

Katz (2003) observed similar 28-day FS for conventional and concrete with coarse RCA and a mixture of fine RCA and natural sand, obtained from 1-day cured concrete; however, the FS of RCAC with RCA obtained from 3- and 28-day concrete was 11.5 % lower than that of conventional concrete. On the other hand, the reduction in FS was respectively 29.9, 20.9 and 31.3 % in concrete with white Portland cement as binder for 1-, 3- and 28-day cured RCA. Yang et al. (2011) observed 3 and 9 % reductions from the control concrete in the 7- and 28-day FS of concrete with RCA as sole coarse aggregate. The addition of crushed brick up to a 50 % substitution level of RCA slightly increased the FS due to the low Young modulus of brick and therefore improved the tensile stress along the matrix-aggregate interface (Fig. 5.32). The failure modes for control and RCA concrete occurred at the aggregate and mortar matrix's interface while both interface and aggregate failure occurred in RCAC containing crushed brick.

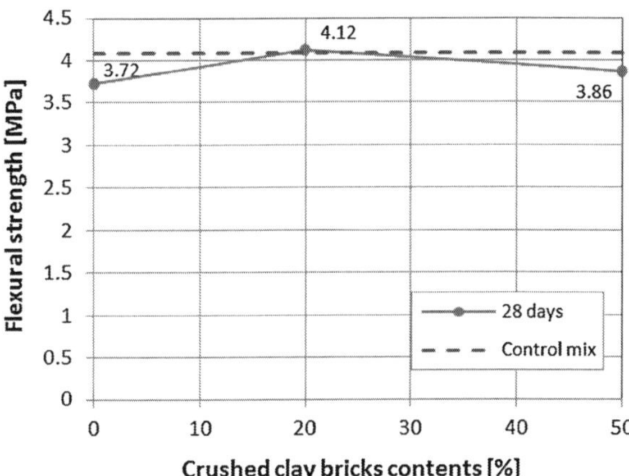

Fig. 5.32 Effect of crushed brick content in coarse RCA on the FS performance of concrete (Yang et al. 2011)

The FS of RCAC in the Chen et al. (2003) study was about 78 % that of conventional concrete when both mixes had a w/c ratio of 0.5. The substitution of 17, 33, 50 and 67 % of RCA by a mixture of bricks and tiles can slightly increase the FS as shown in Fig. 5.33. The differences in FS between conventional and RCA mixes gradually decreased with the w/c ratio. The FS of RCAC was similar to that of conventional concrete at the w/c ratio of 0.67 and significantly higher at w/c of 0.8. The FS of concrete containing RCA with sand-sized particles as well as other impurities like bricks and tiles was much lower than that of conventional concrete especially at lower w/c ratio.

Ahmed (2011) observed that the replacement of 30 % by mass of cement by FA in concrete containing 25 % fine RCA and the replacement of 30 and 40 % by mass of cement by FA in concrete containing 50 % fine RCA could improve the FS (Fig. 5.34). In the Jemas et al. study, the incorporation of FA to replace 10 and 15 % of OPC improved the FS of RCAC and the FS of RCAC prepared at a w/c ratio of 0.55 was even better than that of conventional concrete (Fig. 5.35).

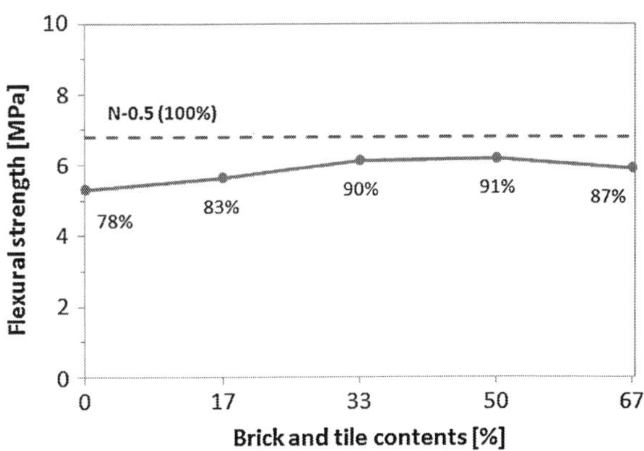

Fig. 5.33 Effect of brick and tile content on the FS of RCAC (Chen et al. 2003)

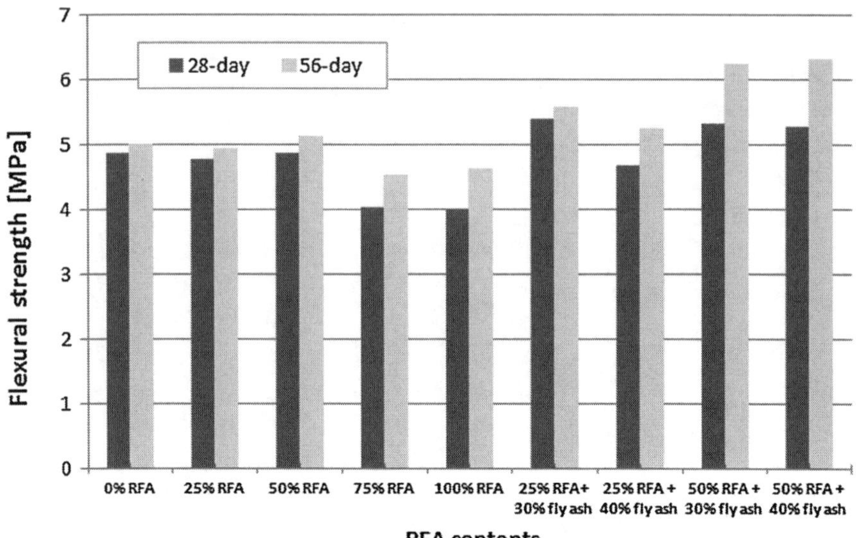

Fig. 5.34 Effect of FA on the FS performance of concrete with fine RCA (Ahmed 2011)

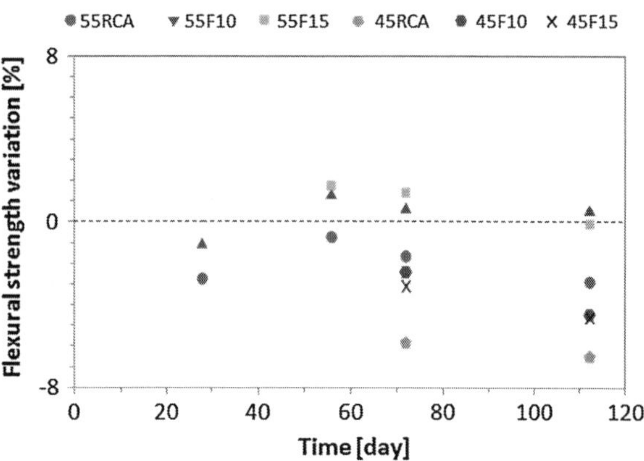

Fig. 5.35 Variation of the FS of conventional concrete and RCAC with and without FA, prepared at w/c of 0.55 and 0.45 (James et al. 2011)

Gupta et al. (2011) observed a reduction of about 6 % in 28-day FS in comparison to a 21.5 % reduction in the corresponding CS, due to the replacement of 100 % coarse NA by RCA. On the other hand, the FS of RCAC with FA as a 10 % replacement of OPC was around 3 % higher than that of conventional concrete and similar to that of conventional concrete with 10 % FA. The FS of RCAC with 20 % FA, however, was about 20 % lower than that of conventional concrete with OPC only. Rao and Khan (2009) observed reductions of about 4 % in FS due to the replacement of 50 % (by mass) of coarse NA by coarse RCA. However, the incorporation of 0.01–0.03 % glass fibre improved the FS of the resulting RCAC.

Tam et al. (2007) observed advantages of the use of a two-stage mixing approach instead of the normal mixing approach to increase several mechanical

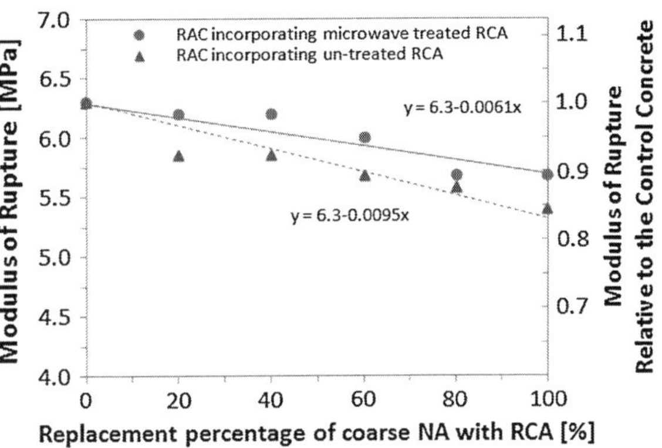

Fig. 5.36 Effect of microwave treatment of RCA on the FS performance of concrete (Akbarnezhad et al. 2011)

properties including the FS of RCAC. The substitution of coarse NA by RCA in the 25–40 % range yielded the optimal FS, along with the other mechanical properties of concrete using the two-stage mixing approach.

Akbarnezhad et al. (2011) observed that the FS of concrete with coarse RCA obtained after a microwave treatment was higher than that observed for concrete with untreated RCA as the microwave treatment can remove adhered mortar content as well as weak RCA particles from concrete (Fig. 5.36). Li et al. (2009) observed an improvement in the FS of RCAC due to the coating of RCA by blast furnace slag, fly ash and silica fume separately or an equal weight mixture of two of these three. These materials were used to replace 20 % of OPC (by weight) in concrete. The improvement was highest for the silica fume and fly ash mixture due to the higher packing density of this mixture.

Yang et al. (2008), after analysing 197 test results of available database along with their own experimental results, observed a decrease of the rupture modulus (indicative FS) of concrete as the water absorption capacity of CDW aggregate increased. The normalized rupture modulus, $f_r/\sqrt{f_c}$, where f_r and f_c are the rupture modulus and CS of concrete against water absorption capacity of RCA respectively, are presented in Fig. 5.37. They observed that the rupture moduli of the control concrete and of the concrete with grade I RCA according to the Korean standard satisfied the expression in the ACI 318-05 norm, regardless of their substitution level of coarse natural aggregate,. On the other hand, the rupture moduli of the concrete with 50 % grade II and of the concrete with grade III RCA were slightly lower than the ACI specified value. The reduction in the rupture modulus due to the incorporation of RCA was due to the weak binding between the components in the concrete matrix owing to the adhered cement paste on the RCA surfaces.

Takavoli and Soroushian (1996) observed lower FS for RCAC than that predicted from the CS of RCAC according to the American ACI Code 318 expression and the difference became larger at higher w/c ratios. Ahmed (2011) reported that the expression used in Australian code, AS3600-2009 could be used for concrete with fine RCA but the same expression for concrete with fine RCA and FA

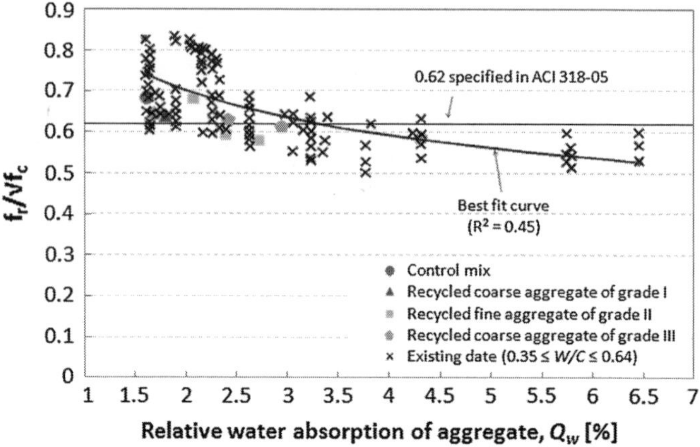

Fig. 5.37 Performance of normalized rupture modulus with the water absorption capacity of natural and CDW aggregate (Yang et al. 2011)

predicts a significantly lower FS value then the experimental value. In the James et al. (2011) study, the existing ACI 318 (2008) code underestimated their FS results. Katz (2003) observed higher values of 28-day FS of conventional concrete and RCAC than the value predicted according to ACI 363R.

5.3.4 Modulus of Elasticity

Similarly to the strength properties, the modulus of elasticity (MO) of concrete containing CDW aggregate is normally lower than that of conventional concrete and it decreases as the content of CDW aggregate in concrete increases. Some typical results are presented in Table 5.16. The causes for reduction in concrete's MO due to the incorporation of CDW aggregate indicated in various references are: (1) the loss of concrete stiffness, which depends on properties such as stiffness of mortar, concrete porosity and aggregate-cement paste bonding; these properties deteriorate due to the addition of RCA to concrete; (2) the lower MO of CDW aggregate than that of natural aggregate, since the concrete's MO is primarily dependent on the MO of the aggregate.

Depending on factors such as substitution ratio, quality and size of aggregate, w/c ratio, the reduction in MO of RCAC may reach 50 % when compared to conventional concrete. The reduction in concrete's MO due to the incorporation of CDW aggregate is generally higher than the corresponding CS reduction. Safiuddin et al. (2011) observed a reduction of about 17.7 % in the 28-day MO in comparison to a 12.2 % reduction in CS of concrete with 100 % coarse RCA when compared to conventional concrete. Chen et al. (2003) observed a 22 % reduction in MO in comparison to a 15 % reduction in CS of concrete due to the inclusion of

Table 5.16 MO of concrete containing various amount of CDW aggregate

Reference	Type of aggregate	Modulus of elasticity (GPa)/substitution level(%)/day	Comment
Evangelista and de Brito 2007	RCA/fine	35.5/0/28; 34.2/30/28; 28.9/100/28	Type of addition: v/v
Etxeberria 2007b	RCA/coarse	32.6/0/28; 31.3/25/28; 28.6/50/28; 27.8/100/28	Type of addition: v/v
Gonzalez-Fonteboa and Martinez-Abella 2002	RCA/coarse	29.7/0/28; 29.1/15/28; 27.8/30/28; 26.6/60/28; 26.7/100/28	Type of addition: v/v
Gonzales–Fonteboa et al. 2011	RA/coarse	29.6/0/28[a]; 28.2/20/28[a]; 26.4/50/28[a]; 24.3/100/28[a] 33.9/0/28[b]; 32.6/20/28[b]; 28.8/50/28[b]; 24.0/100/28[b]	[a] and [b] = water to cement ratios of 0.65 and 0.50 respectively; Type of addition: v/v
Gonzalez-Fonteboa and Martinez-Abella 2008		32.2/0/28[a]; 31.4/0/28[b]; 32.5/0/115[a]; 32.8/0/115[b]; 28.6/100/28[a]; 27.3/100/28[b]; 28.9/100/115[a]; 28.3/100/115[b]	[a] and [b] = concrete containing OPC and OPC-SF as binder respectively; Type of addition: v/v
Rao et al. 2011	RCA/coarse	31.2/0/28; 26.8/25/28; 26.7/50/28; 26.4/100/28	Type of addition: v/v
Domingo-Cabo et al. 2010	RCA/coarse	33.3/0/28; 32.4/20/28; 33.5/50/28; 33.3/100/28[, 1] 36.2/0/28; 32.4/20/28; 34.1/50/28; 31.0/100/28[b, 1] 32.2/0/28; 31.2/20/28; 31.2/50/28; 31.6/100/28[2]	[a] = cured at 100 % humidity condition and at 20 °C; [b] = initial 18 days in 100 % humidity and next 10 days in 65 % humidity and at 23 °C; [1] = prepared at constant w/c; 2: prepared at constant slump by considering absorbed amount of water by RCA; Type of addition: v/v
Thangchirapat et al. 2008	RCA/coarse + fine	34.9/0/28; 31.0/100/28[a]; 29.8/100/28[b]; 26.7/100/28[c]	[a] = replaced coarse NA; [a] = replaced coarse NA + 50 % fine NA; [a] = replaced coarse NA + 100 % fine NA; Type of addition: w/w
Berndt 2009	RCA/coarse	47.2/0/28; 40.1/100/28; 45.6/0/28[a]; 36.2/100/28[a]	[a] = 30 % OPC-70 % slag as binder; Type of addition: v/v

RCA as 100 % (by volume) replacement of coarse NA when the concrete was prepared at w/c of 0.5. In contrast to the CS and STS performance, Etxeberria et al. (2007b) also observed a decrease of the MO caused by the addition of coarse RCA as a replacement of NA due to the lower MO and higher deformation of RCA than NA's. The percentage reduction in 28-day MO due to the replacement of 25, 50 and 100 % of NA by RCA was around 4, 12 and 15 % respectively. Rahal (2007) observed a reduction of only 3 % in the MO for concrete with cylindrical strength between 25 and 30 MPa due to replacement of coarse NA by RCA.

Gonzales-Fonteboa et al. (2011) observed considerable decrease in the MO due to the addition of coarse RCA in concrete, because of the lower MO of RCA than NA's. They observed 4.7, 10.9 and 18.0 % reduction in MO due to the replacement of 20, 50 and 100 % of coarse NA by RCA in concrete at w/c of 0.65. These values were 3.8, 14.9 and 29.2 % at w/c of 0.50. Berndt (2009) observed a reduction of about 15 % in the MO due to a 100 % replacement of coarse NA by RCA. According to them, the low MO of RCAC might have an impact on the structural response (e.g. a low stiffness material is less susceptible to cracking). Corinaldesi (2011) observed a reduction of around 17 % in 28-day MO due to the replacement of 30 % coarse NA by RCA. Xiao et al. (2006a) observed a 45 % reduction in the MO of concrete due to the incorporation of coarse RCA as a 100 % replacement of coarse NA when compared to conventional concrete. Frondistou-Yannas (1977) observed a reduction of up to 40 % in the MO in comparison to a 4–14 % reduction in CS of concrete due to the incorporation of coarse RCA as a full replacement of NA.

In the Rao et al. (2011) study, the MO decreased with the replacement ratio of coarse NA by RCA due to the weaker ITZ between RCA and cement mortar and the lower MO of RCA than NA's. The reduction percentage in 28-day MO of concrete due to the replacement of 25, 50 and 100 % NA by RCA were 14.3, 14.4 and 15.4 % respectively. Evangelista and de Brito (2007) observed a reduction of about 3 % in the 28-day MO of concrete containing fine RCA as a replacement of 30 % by volume of fine NA; however, the MO of concrete containing 100 % fine RCA was 18.6 % lower than that of conventional concrete. The loss of concrete stiffness, which depends on properties such as stiffness of mortar, concrete porosity and aggregate-cement paste bonding was not as significant for smaller incorporation ratios of fine RCA as for higher ones and therefore the MO of concrete was slightly affected for small ratios of fine RCA.

Kou and Poon (2008) observed a 17–23 % reduction in 28-day MO of concrete containing three types of coarse RCA as a 100 % replacement of NA when compared to conventional concrete; however, after 5 years of curing, the reduction dropped to around 10 % indicating higher gain over time in MO of RCAC than that observed in conventional concrete. The increase in MO of the mixes containing three types of RCA as a 100 % replacement of coarse NA between 28 days and 5 years was in the range of 33–40 % in comparison to a 20 % improvement in conventional concrete. Gomez-Soberon (2002) observed a gradual decrease of the MO as the replacement of coarse NA by RCA increased up to 60 % and then became similar at 100 % replacement level (Fig. 5.38). The development of the

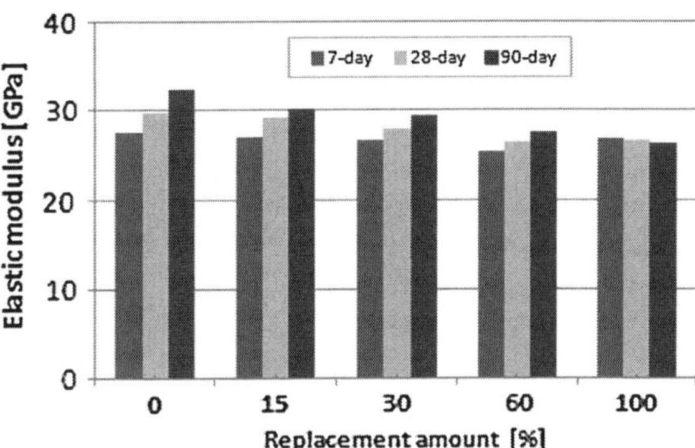

Fig. 5.38 MO of concrete at various ages due to the replacement of coarse NA by coarse RCA (Gomez-Soberon 2002)

MO gradually slowed down with increasing content of RCA in concrete. The MO of concrete containing RCA at 100 % replacement level was almost the same at the various curing ages. The authors could not establish a relationship between total porosity and MO even though the MO decreased as the open porosity increased up to around a 15 % porosity level. Safiuddin et al. (2011) observed a smaller increase in MO as curing time increased than that observed for FS and STS. In this study, the 28-day MO of RCAC was only 11.2 % higher than 7-day value, whereas the FS and STS increases were respectively 40.3 and 17.3 %. The MO increased with the concrete's CS too.

Domingo-Cabo et al. (2010) found a decrease of the 28-day MO due to the incorporation of a good quality coarse RCA as a 0, 20, 50 and 100 % (by volume) replacement of coarse NA in mixes with similar w/c ratio. Unlike most investigations, the RCA used in this one was not pre-saturated before concrete mixing; instead a super-plasticizer was used to prepare a workable mix. After the RCAC were prepared at constant slump by considering the amount of water absorbed by RCA, the 28-day MO was similar to that of conventional concrete. Padmini et al. (2009) observed a significant reduction in MO of concrete with the incorporation of RCA as coarse aggregate owing to the increase of porosity of concrete due to that incorporation. The higher reduction in percentage of the MO was observed for concrete made with smaller coarse RCA due to their higher porosity. However, no effect was detected of the strength of original concrete from which the RCA were generated on the MO of RCAC. Corinaldesi (2010) observed a reduction of around 23 and 13 % in 28-day MO of concrete prepared at w/c of 0.40 and 0.45 respectively, due to the replacement of 30 % of fine and coarse gravel (6–12, FR and 11–22 mm, CR) by similar sized RCA (Fig. 5.39). However, these values for fine and coarse gravel RCA became 22 and 32 % respectively at the w/c ratio of 0.60. Thangchirapat et al. (2008) observed a reduction of around 11 % in the 28-day MO due to a 100 % replacement by weight of coarse NA by RCA. The replacements of fine NA by fine RCA in the concrete with 100 % coarse RCA further lowered the MO. The reduction in MO of mixes with 50 and 100 % fine RCA was respectively around 14 and 24 %.

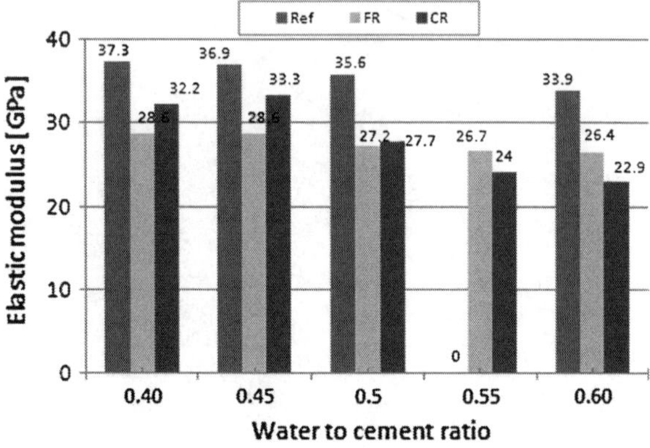

Fig. 5.39 MO of concrete with NA and RCA at different w/c ratios (Corinaldesi 2010)

Table 5.17 MO (E-modulus) of conventional and RCA-containing concrete (Hansen and Boegh 1985)

Item[a]	H	H/H	H/M	H/L	M	M/H	M/M	M/L	L	L/H	L/M	L/L
E- modulus (GPa)	43.4	37.0	36.3	34.8	38.5	33.0	32.0	30.0	30.8	27.5	22.3	22.6
Reduction (%)	–	14	16	20	–	14	17	22	–	11	28	27

[a] Details about H, H/M etc. are in Fig. 5.12

Kou and Poon (2008) observed a decreasing of the MO of conventional as well as of RCA mixes with the water to binder ratio. In comparison to conventional concrete, Hansen and Boegh (1985) observed an 11–28 % reduction in the MO of three different classes of concrete with RCA from three different classes of concrete as a full replacement of coarse NA when the various types of concretes were subjected to 47-day accelerated curing conditions. Depending on the strength of the original concrete, the reduction percentage slightly varied, as presented in Table 5.17.

The reduction in MO of concrete due to the incorporation of RCA was more pronounced in water curing than in steam curing especially at the smaller ratios of RCA (Poon et al. 2006) (Fig. 5.40). Fonseca et al. (2011) reported that the MO of concrete decreased as the incorporation of coarse RCA as a replacement of NA increased, due to the increase in porosity of concrete. The effect of curing conditions on the MO of conventional and RCA concrete was determined in this investigation. The authors used four curing conditions: water immersion, wet chamber, outer environment and laboratory. They observed the lowest MO of both types of concrete when the specimens were cured in laboratory conditions, the driest condition in this study (Fig. 5.41). The variations in MO of both types of concrete were not significant in other curing conditions. The cause of the low MO of concrete observed in laboratory curing condition was the formation of a porous microstructure of cement pastes due to low humidity.

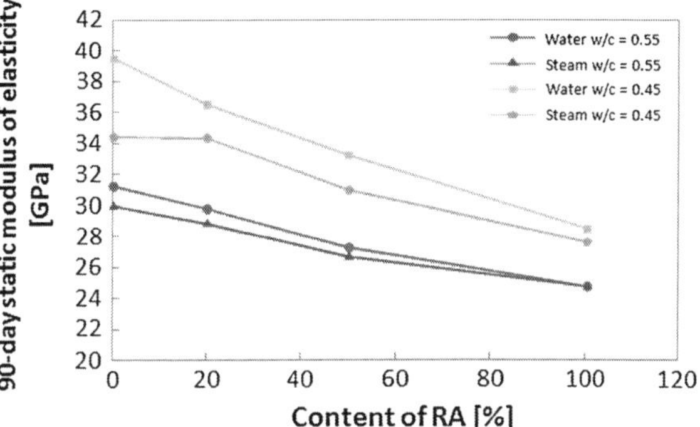

Fig. 5.40 Effect of curing conditions on the MO of concrete (Poon et al.2006)

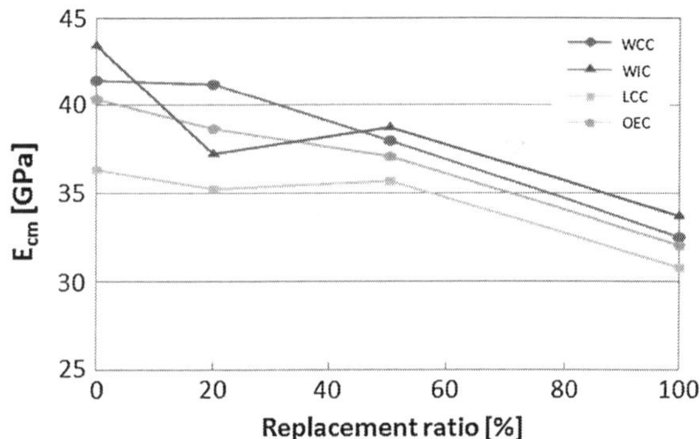

Fig. 5.41 Effect of curing conditions on the MO of concrete containing various ratios of coarse RCA (Fonseca et al. 2011)

Oliveira and Vasquez (1996) observed comparable 28-day MO for concrete containing coarse RCA with various moisture levels; the MO of RCAC was about 75 % that of conventional concrete. Chen et al. (2003) observed insignificant influence of the quality of RCA whether it was washed or unwashed with impurities such as sand particles, bricks and tiles on the MO of the resulting concrete. The differences in MO of conventional and RCA concrete were also similar with the w/c ratio (Fig. 5.42). Chen et al. (2003) also found an insignificant effect of brick and tile contents on the MO of RCAC (Fig. 5.43).

Tam et al. (2007) observed higher MO for concrete prepared using a two-stage mixing approach (in which mixing of the water was divided into two parts: the first one added to the mixed aggregate and the remaining part to the mixed aggregate and cement) than that observed for concrete prepared using a one-step mixing approach (in this approach the whole amount of water was added to the mixed aggregate and cement). In this study, the replacement of 31.3 % of NA by RCA gave the highest improvement in 28-day MO of RCAC prepared by the two-stages mixing when compared to conventional concrete. In another study, Tam and Tam (2008) observed an improvement of about 16 % in the 28-day MO of concrete containing coarse RCA as a 30 % replacement of NA due to the incorporation of silica fume as a 2 % replacement of OPC where the concrete was prepared by a

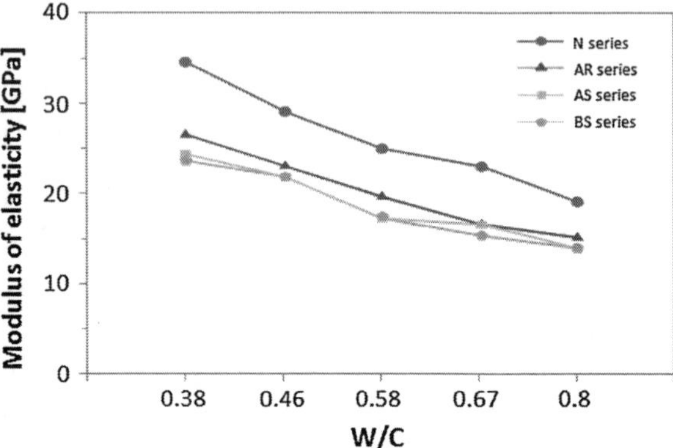

Fig. 5.42 Effect of w/c ratio on the MO of concrete containing NA (N), washed (AR) and unwashed (AS) RCA (Chen et al. 2003)

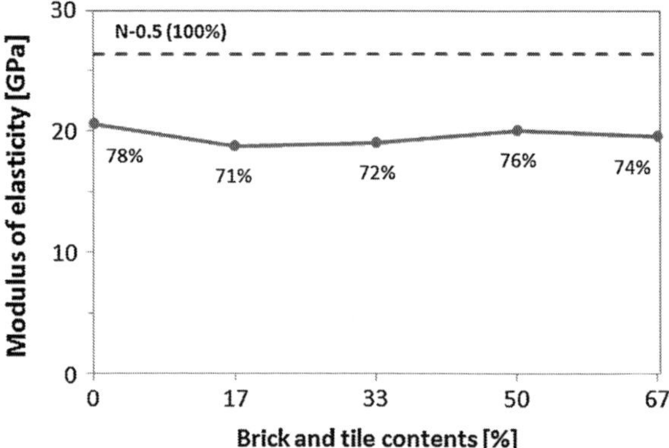

Fig. 5.43 Effect of brick and tile content in RCA on the MO of concrete (Chen et al. 2003)

two-stage mixing approach, when compared to the MO of equivalent RCAC containing 2 % SF prepared by a one-step mixing approach. In comparison to the MO of RCAC prepared by conventional mixing method, Razaqpur et al. (2010) observed 11 and 15 % improvements in the MO of RCAC prepared using the equivalent mortar volume (EMV) method (details are given in the CS section). The MO of RCAC produced by EMV method was comparable or even better than that of conventional concrete because the total natural aggregate volume was the same in both types of concrete. Akbarnezhad et al. (2011) observed a significant improvement in the MO of RCAC due to the use of microwave treated RCA (MRCA). In comparison to conventional concrete, the reduction in MO of RCAC with MRCA as the only coarse aggregate was around 10 % whereas it was around 25 % for RCAC with untreated RCA.

In several references, it was reported that the use of mineral additions could not improve the MO of RCAC although mineral additions normally improve the various strength properties. According to Thangchirapat et al. (2008), the concrete's MO mainly depends on the properties of the aggregate rather than the strength of the cement paste. In their study, they did not observe any beneficial effect of rice husk ash addition into OPC on the MO performance of concrete.

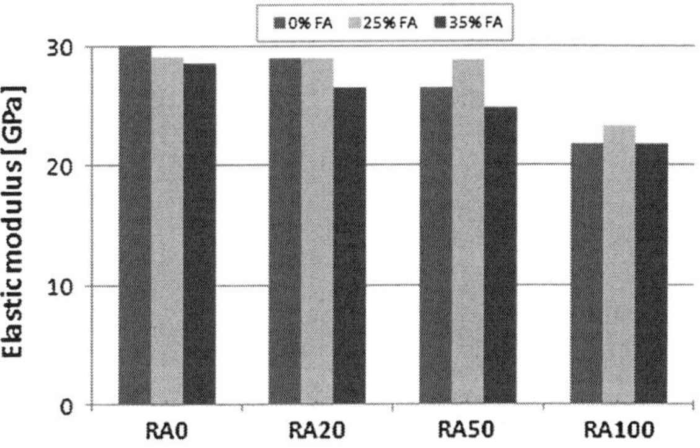

Fig. 5.44 Modulus of elasticity of concrete containing various amount of coarse RCA aggregate due to FA addition Kou and Poon (2006)

Similarly, Gonzalez-Fonteboa and Martinez-Abella (2008) did not observe any improvement of the MO of concrete containing RA as coarse aggregate due to the replacement of 8 % (by weight) of OPC by silica fume. They observed respectively reductions of about 15 and 13 % in the 28- and 115-day MO of concrete with SF while similar reductions of around 11 % were observed for OPC mixes after 28 and 115 days of curing. Berndt (2009) observed a lower MO of RCAC due to the incorporation of blast furnace slag as 50 and 70 % replacement of OPC. At 70 %, the MO of RCAC was around 20 % lower than that of the concrete containing 70 % slag and without RCA.

Kou and Poon (2006) observed a slight improvement of the 28-day MO of concrete containing RCA as a 20, 50 and 100 % replacement of coarse NA due to the incorporation of FA as a 25 % replacement of OPC (Fig. 5.44). However, for 35 % FA addition, the MO of RCAC was lower than the one of RCAC without FA.

Lopez-Gayarre et al. (2009) applied the analysis variance (ANOVA) method to analyse their experimental results on the use of RCA as a replacement of coarse NA in concrete. Experimental variables such as quality of aggregate, replacement ratio, size distribution, declassified content, strength of original concrete and concrete slump value were considered for analysis. Out of these parameters, they observed significant influence of the quality of aggregate on the MO of RCAC when the 100 % coarse NA was replaced by RCA. However, the aggregate's quality had negligible effect at 20 and 50 % replacement levels.

In several studies, the existing relationships between CS and MO (E) proposed in various standard specifications were applied to check the validity of these relations in concrete containing RCA. As for example, the MO of concrete containing various types of CDW aggregate from the Pain et al. (2009) study was around 20 % lower than the MO values estimated by using the Eurocode 2 expression. Rahal (2007) reported that the expression presented in ACI 318-02 for the relationship between CS and MO of concrete overestimated the experimental results for conventional as well as RCA concrete mixes. Kou et al. (2007) and Kou and Poon (2008) also reported that the existing ACI 363R-92 expression overestimated their experimental results. On the other hand, the experimental MO of conventional as well as RCA concrete from the Oliveira and Vasquez (1996) study were consistent with the CEB-FIP model code (The International Federation for Structural Concrete).

Several expressions were also proposed to describe the relationship between MO (E) and cubic CS (f_{cu}) in earlier studies, some of which are presented below:

Ravindraraja and Tam (1985):

$$E = 7770 \times f_{cu}^{0.33} \tag{5.4}$$

Kakizaki et al. (1988):

$$E = 1.9 \times 10^5 \times f_{cu} + \left(\frac{\rho}{2300}\right)^{1.5} \times \sqrt{\frac{f_{cu}}{2000}} (\rho = \text{density}) \tag{5.5}$$

Dhir et al. (1999):

$$E = 370 \times f_{cu} + 13100 \tag{5.6}$$

Some new equations were recently proposed to establish the relationship between MO (E) and cubic CS (f_{cu}) of RCAC. Corinaldesi (2011) observed that the existing relationship between 28-day cubic CD and MO ($f_{cu}$and E respectively) as described in Italian norms, NTC 2008 and presented in Eq. (5.7)) could be applied to the results obtained for conventional concrete but a different relationship (Eq. (5.8)) was necessary for RCAC:

$$E = 22.0\sqrt[3]{\frac{0.83 \cdot f_{cu}}{10}} \tag{5.7}$$

$$E = 18.2\sqrt[3]{\frac{0.83 \cdot f_{cu}}{10}} \tag{5.8}$$

In another study, Corinaldesi (2010) proposed the following expressions to establish relationships between 28-day cubic CS and MO of RCAC containing fine and coarse RCA, respectively:

$$E = 18.8\sqrt[3]{\frac{0.83 \cdot f_{cu}}{10}} \tag{5.9}$$

$$E = 909 \times f_c + 8738 \tag{5.10}$$

From their experimental results, Evangelista and de Brito (2007) observed that the inclusion of concrete's density (ρ) according to the following equation was necessary to establish a relationship between cubic CS (f_c) and MO (E):

$$E = a \times (f_c + 8)^{\frac{1}{3}} \times \left(\frac{\rho}{b}\right)^2 \tag{5.11}$$

Where a, b and s are the regression coefficients whose values are 8917, 2348 and 0.85 respectively.

From the experimental results in various references and also their own, Xiao et al. (2006b) proposed the following expression to relate the CS and the MO of RCAC:

$$E = \frac{10^5}{2.8 + \frac{40.1}{f_{cu}}} \tag{5.12}$$

5.3.5 Flexural and Shear Performances

Several investigations were undertaken to understand the flexural and shear performances of concrete containing RCA. Here, some results will be highlighted from relevant references.

Razaqpur et al. (2010) observed higher ultimate FS in reinforced RCAC beams than in conventional concrete beams regardless of the source of RCA or the tension and steel contents of the beam. The RCAC were prepared using a new mixing method where the total mortar content (new and old) in RCAC and conventional concrete were the same. The flexural failure modes and cracking patterns of both types of concrete were similar. The mid-span deflections of both types of concrete also met the American ACI 318 M-05 specification limit. However, the RCAC beams showed lower cracking moments and slightly smaller crack spacing than the conventional concrete beam.

Sato et al. (2007) observed higher flexural deflection for reinforced concrete beams containing RCA than for conventional concrete under the same moment and w/c value. In these conditions, they observed similar ductility factors, ultimate moments and crack spacing in the control concrete beam as well and the concrete beams containing coarse or fine RCA but the crack width of concrete containing coarse or fine RCA were larger than that observed in the conventional concrete beam. The crack spacing and crack width of the RCAC beam containing coarse RCA were in the ranges of 0.92–1.37 and 0.57–1.3 times those of the conventional concrete beam respectively. The same parameters for RCAC containing fine RCA were 0.74–1.26 and 1.1–1.7 times those of the conventional concrete. They did not detect any cracking or deflection for 1 year under wet conditions but observed many cracks and two times more deflection than in the conventional concrete beam under dry condition when the concrete beams containing fine RCA and conventional concrete were kept under sustained bending moment equivalent to 100 N/mm^2 in tension rebar stress in reinforced concrete sections. Regardless of the type of aggregate, the ultimate moments of the concrete beams can be predicted from the Japanese code, JSCE 2002e.

Tsujino et al. (2007) investigated the flexural performance of concrete containing untreated and oil-coated low and medium qualities RCA as coarse

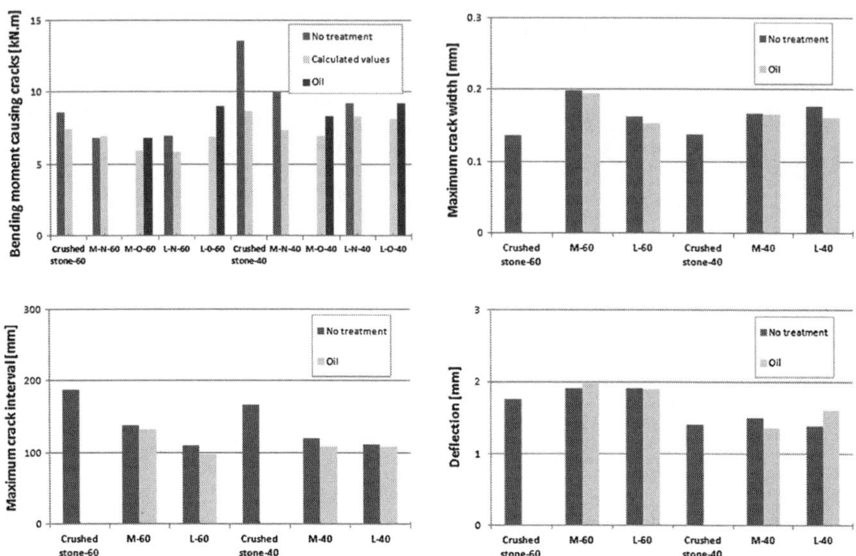

Fig. 5.45 Flexural behaviour of conventional concrete and concrete containing different types of RCA (M and L: medium and low quality RCA; N and O: untreated and oil treated RCA, 60 and 40: w/c of 0.6 and 0.4, respectively) (Tsujino et al. 2007)

aggregates along with conventional concrete at w/c ratios of 0.6 and 0.4 and their results are presented in Fig. 5.45. The cracking moments of concrete containing RCA were slightly lower than that observed for conventional concrete; the differences in cracking moment between conventional concrete and RCAC were higher at low w/c ratio. However, the cracking moment of RCAC can be predicted by conventional equation. The use of RCA as coarse aggregate in concrete also lowered the maximum crack spacing and increased deflection and maximum crack width. However, the maximum crack widths of all types of RCAC were significantly lower than 0.3 mm, the allowable crack width limit for reinforced concrete as specified by the Japanese Architectural Institute. The deflections of RCAC were marginally higher than that of conventional concrete at the w/c ratio of 0.4, indicating problems related to deflection can be overcome by lowering the w/c of RCAC. The authors did not find any beneficial effect on the flexural performance of RCAC due to the use of surface treated RCA.

Etxeberria et al. (2007b) studied the shear behaviour of concrete beams prepared by replacing 0, 25, 50 and 100 % (by volume) of coarse NA by RCA. The beams were prepared with and without transverse reinforcement. They observed a negligible influence of a 25 % replacement of NA by RCA on the shear strength of concrete beam, especially for the beam without transverse reinforcement; however, the shear strength decreased at higher replacement levels (Fig. 5.46). They also concluded that modifications in the concrete composition such as an increase in the cement amount and a decrease of the w/c ratio were necessary to control the shear strength loss due to the incorporation of RCA in concrete.

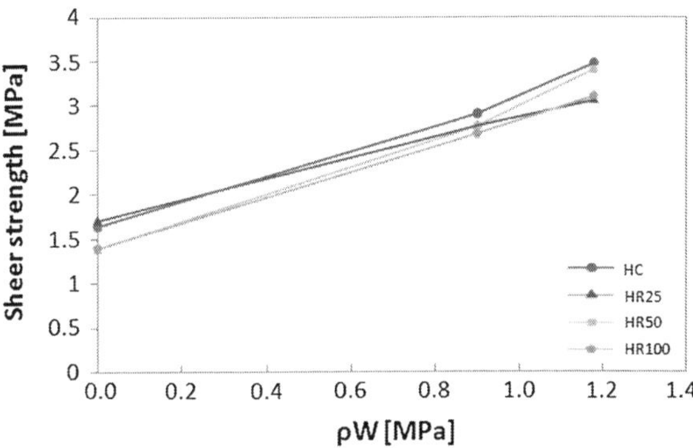

Fig. 5.46 Effect of transverse reinforcement in concrete beams made with NA only (HC) and by replacing 25, 50 and 100 % (HR25, HR50, HR100) of coarse NA by RCA

5.3.6 *Stress–Strain Relationship*

The analysis of the stress–strain curve (SSC) of concrete can yield data such as strength and toughness performance and therefore the evaluation of SSC is essential for structural design of concrete. Several studies were done to evaluate the SSC of concrete with CDW aggregate. The results from a few studies are presented next.

Topçu and Guncan (1995) determined several factors such as toughness, plastic energy capacity and elastic energy capacity from the stress–strain curves of conventional concrete and concrete containing various amount of RCA as a replacement of coarse NA. They observed a gradual decrease of toughness, plastic and elastic energy capacities of the latter as the incorporation of RCA in concrete increased.

Xiao et al. (2006a) observed a significant influence of RCA and replacement ratio of coarse NA by RCA on the stress–strain curve of the resulting concrete (Fig. 5.47). The incorporation of RCA increases the peak strain but significantly decreases the ductility of concrete i.e. ultimate strain. At 100 % replacement of NA by RCA, the increase in the peak strain was 20 %. The higher peak strain was due to the lower stiffness of RCA than that of NA. They also observed that the

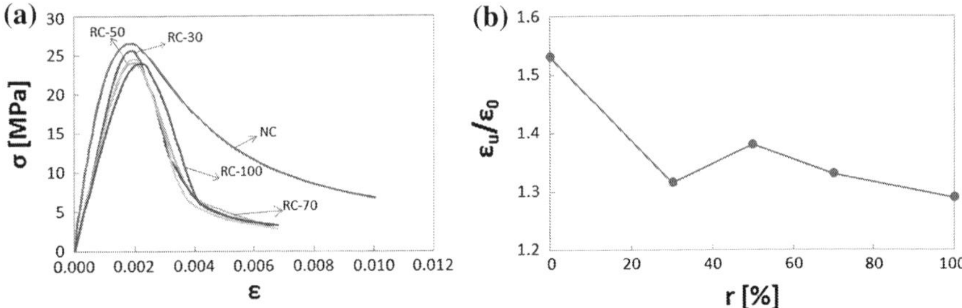

Fig. 5.47 Stress (σ)–strain (ε) curves and ultimate strain of concrete with replacement of coarse NA by coarse RCA (Xiao et al. 2006a). **a** Stress–strain curve. **b** Ultimate strain

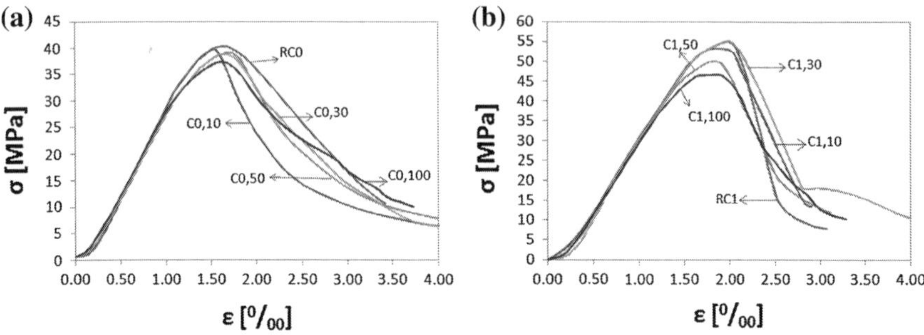

Fig. 5.48 Stress–strain curve of concrete with replacement of fine NA by fine RCA: **a** Without superplasticizers; **b** With superplasticizers (Pereira et al. 2012)

commonly used relationship between CS and MO (E) for NAC was not applicable for RCAC. They also successfully applied the analytical relationship proposed in the Chinese Code GB 50010 for uniaxial compression of NAC after a slight modification to predict the stress–strain curve of RCAC in similar condition.

Tsujino et al. (2006, 2007) studied the load–deflection curves of reinforced concrete beams of conventional concrete as well as of concrete containing untreated and oil treated coarse RCA. They found marginal differences between the NA and the RCA or between the treated and untreated RCA in terms of the plastic behaviour of concrete made with them namely load deflection curves, ultimate concrete strain at compression fibres, ultimate bending moment and toughness ratio or ductility factor. The ultimate concrete strains at compression fibres, bending moment and ductility factors were around 3500 μ, 30 kN.m and 5 respectively for all types of concrete.

Pereira et al. (2012) did not observe any major differences in the stress–strain curve of concrete due to the incorporation of fine RCA as a replacement of fine NA in concrete (Fig. 5.48a). However, the RCAC exhibited earlier stiffness losses than the NAC due to the lower CS of RCAC and the fragile adhered cement paste that facilitates the propagation of cracks. However, the use of superplasticizers increased the yield stress but decreased the yield path length in the stress–strain curves of NAC and RCAC (Fig. 5.48b). Ajdukiewicz and Kliszczewicz (2002) reported that RCAC with recycled basalt aggregate exhibited more brittle behaviour than RCAC with recycled granite aggregate. The stress–strain curves of high-strength conventional concrete and RCAC with a chemical admixture were more linear than that of concrete without the admixture.

5.3.7 *Creep of Concrete*

Creep of concrete is defined as the deformation of structure under sustained load. Creep is dependent on various factors including the properties of aggregate and the

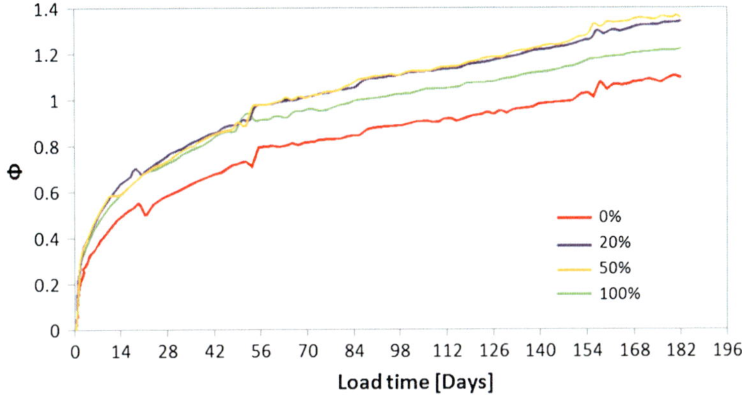

Fig. 5.49 Creep coefficient of concrete with replacement of coarse NA by coarse RCA (Domingo-Cabo et al. 2010)

amount of cement paste. Generally low creep is observed in concrete containing strong aggregate and aggregate with a high stiffness. As the properties of CDW aggregate are different from those of conventional aggregate, the incorporation of CDW aggregate significantly changes the creep of concrete. The incorporation increases the total paste content in concrete too, which also has some effect on the creep performance of concrete. In general the creep of concrete increases with the incorporation of CDW aggregate in concrete. Some results are highlighted next.

Domingo-Cabo et al. (2010) determined total creep deformation, creep coefficient and specific creep deformation of concrete due to the replacement of coarse NA by coarse RCA. They observed a gradual increase of the above parameters with the content of RCA in concrete. They observed a 35, 42 and 51 % higher total creep deformation due to a 20, 50 and 100 % replacement of NA by RCA, respectively, when the specimens were loaded for 180 days. Similarly, the specific creep deformation was also increased by 25, 29 and 32 % due to a 20, 50 and 100 % replacement of NA by RCA. The creep coefficients for various concrete mixes are presented in Fig. 5.49.

After comparing their experimental results with various relationships from ACTM C512-02 Code-2002, RILEM Model B3 Code-1995, CEB-FIP Code-1990 and a model developed by Gardner and Lockman (2001), Domingo-Cabo et al. (2010) concluded that these models were conservative to determine the deformation in the NAC and the RCAC's except the CEB-FIP Code for RCAC with 50 and 100 % coarse RCA. Their results are presented in Fig. 5.50.

The creep of RCAC in the Wesche and Schulz (1982) study, where RCA was used as integral replacement of coarse NA, was 50 % higher than that of conventional concrete. Kou et al. (2007) observed that creep strain increased with the incorporation of coarse RCA in concrete due to the gradual increase of mortar content. The addition of fly ash as a 25 % replacement (by weight) of cement lowers the creep strain of conventional as well as of RCA concrete. The creep

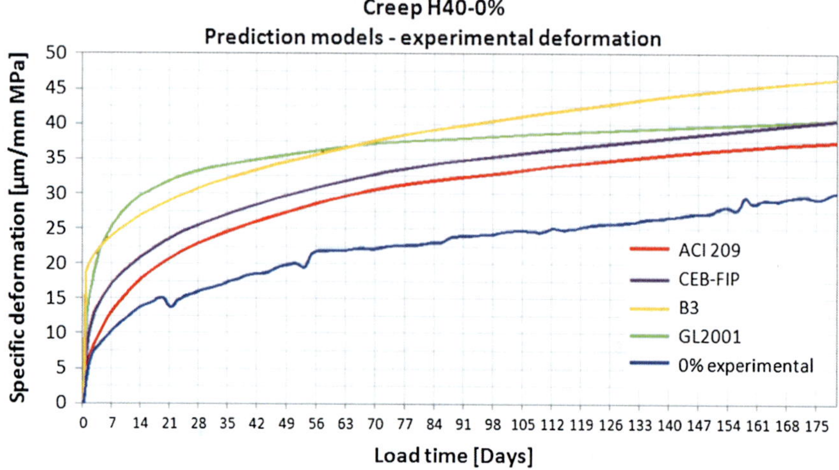

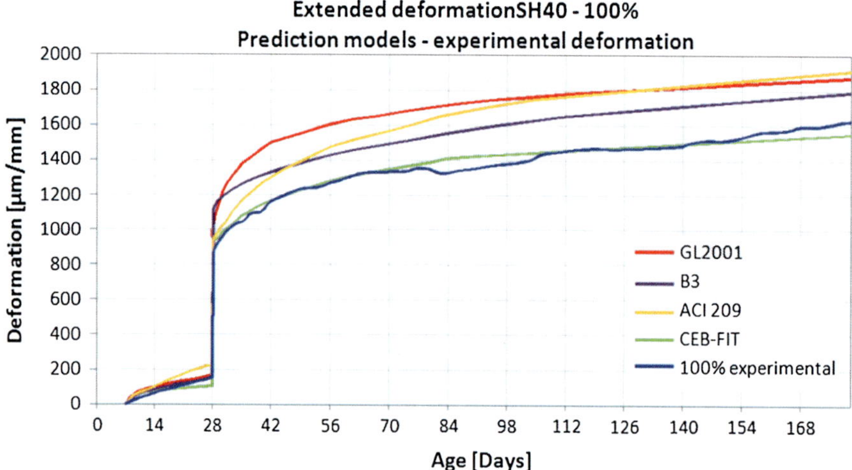

Fig. 5.50 Experimental and predicted creep deformation of NAC and RCAC with 100 % coarse RCA (Domingo-Cabo et al. 2010)

coefficient and specific creep of RCAC's with complete replacement of coarse NA by two types of RCA generated from concrete containing river gravel of limestone gravel in the Fathifazl et al. (2011) study were comparable to or lower than those of the equivalent NAC when the RCAC's were prepared by the EMV method. On the other hand, both parameters for RCAC's prepared by the usual method are higher than those of the NAC. The creep coefficient of RCAC also depended on the residual mortar content in RCA. Concrete with RCA with higher mortar content had higher creep. The authors also proposed a residual mortar factor to fit their experimental results with those predicted by ACI-209 code-1982 and CIB-FIP model code.

Table 5.18 Creep of concrete containing CDW aggregate

Reference	Type of aggregate	Creep coefficient/substitution level(%)/day	Comment
Limbachiya (2010)	RCA/coarse	1.25/0/28; 1.24/30/28; 1.41/50/28; 1.93/100/28;a	a = Concrete with 30 MPa design strength
Limbachiya et al. (2000)	RCA/coarse	1.14/0/90[a]; 1.15/30/90[a]; 1.31/50/90[a]; 1.88/100/90[a]; 0.81/0/90[b]; 0.83/30/90[b]; 0.99/50/90[b]; 1.08/100/90[b];	[a, b] = Concrete with 50 and 60 MPa respectively

Limbachiya (2010) observed almost equal 28-day creep strain for concrete with 30 MPa design strength, when 30 % of coarse NA was replaced by coarse RCA (Table 5.18). However, in comparison with conventional concrete, the creep strain of concrete increased 13 and 54 % when the replacement level was increased to 50 and 100 % of coarse NA, respectively. Limbachiya et al. (2000) observed higher creep in high-strength concrete with coarse RCA as aggregate than that in conventional concrete and creep increased with increasing RCA content (Table 5.18). They identified two reasons for this trend: the increase in cement content to reduce w/c value and achieve the same 90-day CS as conventional concrete and the presence of old cement paste in RCA. Like for conventional concrete, they observed lower creep for RCAC with higher CS (Table 5.18).

Paine et al. (2009) reported that the creep coefficient of concrete containing three types of RA coarse aggregate and a type of RCA coarse aggregate was slightly higher than that of conventional concrete due to differences in the stiffness of NA and CDW aggregates as well as in the CS of concrete containing conventional and CDW aggregates. The creep of RAC and RCAC were in the range of 1.9–2.6 at 100 days. They also observed that the experimentally determined 100-day creep values of concrete containing conventional and CDW aggregates were about 20 % lower than the value predicted by the Eurocode 2 relationship.

Regardless of the w/c ratios and quality of the aggregate, Tsujino et al. (2006) observed higher creep strain in concrete containing RCA as the only coarse aggregate than that in conventional concrete, possibly due to higher paste content in RCAC. The use of oil treated RCA had little effect on the creep strain of the resulting concrete; however, the creeps of concrete containing silane treated RCA were very high at various experimental conditions possibly due to lower bond strength between aggregate and cement paste.

However, in contrast with the above results, Ajdukiewicz and Kliszczzewicz (2002) observed up to 20 % lower creep of 1-year old high-performance/high-strength (hp/hs) concrete containing RCA as full replacement of NA (coarse and fine) or full replacement of coarse aggregate than that of conventional hp/hs concrete. This difference was more visible for concrete prepared using a chemical admixture.

5.3.8 Surface Hardness

The hardness of the surface of concrete has several practical repercussions and therefore several tests are performed to evaluate this property using parameters such as Schmidt hardness and rebound number, abrasion resistance, impact value. In this section, this property of concrete containing CDW aggregate is highlighted.

The Schmidt hardness test, which is a non-destructive method, can yield indirect data on surface hardness and penetration resistance as well as CS of concrete. A few results are available on the evaluation of this property in RCAC too.

Topçu and Sengel (2004) observed decreasing Schmidt rebound hammer with increasing coarse RCA content in two types of concrete with design CS of 16 and 20 MPa. The rebound hammer values of conventional concrete and RCAC were around 20 and 19 respectively for both classes of concrete. The 28-day rebound hammer values of conventional concrete and concrete with RCA as the only fine and coarse aggregate in another study of Topçu (1997) were respectively 21.3 and 11.6.

Rao et al. (2011) observed decreasing rebound numbers with increasing replacement of coarse NA by RCA. The rebound number for conventional concrete and RCAC containing 25, 50 and 100 % coarse RCA were 30.28, 16.80, 16.23 and 14.95, respectively. According to the authors, this may be due to the porous nature of RCA linked to the weak adhered cement paste. They also observed a linear relationship between rebound number and CS of RCAC, which is presented in Fig. 5.51. Sagoe-Crentsil et al. (2001) observed a volume loss about 12 % higher in the abrasion resistance of concrete due to complete replacement of coarse basalt aggregate by RCA aggregate (Fig. 5.52). The use of 35 % slag containing OPC or the increase of 5 % in cement content in the preparation of RCA marginally improved the abrasion behaviour of concrete.

Limbachiya (2010) also observed a gradual increase of abrasion depth of two concrete with design strength class of 35 and 45 MPa with the replacement level of coarse NA by RCA. Like the NAC, the abrasion depth of the RCAC's also decreased with as the CS increased (Table 5.19). Limbachiya et al. (2000) from

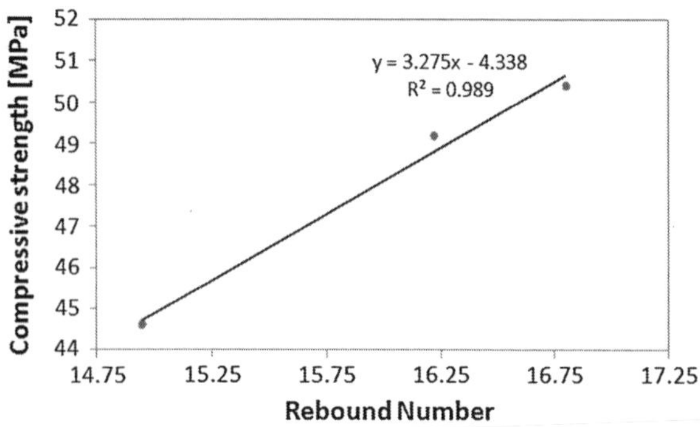

Fig. 5.51 Rebound number *versus* compressive strength for RCAC (Rao et al. 2011)

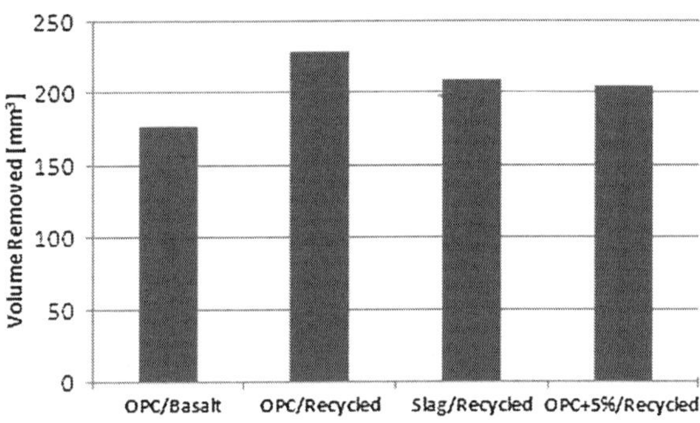

Fig. 5.52 Abrasion resistance of conventional concrete and RCAC's (Sagoe-Crentsil et al. 2001)

their results of concrete with three strength classes (50, 60 and 70 MPa), concluded that the incorporation of coarse RCA as partial or full replacements of NA did not change the abrasion resistance of concrete. The depth of abrasion of RCAC with 100 % coarse RCA at the design strength of 50 and 60 MPa were respectively 0.03 and 0.04 mm lower than that of the conventional concrete. On the other hand, Maas et al. (2012) did not observe any differences in the abrasion behaviour of two types of concrete prepared at w/c ratios of 0.72 and 0.45 by replacing 20 % of coarse NA by low grade RCA. However, the abrasion resistance of concrete deteriorated when 40 % coarse NA was replaced by RCA.

Fonseca et al. (2011) observed comparable abrasion resistance of conventional concrete and RCAC's with a 20, 50 and 100 % replacement by volume of coarse NA by coarse RCA when the concrete specimens were subjected to various curing conditions (Fig. 5.53). However, even though statistically insignificant, concrete containing RCA as the only coarse aggregate had marginally lower wear than the conventional concrete for all curing conditions. According to them, this is due to the better bond between RCA and the cement paste because of the porous nature of RCA.

Poon and Chan (2006) observed similar abrasion resistance but lower skid resistance in concrete paving blocks prepared by using RCA as the only aggregate than in concrete where a part of fine and coarse RCA was replaced by FA. The effect of FA on the behaviour of paving blocks containing crushed brick at 25, 50 and 75 % replacement of RCA was similar. However, the skid and abrasion resistances of paving blocks with or without FA as a partial replacement of RCA normally decrease and increase respectively with the incorporation of crushed brick. The observed improvement of skid resistance of concrete blocks by using FA was due to the formation of a more homogeneous concrete mix along with the smoother surface texture of the concrete blocks.

Table 5.19 Abrasion depth of concrete containing CDW aggregate (Limbachiya 2010)

Type of aggregate	Concrete strength (MPa)	Abrasion depth in mm/substitution level(%)/day
RCA/coarse	35	0.69/0/28; 0.73/30/28; 0.75/50/28; 0.78/100/28;
	45	0.49/0/28; 0.51/30/28; 0.48/50/28; 0.54/100/28;

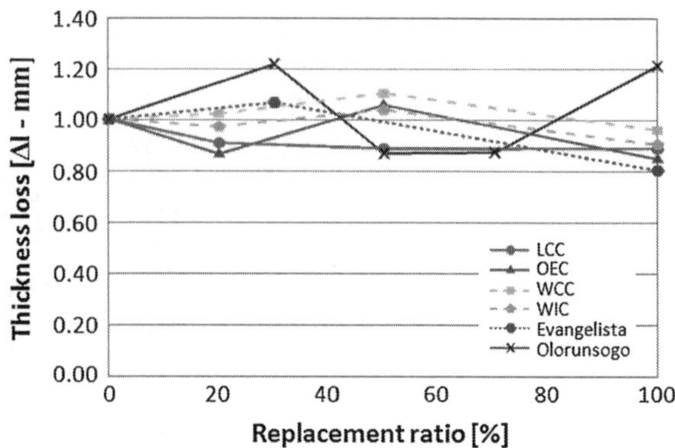

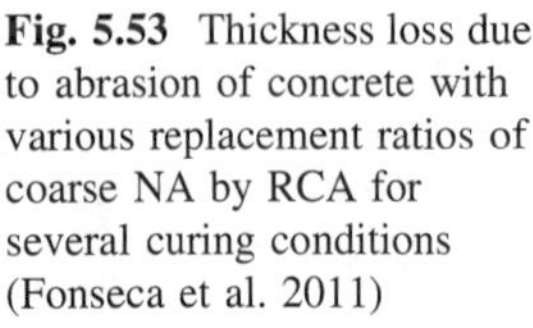
Fig. 5.53 Thickness loss due to abrasion of concrete with various replacement ratios of coarse NA by RCA for several curing conditions (Fonseca et al. 2011)

Poon and Chan (2007), in another study, observed comparable abrasion resistance but higher skid resistance of concrete due to incorporation of impurities such as crushed tiles or mixtures of crushed tiles with glass, brick and wood as 10 % replacement of RCA in concrete paving block. The higher skid resistance was due to the rough surface texture of the contaminating particles. Still all the RCAC paving blocks with and without contaminants met the standard specifications of Hong Kong.

Poon and Lam (2008) observed that the skid resistance of RCAC paving blocks made with an aggregate to cement ratio (A/C) of three was lower than that of blocks made with an A/C of four and six due to higher cement content and therefore smoother surface texture; still the skid resistance of all RCAC paving blocks met the standard specifications of Hong Kong. The quality of RCA did not have any effect on the skid resistance as aggregates were embedded in the cement matrix. Similarly, the abrasion resistance of the RCAC paving blocks made with an A/C of three and four were better and satisfactory than the blocks made with an A/C of six.

Topçu (1997) observed damage depths of 20–30 mm for conventional concrete and 100–130 mm for concrete containing RCA as the only coarse and fine aggregate, when both types of concrete were subjected to an impact test. The higher damage observed in the RCAC concrete was due to the presence of weak adhered mortar.

5.3.9 Other Mechanical Properties

Ajdukiewicz and Kliszczzewicz (2002) observed lower bond stress at failure for high-strength RCAC than for conventional high-strength concrete, which was more prominent for 220 MPa round bars than for 440 MPa ribbed bars. The bond stress was on average about 20 % lower when concrete was prepared with coarse and fine RCA and 8 % lower when RCA was used to replace coarse NA only.

Frondistou-Yannas (1977) observed lower aggregate to mortar bond strength for RCAC than for conventional concrete. The ranking of bond strength *versus* type of coarse aggregate can be arranged as: new granite (56 ± 15 lb) > recycled granite from demolished concrete (49 ± 18 lb) > RCA (39 ± 14 lb) > recycled mortar (3139 ± 8 lb).

Razaqpur et al. (2010) determined the bond performance of conventional concrete and concrete with RCA as the only coarse aggregate prepared by the normal and the EMV methods according to the ASTM A944.99–2003 procedure. They observed lower bond strength in RCAC prepared by the normal method. On the other hand, the bond strength of RCAC prepared by the EMV method was similar or even better than that of conventional concrete. However, they did not observe any effect of aggregate type on the bond performance of the RCAC.

Gonzalez-Fonteboa and Martinez-Abella (2005) observed poorer fatigue performance of RCAC than that of conventional concrete. Concrete fatigue was evaluated via an indirect test where the compressive strength of cylindrical specimens was determined using two loading rates: standard loading rate of 8.66 kN/s and one at 0.06 kN/s. They observed higher strength losses (9.03 %) and therefore poorer fatigue performance for RCAC than for conventional concrete (4.77 %).

In the Xiao and Falkner (2007) study, the shape of the load versus slip curve between RCAC and steel rebars was comparable to that observed for NAC and steel rebars. They also observed a 12 and 6 % reduction in bond strength between RCAC and steel rebars as compared to the strength observed for NAC and steel rebars when 50 and 100 % of NA was replaced by RCA (Fig. 5.54). On the other hand, for deformed rebars, the bond strength was similar for all types of concrete irrespective of the replacement level of coarse NA by RCA. For equivalent CS of NAC and RCAC, they observed higher bond strength between RCAC with 100 % RCA than between NAC and steel rebars.

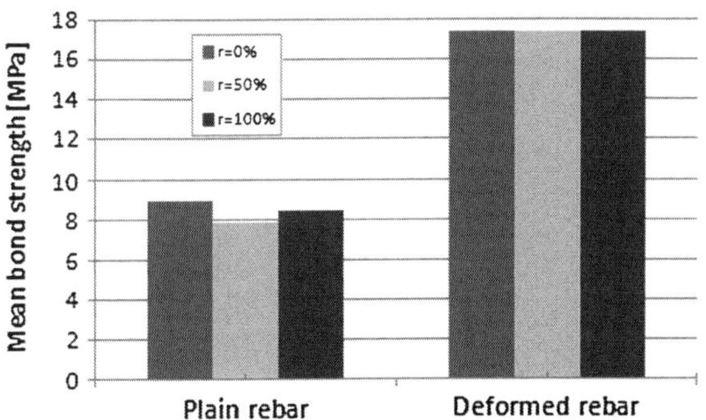

Fig. 5.54 Bond strength between concrete and steel rebars for various types of concrete (Xiao and Falkner 2007)

5.3.10 Failure Mode

Due to adhered mortar in RCA, the failure mode of RCAC is different than that of conventional concrete. The failure of RCAC in the Berndt (2009) study occurred through the old mortar particles and the gravel whereas in conventional concrete (NAC) it occurred through the stone-mortar interface and the gravel .

By examining the surfaces of RCAC after splitting tensile strength tests, Etxeberria et al. (2007b) reported that the failure of concrete (medium strength) mainly occurred through the RCA and it never happened through new interfacial zone. According to them, the weakest point of RCAC was the RCA and in particular the adhered mortar as the new aggregate-paste bond was stronger than the old mortar or the RCA-old cement paste bond due to the use of high quality cement as well as the quality of RCA, since it came from a concrete with low strength and therefore high w/c ratio. Yang et al. (2011) observed that the failure of RCAC made by replacing 50 % coarse RCA by crushed brick aggregate (CBA) during tensile and flexural strength tests occurred within the CBA due to the poor strength of porous CBA. On the other hand, the failure of RCAC with or without the 20 % CBA occurred in the aggregate-mortar interface like in the NAC.

5.3.11 Ultrasonic Pulse Velocity

The assessment of the ultrasonic pulse velocity (UPV) through concrete can give important information such as strength and elastic performance of concrete. This is an inexpensive and quick non-destructive test method to assess the quality of concrete. The influence of CDW aggregate on the UPV of concrete is reported in various references, some of which are highlighted next.

Zega and Di Miao (2009) observed comparable UPV of RCAC's containing three types of coarse RCA as 75 % replacement by volume of NA and their equivalent conventional concrete. Kwan et al. (2012) observed a decreasing UPV of concrete due to the replacement of coarse NA by RCA; however, like in conventional concrete, the UPV of RCAC also increased with curing time. The UPV of all types of concretes was in the 3.66–4.58 km/s range after 28 days of curing and is classed as good category. According to Malhotra (1976), a good category concrete does not contain any large voids or cracks, which would affect the structural integrity of concrete. Rao et al. (2011) observed a decreasing UPV as the replacement of coarse NA by RCA increased. Khatib (2005) observed a systematic decrease of UPV as the replacement of fine NA by fine RCA and fine CBA increased. At the same replacement level, the UPV of concrete containing CBA was higher than that observed for RCA concrete. The authors observed a sharp increase in UPV at the curing age of 1–7 days and then slowed down as the curing time increased. They also proposed an exponential relationship between CS and UPV, which fitted especially well for large replacement levels of fine NA by fine

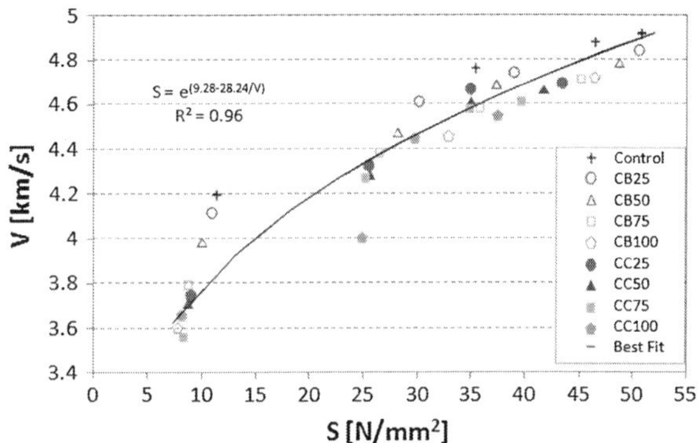

Fig. 5.55 Relationship between the UPV (V) and CS (S) of concrete (Khatib 2005)

RCA and CBA (Fig. 5.55). However, Kwan et al. (2012) commented that the polynomial or exponential relationships were not appropriate to relate UPV with CS of concrete containing RCA as there was a certain UPV value after which an increase in UPV would not necessarily mean an increase in CS. Tu et al. (2006) observed a lower UPV for high-performance (high-strength concrete (hpc/hs) with RCA as replacement of both fine and coarse NA than that observed for hpc/hs with RCA as the only coarse aggregate. The UPV increased with the w/c ratio.

Regardless of the type of mineral addition used as binder, Kou et al. (2011b) observed a decrease in the UPV of 28-day concrete as the replacement of coarse NA by RCA increased. However, the UPV of RCAC at all replacement levels of coarse NA by RCA and with SF and MK as mineral additions was significantly higher than that of conventional concrete with OPC as single binder or RCAC with or without FA and ggbfs. The UPV of RCAC with FA and ggbfs was slightly lower than that of conventional concrete. The gain in UPV between 28 and 90 days of curing for RCAC with mineral additions was higher than that for concrete with OPC as single binder owing to the formation of more hydration products and consequent improvement of the microstructure of RCAC due to a pozzolanic reaction. However, contrary to these results, Topçu (1997) observed a gradual increase of UPV in concrete due to the replacement of all NA by RCA, linked to wider air-voids because of RCA incorporation. The UPV in conventional concrete and concrete with 100 % RCA was around 70 and 93 μs, respectively.

5.4 Durability Performance

The durability of concrete is defined as the ability of concrete to withstand chemical attack and external environmental and physical actions. A concrete with a long service life must have a good durability performance. The durability of concrete depends on various factors such as the properties of concrete's constituents and their proportioning, the curing conditions and external environmental conditions. Several tests are performed to evaluate the durability of concrete. A vast work is already available on the evaluation of various durability behaviours of concrete containing CDW aggregate. In this section these properties are presented based on the collected references.

5.4.1 Drying Shrinkage

Concrete begins to shrink as soon as the hardening process starts by losing unconsumed water (i.e. that does not take part in the cement hydration reaction). The shrinkage of concrete can affect several mechanical and other durability properties of concrete due to the formation of micro cracks. In this section, the shrinkage (especially drying shrinkage) performance of concrete containing CDW aggregate is presented based on the literature. Normally, the incorporation of CDW aggregate in concrete increases the drying shrinkage. The increase in paste content in concrete due to the incorporation of CDW aggregate was identified as the main reason for drying shrinkage of concrete to increase (Kou et al. 2011b; Limbachiya et al. 2000). Table 5.20 shows a few typical examples of drying shrinkage of concrete containing RCA as coarse aggregate.

Limbachiya et al. (2000, Limbachiya (2010) observed higher drying shrinkage in RCAC with 30, 60 and 70 MPa design strength than in equivalent conventional concrete (Table 5.20). The shrinkage of concrete increased with the RCA content and design strength. The reasons for this trend were: the increase in cement content to reduce the w/c value and achieve the same 90-day strength as conventional concrete; and the presence of old cement paste in RCA. Like in conventional concrete, they observed lower shrinkage in RCAC with higher CS. In comparison to conventional concrete, Hansen and Boegh (1985) observed an increase of about 40–60 % in 440-day drying shrinkage of three different classes of structural concrete containing three types of coarse RCA as the only coarse aggregate by comparison with the parent concrete from which the RCA were generated (Table 5.20). The reasons for this trend were the same as in the previous research. Hasaba et al. (1982) observed a drying shrinkage about 70 % higher in concrete with RCA as a full replacement of fine and coarse NA. In comparison to conventional concrete, Poon et al. (2006) observed about 33 and 20 % higher 112-day drying shrinkage of two types of RCAC where 100 % by volume of coarse NA were replaced by RCA and at w/c of 0.55 and 0.45, respectively.

Sagoe-Crentsil et al. (2001) observed a 35 % increase of the 365-day drying shrinkage due to the complete replacement of coarse basalt aggregate by RCA in concrete; however, the decreasing trend of drying shrinkage of both types of concrete with time was similar. The 56-day drying shrinkage strain of both types of concrete was less than the 700 μ, the recommended limit in the Australian standard, AS 3600. Khatib (2005) observed a higher drying shrinkage of conventional concrete and RCAC's containing fine RCA at various levels in the first 10 days of curing (Fig. 5.56). The shrinkage increased with the content of RCA in concrete. Kou and Poon (2009a) observed increasing drying shrinkage of concrete

Table 5.20 Drying shrinkage of concrete containing CDW aggregate

Reference	Type of aggregate	Shrinkage strain/substitution level(%)/day	Comments
Limbachiya et al. (2000)	RCA/coarse	596/0/90[a]; 600/30/90[a]; 625/50/90[a]; 673/100/90[a];d 718/0/90[b]; 728/30/90[b]; 768/50/90[b]; 785/100/90[b];d 769/0/90[c]; 781/30/90[c]; 792/50/90[c]; 818/100/90[c];d	[a, b, c] = concrete with 30, 60 and 70 MPa design strength respectively; d = shrinkage ($\times 10^{-6}$)
Hansen and Boegh (1985)	RCA/coarse	4.0/0/440 [h]; 4.3/0/440 [m]; 5.1/0/440 [l];a 6.4/100/440 [h]; 6.0/100/440 [m]; 5.8/100/440 [l];b 6.1/100/440 [h]; 6.6/100/440 [m]; 6.3/100/440 [l];c 7.9/100/440 [h]; 7.0/100/440 [m]; 7.5/100/440 [l];d	[h, m, l] = aggregate from high, medium and low strength concrete; a, b, c, d = shrinkage ($\times 10^{-4}$) of original concrete, high, medium and low strength RCAC
Limbachiya et al. (2012)	RCA/coarse	280/0/91; 320/30/91; 425/50/91; 810/100/91;a 195/0/91; 250/30/91; 425/50/91; 695/100/91;b	a, b = Concrete with OPC and OPC-FA as binder; Shrinkage in μ strain.

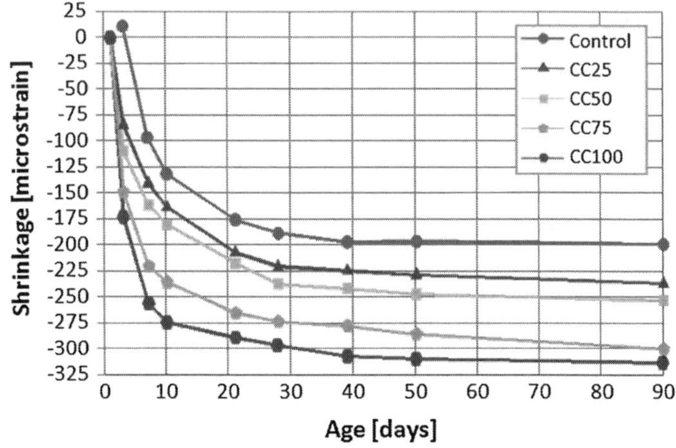

Fig. 5.56 Drying shrinkage of conventional and RCA concrete (CC) *versus* time (Khatib 2005)

at constant w/c ratio and near constant slump due to the replacement of fine NA by fine RCA. Ajdukiewicz and Kliszczzewicz (2002) observed a significant influence of RCA on the shrinkage performance of the resulting high-performance/high-strength concrete. The shrinkage of concrete with RCA generated from two types of concrete with granite and basalt as the coarse aggregates as complete replacement of NA (fine and coarse) was 35–45 % higher than that of conventional concrete. Kou et al. (2011b) reported that the incorporation of RCA as partial or full replacement of coarse natural aggregate in concrete increased the 112-day drying shrinkage of the resulting concrete due to the presence of old cement paste and the low stiffness of RCA.

Domingo-Cabo et al. (2010) measured the drying shrinkage strain of NAC and RCAC with coarse RCA as 20, 50 and 100 % replacement of coarse NA. After 180 days they observed 20 and 70 % higher shrinkage strain of the concrete mixes with 50 and 100 % incorporation of RCA, respectively. The increase in volume of the cement paste and of the porosity of concrete due to the incorporation of RCA was the reason for the higher drying shrinkage of the RCAC. These results are presented in Fig. 5.57.

Zega and Di Miao (2011) observed similar drying shrinkage strains in conventional concrete and RCAC containing fine RCA as 20 % replacement of fine NA when both types of concrete after 180 days of drying shrinkage testing. On the other hand, the shrinkage strain of RCAC containing 30 % fine RCA was slightly lower than that of the NAC due to a lower effective w/c ratio. Regardless of the type of aggregate in concrete, Yang et al. (2008) observed a higher rate of shrinkage strain within the first 10 days of testing and then it gradually slowed down. The shrinkage strain of concrete containing coarse or fine RCA as a 100 % replacement of coarse or fine NA respectively was also lower than that of conventional concrete in the first 10 days due to higher water absorption capacity; however, at a later stage, shrinkage was higher for RCAC due to lower stiffness of

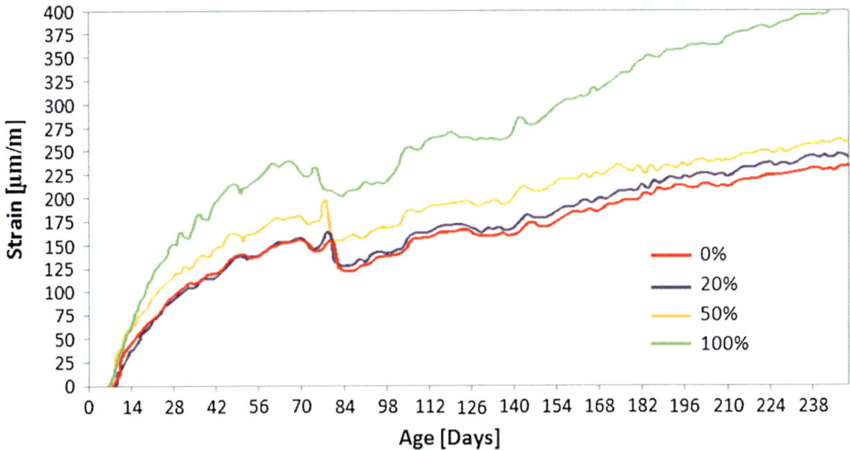

Fig. 5.57 Drying shrinkage of NAC and RCAC *versus* time (Domingo-Cabo et al. 2010)

the RCA than the NA. Moreover, the shrinkage strain of concrete containing coarse RCA with low water absorption capacity was lower than that of concrete containing coarse RCA with higher water absorption capacity or concrete containing fine RCA with higher water absorption capacity. The authors also observed an increasing trend of long-term shrinkage strain of the RCAC as the water absorption capacity of RCA increased (Fig. 5.58).

Corinaldesi (2010) reported that the 180-day shrinkage of concrete containing NA and RCAC prepared by replacing 30 % NA by fine and coarse RCA at the w/c ratios of 0.4–0.5 were almost the same and then shrinkage gradually increased for the w/c ratio of 0.60 (Fig. 5.59). The difference in shrinkage between conventional concrete and RCAC was also higher at low w/c ratio. On the other hand, for equal CS, the 180-day shrinkage of RCAC containing fine and coarse RCA were respectively 23 and 14 % lower than the conventional concrete (Fig. 5.60).

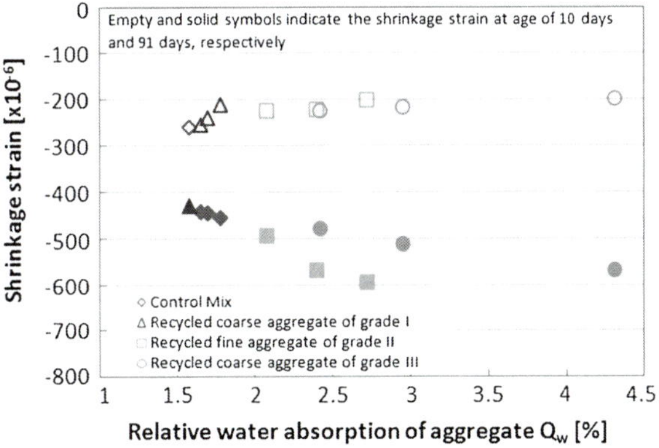

Fig. 5.58 10- and 90-day shrinkage strain of concrete with various w/c ratios (Yang et al. 2008)

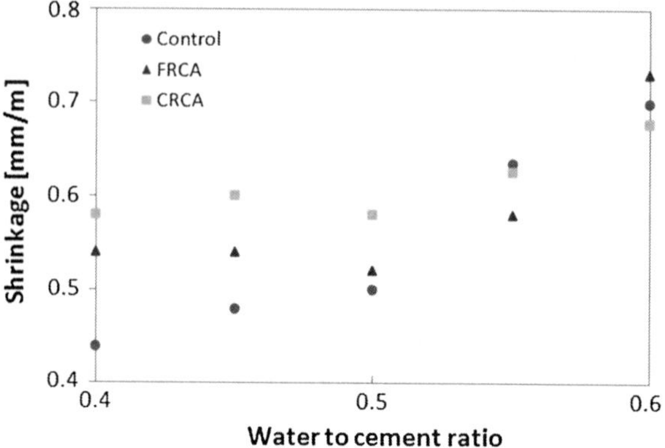

Fig. 5.59 Shrinkage of concrete *versus* w/c ratio (FRCA and CRC: fine and coarse RCA) (Corinaldesi 2010)

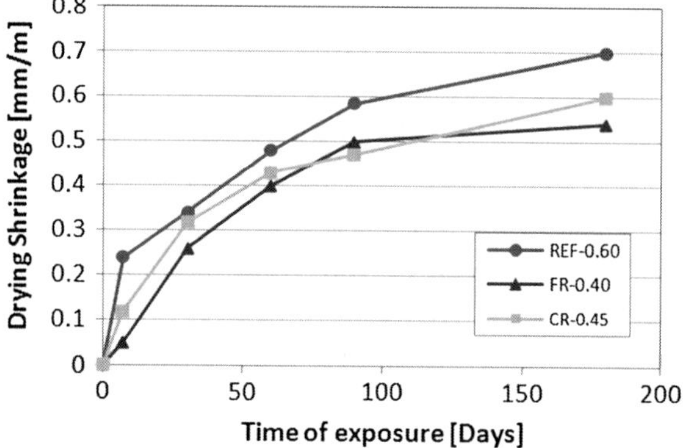

Fig. 5.60 Shrinkage of equal CS conventional concrete and RCAC (FR and CR: fine and coarse RCA, 0.60, 0.45, 0.40 indicate w/c) (Corinaldesi 2010)

Poon et al. (2009) observed increasing drying shrinkage of concrete blocks prepared using fine RA as the only coarse aggregate as the replacement level or the soil content in the RA increased. On the other hand, the drying shrinkages of concrete blocks decreased as the aggregate to cement ratio decreased and the quality of fine RA improved. Lowering of w/c ratio reduced the drying shrinkage of conventional concrete as well as RCAC (Kou et al. 2007, 2008). The drying shrinkage of RCAC increased with the coarse RCA content due to the higher mortar content in concrete (Kou et al. 2007).

In the Kou and Poon (2008) study, reducing the w/c ratio of RCAC from 0.55 to 0.40 was more effective to mitigate drying shrinkage than replacing 25 % of cement by FA. The authors observed an inverse relationship between improvement in CS and increase in drying shrinkage of RCAC (Fig. 5.61). Poon et al. (2006) reported that the drying shrinkage of steam cured concrete containing RCA as the only coarse aggregate and w/c ratios of 0.55 and 0.45 was respectively 14 and 15 % lower than that of the corresponding normal water cured RCAC. Eguchi et al. (2007) observed an approximately linear increase of drying shrinkage strain of RCAC with the water absorption capacity of concrete.

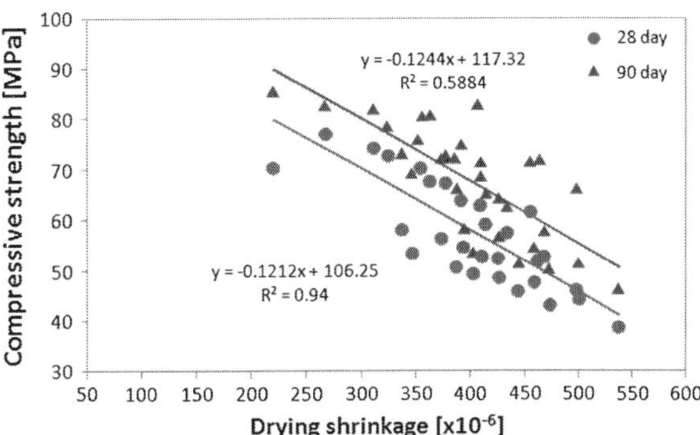

Fig. 5.61 Relationship between CS and drying shrinkage (Kou and Poon 2008)

Depending upon their hydration behaviour, mixing mineral additions with cement has positive or negative effect on the drying shrinkages of concrete containing RCA. Regardless of the replacement ratio of coarse NA by RCA, Kou et al. (2011b) reported that the drying shrinkages of concrete containing SF and MK as 10 and 15 % replacement of OPC was higher than that of the conventional concrete due to the formation of a higher amount of calcium silicate hydrate gel. On the other hand, the drying shrinkage of concrete containing FA and ggbfs was lower than that of the conventional concrete due to the lower hydration rate of FA and ggbfs as well as the restraining effect of unhydrated powder in cement paste. Their results are presented in Fig. 5.62.

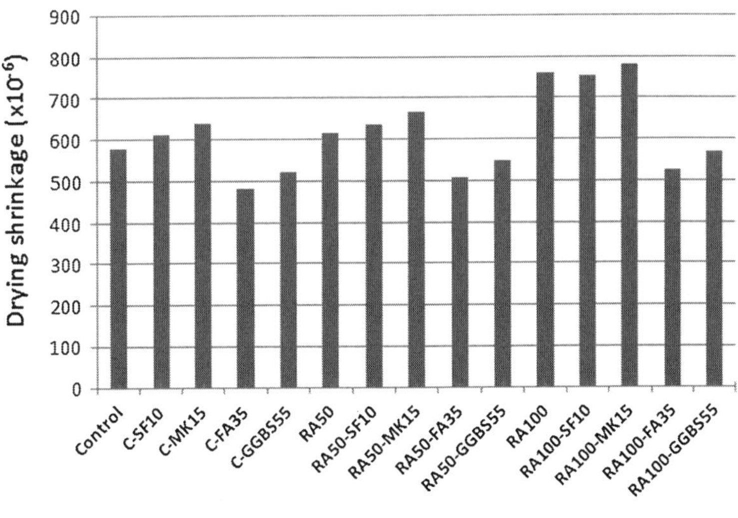

Fig. 5.62 Drying shrinkage of concrete containing 0–100 % of coarse RCA and various mineral additions (Kou et al. 2011b)

Limbachiya et al. (2012) also observed a significant decrease in the drying shrinkage strain of conventional concrete as well as of concrete containing coarse RCA as partial or full replacement of NA when 30 % by weight of OPC was replaced by FA due to the reduction in water demand in the concrete mix because of the lubricating effect of FA as well as the amount of available water the near pore network for any external drying (Table 5.19). However, regardless of the type of binder used, the shrinkage of concrete gradually increased with the RCA content due to the higher water absorption capacity and adhered mortar content of RCA. However, up to a 30 % replacement of coarse NA by RCA in two types of concrete of three different strength classes, the increase in shrinkage strain was not more significant (maximum 36.8 %) than that observed in higher replacement level (maximum 117.9 and 256.4 % at 50 and 100 % replacement level respectively). On the other hand, Sagoe-Crentsil et al. (2001) observed a significant increase of drying shrinkage especially after 91 days due to the replacement of 35 % of OPC by slag in RCAC. An increase in 5 % cement content in RCAC also increases the drying shrinkage but it was not as significant as that observed by slag addition.

The drying shrinkage performance of concrete containing RCA obtained after chemical treatment is reported in several references. Tsujino et al. (2006) stated that the surface coating of medium and low quality coarse RCA by a mineral oil could decrease the drying shrinkages of the resulting concrete, which was more prominent for concrete prepared at w/c of 0.4 than for that prepared at 0.6. In comparison to untreated medium quality RCA, the reduction was 10 % for oil treated medium quality RCA. The drying shrinkage of concrete containing silane coated medium quality RCA prepared at w/c of 0.6 and 0.4 was respectively comparable and lower than that of concrete containing untreated similar quality RCA. On the other hand, for low quality RCA, the drying shrinkage of concrete containing silane treated RCA at both w/c values was higher than that observed for concrete containing untreated RCA. In most of the cases except the concrete containing oil coated medium and low quality RCA prepared at w/c of 0.6, the drying shrinkage of concrete containing treated or untreated RCA was lower than that of conventional concrete.

Kou and Poon (2010) observed higher drying shrinkage of concrete prepared by replacing all coarse NA by untreated or polyvinyl alcohol treated RCA (Fig. 5.63) and cured for 112 days. However, the shrinkage of both concrete mixes with oven dried treated RCA (PI-R(O)-100) and air-dried treated RCA (PI-R(A)-100) was around 15 % lower than that of the concrete with untreated RCA (R-100) due to the lower water absorption capacity of treated RCA than that of the untreated one. Shayan and Xu (2003) observed that the drying shrinkage of concrete with coarse RCA as full replacement of coarse NA or fine RCA as 50 % replacement of fine NA was higher than that of conventional concrete. The use of sodium silicate and lime treated coarse or fine RCA in concrete further increased the drying shrinkage. However, all the concrete compositions met the Australian standard specifications. The drying shrinkage of concrete containing treated coarse and fine RCA as full replacement of coarse NA and as 50 % replacement of fine NA respectively was

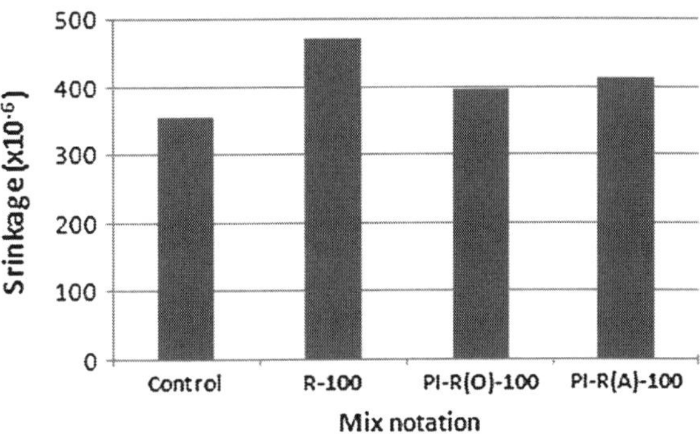

Fig. 5.63 Shrinkage of concrete containing untreated and polyvinyl alcohol treated coarse RCA and conventional concrete (Kou and Poon 2010)

higher than the standard specifications' limit. An OPC blended with 8 % SF was used as binder in this investigation. Fathifazl et al. (2011) observed lower shrinkage strain of NAC and RCAC when both types of concrete were prepared by the EMV method than when they were prepared by the normal method. They also observed lower shrinkage strain in concrete containing RCA originated from a concrete made with river bed gravel than from one made with limestone gravel.

5.4.2 Permeability Properties

The durability of concrete is greatly influenced by its permeability behaviour. A concrete with low permeability has better durability performance. Lower permeability means lower void content in concrete, and therefore water and some other corrosion agents cannot penetrate easily into concrete. Several properties such as water absorption, gas permeability and chloride penetration are evaluated to determine the permeability performance of concrete. Alexander et al. (1999) classified concrete according to its performance in various permeability tests. This classification is presented in Table 5.21.

The types of pores present in concrete also influence the permeability behaviour of concrete: discontinuous capillary pores are desirable for low permeable concrete. Thus the evaluation of porosity of concrete can provide data on the permeability behaviour of concrete. The factors, which improve concrete porosity,

Table 5.21 Concrete durability classification based on permeability parameters (Alexander et al. 1999)

Durability class	Oxygen permeability index (log scale)[a]	Sorptivity (mm/$\sqrt{h}$)	Chloride conductivity (mS/cm)
Excellent	>10.0	<6	<0.75
Good	9.5–10.0	6–10	0.75–1.50
Poor	9.0–9.5	10–15	1.50–2.50
Very poor	<9.0	>15	>2.50

[a] Negative logarithm of oxygen permeability coefficient

can also improve the permeability related durability performance of RCAC. Concrete porosity depends on various factors including aggregate's porosity. The use of CDW aggregate in concrete can change the porosity of concrete due to the higher porosity of CDW aggregate than that of the NA.

Gomez-Soberon (2002) observed an increase in total porosity of concrete with the replacement level of coarse NA by RCA. Total porosity was determined by mercury intrusion porosimetry. After 90 days of curing, total porosity of concrete containing RCA as the only coarse aggregate was around 3.8 % higher than that of conventional concrete. They also observed a significant decrease in total porosity as curing time increased, which was more prominent for RCAC. Kou and Poon (2006) observed that incorporation of coarse RCA in concrete gradually increased the total porosity and average pore's diameter and shifted the pore size distribution to larger pores. The replacement of 25 and 35 % OPC by FA respectively decreased and increased the open porosities of conventional and RCA concrete. On the other hand, decreasing water to binder ratio decreased the open porosities and average pore diameters of both types of concrete. Kou et al. (2011a) observed improvement of total porosity of RCAC as curing time increased.

Properties such as water absorption capacity by immersion and capillarity, chloride and other gas permeation of concrete with CDW aggregate are discussed in this section.

5.4.2.1 Water Absorption

The water absorption capacity of concrete is an important property, which provides data on the water accessible porosity of concrete; concrete with high water absorption capacity is less durable in aggressive environmental conditions. Since the water absorption capacity of CDW aggregate is higher than that of natural aggregate, concrete containing CDW aggregate has higher water absorption capacity than conventional concrete. The water absorption is evaluated by an immersion test, which measures the open porosity of concrete specimens, and by capillarity test, which measures the capillary water absorption due to a difference in pressure occurred between the liquid on the concrete's surface and inside the capillary pores of concrete. Several references are available on the evaluation of water absorption capacity of concrete containing various types of CDW aggregates.

Kwan et al. (2012) observed increasing water absorption capacity of concrete as the replacement level of coarse NA by RCA increased (Fig. 5.64). They state that the replacement of 30 % by weight of coarse NA by RCA led to a water absorption capacity below 3 %, i.e. a concrete considered to have low water absorption capacity. For an 80 % replacement, the water absorption capacity of RCAC was 2.2 times higher than that of conventional concrete. Rao et al. (2011) also observed a gradual increase of water absorption capacity of concrete with the incorporation of coarse RCA to replace coarse NA due to the higher water absorption capacity of RCA, which was about 3.5 times higher than that of NA.

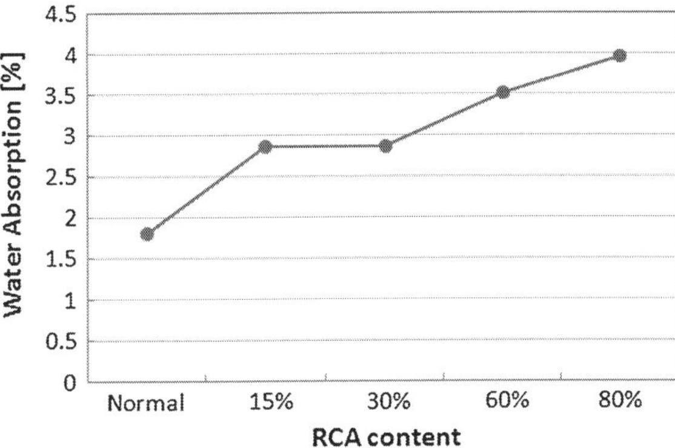

Fig. 5.64 Water absorption capacity of concrete versus coarse RCA content (Kwan et al. 2012)

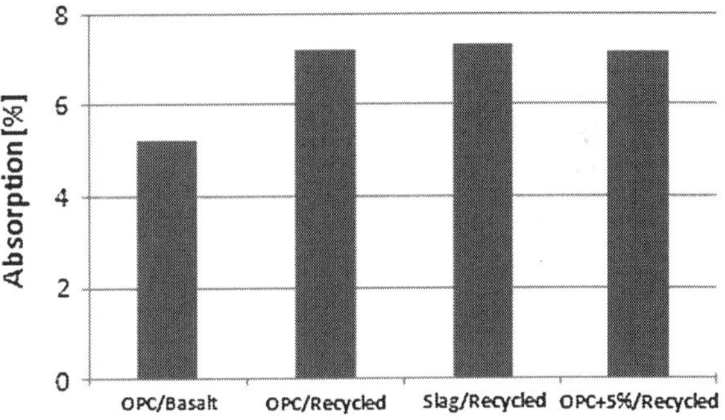

Fig. 5.65 Water absorption due to immersion of conventional concrete and RCAC's (Sagoe-Crentsil et al. 2001)

Sagoe-Crentsil et al. (2001) observed water absorption about 25 % higher due to the complete replacement of coarse basalt aggregate by RCA (Fig. 5.65). The concrete specimens used in this research were moist cured for 6 days after demoulding and then cured at 23 °C at 50 % room humidity before determining the water absorption capacity. Adding slag cement or increasing 5 % OPC content did not have any beneficial effect on the water absorption capacity of RCAC. Grdic (2010) observed water absorption capacity about 0.15–0.37 % higher in self-compacting concrete (SCC) due to a 50–100 % replacement of coarse NA by RCA. They did not observe any water penetration for 50 and 100 % coarse RCA based SCC; on the other hand, the control SCC (with NA only) had a penetration of 10 mm.

Soutsos et al. (2011) observed increasing water absorption capacity of concrete paving blocks as the replacement ratio of coarse and fine NA by similar sized RCA and recycled masonry aggregates (RMA) increased. This increase was higher for concrete with fine recycled aggregate than with coarse one. These results are presented in Fig. 5.66. The authors observed that the replacement of 55 or 25 % of coarse or fine NA by coarse and fine RCA respectively in concrete paving blocks allows maintaining the water absorption capacity below the BS EN1338 specified maximum limit of 6 % along with reasonably good CS and STS. For recycled masonry aggregates, these threshold replacement ratios were respectively 50 and

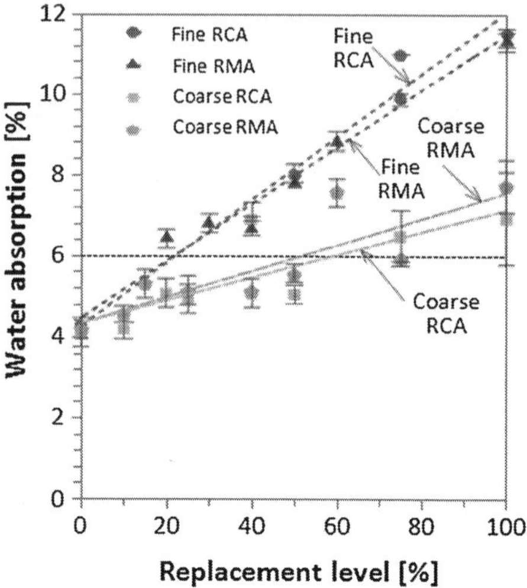

Fig. 5.66 Water absorption capacity of concrete containing various types of CDW aggregate (Soutsos et al. 2011)

20 % for coarse and fine aggregates. The water absorption capacity of conventional concrete and RCAC prepared by replacing 30 and 100 % by volume of fine NA by fine RCA in Evangelista and de Brito (2010) was respectively 11.3, 13.2 and 16.5 %. The water absorption capacity of concrete with 30 and 100 % fine RCA was about 16 and 46 % higher than that of the conventional concrete respectively. Yaprak et al. (2011) also observed a gradual increase of the water absorption capacity of concrete as the replacement of fine NA by RCA increased.

Poon and Lam (2008) observed increasing water absorption capacity of concrete paving blocks containing NA or coarse RCA as the aggregate to cement ratio increased. The water absorption capacity of the concrete also increased with the water absorption capacity of aggregate. A significant reduction in water absorption of the concrete blocks containing coarse RCA was observed when the coarse RCA was partially replaced by low water absorbing aggregates such as coarse NA or coarse recycle glass aggregate. Gomes and de Brito (2009) observed a linear increase of water absorption capacity of concrete with the replacement of coarse NA by RCA and crushed brick and mortar recycled aggregate (CBMRA). The water absorption capacity of concrete containing CBMRA was significantly higher than that observed for RCAC due to the higher porosity of the former type of aggregate.

Poon and Chan (2006) observed a higher cold and hot water absorption capacity of concrete paving blocks due to a 25, 50 and 75 % replacement of RCA by crushed clay brick aggregate and the water absorption increased with the RCA replacement level due to the high water absorption capacity of clay brick aggregate. However, the incorporation of FA to replace part of RCA in both types of concrete (with or without crushed clay brick aggregate) can improve the water absorption capacity of concrete due to the filling of voids within the concrete mixes. In another reference, Poon and Chan (2007) observed higher water absorption capacity of RCAC paving blocks due to the addition of contaminants such as crushed tiles, glass, brick and wood. The water absorption capacity of all types of paving blocks increased with the aggregate to cement ratio.

Mas et al. (2012) observed a gradual increase of water penetration depth in concrete prepared at w/c ratios of 0.45, 0.65 and 0.72 due to the replacement of coarse NA by low quality RCA, which contains 20–30 % of ceramic materials. Yang et al. (2011) observed a water permeability index about 20 % higher in concrete containing RCA as the only coarse aggregate than in conventional concrete. The replacement of RCA by crushed brick gradually increased the water permeability of the resulting concrete due to the higher porosity of crushed brick aggregates than that of RCA (Fig. 5.67). The water permeability of concrete containing RCA as the only coarse aggregate and of concrete containing crushed brick as a 50 % replacement of coarse RCA were 2.31×10^{-7} $m^3/min^{1/2}$ and 3.98×10^{-7} $m^3/min^{1/2}$, respectively. The water permeability coefficient of 84-day cured NAC and RCAC where RCA was used as the only coarse aggregate in the Berndt (2009) study was in the range of 1–1.4 ($\times 10^{-10}$) cm/s and 1.7–1.9 ($\times 10^{-10}$) cm/s, respectively. The marginal increase in water permeability coefficient of RCAC was due to the residual mortar content in RCA. The addition of blast furnace slag to replace 50 and 70 % cement did not have an influence on the permeability performance of concrete. According to the author, the observed permeability of NAC and RCAC's was within the acceptable range for durable concrete, i.e. it was $<3 \times 10^{-10}$ cm/s.

Buyle-Bodin and Zaharieva (2002) observed higher initial water absorption and sorptivity of RCAC's than of conventional concrete (Fig. 5.68). The test was performed in concrete samples obtained after two types of curing, normal water curing and air curing. Both properties of RCAC prepared by replacing only coarse

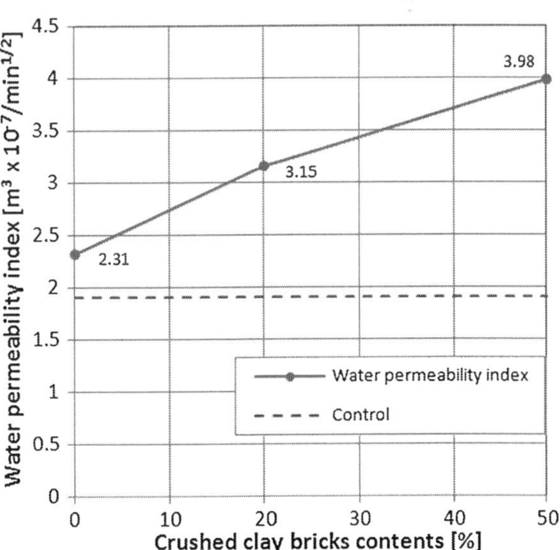

Fig. 5.67 Water permeability of RCAC *c* content of crushed brick as a coarse RCA replacement (Yang et al. 2011)

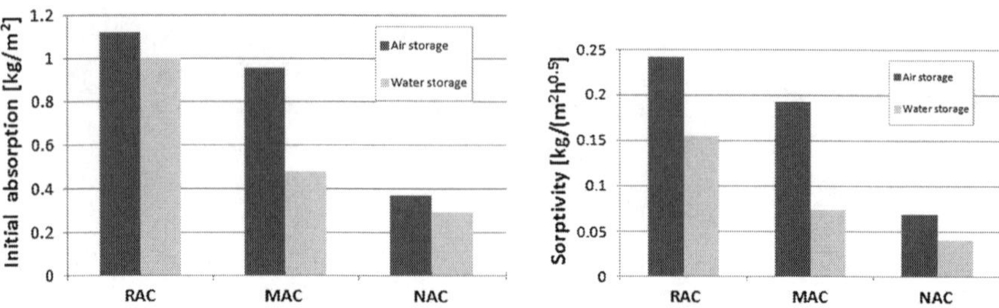

Fig. 5.68 Initial water absorption and sorptivity of conventional and RCA concrete at air- and water-curing conditions (Buyle-Bodin and Zaharieva 2002)

NA (MAC) were considerably better than those of the RCAC prepared by full replacement of fine and coarse NA (RAC), especially when the concrete samples were water cured. In terms of curing conditions, the RCAC's and the conventional concrete (NAC) obtained after water-curing both exhibited better performances due to the finer pore structure.

Zaharieva et al. (2003) observed about two to three times higher water permeability in concrete due to the replacement of coarse and fine NA by RCA and the water permeability of RCAC also increased with the amount of fine RCA. The water permeability of NAC was around 0.8×10^{-20} m^2. According to them, even though the incorporation of RCA in concrete increased the water permeability, the NAC and RCAC's tested could be considered as feebly permeable.

Limbachiya et al. (2000) and Limbachiya (2010) did not observe any effect of a 30 % replacement of coarse NA by RCA on the initial surface water absorption measured at 10 min in five classes of concrete with design compressive strength of 20, 30, 50, 60 and 70 MPa. Their results for the 20 and 30 MPa mixes are presented in Table 5.22. The surface water absorption increased with the RCA content but decreased as the design strength increased. The gradual increase in water absorption of concrete due to the content of RCA was attributed to the raise of cement paste content as more cement was added to reach the design strength of concrete. The same authors (Limbachiya et al. 2012) observed significant reduction in the 10 min initial surface water adsorption of three classes (with 20, 30 and 35 MPa design CS) of conventional concrete as well as mixes with various ratios of coarse RCA as partial or full replacement of NA due to the use of 30 % FA as a mineral addition. This improvement was attributed to the pozzolanic reaction of FA and a refinement of the pore structure and it was especially prominent for the full replacement of coarse NA by RCA.

Table 5.22 Initial surface water absorption at 10 min (ml/m^2/s $\times$ 10^{-2}) of concrete containing RCA (Limbachiya 2010)

Type of aggregate	Strength class (MPa)	Water absorption/substitution level (%)
RCA/coarse	20	50/0; 50.5/30; 55/50; 66/100
	30	32/0; 32/30; 36.5/50; 51/100

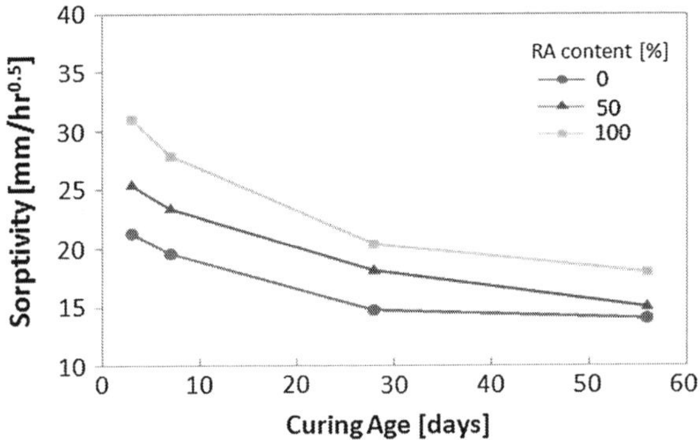

Fig. 5.69 Sorptivity of concrete containing RCA at 0, 50 and 100 % replacement of coarse NA (Olorunsogo and Padayachee 2002)

Olorunsogo and Padayachee (2002) observed a gradual increase in water sorptivity of concrete due to a 50 and 100 % replacement of coarse NA by RCA (Fig. 5.69). The sorptivity of concretes also decreased with increasing curing time. The difference in sorptivity between NAC and RCAC decreased with increasing curing time, indicating more improvement in RCAC than NAC with increasing curing time. However, according to Alexander et al. (1999) classification (Table 5.21), the concrete containing NA and RCA could be classified as poor and very poor respectively.

Evangelista and de Brito (2010) observed an increase of around 34 and 70 % in capillary water sorptivity due to the replacement of 30 and 100 % by volume of fine NA by fine RCA, respectively, attributed to the formation of more capillary pores due to its high porosity (Fig. 5.70a). Zega and Di Miao (2011) also observed a 13 % increase of capillary water absorption capacity of concrete due to the replacement of 20 and 30 % by volume of fine NA by fine RCA because of the higher porosity of RCA than that of NA (Fig. 5.70b). However, the capillary water

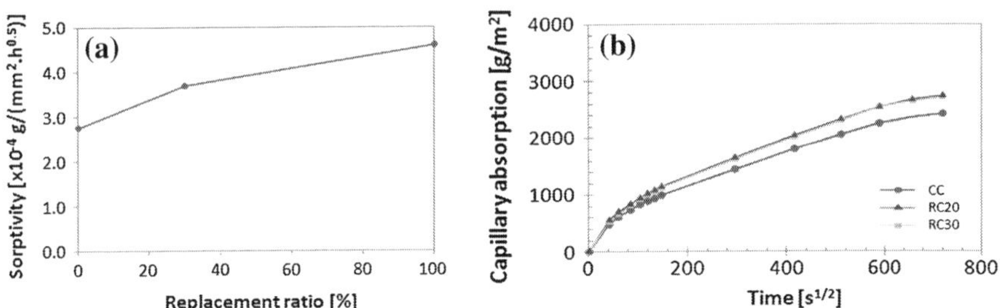

Fig. 5.70 **a** Capillary water sorptivity (Evangelista and de Brito, 2010) and **b** capillary water absorption (Zega and Di Miao 2011) of conventional concrete and RCAC containing RCA as a replacement of 30 and 100 % fine NA

sorptivity of RCAC was below the Argentinean specified maximum value, i.e. lower than 4 $g/m^2/s^{1/2}$.

Gomes and de Brito (2009) observed a progressive increase of the capillary water absorption coefficient with the content of RCA and CBMRA; the maximum increase of capillary water absorption coefficient for concrete with RCA as the only coarse aggregate and for concrete with CMBRA as a 50 % replacement of coarse NA was respectively 16.6 and 71.5 % as compared to conventional concrete. Gonçalves et al. (2004) also observed higher capillary water absorption coefficient of concrete containing RCA than that of conventional concrete; however, they did not observe any effect of increasing the coarse RCA content in concrete on the capillary water absorption coefficient.

5.4.2.2 Chloride Permeability

The determination of diffusivity of chloride ions through concrete can provide data on the permeability performance of concrete. Lower chloride permeability is desirable for durable concrete structures. Another important property that needs to be evaluated for reinforced concrete structure is chloride-induced corrosion. Several studies were undertaken to understand the chloride permeation and chloride-induced corrosion performance of concrete containing CDW aggregate. In several researches, it was reported that the incorporation of CDW aggregate in concrete increases the chloride permeability of concrete; however, some results also indicate negligible influence of CDW aggregate incorporation on the chloride permeability and chloride-induced corrosion performance of concrete.

Olorunsogo and Padayachee (2002) observed a significant increase in chloride conductivity with the replacement of coarse NA by RCA. The chloride conductivity of concrete containing RCA as the only coarse aggregate was 86.5 % higher than that of the conventional concrete. However, the chloride conductivity of RCAC also decreased with curing time (Fig. 5.71). The conventional concrete and the RCA concrete after 56 days of curing are classified as good and very poor, respectively in terms of chloride ion conductivity as presented in Table 5.21. Poon

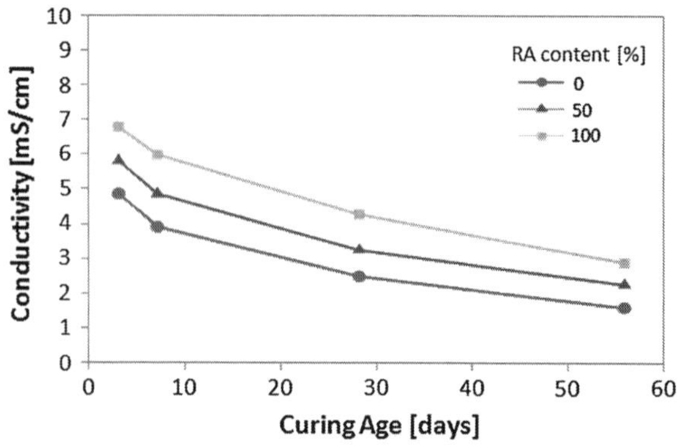

Fig. 5.71 Chloride ion conductivity of conventional and RCA concrete (Olorunsogo and Padayachee 2002)

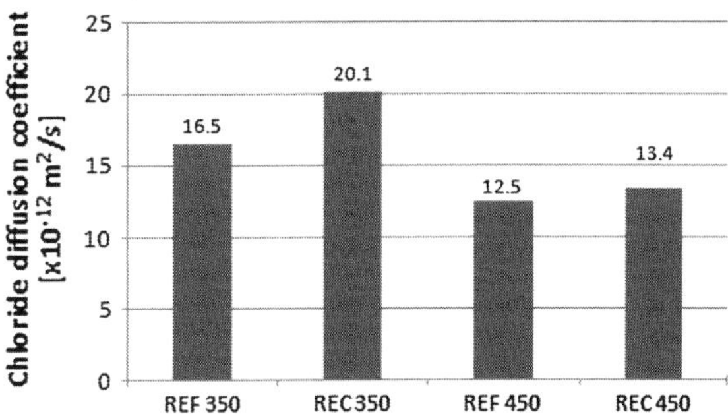

Fig. 5.72 Chloride diffusion coefficient of conventional concrete and RCAC with various cement contents (Gonçalves et al. 2004)

et al. (2006) observed higher resistance against chloride penetration of steam cured RCAC than that observed in normal water cured RCAC, which might be due to the formation of tortuous interconnecting capillary pores because of non-uniform calcium silicate gel formation. The chloride ion penetration resistivity decreased as the coarse RCA content in concrete increased.

Gonçalves et al. (2004) observed a lower chloride ion penetration coefficient with full replacement of coarse NA by RCA. The difference in chloride ion penetration coefficient between conventional concrete and RCAC was lower at higher cement contents indicating a higher influence of paste quality than that of aggregate porosity (Fig. 5.72).

Kou and Poon (2009a) observed lower chloride penetration resistance of concrete due to the incorporation of fine RCA as a replacement of fine NA at constant w/c ratio, which might be due to the poor microstructure formation because of adhered mortar content in RCA. On the other hand, at constant slump, the resistance to chloride penetration of RCAC was comparable to that of conventional concrete. The resistance to chloride penetration of RCAC decreased as the RCA content in concrete increased. Evangelista and de Brito (2010) also observed a linear increase of chloride migration coefficient with the replacement of fine NA by fine RCA because of increasing porosity of concrete (Fig. 5.73). The chloride migration coefficient of RCAC containing fine RCA at 30 and 100 % by volume replacements of fine NA was around 12 and 34 %, respectively.

Rao et al. (2011) observed an increase in chloride penetration depth with the replacement of coarse NA by coarse RCA. At 100 % replacement level, the increase in chloride penetration depth was 14 % as compared to that of conventional concrete. According to the authors, the presence of old porous mortar as well as interfacial transition zone in RCA formed a permeable concrete internal structure. Limbachiya et al. (2012) observed a marginal difference in the resistance to chloride ingress due to the replacement of 30 % of coarse NA by RCA; however, this difference became significant at 50 and 100 % replacement levels. Significant improvement of chloride resistance performance of RCAC was observed, when 30 % of OPC was replaced by FA. Gomes and de Brito (2009) observed higher chloride ion penetration depths and chloride permeability for concrete mixes containing RCA or CBMRA than for conventional concrete

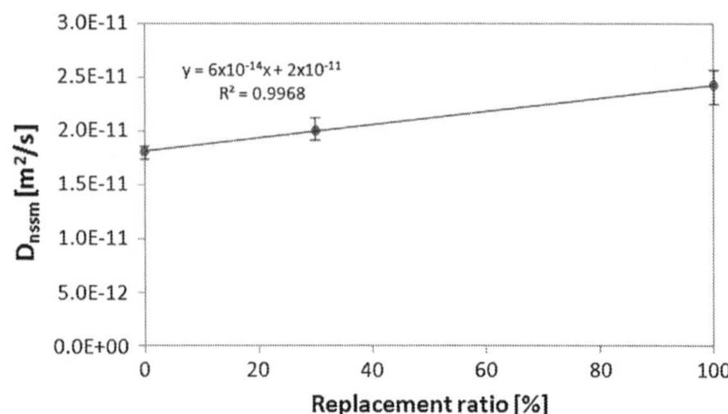

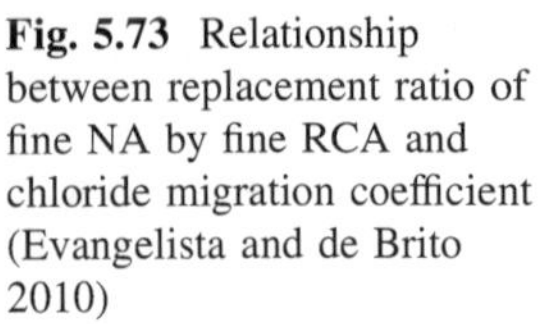
Fig. 5.73 Relationship between replacement ratio of fine NA by fine RCA and chloride migration coefficient (Evangelista and de Brito 2010)

(Table 5.23). The values for the concrete containing CBMRA were significantly higher than those observed for the RCA concrete due to the higher porosity of CBMRA than that of RCA. They also observed an increase in the chloride penetration depth with the water absorption capacity of concrete. The chloride permeability coefficient of concrete mixes containing 50 % RCA and 25 % CBMRA was respectively 5.6 and 18.8 % higher than that of the conventional concrete.

Limbachiya et al. (2000) did not observe any negative effect on the chloride diffusion performance of concrete due to the incorporation of coarse RCA as partial or full replacement of NA in three classes of high-strength concrete with design strength of 50, 60 and 70 MPa; the difference in chloride diffusion coefficients between the conventional and the RCA concrete mixes was below 1×10^{-11} m^2/s. Limbachiya et al. (2000) observed similar chloride-induced corrosion of conventional concrete and RCAC with design strength of 50 MPa except for RCAC containing RCA as the only coarse aggregate when the conventional concrete and RCAC cubes were exposed to a 2.5 M NaCl solution at 20 °C. However, the corrosion current of steel in RCAC with RCA as the only coarse aggregate and the corrosion initiation time were slightly higher and slightly shorter than for the conventional concrete and the RCAC with 50 % RCA, respectively.

Tu et al. (2006) observed higher chloride penetration (CP) for high-performance concrete (hpc) with RCA as replacement of both fine and coarse NA than that observed for hpc with RCA as the only coarse aggregate and NA as the only

Table 5.23 Chloride ion penetration depth of conventional concrete and RCAC's (Gomes and de Brito 2009)

Concrete mix	Porosity of aggregate (%)	Water absorption capacity of concrete (%)	Chloride penetration depth (mm)	Permeability coefficient (m^2/s) ($\times 10^{-12}$)
Conventional	2.29	~13.0	12.50	6.31
RCAC1	8.49	~17.0	13.21	6.66
RCAC2	16.34	~15.5	24.30	7.50
RCAC3	–	~17.0	14.39	7.26

RCAC1, RCAC2, RCAC3: concrete prepared by replacing 50, 25 and 37.5 % by volume of coarse NA by RCA, CBMRA and a 2:1 mixture of RCA and CBMRA, respectively

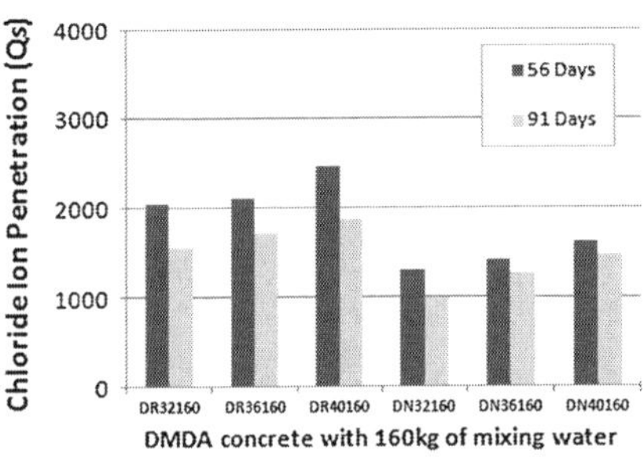

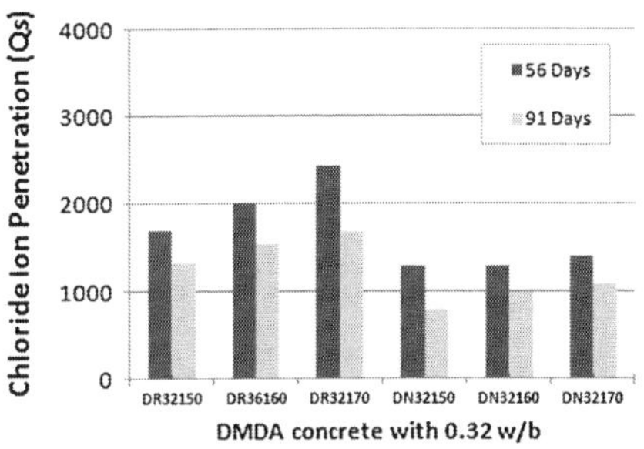

Fig. 5.74 Chloride ion resistivity of RCAC (*DR* Concrete with RCA as fine and coarse aggregate; *DN* Concrete with coarse RCA and fine NA) (Tu et al.2006)

fine aggregate due to the lowering of residual mortar content (Fig. 5.74). At fixed water content, the CP increased with the water to binder ratio. On the other hand, at fixed water to binder ratio, the lower CP was observed for RCAC containing lower water content. The CP of concrete with RCA as coarse NA replacement or as a replacement of fine and coarse NA after 91-day of curing was below 2000 Coulombs, the specified amount according to ASTM C 1202.

Kou et al. (2008) observed a decrease in resistance to chloride ion penetration with the coarse RCA content due to the porous nature of RCA. In the Kou and Poon (2010) study, the chloride permeability of concrete containing RCA as complete replacement of coarse NA was significantly higher than that of the conventional concrete, especially after 28 days of curing. However, the resistance to chloride permeability of concrete with oven dry and air-dried polyvinyl alcohol treated RCA was respectively 32 and 35 % higher than that of concrete with untreated RCA and comparable to that of conventional concrete. The chloride ion permeability of 90-day cured of both types of concrete (conventional and RCAC) was lower than that of 28-day cured concrete and was more significant for concrete with untreated RCA. Kou et al. (2007, 2008) also observed an increase in resistance to chloride ion penetration of normal water cured or steam cured RCAC with decreasing water to binder ratio and increasing curing time.

Like in conventional concrete, the resistance to chloride permeability of concrete with RCA can be improved by using mineral additions (Kou and Poon 2006; Kou et al. 2011b; Poon et al. 2007; Ann et al. 2008; Berndt 2009). This is due to the improvement of the watertightness of concrete (due to a microstructure improvement) and of the chloride binding capacity of cement paste because of the formation of high amounts of calcium silicate hydrate and calcium alumina silicates (Kou et al. 2011b). Kou et al. (2011b) observed a significant improvement of chloride permeability resistance in conventional concrete as well as in RCAC due to the replacement of 10, 15, 35 or 55 % of OPC by SF, MK, FA or ground granulated blast furnace slag (ggbfs), respectively and the ranking of improvement was: ggbfs > FA > MK > SF. The measurement was done for concrete specimens obtained after 28 and 90 days of conventional curing. Berndt (2009) also observed a significant decrease in the chloride diffusion coefficient of NAC and

RCAC after exposure for 1-year to artificial sea water due to the replacement of 50 and 70 % OPC by ggbfs. The chloride diffusion coefficient of NAC and RCAC containing OPC and OPC-slag as binder was in the range of 2×10^{-12} and 8×10^{-13} m^2/s, respectively. Ann et al. (2008) observed a significant increase in chloride ion permeability of concrete due to the incorporation of RCA as complete replacement of coarse NA when both types of concrete (conventional and RCAC) were subjected to rapid chloride ion penetration test (Fig. 5.75). However, at a replacement of 30 or 60 % of OPC by pulverised fuel ash (PFA) and ggbfs, respectively, the chloride ion permeability of RCAC decreased considerably and was even lower than that of the conventional concrete. They also observed improved resistance against chloride-induced corrosion of RCAC containing PFA and ggbfs due to the refinement of the pore structure as well as the increased chloride binding capacity of the cement paste.

Shayan and Xu (2003) reported that concrete with RCA had lesser chloride resistivity and higher negative half-cell potential than conventional concrete although both parameters for RCAC were within acceptable limits. The corrosion current densities of conventional concrete and RCAC were also similar and very low indicating that both types of concrete were resistant to chloride induced corrosion. The chloride ion penetration depth of concrete containing RCA aggregate was also higher than that of the conventional concrete. The depth of penetration of RCAC containing coarse RCA as a full replacement of coarse NA was slightly higher than that of RCAC containing fine RCA as a 50 % replacement of fine NA. On the other hand, the depth of chloride penetration for RCAC containing sodium silicate plus lime treated RCA was higher than that of the RCAC containing untreated RCA. The RCAC had inferior performance than the conventional concrete in the rapid chloride permeability test too; however, the corrosion protection of all types of concrete can be categories as high. The authors did not find any advantages of sodium chloride treatment of RCA on the corrosion

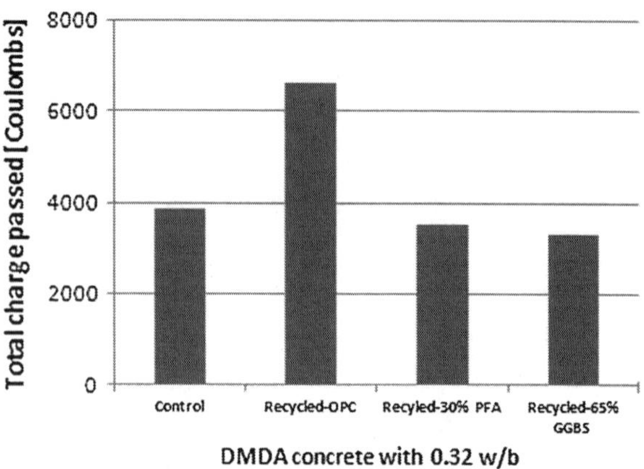

Fig. 5.75 Chloride ion permeability of conventional concrete containing OPC and RCAC containing OPC and PFA and ggbfs (Ann et al. 2008)

resistance of the resulting concrete; however, they found an important role of pre-coating with silica fume.

Kong et al. (2010) observed comparable resistances to chloride ion penetrations of concrete with RCA as the only coarse aggregate and conventional concrete when coarse RCA was pre-coated with FA or slag before mixing. They applied three mixing methods to prepare concrete: normal mixing (NM); double mixing (DM), which is described in Sect. 5.2.1. (for details, see Tam and Tam 2008); triple mixing (TM) in which fine and coarse aggregates were initially mixed for 15 s with part of the mixing water, then the mineral addition was added and mixed for another 15 s to coat the surface of aggregate, after this cement was added and mixed for another 30 s; finally, the remaining water along with the superplasticizer were added. Their results are presented in Fig. 5.76. Razaqpur et al. (2010) observed marginally higher apparent chloride diffusion coefficient of RCAC prepared by the EMV method than by the conventional mixing method; however, the chloride diffusion coefficient of both types of RCAC was lower than or comparable to that observed for conventional concrete. Abbas et al. 2009 reported that the apparent chloride diffusion coefficients of RCAC's prepared by the EMV method were found to be of the same order of magnitude of 10^{-12} m^2/s as the conventional structural grade concrete. They observed higher resistance to chloride diffusion of RCAC prepared by the EMV method due to the addition of FA or ggbfs with OPC and this resistance was significantly better for slag-based RCAC than FA-based RCAC.

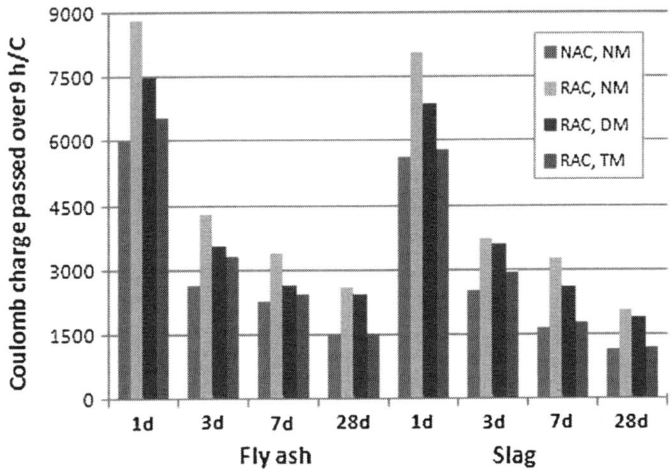

Fig. 5.76 Effect of the mixing method on the resistance to chloride penetration of conventional concrete and RCAC (Kong et al. 2010)

5.4.2.3 Gas Permeability

The gas permeability performance of concrete with CDW aggregate is also reported in the literature. Results are available for several types of gases: air, oxygen, nitrogen.

Kwan et al. (2012) determined the intrinsic permeability of concrete containing various percentages of coarse RCA by using nitrogen gas penetration. They observed a gradual increase in intrinsic permeability of concrete with the content of coarse RCA and the curing time. The difference in permeability between conventional concrete and RCAC decreased with the curing period. They also observed a parabolic inverse relationship between CS and intrinsic permeability, from which they concluded that concrete would achieve a constant permeability as the maturity of concrete increased. Limbachiya et al. (2000) did not observe any effect of 30 % replacement of coarse NA by RCA on the intrinsic air-permeability performance of three classes of high-strength concrete with design strength of 50, 60 and 70 MPa. Air permeability increased with RCA content but decreased with design strength. The increase of air permeability of concrete due to the incorporation of RCA was attributed to the increase of cement paste content as more cement was added to reach the design strength of concrete.

Buyle-Bodin and Zaharieva (2002) observed lower air permeability of concrete with complete replacement of coarse NA by RCA than that of concrete with RCA as complete replacement of fine and coarse aggregates (Table 5.24). Similarly, both types of concrete and conventional concrete as well exhibited considerably lower air permeability after water curing than after air curing. Zaharieva et al. (2003) observed that pre-soaking of RCA can improve the air permeability of concrete as dry aggregate can absorb hydrating water and therefore can hinder the hydration reaction. The air permeability of NAC and RCAC increased significantly due to an increase in pre-treatment temperature i.e. oven drying of concrete specimens at different temperatures after curing (Fig. 5.77).

Gonçalves et al. (2004) found a significant increase in oxygen permeability of concrete due to the replacement of coarse NA by RCA. Permeability as well as the difference of permeability between conventional concrete and RCAC decreased with the cement content in concrete due to a cut in porosity. These results are presented in Fig. 5.78.

Olorunsogo and Padayachee (2002) evaluated the oxygen permeability of concrete various replacement ratios of coarse NA by RCA. They presented their results in terms of oxygen permeability index (OPI), which is defined as the

Table 5.24 Air permeability of concrete containing RCA and conventional concrete at various experimental conditions (Buyle-Bodin and Zaharieva 2002)

Concrete type	Conventional		RCAC with RCA as the only coarse aggregate		RCAC with RCA only (coarse and fine)	
Curing condition	Water	Air	Water	Air	Water	Air
Air permeability ($\times 10^{-18}$) m^2	6.00	20.0	2.80	3.10	1.04	1.45

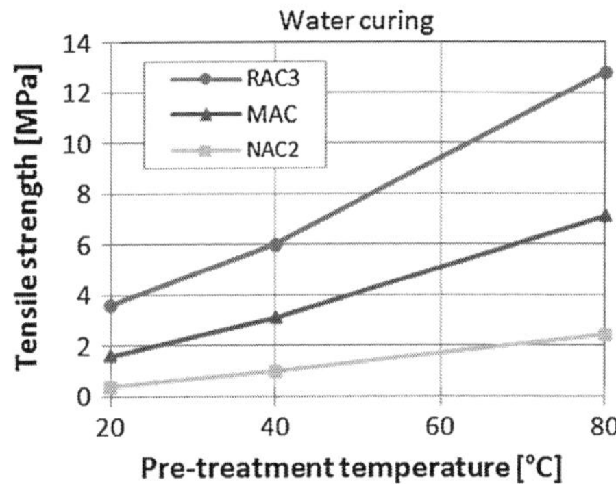

Fig. 5.77 Air permeability of NAC and RCAC at various pre-treatment temperatures (Zaharieva et al. 2003)

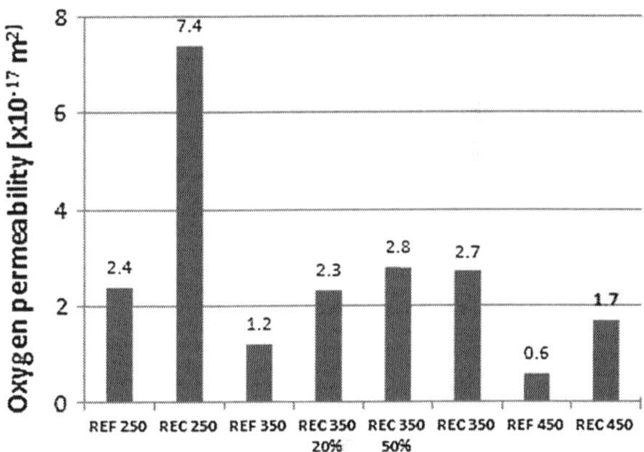

Fig. 5.78 Oxygen permeability of conventional and RCAC's with various cement contents (Gonçalves et al. 2004)

negative logarithm of oxygen permeability coefficient. They observed a gradual decrease of OPI as the replacement of NA by RCA increased; however, the increase in OPI of RCAC's with respect to the curing time was similar to that of conventional concrete. After 56 days of curing, the OPI of conventional concrete was 10 % higher than that of the RCAC with 100 % RCA. These results are presented in Fig. 5.79. The concrete mixes containing 0, 50 and 100 % coarse RCA after 56 days of curing can be considered as excellent, good and poor, respectively, according to the Alexander et al.'s (1999) classification (Table 5.21).

5.4.3 Depth of Carbonation

In a reinforced concrete structure, the steel reinforcement is chemically protected from corrosion by a passive oxide layer due to the presence of the surrounding alkaline environment. However, with time and in the presence of other chemical and physical factors, the alkali content in concrete gradually decreases due to the carbonation of concrete by atmospheric carbon dioxide and therefore the corrosion

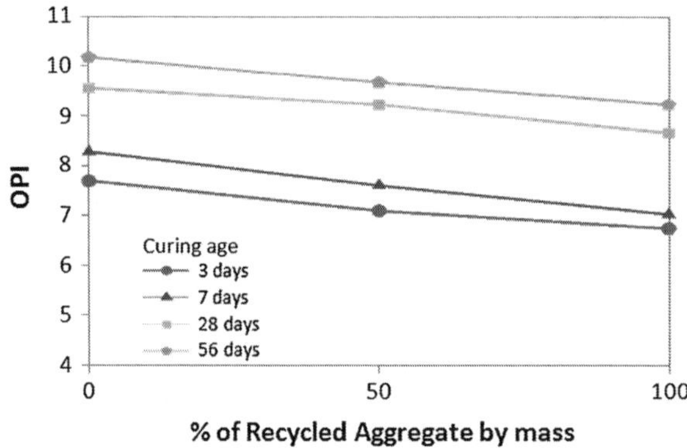

Fig. 5.79 Oxygen permeability index of concrete versus replacement ratio of coarse NA by RCA (Olorunsogo and Padayachee 2002)

Table 5.25 Carbonation depth (in mm) of conventional concrete and RCAC's

References	Type of aggregate	Concrete type	Carbonation depth/substitution level (%, volume)
Limbachiya 2010	RCA/coarse	C30	21/0; 21/30; 20/50; 18.5/100
		C35	18/0; 18/30; 17.5/50; 16.5/100
Gomes and de Brito 2009	RCA/coarse	Conventional	5.13/0
		RCAC1	5.63/50
		RCAC2	5.56/25
		RCAC3	6.57/37.5

C30, C35 Concrete with design strength of 30 and 35 MPa; RCAC1, RCAC2, RCAC3: concrete prepared by replacing coarse NA by RCA, CBMRA and a 2:1 mixture of RCA and CBMRA, respectively

resistance of reinforced structures goes down. The presence of micro cracks and pores in concrete generally enhances the rate of carbonation. Several reports are available on the evaluation of carbonation resistance of concrete. Normally RCAC has higher rate of carbonation than conventional concrete. Hansen (1992), after analysing various studies, concluded that RCAC had 4 times faster carbonation rate than that of conventional concrete. Two typical examples of the effect of RCA on the carbonation depth of concrete are presented in Table 5.25.

Limbachiya (2010) observed similar carbonation depth of two classes of air-entrained conventional concrete and concrete with a 30 % replacement of coarse NA by RCA. However, the carbonation depth of concrete decreased when the replacement level of coarse NA increased to 50 and 100 %. The author gave two reasons for resistance against carbonation to improve due to the incorporation of coarse RCA: increase in calcium hydroxide content with more attached cement paste content and increase in alkalinity due to increased cement content in RCAC to reach equal strength of concrete as well as to reduce the w/c ratio. These results are presented in Table 5.24. In this study, the concrete samples were exposed for

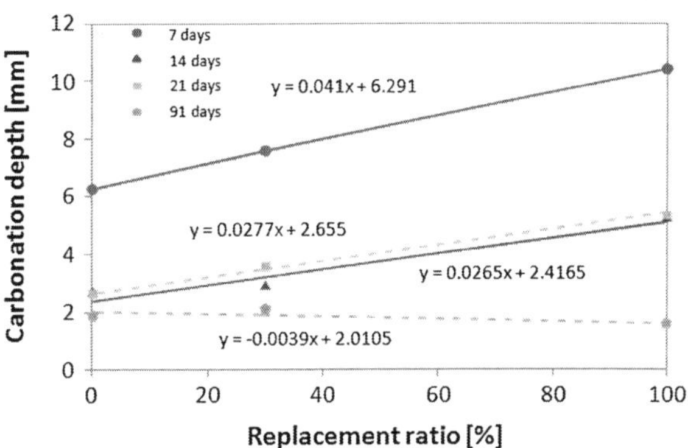

Fig. 5.80 Carbonation depth of concrete *versus* replacement ratio of fine NA by fine RCA (Evangelista and de Brito 2010)

20 weeks to a carbon dioxide atmosphere at 20 °C and 55 % room humidity. Gomes and de Brito (2009) observed higher carbon dioxide penetration depth in concrete with coarse RCA or coarse CBMRA than in conventional concrete (Table 5.24). Evangelista and de Brito (2010) observed a linear increase of carbonation depth with the replacement ratio of fine NA by fine RCA similarly to the capillary water absorption and chloride permeability performances (Fig. 5.80). Zega and Di Miao (2011) observed similar carbonation depth of NAC and RCAC's prepared by replacing 20 and 30 % by volume of fine NA by fine RCA, when concrete was exposed for 620 days to urban-industrial environmental conditions. Shayan and Xu (2003) observed comparable depth of carbonation of conventional concrete and RCAC with coarse RCA as full replacement of coarse NA or fine RCA as 50 % replacement of fine NA, even though a marginally higher carbonation depth was observed in concrete containing fine RCA. However, the use of sodium silicate and lime treated coarse or fine RCA significantly increased the depth of carbonation of the resulting concrete.

Sagoe-Crentsil et al. (2001) observed higher carbonation depth in concrete with RCA as complete replacement of coarse NA than in conventional concrete. The use of slag cement or a 5 % increase in cement content can decrease the carbonation depth of the RCAC, which was more pronounced for RCAC with higher cement content (Fig. 5.81).

Buyle-Bodin and Zaharieva (2002) observed significantly higher carbon penetration depth of concrete due to the complete replacement of coarse and fine NA by RCA (Fig. 5.82a). The carbonation depth of water cured RCAC was around half that of air-cured RCAC. They also observed that the kinetics of carbonation for conventional concrete and RCAC can both be designed according to basic law of diffusion (Fig. 5.82b):

$$x = C \cdot \sqrt{t}$$

where, x, C and t are depth, rate and time of carbonation.

Razaqpur et al. (2010) observed comparable or even lower carbonation depth in RCAC prepared by mixing two methods (conventional and EMV) than in

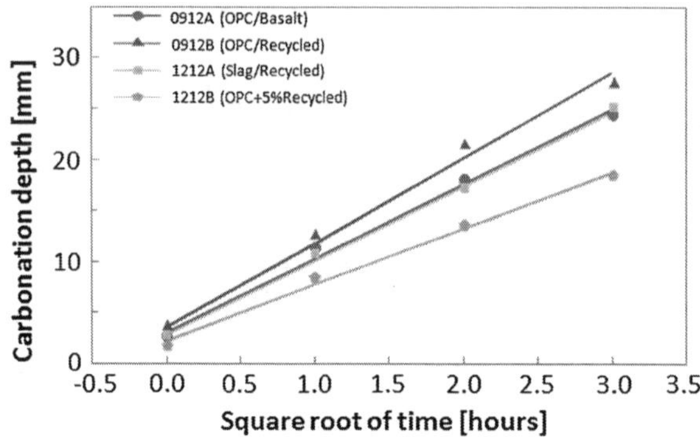

Fig. 5.81 Depth of carbonation versus square root of time of conventional concrete as well as RCAC's during accelerated carbonation test (Sagoe-Crentsil et al. 2001)

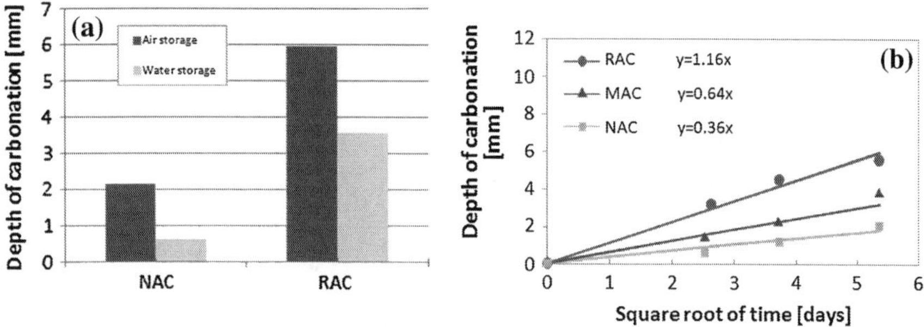

Fig. 5.82 **a** Depth of carbonation and **b** relationship between depth of carbonation and time (Buyle-Bodin and Zaharieva 2002)

conventional concrete, due to the difference in composition of cement of residual and fresh mortars. On the other hand, the carbonation coefficient of RCAC prepared by the conventional method was lower than that prepared by the EMV method due to lower fresh cement content in the later mix. Abbas et al. (2009) reported that the addition of FA and slag increases the carbonation depth due to the consumption of calcium hydroxide because of the pozzolanic reaction. The depth of carbonation of RCAC with or without FA and slag prepared by the conventional method as well as the EMV method and with 140 days of exposure fell in the range of structural grade concrete, i.e. about 0–7 for conventional concrete and 7–15 mm for concrete containing FA and slag.

Limbachiya et al. (2012) did not observe significant differences in the carbonation of conventional concrete and RCAC with various RCA contents when the design strength was 20 MPa. However, the depth of carbonation increased with the replacement ratio of coarse NA by RCA for the 30 and 35 MPa concrete classes and this was more prominent for concrete prepared with 80 % OPC and 20 % FA due to the pozzolanic reaction of FA, which lowered the Portlandite content and thus the pH of the pore solution. The increase in carbonation in RCAC was due to

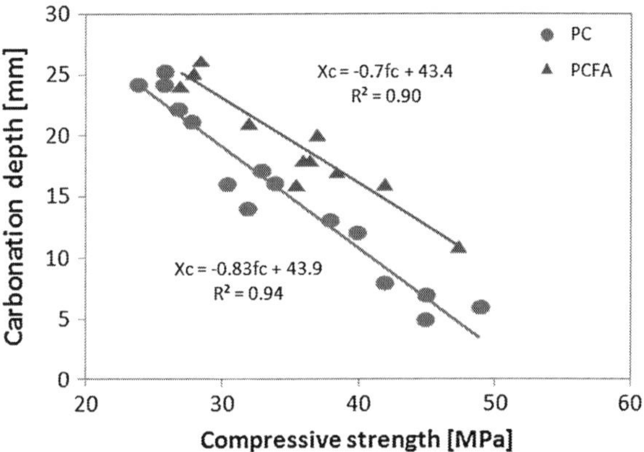

Fig. 5.83 Relationship between CS and carbonation depth (Limbachiya et al. 2012)

the higher water absorption capacity of RCA, which releases water throughout the hydration period and increase the humidity level of concrete. They also observed a linear inverse relationship between CS and carbonation depth of concrete (Fig. 5.83). Like Buyle-Bodin and Zaharieva (2002), they observed that the carbonation depth of RCAC can be predicted by a basic diffusion law.

5.4.4 Freeze–Thaw Durability

The freeze–thaw phenomenon is an important issue for cold region concrete. It occurs due to development of stress in a closed space such as pores in cement paste due to the expansion of water when it freezes. Some cracks may be formed if the stress is higher than the cement paste's strength; the damage can further increase if freezing and thaw cycles continue. The resistance of concrete containing CDW aggregate against freeze–thaw cycle is reported in several references.

Nagataki and Iida (2001) observed decreasing freeze–thaw resistance of concrete due to the incorporation of coarse RCA as a replacement of NA. However, the freeze–thaw resistance was satisfactory as the freezing–thawing factor for RCAC after 300 cycles was >70. Gokce et al. (2004) observed poor freeze–thaw durability of air-entrained concrete with coarse RCA produced from non-air-entrained concrete as the relative modulus of elasticity RCAC was below 60 % after 30 cycles of freeze–thaw. According to them, the presence of a small amount of non-air-entrained RCA can drastically deteriorate the freeze–thaw resistance of air-entrained concrete. The poor freeze–thaw resistance was due to the conversion of overall pore system of concrete containing coarse RCA with adhered mortar and air voids into a partial non-air-entrained void system. On the other hand, regardless of the adhered mortar content in RCA, the freeze–thaw resistance of concrete with coarse RCA originated from air-entrained concrete was even better than that of the conventional concrete after 500 cycles of freeze–thaw cycles even though the RCA has higher permeable voids and water absorption capacity.

Gokce et al. (2004) observed a marginal improvement in the freeze–thaw resistance of concrete due to the use of coarse RCA with small adhered mortar content. The decrease in w/c ratio from 0.55 to 0.33 can significantly improve the freeze–thaw resistance of the resulting concrete with coarse RCA and small adhered mortar content even though still unsatisfactory for long-term exposure. The addition of metakaolin to the above mix (i.e. with RCA with small adhered mortar content and at w/c of 0.3) can lead to a resistance over the standard durability limit of 300 cycles of freeze–thaw; however, a similar RCAC containing silica fume had less resistance than the conventional concrete.

Nagataki and Iida (2001) observed that the freeze–thaw resistance of medium and low strength RCAC as well as of concrete with RCA obtained by primary crushing only (poor quality) was lower than that of high-strength RCAC and RCAC with RCA by a two-stage crushing process. The freeze–thaw resistance of RCAC improved with curing time. After 1 year of curing, the freeze–thaw resistance of RCAC's was similar to that of conventional concrete. Limbachiya et al. (2000) and Limbachiya (2010) observed similar freeze–thaw resistance in air-entrained conventional and high-strength concrete with design strength in the range of 30–50 MPa and in an equivalent type of concrete with RCA at 20–100 % replacement of coarse NA. Figure 5.84 shows the durability factor of 50 MPa NAC and RCAC evaluated with the British standard test where concrete samples are exposed to 300 freeze–thaw cycles. The NAC and RCAC with design strength of 50 MPa also met the British specification (BS 5328 Part 1–1991) for heavy duty external paving blocks.

Oliveira and Vasquez (1996) observed significant influence of moisture content in RCA on the freeze–thaw durability of the resulting concrete. The conventional concrete and the concrete containing RCA with 89.5 % moisture content as the only coarse aggregate resisted more than 100 cycles when both were exposed to freeze–thaw cycles. The RCAC with coarse RCA with 100, 0 and 88 % moisture contents failed after 20, 40 and 80 cycles respectively. Topçu and Sengel (2004) observed marginal deterioration of Schmidt hardness, CS and flexural strength when the conventional concrete and RCAC with 16 and 20 MPa design strength

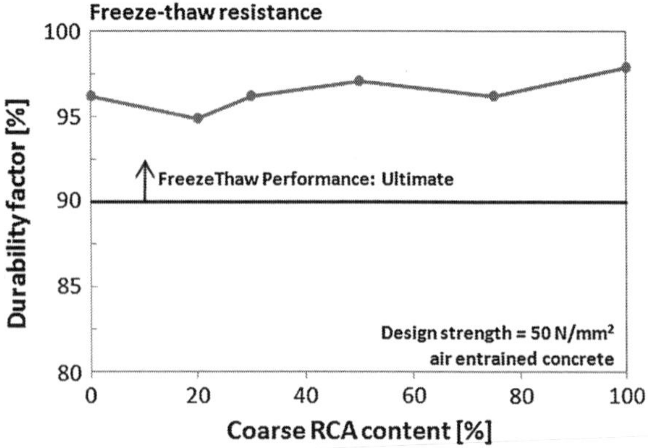

Fig. 5.84 Durability factor evaluated from freeze–thaw resistance tests of concrete *versus* replacement ratio of NA by coarse RCA (Limbachiya et al. 2000)

were exposed to cycles of freeze–thaw (−20 °C for 8 h and then 20 °C for 16 h) for 8 days. Ajdukiewicz and Kliszczzewicz (2002) observed similar or even better freeze–thaw resistivity in high-performance concrete with RCA than in conventional hpc.

Razaqpur et al. (2010) observed similar freeze–thaw resistance of conventional concrete and concrete with RCA as full replacement of coarse NA and prepared by the conventional and EMV methods. The RCAC prepared by the EMV method on the other hand exhibited better freeze–thaw resistance than the RCAC prepared by the conventional method due to the lower mortar content in the former RCAC (Abbas et al. 2009).

5.4.5 Alkali-Aggregate Reactivity and Resistance to Harsh Chemical Substances

A few references are available on the evaluation of the resistance of RCAC to several harmful chemical reactions or chemical environment such as alkali-aggregate reaction and sulphate resistance. Shayan and Xu (2003) observed marginally higher expansion of concrete prisms with replacement of coarse or fine NA by untreated or sodium silicate plus lime treated coarse or fine RCA than of conventional concrete when the specimens of all the types of concrete were subjected to the alkali-aggregate reactivity test for 1 year; however, the expansion of all types of concrete was below 0.024 % and well within the limit, 0.04 %, considered to be indicative of deleterious alkali-aggregate reaction.

Shayan and Xu (2003) observed satisfactory sulphate resistance of concrete with untreated or sodium silicate plus lime treated coarse or fine RCA concrete along with conventional concrete when concrete specimens were stored in a 5 % sodium sulphate solution for 1 year. Limbachiya (2010) observed a comparable expansion of two classes of conventional concrete and concrete with a 30 % replacement of coarse NA by RCA and design strength of 10 and 20 MPa, when both were immersed in a 3 % sodium sulphate solution for 6 months. However, the sulphate-induced expansion of RCAC increased as the replacement level of NA by RCA increased to 50 and 100 %. Limbachiya et al. (2012) observed lower sulphate resistance potential of RCAC than of conventional concrete when both were exposed to a 3 % sodium sulphate solution for 60 days. They observed gradually higher expansion of concrete as the replacement ratio of coarse NA by RCA increased (Fig. 5.85a). However, the addition of FA as a 30 % replacement of OPC slightly improved the sulphate resistance of RCAC due to the reduction in mono-sulphoaluminate and Portlandite contents in the cement paste and the prevention of reaction between free lime and sodium sulphate because of the pozzolanic property of FA (Fig. 5.85b).

Berndt (2009) observed lower dynamic elastic modulus of concrete with RCA as the only coarse aggregate after 12 months exposure into a 5 % sodium sulphate

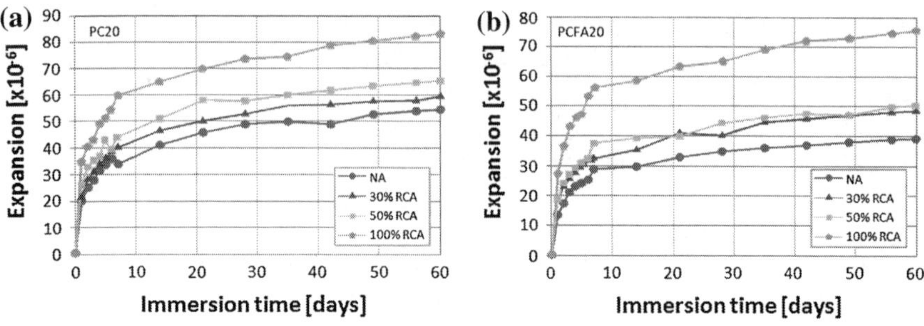

Fig. 5.85 Expansion of concrete with 20 MPa design strength: **a** OPC as binder **b** OPC-30 % FA as binder (Limbachiya et al. 2012)

solution due to the replacement of 50 and 70 % of OPC by ggbfs (Fig. 5.86). Lee et al. (2008) observed lower expansion of cement mortar due to the replacement of 50 % of fine NA by two types of fine RCA with different water absorption capacity when the hardened specimens were kept under sodium sulphate and magnesium sulphate solutions up to 15 months. On the other hand, at 100 % replacement level, depending upon the quality of RCA, the expansion was comparable or significantly higher than that of the conventional mortar. The mortar with higher water absorption capacity has higher expansion than the other one (Table 5.26).

Lee (2009) reported that the magnesium sulphate resistance of cement mortar with replacement of NA by fine RCA depended on the replacement ratio; the loss of CS and the expansion of RCA mortar (RCAM) with 25 and 50 % replacement of NA by RCA were lower and those of the RCAM with 75 and 100 % RCA were higher than the corresponding values of the conventional cement mortar when the specimens were cured in a 4.24 % magnesium sulphate solution for 1 year

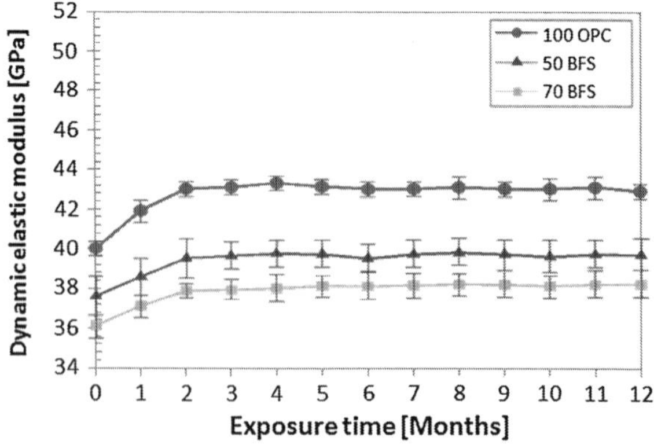

Fig. 5.86 Dynamic elastic modulus of RCAC containing OPC and OPC-slag binder up to 12 months exposure in a 5 % sulphate solution (Berndt 2009)

Table 5.26 Expansion of cement mortar immersed for 9 and 15 months in sodium and magnesium sulphate solutions (Lee et al. 2008)

Type of solution	Replacement amount	Expansion (%) of					
		Conventional mortar		RCA-A		RCA-B	
		9 months	15 months	9 months	15 months	9 months	15 months
Sodium sulphate	0	0.192	1.032				
	50			0.118	0.696	0.134	0.517
	100			0.954	Collapsed	0.287	0.974
Magnesium sulphate	0	0.105	0.523				
	50			0.070	0.386	0.085	0.303
	100			0.202	1.122	0.205	0.757

RCA-A and RCA-B Mortar with RCA with water absorption capacity of 10.35 and 6.59 % respectively

(Fig. 5.87). They also reported that the RCAM with less porous fine RCA has higher resistance to magnesium sulphate attack.

5.4.6 Other Durability Properties

Regardless of type of aggregate, in the Kwan et al. (2012) study, conventional concrete and RCAC slightly shrunk in the initial 24 h of wet curing and then expanded with further curing as well as with increasing replacement level of coarse NA by RCA (Fig. 5.88). The higher expansion in RCAC than in conventional concrete was due to the development of high hydrostatic pressure in the specimen because of the higher water absorption capacity of RCA than of NA.

Tu et al. (2006) observed significantly higher resistivity of high-performance concrete with RCA as complete replacement of fine and coarse NA or RCA as complete replacement of coarse NA than the minimum value for durable concrete, 20 kΩ-cm on or after 28 days of curing. The resistivity of concrete containing

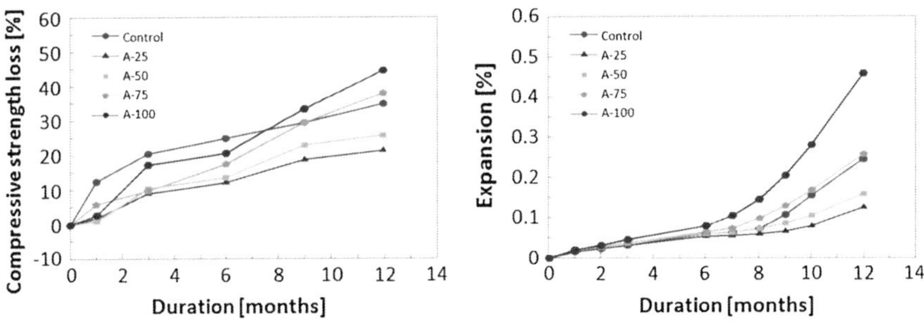

Fig. 5.87 CS loss and expansion due to magnesium sulphate attack of cement mortar with various contents of fine RCA (Lee 2009)

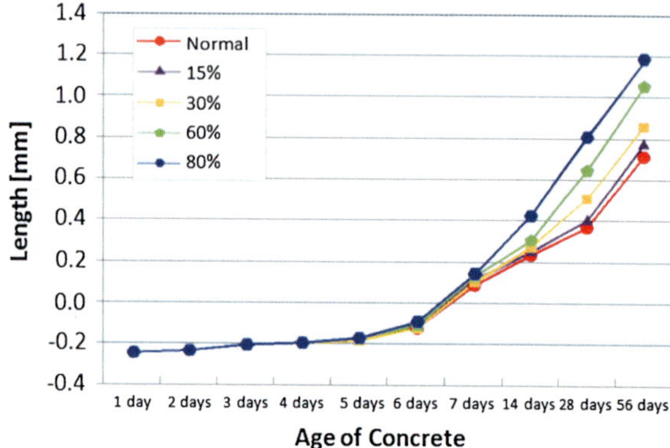

Fig. 5.88 Shrinkage followed by expansion of concrete due to water curing of concrete (Kwan et al. 2012)

RCA as a complete replacement of coarse NA was higher than that of the concrete containing RCA as a complete replacement of fine and coarse NA at different w/c ratios. Sani et al. (2005) observed a lower calcium leaching rate in RCAC prepared by completely replacing coarse NA and a part of fine NA by RCA when water was percolated through both types of concrete, despite the higher porosity of RCAC than of NAC (Fig. 5.89). The addition of FA further improved the leachability of ions for both types of concrete due to the pozzolanic activity.

Vieira et al. (2011) observed no significant differences in thermal response and mechanical properties namely CS, STS and MO of NAC and RCAC with replacement of 20, 50 and 100 % by volume of coarse NA by RCA when they were exposed for 1 h to temperatures of 400, 600 and 800 °C. On the other hand, Zega and Di Miao (2009) observed marginally good post-fire performances of CS, MO (E) and UPV for three types of RCAC's with a replacement of 75 % by volume of coarse NA by RCA from concrete containing three types of coarse aggregate (granitic crushed stone, quartzite crushed stone and siliceous gravel), when compared to equivalent conventional concrete with the same types of coarse NA, when the concrete specimens were exposed to a temperature of 500 °C for 1 h. The losses in percentage of various properties of the NAC and RCAC's due to

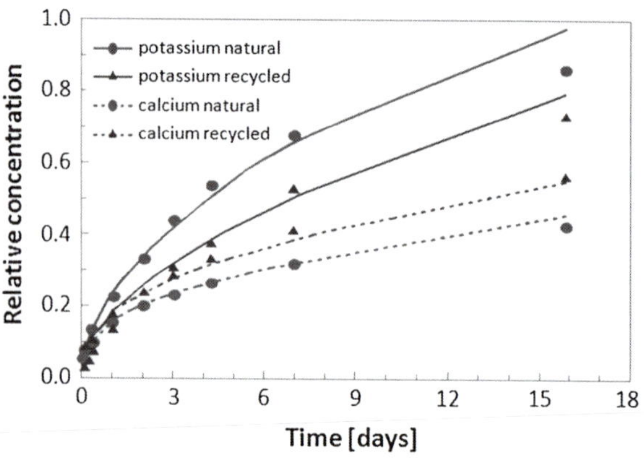

Fig. 5.89 Calcium and potassium leaching for NAC and RCAC (Sani et al. 2005)

Table 5.27 Loss in some mechanical properties (%) of NAC and RCAC due to exposure to 500 °C (Zega and Di Miao 2009)

Property	w/c	Loss (%)					
		GG		SG		QG	
		NAC	RCAC	NAC	RCAC	NAC	RCAC
CS	0.4	22	10	23	20	6	5
	0.7	15	13	24	21	24	16
MO(E)_static	0.4	46	28	46	28	26	23
	0.7	52	48	55	49	47	47
UPV	0.4	34	27	41	35	27	21
	0.7	41	38	47	45	39	38
MO(E)_dynamic	0.4	61	56	72	73	50	48
	0.7	68	69	79	77	72	67

GG, SG and QG Granitic gravel, siliceous gravel and quartzite gravel, respectively

the exposure to high temperatures are presented in Table 5.27. The authors also observed a better performance of RCAC's than NAC when concrete were prepared at w/c of 0.4 than at 0.7. Moreover, the RCAC containing recycled quartzite coarse aggregate exhibited better performance than the RCAC's with the other two types of recycled coarse aggregate at w/c of 0.4 but behaved similarly at w/c of 0.7.

Poon et al. (2009) studied the high temperature performance of concrete with two types of RA as the only coarse aggregate (CRA1 and CRA2) by varying several parameters during the preparation of the concrete mixes. The residual CS of various series of concrete after heating at 800 °C for 1 h is presented in Fig. 5.90. The authors observed a gradual increase of residual CS of concrete due to the variation of these parameters: gradual simultaneous increase of fine RA (FRA1) at aggregate to cement ratio of 10:1; gradual increase of aggregate to cement ratio at 100 % CRA2 and FRA2 contents and gradual increase of FRA1 content in concrete having CRA1 as the only coarse aggregate at aggregate to cement ratio of 12:1 (A, B, and C series in Fig. 5.90 respectively). The soil content

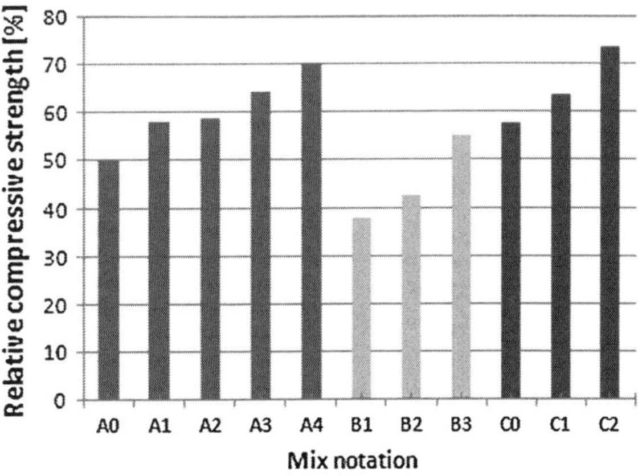

Fig. 5.90 Residual CS of RCAC after heating at 800 °C (Poon et al. 2009)

in these series also increased with the FRA1 content. According to the authors, the low residual CS of the B series in comparison to the other two series was due to relatively low soil content and high residual mortar content in FRA2 than in FRA1. The gradual improvement of the performance of concrete due to the increase of soil content was due to the formation of crystalline calcium aluminate silicates at high temperature.

5.5 Conclusions

The recent developments on the properties of concrete with CDW as aggregate were discussed thoroughly in this chapter. The results are presented in three different sections to present fresh and hardened properties of concrete. Aggregates generated from recycled concrete or other ceramic-based waste materials are reported to be used as partial or full replacements of fine and coarse aggregates in concrete. The use of these aggregates as coarse aggregate is more versatile than that as fine aggregates. Both normal and high-strength concrete can be produced with CDW aggregates. In several investigations, it was reported that 30 % replacement of natural aggregates by aggregates generated from waste concrete did not substantially deteriorate the mechanical and durability performances of the resulting concrete. However, the presence of impurities such as clay brick, tiles and other ceramics can strongly jeopardize the performance of concrete and therefore in this case replacement ratios should be lower than the one mentioned above. Some important conclusions are indicated below:

1. The use of any type of CDW aggregates substantially lowers the workability of concrete. This is mainly due to the higher water absorption capacity of these types of aggregates than that of NA because of adhered porous mortar. Therefore several mixing procedures have been developed to improve the workability performance of CDW aggregates-based concrete; out of these, pre-saturating aggregates for 10 min before mixing or using 85–90 % humid CDW aggregate are the most widely used techniques. The density of concrete also mildly decreases with the replacement of NA by CDW aggregates;
2. All the strength properties and modulus of elasticity of concrete are also deteriorated by the incorporation of CDW aggregates as partial or full replacement of NA. However, in comparison to the reduction in compressive strength, the reduction in flexural and splitting tensile strength is not so prominent. The reduction in these properties of concrete is due mainly to the presence of adhered mortar in CDW aggregate. The reduction in modulus of elasticity is also very high due to the lower modulus of elasticity of CDW aggregates than that of NA. However, pre-saturation of the aggregates before mixing can increase the strength properties and the modulus of elasticity of concrete;

3. The load-deflection curve of concrete containing CDW aggregate with NA is marginally different from the load-deflection curve of conventional concrete. Concrete with CDW aggregates normally has lower toughness and ductility performances but higher creep than those of conventional concrete. The surface hardness properties such as Schmidt hardness, skid resistance, impact and abrasion resistances of concrete with CDW aggregates are normally inferior to those of conventional concrete, primarily due to the presence of adhered mortar. However, several researchers observed similar abrasion behaviour of CDW aggregates-based and natural aggregates-based concrete;
4. The incorporation of CDW aggregates in concrete increases its drying shrinkage due to the higher paste content. This incorporation increases total porosity and therefore increases various permeability properties such as water absorption, chloride migration, depth of carbonation, various gas permeability of the resulting concrete. Concrete containing CDW aggregates has poorer freeze–thaw resistance than that of conventional concrete. The resistance of CDW aggregates-based concrete to some harsh chemical solutions is also poorer than that of NA-based concrete;
5. However, by improving the concrete mixing preparation techniques, almost all properties of concrete with CDW aggregates as a replacement of NA can be substantially improved and in some cases property results are comparable to those of conventional properties. The replacement of cement up to given amount by several mineral additions such as fly ash, blast furnace slag, silica fume, metakaolin can improve almost all properties of concrete with CDW aggregate at various stages of curing depending on their reactivity. The properties of concrete with CDW aggregate as partial replacement of NA up to given replacement ratios (30 %, in most of the cases) are similar to the equivalent properties of conventional concrete.

References

Abbas A, Fathifazl G, Isgor OB, Razaqpur AG, Fournier B, Foo S (2009) Durability of recycled aggregate concrete designed with equivalent mortar volume method. Cement Concr Compos 31(8):555–563

Ahmed SFU (2011) Properties of concrete containing recycled fine aggregate and fly ash, Concrete 2011 conference, The concrete Institute of Australia, 2011, Perth

Ajdukiewicz A, Kliszczzewicz A (2002) Influence of recycled aggregates on mechanical properties of HS/HPC. Cement Concr Compos 24(2):269–279

Akbarnezhad A, Ong KCG, Zhang MH, Tam CT, Foo TWJ (2011) Microwave-assisted beneficiation of recycled concrete aggregates. Constr Build Mater 25(8):3469–3479

Alexander MG, Ballim Y, Maketchine JR (1999) Guide to use of durability indexes for achieving durability in concrete structures, Research monograph no. 2, collaborative research by Universities of Cape Town and Witwatersrand, p 33

Ann KY, Moon YB, Kim HY, Royu J (2008) Durability of recycled aggregate concrete using pozzolanic materials. Waste Manage (Oxford) 28(6):993–999

Berndt ML (2009) Properties of sustainable concrete containing fly ash, slag and recycled concrete aggregate. Constr Build Mater 23(7):2606–2613

Buck AD (1977) Recycled aggregate as a source of aggregate. ACI J 74(5):212–219

Buyle-Bodin F, Zaharieva RH (2002) Influence of industrially produced recycled aggregates on flow properties of concrete. Mater Struct 35(8):504–509

Casuccio M, Torrijos MC, Giaccio G, Zerbino R (2008) Failure mechanism of recycled aggregate concrete. Constr Build Mater 22(7):1500–1506

Chen ZP, Huang KW, Zhang XG, Xue JY (2010) Experimental research on the flexural strength of recycled coarse aggregate concrete, International conference on mechanic automation and control engineering (MACE) 2010, Wuhan, PR China, pp 1041–1043

Chen H-J, Yen T, Chen K-H (2003) Use of building rubbles as recycled aggregates. Cem Concr Res 33(1):125–132

Corinaldesi V, Moriconi G (2009) Influence of mineral additions on the performance of 100 % recycled aggregate concrete. Constr Build Mater 23(8):2869–2876

Corinaldesi V, Moriconi G (2010) Mechanical and elastic behaviour of concretes made of recycled-concrete. Constr Build Mater 24(9):1616–1620

Corinaldesi V (2011) Structural concrete prepared with coarse recycled concrete aggregate: from investigation to design, advances in civil engineering, advances in civil engineering, vol. 2011, Article ID 283984, p 6

Courard L, Michel F, Delhez P (2010) Use of concrete road recycled aggregates for roller compacted concrete. Constr Build Mater 24(3):390–395

De Brito J, Alves F (2008) Concrete with recycled aggregates: the Portuguese experimental research. Mater Struct 43(1):35–51

De Brito J, Robles R (2010) Recycled aggregate concretes methodology for estimating its long term properties. Indian J Eng Mater Sci 17(6):449–462

Dapena E, Alaejos P, Lobet A, Pérez D (2011) Effect of recycled sand content on characteristics of mortars and concretes. J Mater Civ Eng 23(4):414–422

Dhir RK, Limbachiya MC, Leelawat T (1999) Suitability of recycled concrete aggregate for use in BS 5328 designated mixes. Struct Buildings 134(3):257–274

Domingo-Cabo A, Lazaro C, Gayarre FL, Serrano MA, Lopez-Colina C (2010) Long term deformations by creep and shrinkage in recycled aggregate concrete. Mater Struct 43(8):1147–1160

Eguchi K, Teranishi K, Nakagome A, Kishimoto H, Sinozaki K, Narikawa M (2007) Application of recycled coarse aggregate by mixture to concrete construction. Constr Build Mater 21(7):1542–1551

Etxeberria M, Vazquez E, Mari A, Barra M (2007a) Influence of amount of recycled coarse aggregates and production process on properties of recycled aggregate concrete. Cem Concr Res 37(5):735–742

Etxeberria M, Mari A, Vazquez E (2007b) Recycled aggregate concrete as structural material. Mater Struct 40(5):529–541

Evangelista L, de Brito J (2007) Mechanical behaviour of concrete made with fine recycled concrete aggregates. Cem Concr Compos29(5):397–401

Evangelista L, de Brito J (2010) Durability performance of concrete made with fine recycled Concrete Aggregates. Cem Concr Compos 32(1):9–14

Fathifazl G, Razaqpur AG, Isgor OB, Abbas A, Fournier B, Foo S (2011) Creep and drying shrinkage characteristics of concrete produced with coarse recycled concrete aggregate. Cement Concr Compos 33(10):1026–1037

Ferreira L, de Brito J, Barra M (2011) Influence of the pre-saturation of recycled concrete aggregates on the mechanical strength and durability of structural concrete. Mag Concr Res 63(8):617–627

Fonseca N, de Brito J, Evangelista L (2011) The influence of curing conditions on the mechanical performance of concrete made with coarse recycled concrete aggregates. Cement Concr Compos 33(6):637–643

Frondistou-Yannas S (1977) Waste concrete as aggregate for new concrete. ACI J Proc 74(8):373–374 (title no. 74–37)

Gardner NJ, Lockman LJ (2001) Design provision for drying shrinkage and creep of normal strength concrete. ACI Mater J 98(2):159–167

Gokce A, Nagataki S, Saeki T, Hisda M (2004) Freezing and thawing resistance of air-entrained concrete incorporating recycled coarse aggregate: the role of air-content in demolished concrete. Cem Concr Res 34(5):799–806

Gomes M, de Brito J (2009) Structural concrete with incorporation of coarse recycled concrete and ceramic aggregates: durability performance. Mater Struct 42(5):663–675

Gomez-Soberon JMV (2002) Porosity of recycled concrete with substitution of recycled concrete aggregate: an experimental study. Cem Concr Res 32(8):1301–1311

Gonçalves A, Esteves A, Vieira M (2004) Influence of recycled concrete aggregates on concrete durability, Abstract ID number: 279, http://congreso.cimne.com/rilem04/admin/Files/p279.pdf. Accessed 11 April 2012

Gonzalez-Fonteboa B, Martinez-Abella F (2008) Concretes with aggregates from demolition waste and silica fume: materials and mechanical properties. Build Environ 43(4):429–437

Gonzalez-Fonteboa B, Martinez-Abella F (2005) Recycled aggregates concrete: aggregates and mix properties. Materiales de Construcción 55(279):53–66

Gonzalez-Fonteboa B, Martinez-Abella F, Eiras-Lopez J, Seara-Paz S (2011) Effect of recycled coarse aggregate on damage of recycled concrete. Mater Struct 44(10):1759–1770

Grdic ZJ (2010) Properties of self-compacting concrete prepared with coarse recycled concrete aggregate. Constr Build Mater 24(7):1129–1133

Gull I (2011) Testing of strength of recycled waste concrete and its applicability. J Constr Eng Manag 137(1):1–5

Gupta A, Mandal S, Ghosh S (2010) Direct compressive strength and elastic modulus of recycled aggregate concrete. Int J Civil Struct Eng 2(1):292–304

Hansen TC (1992) Recycling of demolished concrete and masonry. RILEM report no. 6, E&FN Spon, London

Hansen TC, Boegh E (1985) Elasticity and drying shrinkage of recycled-aggregate concrete. ACI Journal 82(5):648–652

Hansen TC, Narud H (1983) Strength of recycled concrete made from crushed concrete coarse aggregate. Concr Int 5(1):79–83

Hasaba S, Kawamura M, Torli K, Takemoto K (1982) Drying shrinkage and durability of concrete made from recycled concrete aggregates. Trans Jpn Concr Inst 3:55–60

Heeralal M, Kumar PR, Rao YV (2009) Flexural fatigue characteristics of steel fiber reinforced recycled aggregate concrete (SFRRAC). Facta Universitatis, Ser Archit Civil Eng 7(1):19–33

James MN, Choi W, Abu-Lebdeh T (2011) Use of recycled aggregate and fly ash in concrete Pavement. Am J Eng Appl Sci 4(2):201–208

Kakizaki M, Harada M, Soshiroda T, Kubota S, Ikeda T, Kasai Y (1988) Strength and elastic modulus of recycled aggregate concrete, 2nd international RILEM symposium on demolition and reuse of concrete and masonry. Tokyo, Japan, pp 565–574

Katz A (2003) Properties of concrete made with recycled aggregate from partially hydrated old concrete. Cem Concr Res 33(5):703–711

Katz A (2004) Treatments for the improvement of recycled aggregate. J Mater Civ Eng 16(4):597–603

Khatib JM (2005) Properties of concrete incorporating fine recycled aggregate. Cem Concr Res 35(4):763–769

Kim MH, Nam SI, Kim JM (1993) An experimental study on the workability and engineering properties of recycled aggregate concrete according to the combination condition of recycled aggregate. J Archit Inst Korea 9(11):109–120

Kong D, Lei T, Zheng J, Ma C, Jiang J, Jiang J (2010) Effect and mechanism of surface-coating pozzolanic materials around aggregate on properties and ITZ microstructure of recycled aggregate concrete. Constr Build Mater 24(5):701–708

Kou S, Poon C (2006) Compressive strength, pore size distribution and chloride-ion penetration of recycled aggregate concrete incorporating class-F fly ash. J Wuhan Univ Technol Mater Sci Edition 21(4):130–136

Kou SC, Poon CS, Chan D (2007) Influence of fly ash as cement replacement on the properties of recycled aggregate concrete. J Mater Civ Eng 19(9):709–717

Kou SC, Poon CH, Chan D (2008) Influence of fly ash as a cement addition on the hardened properties of recycled aggregate concrete. Mater and Struct 41(7):1191–1201

Kou S, Poon C (2009a) Properties of concrete prepared with crushed fine stone, furnace bottom ash and fine recycled aggregate as fine aggregates. Constr Build Mater 23(2):2877–2886

Kou SC, Poon CS (2009b) Properties of self-compacting concrete prepared with coarse and fine recycled concrete aggregates. Cement Concr Compos 31(9):622–627

Kou S, Poon C, Etxeberria M, (2011a) Influence of recycled aggregate on the long term mechanical properties and pore size distribution of concrete, Cement Concr Compos 33(2):286–291

Kou S, Poon C, Agrela F (2011b) Comparisons of natural and recycled aggregate concretes prepared with the addition of different mineral admixtures. Cement Concr Compos 33(8):788–795

Kou S, Poon C (2008) Mechanical properties of 5-year-old concrete prepared with recycled aggregates obtained from three different sources. Mag Concr Res 61(1):57–64

Kou S, Poon C (2010) Properties of concrete prepared with PVA-impregnated recycled concrete aggregates. Cement Concr Compos 32(8):649–654

Kutcharlapatty S, Sarkar AK, Rajamane NP (2011) Nanosilica improves recycled concrete aggregates. New Build Mate Constr Worlds 16(1):190–199

Kwan WH, Ramli M, Kam KJ, Sulieman MZ (2012) Influence of the amount of recycled coarse aggregate in concrete design and durability properties. Constr Build Mater 26(1):565–573

Lee ST, Swamy RN, Kim SS, Park YG, (2008). Durability of Mortars Made with Recycled Fine Aggregates Exposed to Sulfate Solutions. J Mater Civ Eng 20 (1): 63−70

Lee ST (2009) Influence of recycled fine aggregates on the resistance of mortars to magnesium sulfate attack. Waste Manag 29(8):2385–2391

Leite MB (2001) Evaluation of the mechanical properties of concretes made with recycled aggregates form construction and demolition waste (in Portuguese), PhD Thesis in civil engineering, School of Engineering of the Federal University of Rio Grande do Sul, Porto Alegre, Brazil

Li J, Xiao H, Zhou Y (2009) Influence of coating recycled aggregate surface with pozzolanic powder on properties of recycled aggregate concrete. Constr Build Mater 23(3):1287–1291

Limbachiya MC (2010) Recycled aggregate: production, properties and value added sustainable applications. J Wuhan Univ Technol Mater Sci Edition 27(6):1011–1016

Limbachiya MC, Koulouris A, Roberts JJ, Fried AN (2004) Properties of recycled aggregate concrete, in RILEM international symposium on environment-conscious materials and systems for sustainable development, RILEM publication SARL, pp 127–136

Limbachiya MC, Leelawat T, Dhir RK (2000) Use of recycled concrete aggregate in high-strength concrete. Mater Struct 33(9):574–580

Limbachiya M, Meddah MS, Ouchagour Y (2012) Use of recycled concrete aggregate in fly-ash concrete. Constr Build Mater 27(1):439–449

Lin Y-H, Tyan Y-Y, Chang T-P, Chang C-Y (2004) An assessment of optimal mixture for concrete made with recycled concrete aggregates, Cement Concrete Res 34(8):1373–1380

López-Gayarre F, Serna P, Domingo-Cabo A, Serrano-López MA, López-Colina C (2009) Influence of recycled aggregate quality and proportioning criteria on recycled concrete properties. Waste Manage 29(12):3022–3028

Malesev M, Radonjanin V, Marinkovic S (2010) Recycled concrete as aggregate for structural concrete production. Sustainability 2(5):1204–1225

Malhotra VM (1976) Testing hardened concrete: non-destructive methods. American Concrete Institute, Monograph No. 9, Reston

Malhotra VM (1978) Use of recycled concrete as new aggregate. Symposium on energy and resource conservation in the concrete industry, CANMET report no. 76–8, CANMET, Ottawa, Canada, pp 4–16

Mas B, Cladera A, Olmo T, Pitarch F (2012) Influence of the amount of mixed recycled aggregates on the properties of concrete for non-structural use. Constr Build Mater 27(1):612–622

Merlet JD, Pimienta P (1993) Mechanical and physical-chemical properties of concrete produced with coarse and fine recycled aggregates' in 'Demolition and reuse of concrete and masonry, Odense, pp 343–353

Mukai T, Kikuchi M (1978) Studies on utilization of recycled concrete for structural members (Parts 1 and 2). Summaries of technical papers of annual meeting (in Japanese). Archit Inst Jpn 85–86:87–88

Nagataki S, Gokce A, Saeki T, Hisada M (2004) Assessment of recycling process induced damage sensitivity of recycled concrete aggregates. Cem Concr Res 34(6):965–971

Nagataki S, Iida K (2001) Recycling of demolished concrete, ACI Special Publication, vol. 200. Reston, USA, pp 1–20

Oliveira M, Vasquez E (1996) The influence of retained moisture in aggregates from recycling on the properties of new hardened concrete. Waste Manage 16(1–3):113–117

Olorunsogo FT, Padayachee N (2002) Performance of recycled aggregate concrete monitored by durability indexes. Cem Concr Res 32(2):179–185

Padmini AK, Ramamurthy K, Mathews MS (2009) Influence of parent concrete on the properties of recycled aggregate concrete. Constr Build Mater 23(2):829–836

Padmini AK, Ramamurthy K, Mathews MS (2002) Relative moisture movement through recycled aggregate concrete. Maga Concr Res 54(5):377–384

Paine KA, Collery DJ, Dhir RK (2009) Strength and deformation characteristics of concrete containing coarse recycled and manufactured aggregates, proceedings of the 11th international conference on non-conventional materials and technologies (NOCMAT 2009), 2009, Bath, pp 1–9

Park SB, Seo DS, Lee J (2005) Studies on the sound absorption characteristics of porous concrete based on the content of recycled aggregate and target void ratio. Cem Concr Res 35(9):1846–1854

Pereira P, Evangelista L, de Brito J (2012) The effect of superplasticizers on the workability and compressive strength of concrete made with fine recycled concrete aggregates. Constr Build Mater 28(1):722–729

Poon CS, Chan D (2006) Paving blocks made with recycled concrete aggregate and crushed clay brick. Constr Build Mater 20(8):569–577

Poon CS, Chan D (2007) Effects of contaminants on the properties of concrete paving blocks prepared with recycled concrete aggregates. Constr Build Mater 21(1):164–175

Poon C, Kou S (2010) Effects of fly ash on the mechanical properties of 10- year-old concrete prepared with recycled concrete aggregates, proceedings, 2nd international conference on waste engineering and management-ICWEM 2010, Shanghai, China, pp 46–59

Poon C, Kou S, Chan D (2006) Influence of stream curing on hardened properties of recycled aggregate concrete. Magaz Concr Res 58(5):289–299

Poon CS, Kou SC, Wan H-W, Etxeberria M (2009) Properties of concrete blocks prepared with low grade recycled aggregates. Waste Manage 29(8):2369–2377

Poon CS, Lam L (2008) The effect of aggregate-to-cement ratio and types of aggregates on the properties of pre-cast concrete blocks. Cement Concr Compos 30(4):283–289

Poon CS, Shui ZH, Lam L (2004a) Effect of microstructure of ITZ on compressive strength of concrete prepared with recycled aggregates. Constr Build Mater 18(6):461–468

Poon CS, Shui ZH, Lam L, Fok H, Kou SC (2004b) Influence of moisture states of natural and recycled aggregates on the slump and compressive strength of concrete. Cem Concr Res 34(1):31–36

Poon CS, Kou SC, Lam L (2007) Influence of recycled aggregate on slump and bleeding of fresh concrete. Mater Struct 40(9):981–986

Ravindrarajah RS, Tam TC (1985) Properties of concrete made with crushed concrete as coarse aggregate. Maga Concr Res 37(130):29–38

Rahal K (2007) Mechanical properties of concrete with recycled coarse aggregate. Build Environ 42(1):407–415

Rao CM, Bhattacharyya SK, Barai SV (2011) Influence of field recycled aggregate on properties of concrete. Mater Struct 44(1):205–220

Rao KJ, Khan TA (2009) Suitability of glass fibers in high strength recycled aggregate concrete-An experimental investigation. Asian J Civil Eng (Building and Housing) 10(6):681–689

Rasheeduzzafar AK, Khan A (1984) Recycled concrete-a source of new aggregate. Cem Concr Aggreg 6(1):17–27

Razaqpur AG, Fathifazal G, Isgor B, Abbas A, Fournier B, Foo S (2010) How to produce high quality concrete mixes with recycled concrete aggregate, 2nd international conference on waste engineering and management—ICWEM 2010, RILEM publications, SARL, pp 11–35

Rustom R, Taha S, Badarnah A, Barahma H (2007) Properties of recycled aggregate in concrete and road pavement applications. Islamic Univ J (Ser Nat Stud Eng) 5(2):247–264

Ryu JS (2002) An experimental study on the effect of recycled aggregate on concrete properties. Maga Concr Res 54(1):7–12

Safiuddin M, Alengaram UJ, Salam MA, Jumaat MZ, Jaafar FF, Saad HB (2011) Properties of high-workability concrete with recycled concrete aggregate. Mater Res 14(2):248–255

Sagoe-Crentsil KK, Brown T, Taylor AH (2001) Performance of concrete made with commercially produced coarse recycled concrete aggregate. Cem Concr Res 31(5):707–712

Santos JR, Branco F de Brito J (2002) Mechanical properties of concrete with coarse recycled concrete aggregates. Sustainable building 2002, Conference proceedings, Oslo

Sani D, Moriconi G, Fava G, Corinaldesi V (2005) Leaching and mechanical behaviour of concrete manufactured with recycled aggregates. Waste Manage 25(2):177–182

Sato R, Maruyama I, Takahisa S, Sogo M (2007) Flexural behaviour of reinforced recycled concrete beams. J Adv Concr Technol 5(1):43–61

Shayan A, Xu A (2003) Performance and properties of structural concrete made with recycled concrete aggregate. ACI Mater J 100(5):371–380

Singh SK, Sharma PC (2007) Use of recycled aggregates in concrete—a paradigm shift, NBM&CW October (no page) (http://nbmcw.com/articles/concrete/waste-material-by-product/576-use-of-recycled-aggregates-in-concrete-a-paradigm-shift.html). Accessed 28 Feb 2012

Soutsos M, Tang K, Millard S (2011) Use of recycled demolition aggregate in precast products, phase II: concrete paving blocks. Constr Build Mater 25(7):3131–3143

Tabsh SW, Abdelfatah AS (2009) Influence of recycled concrete aggregates on strength properties of concrete. Constr Build Mater 23(2):1163–1167

Tam VWY, Tam CM (2008) Diversifying two-stage mixing approach (TSMA) for recycled aggregate concrete: TSMAs and TSMAsc. Constr Build Mater 22(10):2068–2077

Tam VWY, Tam CM, Wang Y (2007) Optimization on proportion for recycled aggregate in concrete using two-stage mixing approach. Constr Build Mater 21(10):1928–1939

Tavakoli M, Soroushian P (1996) Strength of recycled aggregate concrete made using field demolished concrete as aggregate. ACI Mater J 93(2):178–181

Thangchirapat W, Buranasing R, Jaturapitakkul C, Chindaprasirt P (2008) Influence of rice husk–bark ash on mechanical properties of concrete containing high amount of recycled aggregates. Constr Build Mater 22(8):1812–1819

Topçu IB (1997) Physical and mechanical properties of concretes produced with waste concrete. Cem Concr Res 27(12):1817–1823

Topçu IB, Guncan NF, (1995) Using waste concrete as aggregate, Cem Concr Res 25 (7): 1385–1390

Topçu IB, Sardimir M (2008) Prediction of mechanical properties of recycled aggregate concretes containing silica fume using artificial neural networks and fuzzy logic. Comput Mater Sci 42(1):74–82

Topçu IB, Sengel S (2004) Properties of concretes produced with waste concrete aggregate. Cem Concr Res 34(8):1307–1312

Tsujino M, Noguchi T, Tamura M, Kanematsu M, Maruyama I (2007) Application of conventionally recycled coarse aggregate to concrete structure by surface modification treatment. J Adv Concr Technol 5(1):13–25

Tsujino M, Noguchi T, Tamura M, Kanematsu M, Maruyama I, Nagai H (2006) Study on the application of low-quality recycled coarse aggregate to concrete structure by surface modification treatment, proceedings, 2nd Asian concrete federation conference, Bali, Indonesia, 20–21 Nov 2006, pp 36–45

Tu T-Y, Chen Y–Y, Hwang C-L (2006) Properties of HPC with recycled aggregates. Cem Concr Res 36(5):943–950

Vieira JPB, Correia JR, de Brito J (2011) Post-fire residual mechanical properties of concrete made with recycled concrete coarse aggregates. Cem Concr Res 41(5):533–541

Wesche K, Schulz R (1982) Benton aus afbereitetem altbeton. Technologie und eigenschaften, Benton (Dusseldorf), no. 2, Feb, pp 64–68; no. 3, Mar, pp 108–112

Xiao J, Falkner H (2007) Bond behaviour between recycled aggregate and steel rebars. Constr Build Mater 21(2):395–401

Xiao JZ, Li JB, Zhang C (2006a) Mechanical properties of recycled aggregate concrete under uniaxial loading. Cem Concr Res 25(6):1187–1194

Xiao JZ, Li JB, Zhang C (2006b) On relationships between the mechanical properties of recycled aggregate concrete: an overview. Mater Struct 39(6):655–664

Yang J, Chung H, Ashour AF (2008) Influence of type and replacement level of recycled aggregates on concrete properties. ACI Mater J 105(3):289–296

Yang J, Du Q, Bao Y (2011) Concrete with recycled concrete aggregate and crushed clay bricks. Constr Build Mater 25(4):1935–1945

Yaprak H, Aruntas HY, Demir I, Simsek O, Durmus D (2011) Effects of the fine recycled concrete aggregates on the concrete properties. Int J Phys Sci 6(10):2455–2461

Yong PC, Teo DCL (2009) Utilization of recycled aggregate as coarse aggregate in concrete. UNIMAS E-J Civil Eng 1(1):1–6

Zaharieva R, Buyle-Bodin F, Skoczylas F, Wirquin E (2003) Assessment of the surface permeation properties of recycled aggregate concrete. Cement Concr Compos 25(2):223–232

Zega CJ, Di Miao AA (2009) Recycled concrete made with different natural coarse aggregates exposed to high temperature. Constr Build Mater 23(5):2047–2052

Zega CJ, Di Miao AA (2011) Use of recycled fine aggregate in concretes with durable requirements. Waste Manage 31(11):2336–2340

Chapter 6
Methodologies for Estimating Properties of Concrete Containing Recycled Aggregates: Analyses of Experimental Research

Symbols

Ab_cap_{RAC}	Capillary water absorption of the recycled aggregates concrete
Ab_cap_{RC}	Capillary water absorption of the reference concrete
Ab_im_{RAC}	Water absorption by immersion of the recycled aggregates concrete
Ab_im_{RC}	Water absorption by immersion of the reference concrete
$carbonation_{RAC}$	Carbonation depth of the reference concrete
$carbonation_{RC}$	Carbonation depth of the recycled aggregates concrete
CDW	Construction and demolition waste
chloride $pen._{RAC}$	Chloride penetration depth of the reference concrete
chloride $pen._{RC}$	Chloride penetration depth of the recycled aggregates concrete
D_{CNA}	Density of the coarse natural aggregates
D_{CRA}	Density of the coarse recycled aggregates
D_{FNA}	Density of the fine natural aggregates
D_{FRA}	Density of the fine recycled aggregates
D_{mix}	Weighed density of the aggregates in the concrete mix
D_{RAC}	Weighed density of the aggregates in the recycled aggregates concrete
D_{RC}	Weighed density of the aggregates in the reference concrete
D'_{RAC}	Density of the recycled aggregates concrete
D'_{RC}	Density of the reference concrete
E_{cRAC}	Modulus of elasticity of the recycled aggregates concrete
E_{cRC}	Modulus of elasticity of the reference concrete
FA	Percentage of fine aggregates in the mix
f_{c7RAC}	7-day compressive strength of the recycled aggregates concrete
f_{c7RC}	7-day compressive strength of the reference concrete
f_{cRAC}	28-day compressive strength of the recycled aggregates concrete

J. de Brito and N. Saikia, *Recycled Aggregate in Concrete*,
Green Energy and Technology, DOI: 10.1007/978-1-4471-4540-0_6,

f_{cRC}	28-day compressive strength of the reference concrete
f_{spRAC}	Splitting tensile strength of the recycled aggregates concrete
f_{spRC}	Splitting tensile strength of the reference concrete
f_{tRAC}	Flexural strength of the recycled aggregates concrete
f_{tRC}	Flexural strength of the reference concrete
NA	Natural aggregates
RA	Recycled aggregates
RAC	Recycled aggregates concrete(s)
RC	Reference concrete (mix without recycled aggregates)
$shrinkage_{RAC}$	90-day shrinkage of the reference concrete
$shrinkage_{RC}$	90-day shrinkage of the recycled aggregates concrete
$subst_{FRA}$	Substitution rate of fine recycled aggregates with fine natural aggregates
$subst_{CRA}$	Substitution rate of coarse recycled aggregates with coarse natural aggregates
w/c	Water/cement
wa_{RAC}	Weighed water absorption of the aggregates in the recycled aggregates concrete
wa_{RC}	Weighed water absorption of the aggregates in the reference concrete
Δ_{lRAC}	Abrasion loss of mass of the reference concrete
Δ_{lRC}	Abrasion loss of mass of the recycled aggregates concrete

6.1 Introduction

The use of concrete now accounts for such a high level of consumption of non-renewable natural materials that, in some countries, there is already a shortage of these products. The demolition of concrete structures produces waste that is difficult to store owing to the lack of proper dumping places and high transportation and storage costs. The concern with the need of raw materials and the production of enormous quantities of waste has led to several studies on solutions to these problems. A number of studies have recently been published on the properties of recycled aggregates and this aggregate-based various concretes. A comprehensive discussion on various aspects of these materials is already presented in Chaps. 3 and 5 of this book. In Chap. 5, relationships of compressive strength with various other properties of concrete containing CDW aggregate (here RA and RCA will be used to indicate recycled aggregate and recycled concrete aggregate respectively), proposed in various studies, were also presented during the discussion of those properties. The relationships with compressive strength, the most important and extensively studied property of concrete containing recycled aggregate (RAC), with some other properties with sufficiently high correlation coefficients as indicated by researches, are summarised in Table 6.1.

Table 6.1 Relationship between various properties and compressive strength of concrete containing CDW aggregates, proposed in various studies

Property	Relationship	Reference
Splitting tensile strength (f_{sp}, f_{ctm})	$f_{sp} = 0.0931\, f_{cu}^{0.8842}$	Kou and Poon (2008)
	$f_{sp} = 0.24\, f_{cu}^{0.65}$	Xiao et al. (2006)
	$f_{ctm} = 0.3\, f_{ck}^{2/3}$ ($\leq$C50/60) $f_{ctm} = 2.12\ \ln(1 + (f_{cm}/10)) >$ C50/60	Paine et al. (2009)
Flexural strength (f_f)	$f_f = 0.75\, \sqrt{f_{cu}}$	Xiao et al. (2006)
Modulus of elasticity (E)	$E = 7770 \times f_{cu}^{0.33}$	Ravindrarajah and Tam (1985)
	$E = 1.9 \times 10^5 \times f_{cu} + \left(\frac{\rho}{2300}\right)^{1.5} \times \sqrt{\frac{f_{cu}}{2000}}$	Kakizaki et al. (1988)
	$E = 370 \times f_{cu} + 13100$	Dhir et al. (1999)
	$E = 18.2\sqrt[3]{\frac{0.83 f_{cu}}{10}}$	Corinaldesi (2011)
	$E = 18.8\sqrt[3]{\frac{0.83 f_{cu}}{10}}$ $E = 909 \times f_{cu} + 8738$	Corinaldesi (2010)
	$E = 8917 \times (f_{cu} + 8)^{\frac{1}{0.85}} \times \left(\frac{\rho}{2345}\right)^2$	Evangelista and de Brito (2007)
	$E = \frac{10^5}{2.8 + \frac{40.1}{f_{cu}}}$	Xiao et al. (2006)
Dry density (ρ)	$f_{cu} = 0.069 \times \rho - 116.1$	Xiao et al. (2006)

f_{cu}, f_{ck}, f_{cm} cubic compressive strength (ultimate, characteristic and mean values)

In this chapter, a methodology developed and patented by de Brito (2007) will be applied to analyse experimental studies reported in the various literatures done in various countries as well as in Portuguese researches on concrete with recycled concrete aggregates. The description of this methodology is presented in the following section. The results are presented in two further sections. The following section of this chapter summarises the knowledge acquired through past experimental research on RAC performed by researchers from various countries and statistically and graphically processes the published data, aiming at correlating the RAC properties with the properties of the RA used to replace natural aggregates (NA). The data collected are then systemised and some of the results in the field of structural concrete are interpreted. A parallel study was performed using experimental results done by Portuguese researchers and included in the following section. A similar methodology to that adopted in the next international review section was used in the Portuguese review section.

This chapter therefore describes the methodology adopted in the data processing that leads to the estimation of the long-term behaviour of RAC. This innovative methodology provides the building owner, the structural designer and the builder with reliable information to render viable a process (the reuse of inert waste in concrete production) that at the moment faces practical technological limitations. It shifts the reuse of these materials from the currently dominant practices that in effect lead to their down-cycling.

6.2 Methodology

This section presents a methodology to estimate the long-term properties of structural concrete made with recycled aggregates [such as those resulting from construction and demolition waste (CDW)] based on results that can be obtained at a very early stage of the construction process. In order to establish correlations between the RAC properties and the density and water absorption of the aggregates used in the mix and the 7-day compressive strength of the resulting concrete, a graphic analysis methodology was created and patented in Portugal (de Brito 2007), involving the following steps:

1. Analysis and organisation of the data available from each experimental campaign, including the properties of the NA, the RA, the RAC and the reference concrete (RC);
2. Study of the composition of the concrete (RC and RAC) and of the properties of the aggregates to determine the values for density and water absorption of the aggregates (RA and NA) used in the mixes; in the case of the Portuguese experimental research, calculation of the exact value of the density and water absorption of the aggregates used in the mix, through the mix proportions of the concrete mixes (with NA only and with RA and the individual density and water absorption of the aggregates (natural and recycled);
3. Graphical analysis of the relationship between the replacement rate of NA by RA and each property of concrete;
4. Graphical analysis of the variation of the ratio between the properties of concrete with RA and the one with NA only (reference conventional concrete) and the replacement rate of NA by RA;
5. Graphical analysis of the variation of the ratio between the properties of concrete with RA and the RC and the ratio between the weighed density of aggregates in the concrete mix with RA and the RC (as in Eq. 6.1);

$$D = \frac{\text{FA}}{100} \times \left[\frac{\text{subst}_{\text{FRA}} \times d_{\text{FRA}} + (100 - \text{subst}_{\text{FRA}}) \times d_{\text{FPA}}}{100}\right] + \frac{(100 - \text{FA})}{100} \times \left[\frac{\text{subst}_{\text{CRA}} \times d_{\text{CRA}} + (100 - \text{subst}_{\text{CRA}}) \times d_{\text{CPA}}}{100}\right] \quad (6.1)$$

   where D, weighed density of the mixture of aggregates in the concrete mix; FA, percentage of fine aggregates used in the mixture; $\text{subst}_{\text{FRA}}$, replacement ratio (in percentage) of fine primary aggregates with fine recycled aggregates; $\text{subst}_{\text{CRA}}$, replacement ratio (in percentage) of coarse primary aggregates with coarse recycled aggregates; d_{FRA}, density of the fine recycled aggregates; d_{FPA}, density of the fine primary aggregates; d_{CRA}, density of the coarse recycled aggregates; d_{CPA}, density of the coarse primary aggregates.
6. Graphical analysis of the variation of the ratio between the properties of concrete with RA and the RC and the ratio between the weighed water absorption of aggregates in the concrete mix with RA and the RC (similar to Eq. 6.1);

Table 6.2 Qualitative rating of the correlation coefficients

Rating	Values range
Very good	$R^2 \geq 0.95$
Good	$0.80 \leq R^2 < 0.95$
Acceptable	$0.65 \leq R^2 < 0.80$
Non-acceptable	$R^2 < 0.65$

7. Graphical analysis of the variation of the ratio between the properties of concrete with RA and the reference concrete and the ratio between the compressive strength at 7 days of concrete with RA and the reference concrete (similar to Eq. 6.1);
8. Compilation of the data in a table, including the slope of linear regression line and the respective correlation coefficient.

In the case of the methodology applied for analysis of the literature data (presented above), the following two additional steps were adopted to complete the methodology:

9. Superposition of the graphical results of each concrete property for the various campaigns analysed and determination of the linear regression lines with the respective correlation coefficient;
10. Correction of the linear regression lines obtained, to make them representative of the physical behaviour under analysis, forcing them to pass through the point that corresponds to the RC, with the inconvenience of lowering the corresponding correlation coefficient.

Table 6.2 gives the qualitative criteria defined in terms of the correlation coefficient R^2 and applied in both sections.

The methodology's feasibility is explained by the fact that, for conventional concrete as well, it is possible to establish reliable correlations between the various properties of hardened concrete and the density (of which the water absorption is an indirect measure) of the aggregates or the compressive strength at early age.

6.3 Analysis of the Literature Data: Estimation of Long-Term Properties

This study started with a bibliographic Internet-based survey within the RAC topic, with emphasis on the published results of international experimental campaigns. The survey also included scientific journals' papers, article compilations in conferences and seminars, graduate, master's and PhD theses. To better understand the behaviour of RAC in the fresh and hardened states, this study set out to collect experimental data from various international campaigns, and use it to correlate some of the properties of the aggregates (density and water absorption) and the 7-day compressive strength of concrete with the most relevant properties of the RAC.

This survey was guided by the following criteria: availability of the test results performed on RA, especially for water absorption and density; availability of the experimental values relative to the greatest possible number of concrete properties in the fresh and hardened states (mechanical and durability-related), especially the 7-day compressive strength; the greatest possible NA/RA replacement ratio options; the greatest possible number of fixed parameters (e.g. water/cement (w/c) ratio, grading curve, workability, curing method) during the experimental production of RAC; existence of an RC, without which the corresponding campaign was eliminated from the survey. Of the numerous campaigns analysed, only a small group was considered apt for data processing, since most studies did not comply with some, and in the majority of cases, most of the conditions stated above.

6.3.1 Properties of the RA Mixture

In order to obtain the weighed value of the density of the aggregates present in a concrete mix, Eq. 6.1 was used. Through this general equation, the weighed density value of the aggregates in the concrete mixes with the various replacement ratios of NA with RA (from 0—reference concrete—to 100 %) was determined. Although in most of the experimental campaigns, there was no replacement of the fine fraction, i.e. its density is kept constant for all the mixes, the weighed density of the mixture considers them in its calculation.

The transformation of the experimental absolute results into relative values by comparison with the RC allows the comparison between the different campaigns performed. The analysis procedure adopted for the density was also applied to the water absorption of the mixture of aggregates by replacing all the density values in Eq. 6.1 with the corresponding water absorption values. In both equations, it should be noted that: (i) both the fine and coarse fraction are taken into account according to their relative importance in the aggregates' mixture; (ii) both the natural and recycled aggregates are taken into account according to their relative importance in the same mixture; (iii) it is not necessary to classify or in some other way thoroughly identify the nature of the recycled aggregates, as long as it is possible to homogenise the characteristics of each batch, which makes it much easier in practice to use recycled aggregates in concrete production.

6.3.2 Conclusions Drawn from the Bibliographic Survey

The present study has revealed the high heterogeneity of the procedures adopted by international researchers, which sometimes led to difficulties when performing comparative studies such as this. It has been found that the results of experimental campaigns, even in leading international publications, are not always presented in such a way as to allow their in-depth analysis due to lack of specific data.

The survey of international experimental campaigns has revealed great differences at the level of procedures and of organisation/presentation of the results. In this process, the analysis of several campaigns had to be abandoned because of unavailability of data. This is either due to existing information not being included in the description, or sometimes because relevant data were simply not determined. Important examples of this type of limitation are the absence of data on the aggregates' properties, both RA and NA, or on the composition of the concrete mixes tested. Another situation concerned the w/c ratio, since in some of the campaigns, it was not specified whether the ratio concerned all the water in the mix or just the effective water (the first one minus the water directly absorbed by the recycled aggregates during mixing).

Another data collection problem was the variability of the factors introduced in each campaign. In order to allow a scientifically valid comparison, the campaigns analysed should be similar in the greatest number of factors that affect the production of concrete. In order to improve the quality of the comparison, the following parameters are the ones that are more important to keep constant in each family of RC and RAC: effective w/c ratio (differentiating the total amount of water introduced into the mix from that which effectively contributes to the hydration of the cement and the workability of fresh concrete); workability (this property must be maintained by using plasticizers or increasing the total quantity of water without increasing the effective w/c ratio, for example by pre-saturating the recycled aggregates); grading curve of the aggregates (when replacing PA with RA this curve should be kept exactly constant because any change leads to uncontrolled shifts in almost every relevant property of concrete).

This variability of the criteria from one campaign to another is translated into a decrease in the linearity of the trends detected (and of the respective correlation coefficients) when the analysis of the results progresses from the individual campaigns to the summation of results from several of them. The process of comparison is thus rendered difficult, leading to an artificial scattering of the results that should not occur if the conditions stated above are ensured in every campaign.

6.3.3 Experimental Campaigns Selected for Analysis

In the Carrijo (2005) campaign, river sand and basaltic gravel were used as NA in the production of RC, and coarse ceramic and recycled concrete as RA in the production of RCA. Their water absorption was simplistically considered nil (for comparative purposes, in the present study the value was changed to 1 %). Three different values of the w/c ratio were defined in the production of concrete: 0.4, 0.5 and 0.67). Only the coarse aggregates were always replaced with a replacement ratio of 100 %. This strongly impairs a better understanding of the gradual evolution of the properties as NA is replaced with RA. Another parameter that was kept constant was the quantity of water in the different concrete mixes. The families of RAC were defined by the RA density, in four categories. The hardened concrete was tested for compressive strength, water absorption and modulus of elasticity.

In the Kou et al. (2004) campaign, where the RA used in the production of RCA were coarse and undifferentiated, three families of concrete were considered according to the amount of fly ash added (0, 25 and 35 % of the initial amount of cement) to replace the cement, leading to three different types of RC. The influence of the method used to cure the concrete was also analysed. The traditional method was immersion in a water tank at 27 °C (80.6 °F) after 24 h of cure in natural conditions. In an alternative process the specimens were exposed to water vapour at 65 °C (149 °F) for 8 h and afterwards immersed in a water tank until they were tested. The present study does not include the comparison of results from the different curing processes. The effective w/c ratio was kept constant at 0.45. The hardened concrete was tested for compressive strength, modulus of elasticity, chloride penetration and shrinkage.

In the Leite (2001) campaign, where the RA used in the production of RCA were coarse and fine ceramic and recycled concrete, several replacement NA/RA ratios were selected within each family of RCA, defined by a predetermined w/c ratio (0.4, 0.45, 0.60, 0.75, and 0.80), leading to multiple linear regression equations representative of each concrete property under analysis. For comparative purposes with the other campaigns analysed within this study, only the families corresponding to w/c of 0.40 and 0.45 were selected, due to the contradictory nature of many of the results of the other families where a high percentage of fine RA was used. The hardened concrete was tested for compressive strength, splitting and flexural tensile strength and modulus of elasticity.

In the Gomez-Soberón (2002) campaign, the RA were obtained in the laboratory by crushing a concrete produced for that purpose. Two size distributions were obtained for coarse RA, not strictly the same as for the NA. Each family of RCA was defined in terms of age when the tests were performed, since every other parameter remained constant except for the replacement ratio (0, 15, 30, 60 and 100 %) of coarse NA with equivalent recycled concrete RA. The greater water absorption capacity of RA compared with NA was taken into account by pre-moistening the RA before mixing. The effective w/c ratio was kept constant at 0.52. The hardened concrete was tested for porosity, density, permeability, compressive strength, splitting tensile strength, modulus of elasticity, water absorption, creep and shrinkage.

In the Cervantes et al. (2007) campaign, the RCA families were defined in terms of the addition of synthetic fibres in the concrete production. Only the coarse fraction of NA was replaced with different ratios of recycled concrete RA (0, 50 and 100 %). In order to maximise the number of valid results, the family with 0.2 % of synthetic fibres was also considered in the present study. The effective w/c ratio remained constant at 0.51. The hardened concrete was tested for compressive strength, splitting tensile strength, modulus of elasticity and shrinkage.

In the Katz (2003) campaign, the RA was produced in the laboratory by crushing concrete specimens 1, 3 and 28 days old. Three grades of aggregate groups were used. The PA was replaced with RA for the following fractions: 2.36–9.5 mm and 9.5–25 mm (both coarse), and 0–2.36 mm. Fine RCA was used in small amounts and only to improve workability. Families were defined in terms of the type of cement used (traditional and white Portland) and the age of RA at crushing. The traditional cement class was lower than the white cement class, and that had some influence on the results. The hardened concrete was tested for compressive strength, splitting and flexural tensile strength, modulus of elasticity, water absorption, carbonation penetration and shrinkage.

6.3.4 Properties

6.3.4.1 Compressive Strength

Compressive strength is the most common tested property of hardened concrete, and for this reason it was possible to obtain results from four campaigns: Carrijo (2005), Leite (2001), Kou et al. (2004) and Gomez-Soberón (2002). The general trend identified for this property indicates a reduction of strength with the increase of the PA/RA replacement ratio.

Figure 6.1 (top) shows the variation of the ratio between the 28- and 90-day compressive strengths of concrete (f_c) and the ratio between the densities (D) of the mixture of aggregates. The correlation coefficient is considered good (according to the criteria defined in Eq. 6.1) and a linear relation between the parameters can be identified. The reduction of the density and mechanical strength of RA compared with NA, due to the higher percentage of attached mortar in RA, contributes to the reduction of the ratio between the compressive strengths of concrete.

The same analysis was performed with the variation of the ratio between the compressive strengths of concrete and the ratio between the water absorptions (w_a) of the mixture of aggregates and is presented in Fig. 6.1 (centre). The correlation coefficient is not acceptable, indicating however a trend for a linear behaviour between the ratios.

Figure 6.1 (bottom) shows the variation of the ratio between the 28- and 90-day compressive strengths of concrete and the ratio between the 7-day compressive strengths of concrete (f_{c7}). The lack of data about the 7-day compressive strength of the concrete in the campaign of Carrijo (2005) excluded the author from this particular analysis. The negative values in the abscissa axis mean that in the campaign of Gomez-Soberón (2002), some of the results for the 7-day compressive strength of the concrete with RA were higher than the conventional concrete. This particular behaviour is not to be expected and contradicts most of the research; the values were nevertheless included in the analysis, contributing to the reduction of the correlation coefficient, which was, however, acceptable.

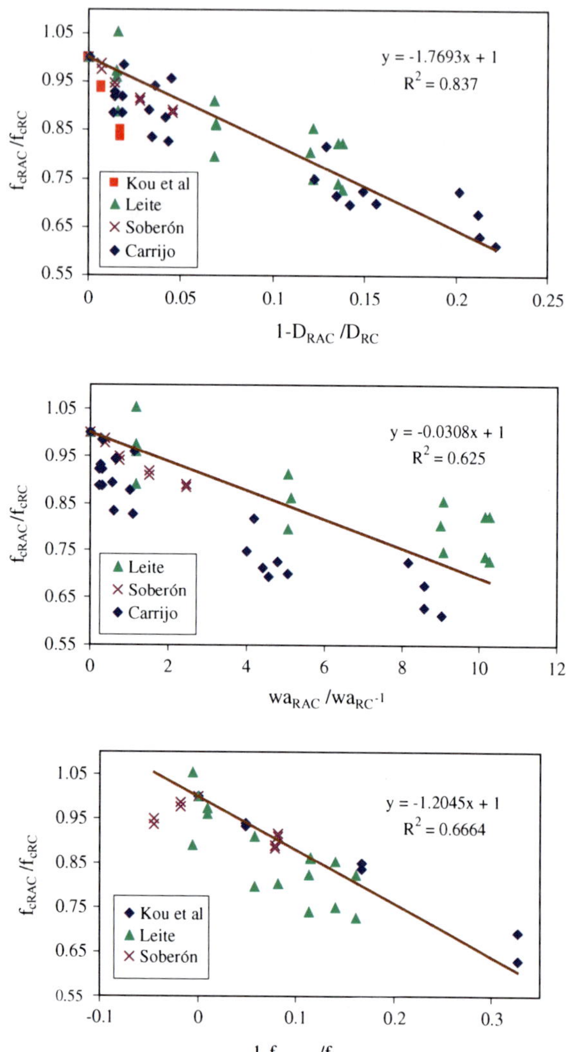

Fig. 6.1 Ratio between the 28- and 90-day compressive strengths of concrete versus the ratio between the densities (*top*), the water absorptions (*centre*) of the mixture of aggregates and the 7-day compressive strengths of concrete (*bottom*)

6.3.4.2 Modulus of Elasticity

Modulus of elasticity results were obtained from the campaigns of Carrijo (2005), Leite (2001), Kou et al. (2004) and Gomez-Soberón (2002). In every case, the modulus of elasticity decreases when the NA/RA replacement ratio increases. This behaviour is mostly attributed to the lower stiffness of RA compared with NA. The higher porosity of RA is responsible for the higher deformation of these aggregates when compared with NA, and this effect is also reflected in concrete with RA when compared to conventional concrete.

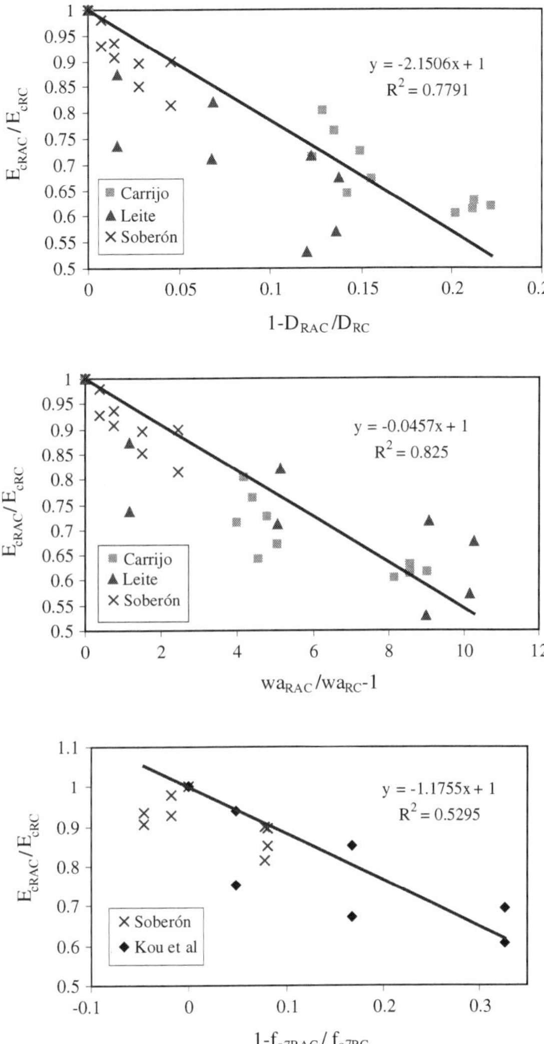

Fig. 6.2 Ratio between the 28- and 90-day moduli of elasticity of concrete versus the ratio between the densities (*top*), the water absorptions (*centre*) of the mixture of aggregates and the 7-day compressive strengths of concrete (*bottom*)

Figure 6.2 (top) shows the variation of the ratio between the 28- and 90-day moduli of elasticity of concrete (E_c) and the ratio between the densities of the mixture of aggregates. The correlation coefficient is acceptable. The same variation for the ratio between the water absorptions of the mixture of aggregates is presented in Fig. 6.2 (centre). The correlation coefficient is good. Figure 6.2 (bottom) shows the variation of the ratio between the 28- and 90-day moduli of elasticity of concrete and the ratio between the 7-day compressive strengths of concrete. The correlation coefficient is not acceptable.

The comparison of the slopes of the correlation lines between the moduli of elasticity and the compressive strengths show similar values, despite the fact that in the literature (Angulo 1998; Azzouz et al. 2002; de Brito 2005; Etxeberria et al. 2007), it is often stated that the inclusion of recycled aggregates has a much stronger influence on stiffness than on mechanical strength.

6.3.4.3 Splitting Tensile Strength

The campaigns of Kou et al. (2004), Gomez-Soberón (2002) and Leite (2001) tested concrete splitting tensile strength at different ages (28 and 90 days). The results of Leite (2001) were only for the age of 28 days and were very scattered. Generally, the results of splitting tensile strength graphs obtained confirm the scatter of test results for this property in different campaigns. The general trend identified for this property indicates a reduction of strength with the increase of the NA/RA replacement ratio. This is explained, as for compressive strength, by the lower mechanical characteristics of the mortar adhering to the primary aggregates (in the case of recycled concrete RA) and of the ceramics (when they were used as RA).

Figure 6.3 (top) shows the variation of the ratio between the 90-day splitting tensile strengths of concrete (f_{sp}) and the ratio between the densities of the mixture of aggregates only for the campaign of Gomez-Soberón (2002). The correlation coefficient is good. Figure 6.3 (centre) shows the same correlation but for the water absorption of mixture of aggregates. The correlation coefficient obtained is also good. The variation of the ratio between the 90-day tensile strengths of concrete and the ratio between the 7-day compressive strengths is shown in Fig. 6.3 (bottom). The correlation factor is good.

Judging by the slope of the correlation lines for all three parameters, the splitting tensile strength is slightly less affected by the NA/RA replacement than the compressive strength, in accordance with the existing literature (Buttler 2003; Chen et al. 2003; de Brito 2005; Larrañaga 2004).

6.3.4.4 Flexural Tensile Strength

For flexural tensile strength, the test results of Leite (2001) at 28 and 90 days are used. The general trend is the same as for the splitting tensile strength and for the same reasons.

Figure 6.4, representing the variation of the ratio between the 28 and 90-day flexural strengths of concrete (f_t) and the ratio between the three parameters, indicates the existence of a linear relationship of the variation. In the three graphs the correlation coefficients are all good. However, the slope of the correlation lines is much higher for all the reference parameters than for the splitting tensile strength and even the compressive strength, which contradicts the results of most experimental campaigns reviewed in this study (Chen et al. 2003; Latterza and Machado 2003; Leite 2001).

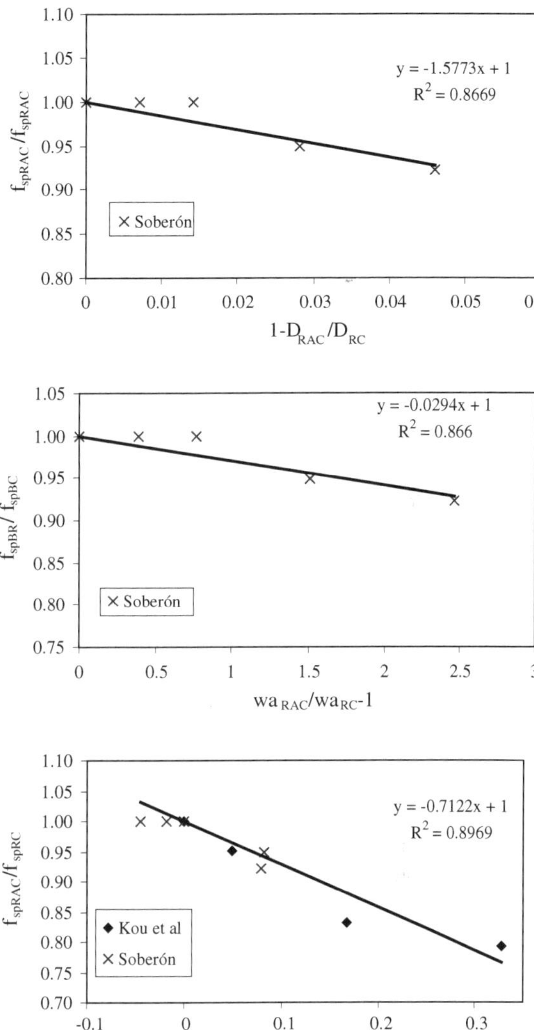

Fig. 6.3 Ratio between the 90-day splitting tensile strengths of concrete versus the ratio between the densities (*top*), the water absorptions (*centre*) of the mixture of aggregates and the 7-day compressive strengths of concrete (*bottom*)

6.3.4.5 Shrinkage

Shrinkage results were obtained from the campaigns of Cervantes et al. (2007) and Gomez-Soberón (2002) at the age of 28 and 90 days, respectively. The general trend is that of a progressive increase of shrinkage with the inclusion of recycled aggregates in the concrete mixes, which is explained by the higher absorption and lower stiffness of the RA compared with the NA.

Figure 6.5 (top) shows the variation of the ratio between the 28- and 90-day shrinkages of concrete and the ratio between the densities of the mixture of

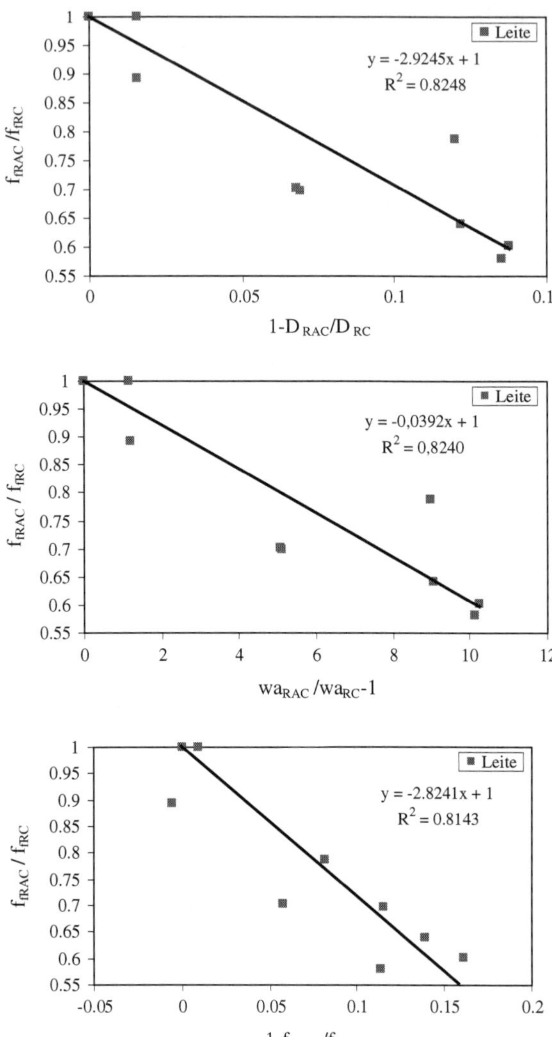

Fig. 6.4 Ratio between the 28- and 90-day flexural tensile strengths of concrete versus the ratio between the densities (*top*), the water absorptions (*centre*) of the mixture of aggregates and the 7-day compressive strengths of concrete (*bottom*)

aggregates. The correlation coefficient obtained is not acceptable. Figure 6.5 (centre) shows the same correlation but for the water absorption of the mixture of aggregates. The correlation coefficient is not acceptable. Since the 7-day compressive strength results for the campaign of Cervantes et al. (2007) are considered inconsistent, the analysis of the correlation with this parameter is not presented. Therefore, Fig. 6.5 (bottom) shows the variation of the ratio between the 90-day shrinkages of concrete and the ratio between the compressive strengths of concrete for the campaign of Gomez-Soberón (2002). The correlation coefficient obtained is good.

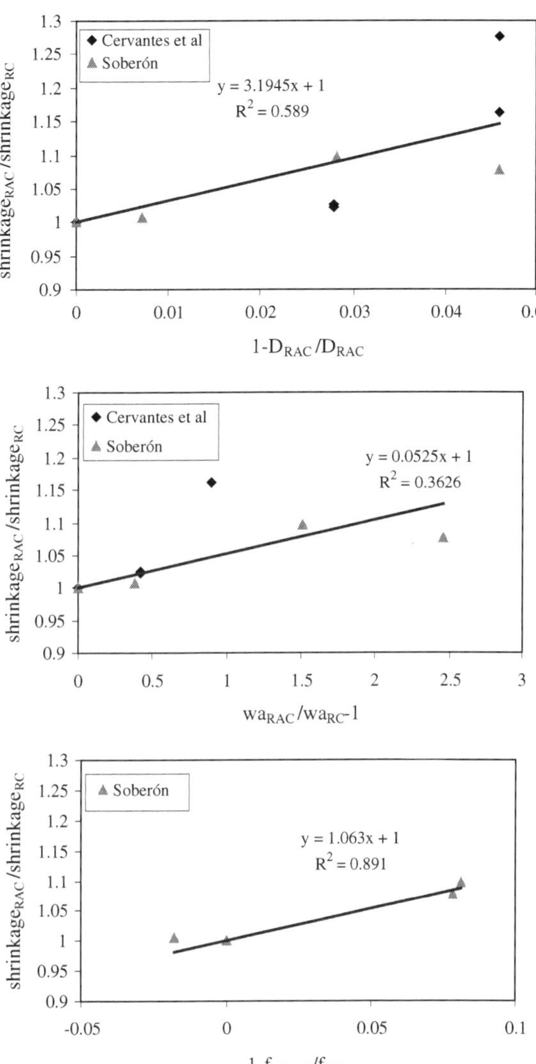

Fig. 6.5 Ratio between the 28- and 90-day shrinkages of concrete versus the ratio between the densities (*top*), the water absorptions (*centre*) of the mixture of aggregates and the 7-day compressive strengths of concrete (*bottom*)

In the literature (Azzouz et al. 2002; Buyle-Bodin and Zaharieva 2002; de Brito 2005; Fraaij et al. 2002), shrinkage is often considered as the single property of concrete most affected by the replacement of NA with RA. However, in this study the slopes of the correlation lines for the 90-day shrinkage are similar to those of the modulus of elasticity and of the compressive strength, even though they significantly increase for the 28-day shrinkage (see Table 6.2). The results are not conclusive because they are not enough to be representative.

6.3.4.6 Creep

Gomez-Soberón (2002) tested the 90-day creep of concrete. The reduction of the stiffness of RA compared with NA contributes to a higher creep with the increase of the NA/RA replacement ratio. Furthermore, a hypothetical increment of the w/c ratio, to balance the higher water absorption of RA compared with NA, can contribute to higher values of creep in the concrete with RA.

Figure 6.6 summarises the ratio between the 90-day creep values of concrete and the ratio of the densities and water absorptions of the mixture of aggregates and the 7-day compressive strengths of concrete. The correlation coefficients obtained are very good for the variation with the ratio between the properties of the aggregates in the mixture and acceptable for the variation with the ratio between the 7-day compressive strengths of concrete.

There are very few studies on creep of RCA, and these point to a slightly better behaviour than for shrinkage (Fraaij et al. 2002; Limbachiya et al. 2000; Mendes et al. 2004; Roos 1998). However, the slopes of the correlation lines obtained from the Gomez-Soberón (2002) study are higher than those in Fig. 6.6 (top and centre) (the exception is Fig. 6.6 (bottom) concerning the 7-day compressive strength of concrete). The small number of valid results for both shrinkage and creep does not allow definitive conclusions.

6.3.4.7 Water Absorption by Immersion

The water absorption by immersion of concrete was tested by Gomez-Soberón (2002) according to the UNE 83-310-90 norm. It is expected that increasing the PA/RA replacement ratio will increase the water absorption of concrete, mostly because of the much higher water absorption of RA compared with NA (due to the porosity of the mortar attached to the first, in the case of recycled aggregates RA, or the intrinsic porosity of the ceramic RA).

Figure 6.7 summarises the variation of the ratio between the 28-day water absorptions by immersion of concrete and the ratio of the densities and water absorptions of the mixture of aggregates and the 7-day compressive strengths of concrete. The correlation coefficients are very good for the variation with the ratio between the properties of the mixture of aggregates and not acceptable for the variation with the ratio between the 7-day compressive strengths of concrete.

As stated above, concrete water absorption by immersion is very strongly linked to the porosity of the aggregates and thus is expected to be one of the characteristics of RCA most affected, in relation to RC (Angulo 1998; Azzouz et al. 2002; Barra and Vazquez 1998; Buttler 2003; de Brito 2005). The slopes of the correlation lines in Fig. 6.7 are similar to those obtained for the corresponding parameters for creep (Fig. 6.7) thus corroborating the point.

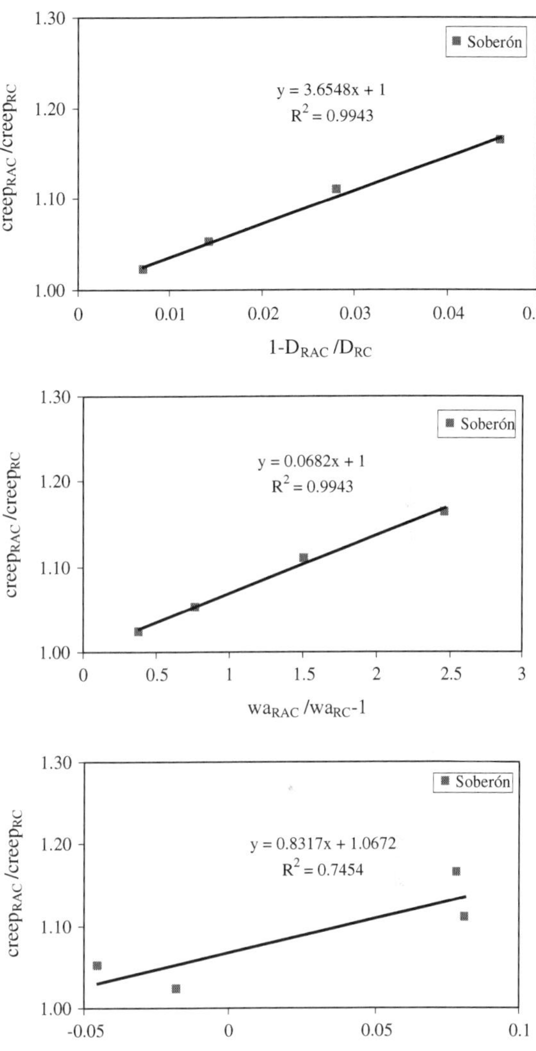

Fig. 6.6 Ratio between the 90-day creep values of concrete versus the ratio between the densities (*top*), the water absorptions (*centre*) of the mixture of aggregates and the 7-day compressive strengths of concrete (*bottom*)

6.3.4.8 Carbonation Penetration Depth

The experimental campaign of Katz (2003) analysed the carbonation effect on concrete for a conventional concrete and a concrete with recycled aggregates only. For this study, and in order to collect the maximum amount of results, values obtained in the three areas of the concrete specimen tested (top, bottom and sides) were used. Like the resistance to chloride penetration, the carbonation penetration

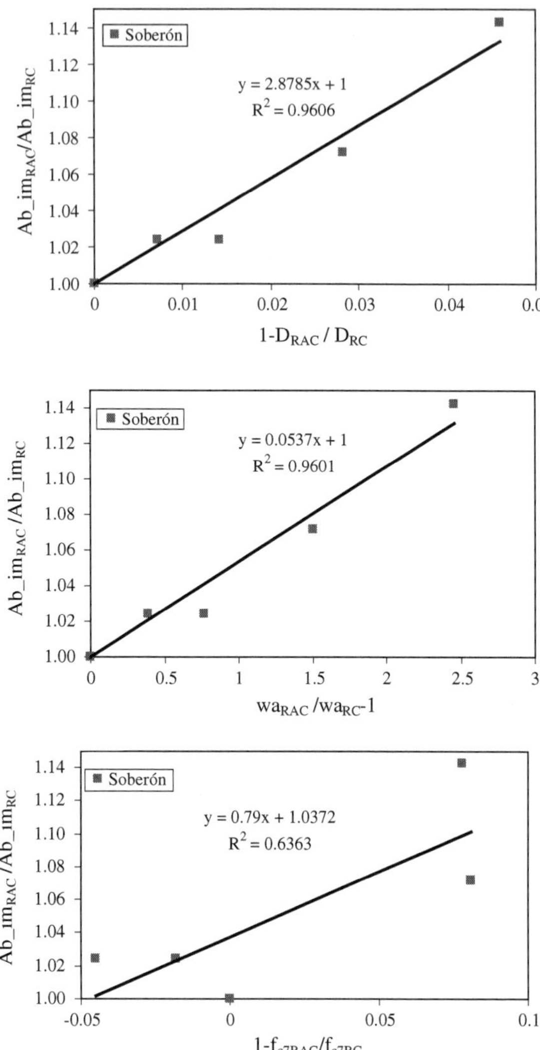

Fig. 6.7 Ratio between the 28-day water absorptions by immersion of concrete versus the ratio between the densities (*top*), the water absorptions (*centre*) of the mixture of aggregates and the 7-day compressive strengths of concrete (*bottom*)

resistance decreases with an increase in the replacement ratio of RA for NA. This behaviour is mostly justified by the higher porosity of RA compared with NA.

Figure 6.8 shows the variation of the ratio between the 7-day carbonation depths of concrete and the ratio of the densities and water absorptions of the mixture of aggregates and the 7-day compressive strengths of concrete. The correlation coefficients are acceptable for the variation with the ratio between the properties of the mixture of aggregates and good for the variation with the ratio between the 7-day compressive strengths of concrete.

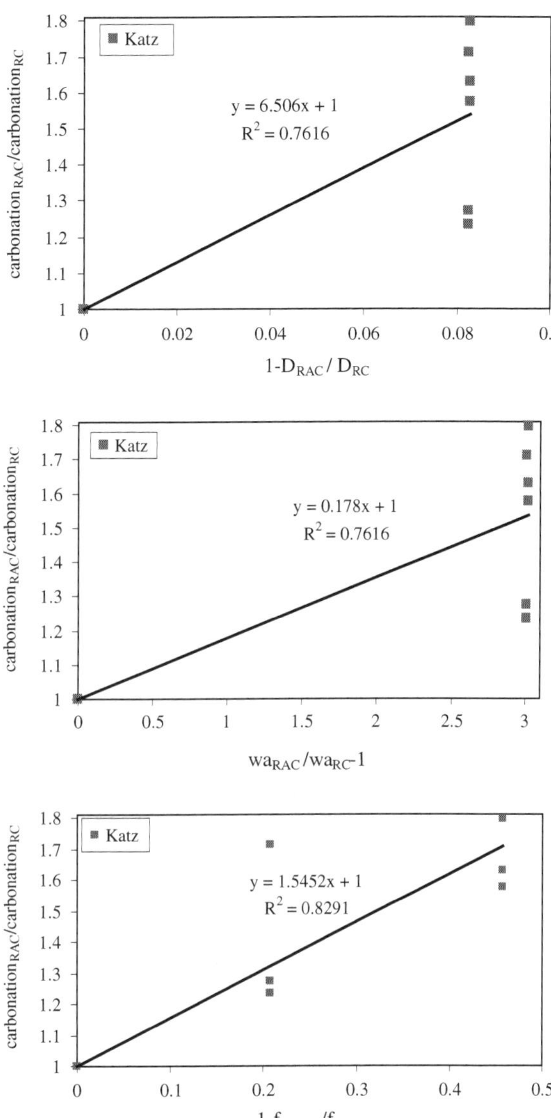

Fig. 6.8 Ratio between the 7-day carbonation depths of concrete versus the ratio between the densities (*top*), the water absorptions (*centre*) of the mixture of aggregates and the 7-day compressive strengths of concrete (*bottom*)

Although this conclusion is based on the results of a single campaign, the slopes of the correlation lines in Fig. 6.8 are the highest for all the properties analysed so far, which gives a good indication that RCA are especially susceptible to this mechanism of degradation that initiates the steel reinforcement corrosion process, as reported in the existing literature (Buyle-Bodin and Zaharieva 2002; de Brito 2005; Levy 2001; Merlet and Pimienta 1993).

6.3.4.9 Chloride Penetration Depth

The chloride penetration results were obtained by Kou et al. (2004) using the test defined in the ASTM C1202-94. This norm establishes the relationship between the electric charge across concrete over a certain period of time and the chloride penetration resistance of concrete. The higher values of the electric charge correspond to a lower resistance to chloride penetration. The chloride penetration resistance is expected to decrease with the increase in the NA/RA replacement ratio, for the same reason as for carbonation penetration, i.e. the higher porosity of RA compared with NA.

Figure 6.9 summarises the variation of the ratio between the 28- and 90-day electric charges measured in concrete (Elect) and the ratio between the densities and water absorptions of the mixture of aggregates and the 7-day compressive strengths of concrete. The correlation coefficients obtained are all good, expressing a linear trend.

The slopes of the correlation lines in Fig. 6.9 are not directly comparable with the ones in the equivalent graphs for the other properties (because the property analysed is not the one directly measured). However, it is very clear that, as for carbonation penetration, RCA are very susceptible to degradation phenomena related to the presence of chlorides in the environment surrounding concrete elements, as reported in the literature (de Brito 2005; Fraaij et al. 2002; Levy 2001; Limbachiya et al. 2000; Olorunsogo and Padayachee 2001).

Table 6.3 shows all the correlation trend lines presented in this paper. The correlation coefficient classification is identified by different colours.

6.4 Analyses of Results from Portuguese Experimental Results

This research started with the search for the experimental studies conducted in Portugal that used recycled CDW as aggregate in the production of structural concrete. The results obtained for the properties of the recycled aggregates and for fresh and hardened concrete were noted and then related to other properties of the aggregates (density and water absorption) and the concrete compressive strength at the age of 7 days. Summary forms were prepared for each campaign, indicating materials used, properties tested and other relevant details. The methodology described in the previous section along with Eq. (6.1) to calculate the weighed density value was also used here and therefore is not described.

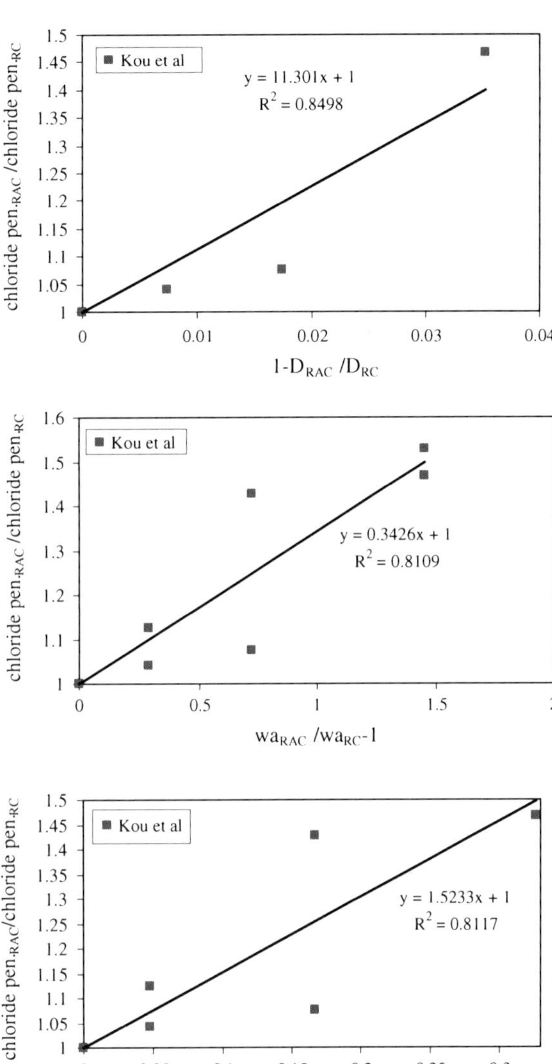

Fig. 6.9 Ratio between 28- and 90-day electric charges measured in concrete versus the ratio between the densities (*top*), the water absorptions (*centre*) of the mixture of aggregates and the 7-day compressive strengths of concrete (*bottom*)

6.4.1 Short Description of the Analysed Experimental Studies

The experimental studies described next were analysed to develop the methodology.

Rosa (2002) tested replacement rates of 1/3, 2/3 and 3/3 of coarse limestone aggregate with coarse recycled ceramic aggregate (from crushed standard hollow red clay wall bricks from a single batch) to determine the compressive and tensile

Table 6.3 Summary of the correlation trend lines between the different concrete properties and the densities and water absorptions of the mixture of aggregates and the 7-day compressive strengths of concrete (Robles 2003)

Property	Campaigns	Aggregates' density		Aggregates' water absorption		Concrete 7-daycompressive strength	
		R^2	slope	R^2	slope	R^2	slope
f_{c28}	Carrijo/Kou/Leite/Gomez-Soberón	0.8697	-1.8354	0.6692	-0.0369	0.6339	-1.3551
f_{c90}	Kou/Leite/Gomez-Soberón	-	-	-	-	0.7616	-1.0539
	Leite/Gomez-Soberón	0.8152	-1.4169	0.7716	-0.0190	-	-
f_c	Carrijo/Kou/Leite/Gomez-Soberón	0.837	-1.7693	-	-	0.6664	-1.2045
	Carrijo/Leite/Gomez-Soberón	-	-	0.6250	-0.0308	-	-
E_{c28}	Carrijo/Kou/Gomez-Soberón	0.7591	-1.9224	0.7300	-0.0506	0.6356	-1.3738
E_{c90}	Kou/Gomez-Soberón	0.7565	-4.9946	0.6593	-0.0934	0.4215	-0.9771
E_c	Carrijo/Leite/Gomez-Soberón	0.7791	-2.1506	0.8250	-0.0457	-	-
	Carrijo/Kou/Gomez-Soberón	-	-	-	-	0.5295	-1.1755
f_{sp28}	Leite/Gomez-Soberón	0.4906	-1.3441	0.4693	-0.0180	-	-
	Gomez-Soberón/Kou	-	-	-	-	0.6356	-0.6321
f_{sp90}	Gomez-Soberón/Kou	-	-	-	-	0.8969	-0.7122
	Gomez-Soberón	0.8669	-1.5773	0.8660	-0.0294	-	-
f_{sp}	Leite/Gomez-Soberón	0.5358	-1.3530	0.5094	-0.0183	-	-
	Gomez-Soberón/Kou	-	-	-	-	0.7858	-0.6721
f_f	Leite	0.8248	-2.9245	0.8240	-0.0392	0.8143	-2.8241
$Shrinkage_{28}$	Cervantes	0.6280	3.7310	0.7402	0.2117	-	-
$Shrinkage_{90}$	Gomez-Soberón	0.7397	2.1448	0.7402	0.0400	0.8910	1.0630
Shrinkage	Soberón/Cervantes	0.5890	3.1945	0.3626	0.0525	-	-
$Creep_{90}$	Gomez-Soberón	0.9943	3.6548	0.9943	0.0682	0.7454	1.0672
$Absorption_{28}$	Gomez-Soberón	0.9606	2.8785	0.9601	0.0537	0.6363	0.7900
$Carbonation_7$	Katz	0.7616	6.506	0.7616	0.1780	0.8291	1.5452
$Chloride_{28}$	Kou	0.8101	14.0860	0.8888	0.4108	0.8903	1.8266
$Chloride_{90}$	Kou	0.8821	16.8710	0.8455	0.2744	0.8460	1.2199
Chloride	Kou	0.8498	11.3010	0.8109	0.3426	0.8117	1.5233

Colour	Meaning
(yellow-green)	correlation coefficient acceptable ($0.65 \leq R^2 < 0.80$)
(orange)	correlation coefficient good ($0.80 \leq R^2 < 0.95$)
(green)	correlation coefficient very good ($R^2 \geq 0.95$)

strengths, abrasion resistance, water absorption by capillarity and immersion of the hardened concrete (see also de Brito et al. 2005; Correia et al. 2006).

Rocha and Resende (2004) tested replacement rates of 50 and 100 % of coarse granite aggregate with coarse recycled concrete aggregate (from crushed concrete test cubes made in laboratory) to determine the density, compressive and tensile strengths and modulus of elasticity of hardened concrete.

Figueiredo (2005) tested replacement rates of 50 and 100 % of coarse granite aggregate with coarse recycled concrete (from crushed concrete test cubes made in laboratory) and white ceramic aggregate (from crushed standard sanitary elements and wall tiles, from single batches) to determine the compressive and tensile strengths, modulus of elasticity, abrasion resistance, shrinkage, water absorption by capillarity and immersion, carbonation and chloride penetration of the hardened concrete.

Matias and de Brito (2005) tested replacement rates of 25, 50 and 100 % of coarse limestone aggregate with coarse recycled concrete aggregate (from crushed concrete test cubes made in laboratory), with a focus on the use of superplasticizers, to determine the compressive and tensile strength, abrasion resistance, shrinkage, water absorption by capillarity and immersion, carbonation and chloride penetration of the hardened concrete.

Evangelista (2007) tested replacement rates of 10, 20, 30, 50 and 100 % of fine limestone aggregate with fine recycled concrete aggregate (from crushed concrete test cubes made in laboratory) to determine the compressive and tensile strengths, modulus of elasticity, abrasion resistance, shrinkage, water absorption by capillarity and immersion, carbonation and chloride penetration of the hardened concrete (see also Evangelista and de Brito 2007, 2010).

Gomes (2007) tested various combinations of replacement rates of coarse limestone aggregate with coarse recycled concrete (from crushed concrete test cubes made in laboratory), ceramic and mortar aggregates (from demolished standard partition walls made of hollow red clay bricks and cement-based renders of previously known characteristics) to determine the compressive and tensile strengths, modulus of elasticity, shrinkage, water absorption by capillarity and immersion, carbonation and chloride penetration of the hardened concrete (see also Gomes and de Brito 2009);

Ferreira (2007) tested replacement rates of 20, 50 and 100 % of coarse limestone aggregate with coarse recycled concrete aggregate (from crushed concrete from the demolition of a former factory in Barcelona, Spain), with a focus on the recycled aggregate pre-saturation process, to determine the density, compressive strength, modulus of elasticity, shrinkage and water absorption by capillarity and immersion of the hardened concrete.

6.4.2 Results and Discussion

6.4.2.1 General Remarks

In each study, the w/c ratio was one of the properties kept constant in the RC (the mix without recycled aggregates, RC) and the various concrete mixes made with recycled aggregates (RAC). For the comparison to be successful (i.e. for the influence of the recycled aggregates to stand out without any entropy from other parameters), these properties should be kept unaltered: (1) Effective w/c ratio; (2)

Slump (according to the experience gained in these various experimental studies, keeping the effective w/c ratio constant usually leads to equal slump values, regardless of the type of aggregates used); (3) Aggregate size distribution.

As expected, the results presented by each experimental study individually exhibit a far better linearity and bigger correlation values compared with the juxtaposed campaigns' values presented below. Various hardened and fresh concrete properties were analysed in this study. The primary property analysed for fresh concrete was density. It showed a proportional decrease as the substitution rate increased, which can be linked to the lower density presented by recycled aggregates. The hardened concrete mixes were analysed for their mechanical properties and durability. Most of these properties had a poorer performance with the use of recycled aggregates, as expected, and in agreement with the international references cited above.

Each property is graphically analysed, with the introduction of a regression line, corrected so as to pass through the point corresponding to the RC.

6.4.2.2 Concrete Density

The ratio between the hardened concrete densities and the ratio between the densities of the aggregates in the mixes showed a linear trend with a good correlation coefficient.

Figure 6.10 shows the relationship between the ratio between the hardened concrete densities and the ratio between the densities and the water absorptions of the aggregates in the mix and the 7-day compressive strengths of concrete, respectively. The correlation coefficients (see Table 6.2) are considered, respectively, good, acceptable and not acceptable.

For every parameter used as an indicator, the concrete density decreases with the substitution rate because of the lower density of the recycled aggregates, due to their lower intrinsic porosity (in ceramic and mortar aggregates) or the mortar adhering to the original aggregates (in recycled concrete aggregates).

6.4.2.3 Compressive Strength

Figure 6.11 depicts the relationship between the ratio between the 28-day compressive strengths of the hardened concrete and the ratio between the densities and the water absorptions of the aggregates in the mix and the 7-day compressive strengths of concrete, respectively. The correlation coefficients are considered, respectively, acceptable, not acceptable and not acceptable. Although the correlation coefficients were disappointingly low (due to the large number and variety of studies juxtaposed in the same graph), it is still possible to see a clear trend towards a fall in compressive strength as the substitution rate increases, explained by the lower mechanical characteristics of the ceramics and of the mortar adhering to the NA (in recycled concrete).

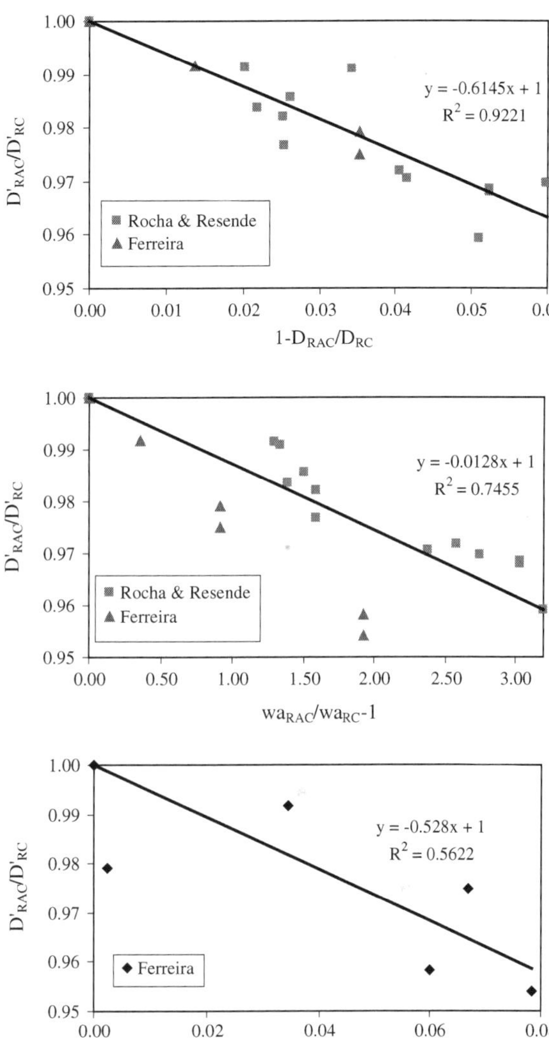

Fig. 6.10 Ratio between hardened concrete densities versus ratio between densities of aggregates in mix (*top*), water absorptions of aggregates in mix (*middle*) and 7-day compressive strengths (*bottom*) of concrete

6.4.2.4 Modulus of Elasticity

Figure 6.12 shows the relationship between the ratio between the moduli of elasticity of the hardened concrete and the ratio between the densities and the water absorptions of the aggregates in the mix and the 7-day compressive strengths of concrete, respectively. The correlation coefficients are considered, respectively, good, good and acceptable.

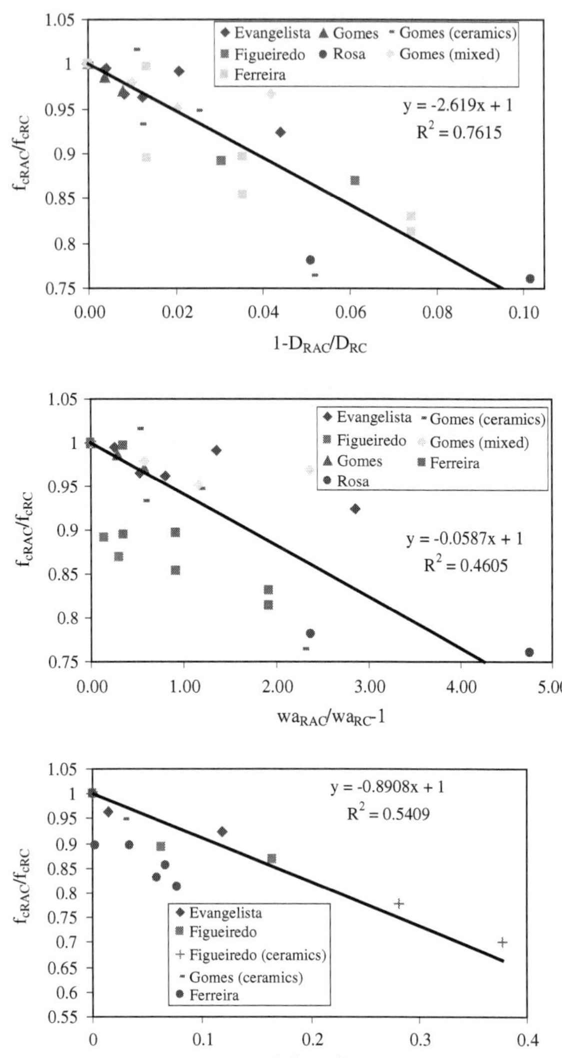

Fig. 6.11 Ratio between concrete 28-day compressive strengths versus ratio between densities of aggregates in mix (*top*), water absorptions of aggregates in mix (*middle*) and 7-day compressive strengths (*bottom*) of concrete

Again, there is a clear trend towards a decrease of a mechanical property of the hardened concrete (in this case the modulus of elasticity) as the replacement rate increases. The reason is the same as for the compressive strength, though related to the lower stiffness of the recycled aggregates. The slope of the correlation lines is higher for the modulus of elasticity than for the compressive strength in every parameter, confirming a trend of greater loss of stiffness for the concrete containing recycled aggregates (see Chap. 5).

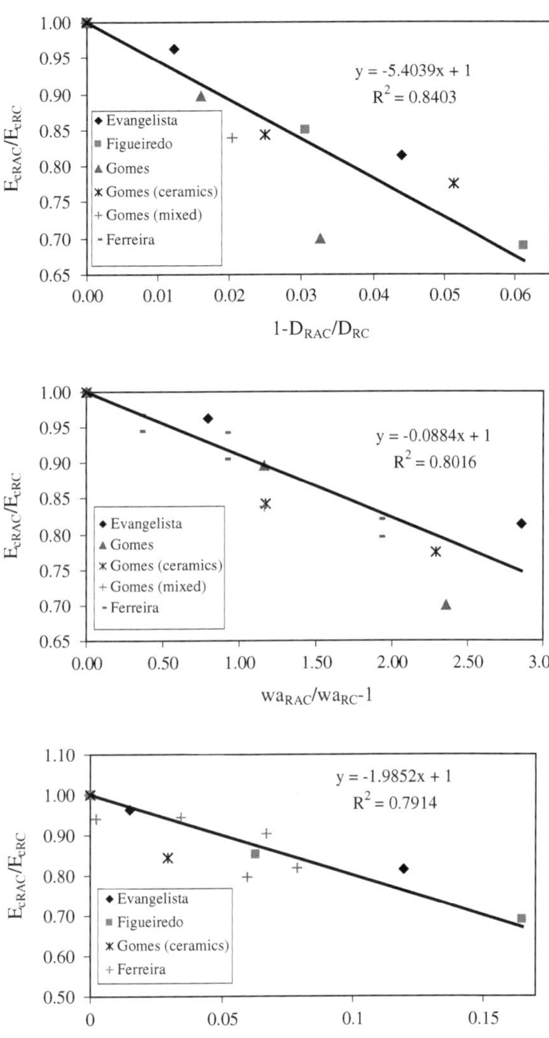

Fig. 6.12 Ratio between concrete moduli of elasticity versus ratio between densities of aggregates in mix (*top*), water absorptions of aggregates in mix (*middle*) and 7-day compressive strengths (*bottom*) of concrete

6.4.2.5 Splitting Tensile Strength

Figure 6.13 shows the relationship between the ratio between the splitting tensile strengths of the hardened concrete and the ratio between the densities and the water absorptions of the aggregates in the mix and the 7-day compressive strengths of concrete, respectively. The correlation coefficients are considered respectively acceptable, not acceptable and good.

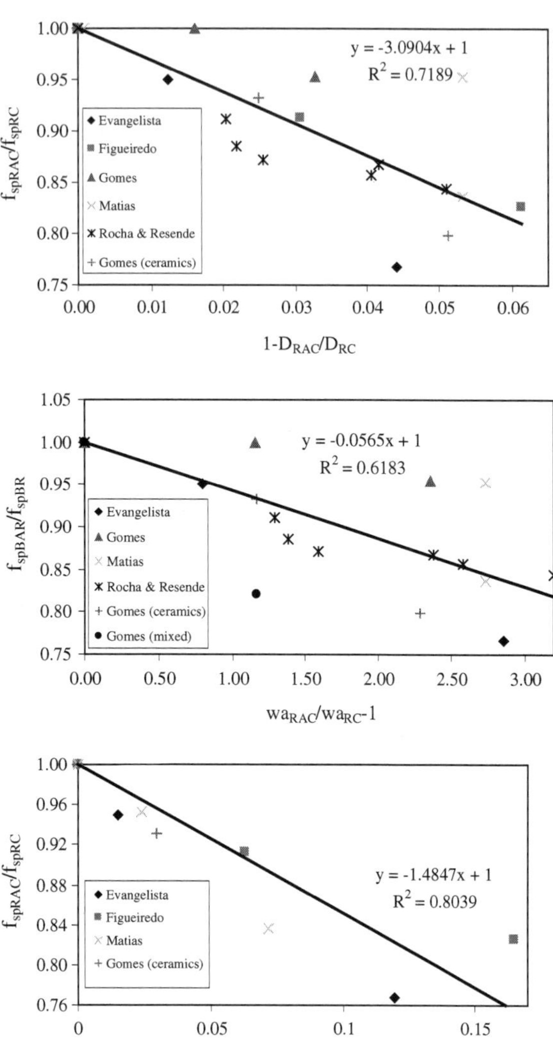

Fig. 6.13 Ratio between concrete splitting tensile strengths versus ratio between densities of aggregates in mix (*top*), water absorptions of aggregates in mix (*middle*) and 7-day compressive strengths (*bottom*) of concrete

Once again, the lower mechanical properties of the recycled aggregates when compared with the natural ones lead to a fall in the splitting tensile strength of the former as the replacement rate increases. The slope of the correlation lines, even though lower than for the modulus of elasticity, is generally higher than for the compressive strength, which somehow contradicts the existing literature as presented in Chap. 5, which states that the concrete mixes that use recycled aggregates are slightly less sensitive to the inclusion of these aggregates for tensile strength than for compressive strength.

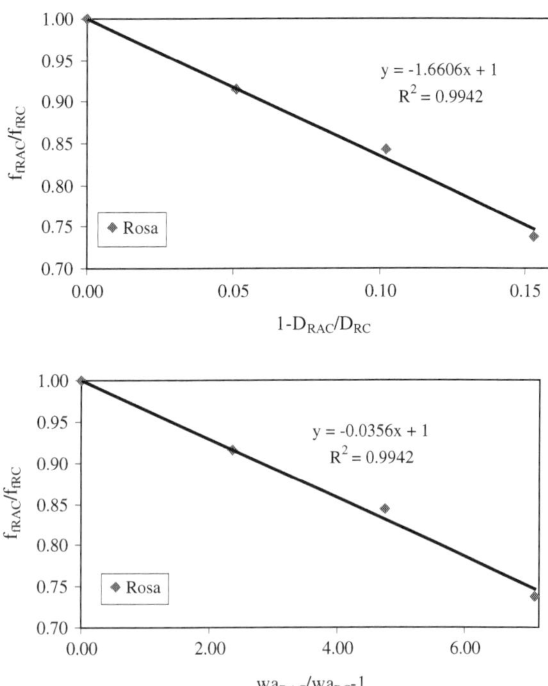

Fig. 6.14 Ratio between concrete flexural strengths versus ratio between densities (*top*) and water absorptions (*bottom*) of aggregates in the mix

6.4.2.6 Flexural Tensile Strength

Figure 6.14 shows the relationship between the ratio between the flexural tensile strengths of the hardened concrete and the ratio between the densities and the water absorptions of the aggregates in the mix, respectively. The correlation coefficients are both considered very good.

The conclusions are the same as for splitting tensile strength, except that this property seems to be less sensitive to the inclusion of recycled aggregates than the one which is in accordance with the discussion presented in Chap. 5.

6.4.2.7 Abrasion Resistance

Figure 6.15 shows the relationship between the ratio between the abrasion resistances (inversely proportional to the loss of mass) of the hardened concrete and the ratio between the densities and the water absorptions of the aggregates in the mix and the 7-day compressive strengths of concrete, respectively. The correlation coefficients are considered, respectively, good, good and not acceptable.

Abrasion resistance of concrete with recycled aggregates tends to increase (less loss of mass) as the replacement ratio increases, because of the greater adherence

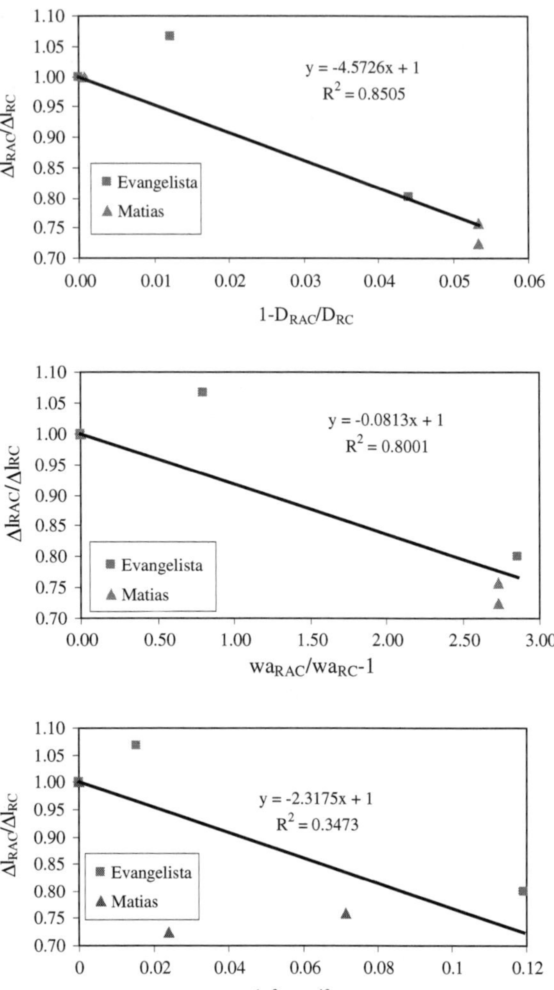

Fig. 6.15 Ratio between concrete abrasion losses of mass versus ratio between densities of aggregates in mix (*top*), water absorptions of aggregates in mix (*middle*) and 7-day compressive strengths (*bottom*) of concrete

between particles provided by the more porous surface of recycled aggregates. It is the only property analysed where concrete with recycled aggregates behaves better than conventional concrete.

6.4.2.8 Shrinkage

Figure 6.16 shows the relationship between the ratio between the shrinkages of the hardened concrete and the ratio between the densities and the water absorptions of the aggregates in the mix and the 7-day compressive strengths of concrete,

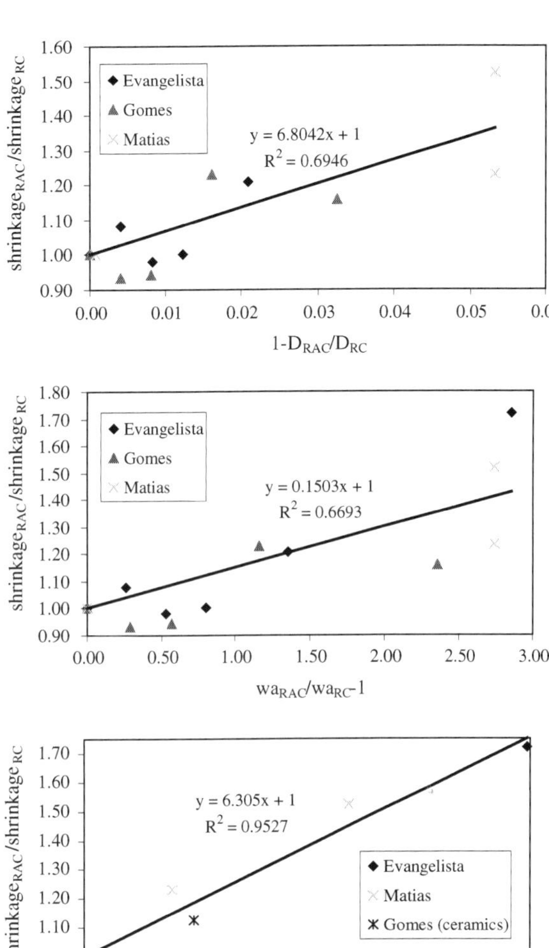

Fig. 6.16 Ratio between concrete shrinkages versus ratio between densities of aggregates in mix (*top*), water absorptions of aggregates in mix (*middle*) and 7-day compressive strengths (*bottom*) of concrete

respectively. The correlation coefficients are considered, respectively, acceptable, acceptable and very good.

Shrinkage of hardened concrete is much affected by the porosity of the aggregates and their stiffness and shows a clear trend towards increasing with the substitution rate, as expected, based on the relative properties of the recycled aggregates and on the literature review. Another conclusion often reported in the literature is that shrinkage is the most sensitive property in terms of the inclusion of recycled aggregates. This is confirmed by the higher slopes of the correlation lines, even higher than those for the modulus of elasticity.

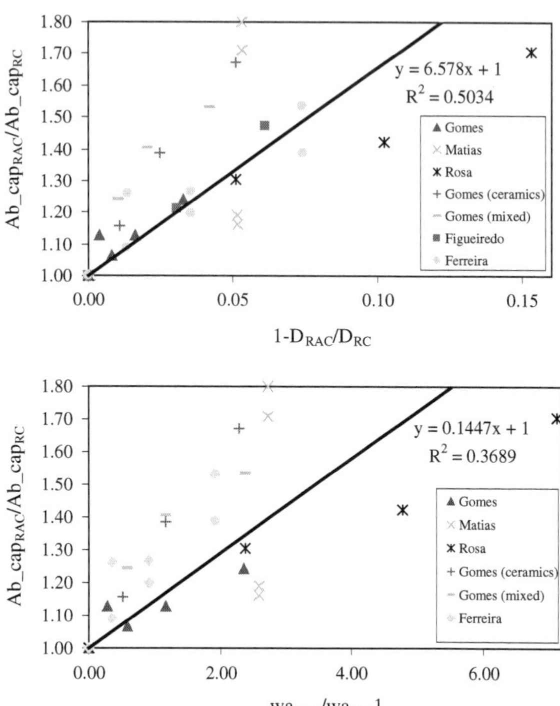

Fig. 6.17 Ratio between concrete capillary water absorptions versus ratio between densities (*top*) and water absorptions (*bottom*) of aggregates in the mix

6.4.2.9 Water Absorption by Capillarity

Figure 6.17 shows the relationship between the ratio between the capillary water absorptions of the hardened concrete and the ratio between the densities and the water absorptions of the aggregates in the mix, respectively. The correlation coefficients are both considered not acceptable.

Although the correlation obtained is poor, the trend is very clear and shows the strong impact of the higher porosity of the recycled aggregates on the water absorption by capillarity, which is only smaller than that found for the shrinkage. Again, this confirms the existing literature reports which single out durability as the performance aspect of concrete with recycled aggregates where the losses are greatest, after volumetric instability.

6.4.2.10 Water Absorption by Immersion

Figure 6.18 shows the relationship between the ratio between the water absorptions by immersion of the hardened concrete and the ratio between the densities and the water absorptions of the aggregates in the mix and the 7-day compressive strengths of concrete, respectively. The correlation coefficients are considered, respectively, not acceptable, not acceptable and good.

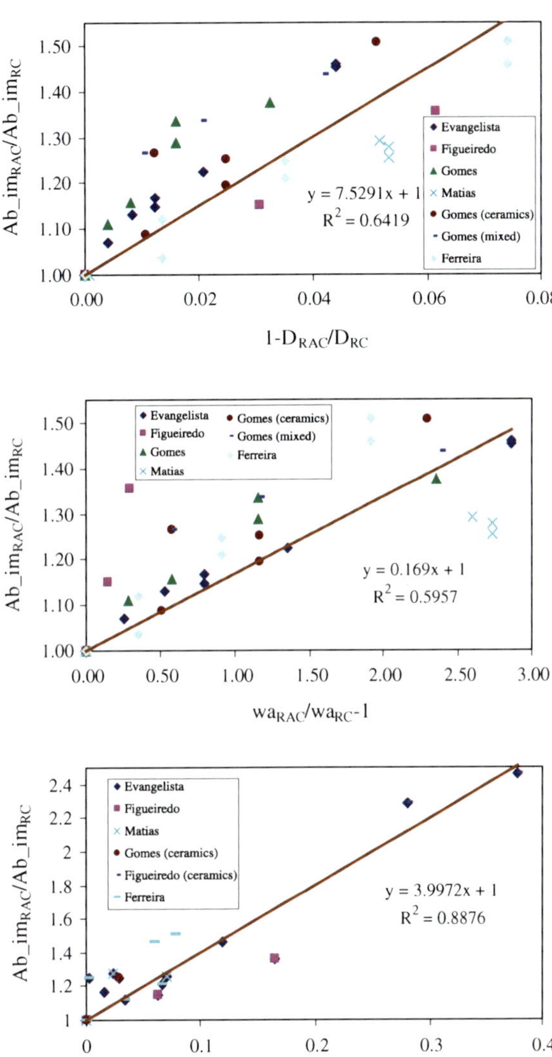

Fig. 6.18 Ratio between concrete water absorptions by immersion versus ratio between densities of aggregates in mix (*top*), water absorptions of aggregates in mix (*middle*) and 7-day compressive strengths (*bottom*) of concrete

The conclusions are the same as for water absorption by capillarity with similar slopes in terms of correlation lines.

6.4.2.11 Carbonation Penetration Depth

Figure 6.19 shows the relationship between the ratio between the carbonation depths of the hardened concrete and the ratio between the densities and the water

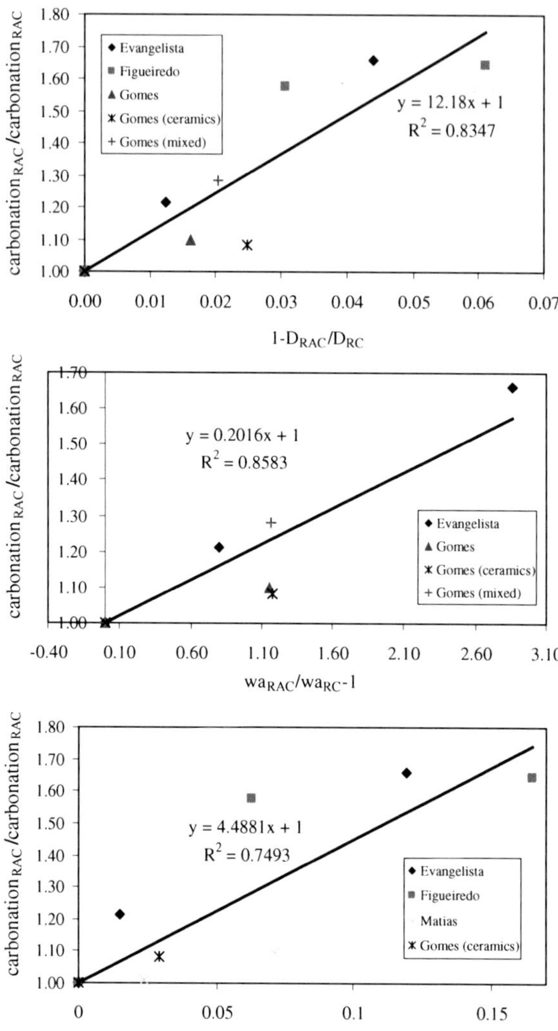

Fig. 6.19 Ratio between concrete carbonation depths versus ratio between densities of aggregates in mix (*top*), water absorptions of aggregates in mix (*middle*) and 7-day compressive strengths (*bottom*) of concrete

absorptions of the aggregates in the mix and the 7-day compressive strengths of concrete, respectively. The correlation coefficients are considered, respectively, good, good and acceptable.

The graphs exhibit a very strong influence of the substitution rate on the depth of the carbonation front which reaches within the concrete, stronger than for any other characteristic so far. This has been identified in the literature review as one of the most vulnerable aspects of introducing recycled aggregates into concrete production and is explained, as for the other durability-related characteristics (e.g.

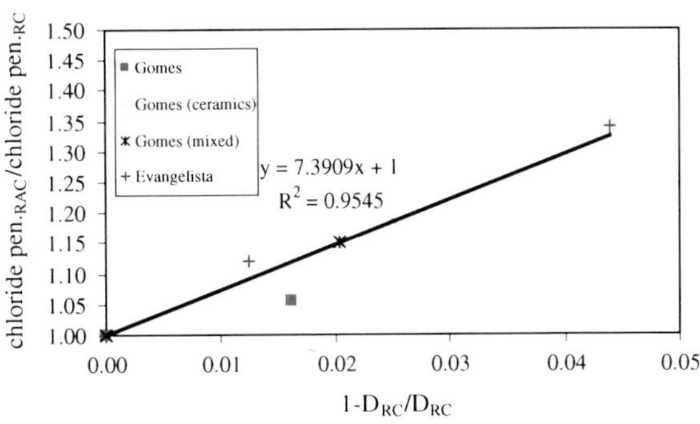

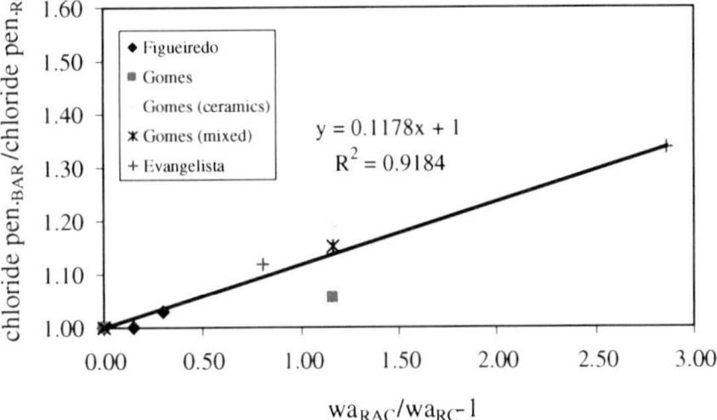

Fig. 6.20 Ratio between concrete chloride penetration depths versus ratio between densities (*top*) and water absorptions (*bottom*) of aggregates in the mix

water absorption), by the greater porosity of these aggregates when compared to that of NA.

6.4.2.12 Chloride Diffusion Coefficients

Figure 6.20 shows the relationship between the ratio between the chloride diffusion coefficients of the hardened concrete and the ratio between the densities and the water absorptions of the aggregates in the mix, respectively. The correlation coefficients are considered, respectively, very good and good.

Although the slope of the correlation lines is smaller than for the carbonation, the trend of increase of vulnerability to chloride penetration arising from the inclusion of recycled aggregates is very clear. The reasons for this trend are still the same (i.e. greater porosity of the recycled aggregates).

Table 6.4 summarises the correlation coefficients for the relationship of each hardened concrete property with the weighed densities and water absorptions of the aggregates in the mix and the 7-day compressive strengths of concrete. The regression lines' slopes are also presented in this table. The correlation coefficient classification was identified by different colours.

Table 6.4 Summary of the correlation of each concrete property with the density and water absorption of the aggregates in the mix and the 7-day compressive strength of concrete

Property	Experimental studies	Aggregate density		Aggregate water absorption		Concrete 7-day compressive strength	
		R^2	Slope	R^2	Slope	R^2	Slope
Density	Rocha and Resende/Ferreira	0.9221	-0.6145	0.7455	-0.0128	-	-
	Ferreira					0.5622	-0.528
Compressive strength	Evangelista/Figueiredo/Gomes/Rosa/Ferreira	0.7615	-2.619	0.4605	-0.0587	-	-
	Evangelista/Figueiredo/Gomes/Ferreira	-	-	-	-	0.5409	-0.88908
Modulus of elasticity	Evangelista/Figueiredo/Gomes/Ferreira	0.8403	-5.4039	-	-		
	Evangelista/Gomes/Ferreira	-	-	0.8016	-0.0884	-	-
	Evangelista/ Figueiredo/ Ferreira	-	-	-	-	0.7914	-1.9852
Splitting tensile strength	Evangelista/Figueiredo/Gomes/Matias/Rocha and Resende	0.7189	-3.0904	-	-	-	-
	Evangelista/Gomes/Matias/Rocha and Resende	-	-	0.6183	-0.0565	-	-
	Evangelista/Figueiredo/Gomes/Matias	-	-	-	-	0.8039	-1.4847
Flexural tensile strength	Rosa	0.9942	-1.6606	0.9942	-0.0356	-	-
Abrasion resistance	Evangelista/Matias	0.8505	-4.5726	0.8001	-0.0813	0.3473	-2.3175
Shrinkage	Evangelista/Gomes/Matias	0.6946	6.8042	0.6693	0.1503	-	-
	Evangelista/Gomes/Matias	-	-	-	-	0.9527	6.305
Water absorption by capillarity	Figueiredo/Gomes/Matias/Rosa/Ferreira	0.5034	6.578	-	-	-	-
	Gomes/Matias/Rosa/Ferreira	-	-	0.3689	0.1447	-	-
Water absorption by immersion	Evangelista/Figueiredo/Gomes/Matias/	-	-	-	-	0.8876	3.9972
	Evangelista/Figueiredo/Gomes/Matias/Ferreira	0.6419	7.5291	0.5957	0.169	-	-
Carbonation depth	Evangelista/Figueiredo/Gomes/Matias	-	-	-	-	0.7493	4.4881
	Evangelista/Figueiredo/Gomes	0.8347	12.18	-	-	-	-
	Evangelista/Gomes	-	-	0.8583	0.2016	-	-
Chloride penetration	Evangelista/Figueiredo/Gomes	-	-	0.9184	0.1178	-	-
	Evangelista/Gomes	0.9545	7.3909	-	-	-	-

(yellow) correlation coefficient acceptable ($0.65 \leq R^2 < 0.80$)
(orange) correlation coefficient good ($0.80 \leq R^2 < 0.95$)
(green) correlation coefficient very good ($R^2 \geq 0.95$)

6.5 Conclusions

A graphical analysis method was developed and successfully used to analyse some of the most important properties of concrete. This analysis proves the viability of a methodology able to anticipate the behaviour of some fresh and hardened concrete properties. The prior knowledge of properties such as density and water absorption of the aggregates in the concrete mix will make it possible to estimate the values of some hardened concrete properties.

The search for international experimental results for this study revealed great differences in the test procedures and organisation of the published information. Most of the campaigns accepted the variation of more than one parameter, including the w/c ratio, making the analysis of the effect of the replacement ratio impracticable. These obstacles meant that a great number of campaigns could not be used in the graphical analysis method used in this study. However, this study managed to analyse some of the most important properties of concrete and to develop the concept initially devised of finding correlations between the relative values of these concrete properties and the weighed density and water absorption of the mixture of aggregates and the relative 7-day compressive strength of concrete.

This study has also shown the development level of Portuguese experimental studies on this subject. The Portuguese studies follow very similar and strict scientific procedures that allowed their results to be used with increased safety and efficiency.

It can be stated that the relationships between the ratio of several of the hardened concrete properties analysed and the ratio between densities and water absorptions of the aggregates in the mix and the ratio between 7-day compressive strengths of concrete revealed some linear trends although some exceptions are observed in some of the properties. The relationships between various hardened properties and the ratio between densities and water absorptions of the mixture of aggregates and the 7-day compressive strengths of concrete allowed the following conclusions to be drawn:

- Generally, the density of the mixture of aggregates showed higher correlation coefficients in the graphical analysis for the hardened concrete properties; on the other hand, the 7-day compressive strength of concrete seems to be the least suitable parameter to estimate the long-term concrete properties, since, in general, the lowest correlation coefficients were obtained with this property; this can be explained by the influence of the variation of mixture procedures from one campaign to the other and by the higher scatter of results for young concrete;
- The lowest results were obtained for the splitting tensile strength and can be justified by the greater variability of this property compared with the compressive strength, for example (a trend common to conventional concrete);
- For every property of hardened concrete analysed the performance of RCA is worse than that of the RC and the difference increases with the value of the replacement ratio;
- Based on the slope of the correlation lines for either of three reference parameters, density and water absorption of the mixture of aggregates and 7-day compressive strength of concrete, the effect on RCA of the inclusion of RA instead of NA grows from the mechanical properties (compressive strength, splitting and flexural tensile strength and modulus of elasticity) to those related to its rheological behaviour (shrinkage and creep), and from these to the durability-related properties (water absorption, carbonation and chloride penetration).

Notwithstanding the variability of factors introduced by each researcher in the experimental procedures, it was still possible to validate this methodology to estimate the properties of the concrete with recycled aggregates. The major advantage of this procedure is related to the low cost and short time needed to obtain the results to estimate the long-term properties of hardened concrete. The generalisation of this methodology could, in the future, allow construction developers to decide cheaply and quickly on the use of RA in the construction of new concrete structures.

A practical example of this application, once it has been fully developed, would be the chance for a building owner, who is going to rebuild after demolishing an existing structure, to decide how to reuse the waste resulting from the demolition in the new structure, and provide the structural designer with the necessary data on the expected evolution of the properties of the RAC to be used in the calculations. With these data, the designer can adjust the reinforced concrete project, allowing a considerable saving in the whole process and a contribution to the effective sustainability of the sector.

References

Angulo S (1998) Production of concrete with recycled aggregates. Diploma Thesis in Civil Engineering, State University of Londrina, Londrina, Brazil (in Portuguese)

Azzouz L, Kenai S, Debied F (2002) Mechanical properties and durability of concrete made with coarse and fine recycled aggregates. International symposium on sustainable concrete construction, University of Dundee, Scotland, pp 383–392

Barra M, Vazquez E (1998) Properties of concretes with recycled aggregates: influence of properties of the aggregates and their interpretation. International symposium on the use of recycled concrete aggregate, University of Dundee, Scotland, pp 19–30

Buttler A (2003) Concrete with coarse recycled concrete aggregates—influence of the age of recycling on the properties of the recycled aggregates and concrete (in Portuguese). Masters Dissertation, University of São Paulo, Brazil

Buyle-Bodin F, Hadjieva-Zaharieva R (2002) Influence of industrially produced recycled aggregates on flow properties of concrete. Mater Struct 35(8):504–509

Carrijo P (2005) Analysis of the influence of the density of coarse aggregates from construction and demolition waste on the mechanical performance of concrete (in Portuguese). Master Dissertation, University of São Paulo, São Paulo, Brazil

Cervantes V, Roesler J, Bordelon A (2007) Fracture and drying shrinkage properties of concrete containing recycled concrete aggregate. Technical note CEAT, University of Illinois, USA

Chen H, Yen T, Chen K (2003) Use of building rubbles as recycled aggregates. Cem Concr Res 33(1):125–132

Corinaldesi V (2010) Mechanical and elastic behaviour of concretes made of recycled-concrete. Constr Build Mater 24(9):1616–1620

Corinaldesi V (2011) Structural concrete prepared with coarse recycled concrete aggregate: from investigation to design. Adv Civil Eng 2011:6 (Article ID 283984)

Correia J, de Brito J, Pereira AS (2006) Effects on concrete durability of using recycled ceramic aggregates. Mater Struct 39(2):151–158

De Brito J (2007) Methodology to estimate the properties of concrete with recycled aggregates (in Portuguese). Patent PT No. 103756, Lisbon Portugal

De Brito J (2005) Recycled aggregates and its influence on concrete properties (in Portuguese). Public Lecture within the Full Professorship in Civil Engineering Pre-Admission Examination, Instituto Superior Técnico, Technical University of Lisbon, Lisbon, Portugal

De Brito J, Pereira AS, Correia J (2005) Mechanical behaviour of non-structural concrete made with recycled ceramic aggregates. Cement Concr Compos 27(4):429–433

Dhir RK, Limbachiya MC, Leelawat T (1999) Suitability of recycled concrete aggregate for use in BS 5328 designated mixes. Struct Build 134(3):257–274

Etxeberria M, Vazquez E, Marí A (2007) Recycled concrete aggregate as a structural material. Mater Struct 40(5):529–541

Evangelista L (2007) Concrete made with fine recycled concrete aggregates (in Portuguese). Masters Dissertation, Instituto Superior Técnico, Technical University of Lisbon, Lisbon, Portugal

Evangelista L, de Brito J (2007) Mechanical behaviour of concrete made with fine recycled concrete aggregates. Cement Concr Compos 29(5):397–401

Evangelista L, de Brito J (2010) Durability performance of concrete made with fine recycled concrete aggregates. Cement Concr Compos 32(1):9–14

Ferreira L (2007) Structural concrete with coarse recycled concrete aggregates—influence of the pre-saturation (in Portuguese). Masters Dissertation, Instituto Superior Técnico, Technical University of Lisbon, Lisbon, Portugal

Figueiredo F (2005) Integrated management of construction and demolition waste (in Portuguese). Report of research project POCTI/ECM/43057/2001, Faculty of Engineering of Porto University, Porto, Portugal

Fraaij AL, Pietersen HS, de Vries J (2002) Performance of concrete with recycled aggregates. International conference on sustainable concrete construction, Dundee, Scotland, pp 187–198

Gomes M (2007) Structural concrete made with coarse ceramic and recycled concrete aggregates (in Portuguese). Masters Dissertation, Instituto Superior Técnico, Technical University of Lisbon, Lisbon, Portugal

Gomes M, de Brito J (2009) Structural concrete with incorporation of coarse recycled concrete and rendered ceramics aggregates: durability performance. Mater Struct 42(5):663–675

Gomez-Soberón J (2002) Porosity of recycled concrete with substitution of recycled concrete aggregate, an experimental study. Cem Concr Res 32(8):1301–1311

Kakizaki M, Harada M, Soshiroda T, Kubota S, Ikeda T, Kasai Y (1988) Strength and elastic modulus of recycled aggregate concrete. 2nd international RILEM symposium on demolition and reuse of concrete and masonry, Tokyo, Japan, pp 565–574

Katz A (2003) Properties of concrete made with recycled aggregate from partially hydrated old concrete. Cem Concr Res 33(5):703–711

Kou S, Poon C (2008) Mechanical properties of 5-year-old concrete prepared with recycled aggregates obtained from three different sources. Mag Concr Res 61(1):57–64

Kou SC, Poon CS, Chan D (2004) Properties of steam cured recycled aggregate fly ash concrete. International RILEM conference on the use of recycled materials in buildings and structures, Barcelona, Spain, pp 590–599

Larrañaga ME (2004) Experimental study on microstructure and structural behaviour of recycled aggregate concrete. PhD Thesis, Polytechnic University of Catalonia, Barcelona, Spain

Latterza L, Machado E (2003) Concrete with coarse recycled aggregate: properties in the fresh and hardened states and application in light precast elements (in Portuguese). Eng Struct J 21:27–58

Leite M (2001) Evaluation of the mechanical properties of concrete made with recycled aggregates from construction and demolition waste (in Portuguese). PhD Thesis in Civil Engineering, Federal University of Rio Grande do Sul, Porto Alegre, Brazil

Levy S (2001) Contribution to the study of the durability of concrete made with concrete and masonry waste (in Portuguese). PhD Thesis in Civil Engineering, Polytechnic School of São Paulo, São Paulo, Brazil

Limbachiya MC, Dhir RK, Leelawat T (2000) Use of recycled concrete aggregate in high-strength concrete. Mater Struct 33(9):574–580

Matias D, de Brito J (2005) Concrete with coarse recycled concrete aggregates resorting to plasticizers—experimental study performed at IST (in Portuguese). Report ICIST DTC 3/05, Instituto Superior Técnico, Technical University of Lisbon, Lisbon, Portugal

Mendes T, Morales G, Carbonari G (2004) Study on ARC's aggregate utilization recycled of concrete. International RILEM conference on the use of recycled materials in buildings and structures, Barcelona, Spain, pp 629–635

Merlet JD, Pimienta P (1993) Mechanical and physico-chemical properties of concrete produced with coarse and fine recycled concrete aggregates. Demolition and reuse of concrete and masonry, RILEM proceedings 23, Odense, pp 343–353

Olorunsogo FT, Padayachee N (2001) Performance of recycled aggregate concrete monitored by durability indexes. Cem Concr Res 32(2):179–185

Paine KA, Collery DJ, Dhir RK (2009) Strength and deformation characteristics of concrete containing coarse recycled and manufactured aggregates. Proceedings of the 11th international conference on non-conventional materials and technologies (NOCMAT 2009), 2009, Bath, UK, pp 1–9

Ravindrarajah RS, Tam TC (1985) Properties of concrete made with crushed concrete as coarse aggregate. Mag Concr Res 37(130):29–38

Robles R (2003) Prediction of the behavior of concrete with coarse recycled aggregates—the international experimental 'state-of-the-art' summary (in Portuguese). Masters Dissertation, Instituto Superior Técnico, Technical University of Lisbon

Rocha B, Resende C (2004) Properties of concrete made with recycled aggregates (in Portuguese). Diploma Thesis, Aveiro University, Aveiro, Portugal

Roos F (1998) Verification of the dimensioning values for concrete with recycled aggregates. International symposium on the use of recycled concrete aggregate, University of Dundee, Scotland, pp 309–319

Rosa AS (2002) Use of coarse ceramic aggregates in concrete production (in Portuguese). Masters Dissertation, Instituto Superior Técnico, Technical University of Lisbon, Lisbon, Portugal

Xiao JZ, Li JB, Zhang C (2006) On relationships between the mechanical properties of recycled aggregate concrete: an overview. Mater Struct 39(6):655–664

Chapter 7
Concrete with Recycled Aggregates in International Codes

7.1 Introduction

Despite being an ancient activity, the management of waste produced in construction activities did not get much attention until the last decade. Construction and demolition waste (CDW) is not subjected to management practices as with municipal solid waste (MSW), perhaps due to the higher toxicity of the latter as compared with the former's. Recently, rapid urban expansion, stringent environmental regulations and the scarcity of land filling areas as well as the natural resources over-exploitation led to the need of using CDW as aggregate for construction purposes. CDW contain significant amounts of inert materials whose properties are being investigated and which have been recognised for use as aggregate, although significant differences exist when compared to conventional natural aggregates (NA). The use of recycled concrete waste-based aggregates in new concrete is a way of maximising the economic benefits of CDW and, even though it has been the subject of study for a long time, opinions are still not consensual (Hansen 1992). As expected, concrete made with recycled aggregates (RA) has different characteristics from those of conventional concrete, and these differences are strongly dependent on the type and quality of the aggregates used. Therefore, it is possible to obtain concrete with no structural use, concrete with strength classes considered average and high strength concrete (Vazquez et al. 2006).

The properties of RA vary widely, largely but not exclusively because of their many possible compositions. For example, concrete that includes concrete aggregates has properties distinct from those of concrete made with ceramic aggregates, and an inferior performance is in principle expected from the latter. Besides composition, the grading curve, the density and the water absorption of the aggregates, are among the properties that influence the mechanical characteristics and durability of concrete (De Brito 2005; De Brito et al. 2005; Etxeberria et al. 2007; Kou 2008; Limbachiya et al. 2000; Xiao et al. 2006).

J. de Brito and N. Saikia, *Recycled Aggregate in Concrete*,
Green Energy and Technology, DOI: 10.1007/978-1-4471-4540-0_7,

However, due to lack of specifications that ensures the maintenance of safety and in-use requirement of built elements, the use of this waste material is still in ordinary applications such as bases and sub-bases of road pavements despite the potential shown by CDW aggregates in various experimental researches performed worldwide. This chapter surveys the greatest possible number of standards and specifications relating to the use of RA in place of natural (primary) aggregates. For some of these, a direct analysis was possible but others were analysed indirectly through scientific papers, dissertations or books (Gonçalves 2007).

This chapter presents an overall review of how the production of concrete with CDW aggregates is treated in normative standards of different countries, as well as its response to the need for a sustainable development. Based on existing norms and specifications that allow the use of RA in the production of concrete, a comparison is made of the parameters involved, such as density, water absorption and contaminants ratio within the RA and maximum strength allowed for RA concrete (RAC).

7.2 Normative Documents

Specifications in use in several countries for the requirements of RA and their quality control, together with their application and conditions of use are given below. The comparison of all the parameters is also made in the following section of this chapter.

7.2.1 Brazil

7.2.1.1 Introduction

In Brazil, specification NBR 15.116 "Recycled aggregates from construction and demolition waste" (2005) allows the use of RA only in non-structural concrete, and both coarse and fine fractions are permitted in concrete production.

7.2.1.2 Classification

CDW is separated into four classes (A, B, C and D) and of these, waste of Class A can be considered as aggregate for use in concrete. The waste types that belong to Class A are the following:

- Construction waste generated in the demolition, renovation and repair of pavements and other infrastructural works including soil from earthworks;

- Construction waste generated in the demolition, renovation and repair of buildings, which includes mortar, concrete and ceramic components such as bricks, blocks, tiles, ceramic board among others;
- Waste manufacturing process and/or demolition of prefabricated concrete produced on-site.

The CDW aggregates classified as class A are divided into two subclasses according to their composition:

- Recycled concrete aggregates (RCA): RA composed of more than 90 % concrete and natural resources;
- Mixed aggregates (MA): RA composed of less than 90 % concrete and natural resources.

Again, to distinguish RCA and MA, CDW is separated into four groups according to their composition:

- Group 1: CDW consisting of more than 50 % by volume of hardened cement paste;
- Group 2: CDW consisting of more than 50 % by volume of rock particles;
- Group 3: CDW containing red or white ceramics with polished surfaces not more than 50 % by volume;
- Group 4: CDW containing non-mineral organic materials such as wood, plastic, bitumen and charred materials, and contaminants like glass, ceramic tiles or gypsum.

If the proportion of the first two groups is more than 90 %, then the CDW can be defined as RCA; otherwise it is an MA.

7.2.1.3 Requirements for CDW Aggregates

Aggregates from class A can be used to replace part of the NA in non-structural concrete if they comply with the properties presented in Table 7.1.

7.2.1.4 Final Assessment

Contrary to what is observed in other specifications, this standard does not impose requirements on the minimum density of the aggregates. However, the density can be derived from the water absorption capacity of the CDW aggregate. The requirements in terms of water absorption are quite demanding, and limit the use of masonry rubble, even in the non-structural solutions envisaged in this specification.

Given that this type of aggregate consists essentially of fragments of rock or concrete, the standard assumes a very conservative position with respect to use in concrete, since it only allows the production of concrete for non-structural

Table 7.1 Requirements for recycled aggregates (NBR 15.116)

Properties		Class A CDW aggregates			
		RCA		MA	
		Coarse	Fine	Coarse	Fine
Contents of fragments based on cement and rocks (%)		≥90	–	<90	–
Water absorption (%)		≤7	≤12	≤12	≤17
Contaminants—maximum levels in relation to the mass of CDW aggregate (%)	Chlorides	1			
	Sulphates	1			
	Non-mineral materials[a]	2			
	Clay lumps	2			
	Total	3			
Contents of materials lower than 75 μm (%)		≤7	≤15	≤10	≤20

[a] Plastics, bitumen, charred materials, glass and ceramic tiles

applications. These aggregates, being mainly composed of concrete and stone fragments, could be considered for using in preparation of high-quality concrete classes. This statement is supported by the norm itself when CDW contains more than 90 % concrete and stone fragments and by comparison with other normative references, such as RILEM or BS 8500-2, or even experimental results reported by Hansen (1992) or Angulo (2005). However, it is anticipated that, compared to conventional concrete, some modification like a higher cement content and/or effective aggregate to cement ratio or even the use of superplasticizers may be necessary.

Regarding MA, the standard does not provide a very clear picture since, to belong to this class, all it takes is that less than 90 % of its constituents are derived from fragments of concrete or rock. This makes the limits short when the proportion of ceramic material is large, since 12 % water absorption for coarse ceramic aggregates is easily surpassed, and the same holds for the threshold for fine aggregates. Thus, it is assumed that even within the MA, the standard tries to reduce the presence of ceramic aggregates because of the negative effects they have on the properties of concrete.

Concerning the percentage of particles smaller than 75 μ, this standard is very permissive, allowing percentages of 7 and 10 % for coarse RCA and MA, respectively. These values may influence properties of concrete such as compressive strength or modulus of elasticity, since, as mentioned, micro-fines are likely to contain contaminants such as clay. Thus, a lower value for this requirement might allow this standard to be more permissible towards the use of RA in structural concrete. The standard does not restrict the use of fine aggregates in concrete production, because it has been limited to non-structural purposes, where the possible damage resulting from the application of the fine fraction is not as significant.

Angulo (2005) explains the fact that the standard does not control the reactivity of the aggregates to alkali–silica reactions in this regard that these aggregates have a porosity that allows them to accommodate the product expanding into the cementitious matrix without causing cracking or affecting the mechanical properties of concrete.

Finally, it is observed that this standard specifies a maximum content of water-soluble chlorides, unlike, for example, in Germany or Hong Kong, which suggest limitations on acid soluble chlorides. Such complacency is explained by the low level of demand to which the concrete is subjected, where the chlorides concentration and its effects are not so burdensome. However, it should be noted that the suggested value of 1 % appears very high, considering that it is a percentage by weight of aggregates.

7.2.2 Germany

In Germany, there are two standards, presented below, the first of which refers to the requirements of CDW aggregate, while the second proposes rules for the implementation of these aggregates in concrete.

7.2.2.1 DIN 4226-100 "Aggregates for Mortar and Concrete—Recycled Aggregates"

This standard specifies requirements for aggregates with particle density higher than 1,500 kg/m^3 for use in mortar and concrete. It also specifies the system of production control and conformity assessment. The particle size must follow the requirements specified in DIN 4226-1 "Aggregates for concrete and mortar—normal density and high density aggregates".

Classification of Aggregate

In the German norm DIN 4226-100, CDW aggregates are classed into four types: (1) waste concrete, (2) CDW, (3) Masonry waste, (4) mixed material.

The classification of CDW aggregates is based on their composition and origin. The first type is mainly composed of waste concrete. Type 2 is slightly more embracing than type 1, i.e. a scenario of demolition of a concrete building, where concrete is the major constituent but a portion of masonry may also be included. Type 3 includes those cases in which masonry is the major constituent, but some concrete may also be included. Type 4 is somewhat restrictive and can more easily represent a scenario of a recycling centre where one can see a bit of all materials, but where concrete and masonry are dominant.

Table 7.2 Composition of recycled aggregates (DIN 4226-100)

Constituents	Limits of mass of CDW aggregate by aggregate type[a] (%)			
	1	2	3	4
Concrete	≥90	≥70	≤20	≥80
Clinker and solid brick	≤10	≤30	≥80	
Stone masonry			≤5	
Other materials	≤2	≤3	≤5	≤20
Bituminous	≤1	≤1	≤1	
Contaminants	≤1	≤1	≤1	≤1

[a] Classified in "Classification of Aggregate"

Composition

To determine the composition, individual components must be separated manually and weighed. Each representative sample should weigh no less than 1,000 g for particle sizes up to 8 mm and no less than 2,500 g for those exceeding 8 mm. The composition of RA shall meet the requirements shown in Table 7.2.

Density of the Particles and Water Absorption

Table 7.3 shows the bulk density and water absorption capacity of various types of CDW aggregates as classified in "Classification of Aggregate". The range of densities and water absorptions is consistent with the composition of the various types of aggregates defined by the standard. Types 1 and 2, predominantly consisting of concrete, have densities closer to the density of concrete, while, as the percentage of masonry increases, the minimum allowable density decreases, as the water absorption increases (for type 3 aggregates). This is because concrete has lower water absorption than masonry, so that the greater the percentage of concrete, the lower the water absorption of the respective aggregates. For type 4 aggregates, the required minimum density is much lower due to the possible existence of up to 20 % of materials with densities much lower than those of the

Table 7.3 Requirements for CDW aggregates—bulk density and water absorption (DIN 4226-100)

Properties	Type of aggregate			
	1	2	3	4
Minimum bulk density (kg/m^3)	2,000		1,800	1,500
Maximum permissible deviation of bulk density from average value declared by the manufacturer (kg/m^3)	±150			n/a
Maximum water absorption capacity after 10 min (%)	10	15	20	n/a

n/a Not allowed

ceramics or concrete, such as mortars, aerated concrete or lightweight concrete. The requirements in terms of density concern the oven-dry density.

If, for type 1, the relationship between the density and minimum water absorption is very close to what is expected for concrete aggregates, for type 2 the maximum allowed water absorption is quite high. It is common to have exclusively ceramic aggregates with water absorption values of this magnitude, but with aggregates mainly of concrete, where the brick portion is lower than 30 %, this criterion does not seem conditioning. The same line of reasoning can be assumed for type 3 aggregates, which, despite the low permissible minimum density, is associated with very high water absorption. No limit is set in terms of water absorption for type 4 aggregates, because, despite having more materials such as mortars, plasters or aerated concrete, they should not be used for demanding purposes, where this property can be quite important.

De Brito (2005) and Hansen (1992) reported a series of experimental studies where densities of around 2,300 kg/m^3 for concrete aggregates and 2,000 kg/m^3 for masonry aggregates were obtained. Thus, it is expected that for the minimum values of allowable densities and based on the compositions imposed, there are great difficulties in complying with the code, also because there is still a tolerance of 150 kg/m^3, and one can thus obtain densities of 1,850 kg/m^3 (type 1 and 2 aggregates) and 1,650 kg/m^3 (type 3 aggregate). Even for type 4 aggregates, in which quite lighter materials than concrete can be found, the density of 1,500 kg/m^3 is not very restrictive, since in the most unfavourable case 80 % of the constituents is masonry.

Chlorides and Sulphates

The maximum allowed levels of chlorides and sulphates in RA are shown in Table 7.4. Regarding chlorides content in RA, the German standard is quite demanding. By considering the use of RA in structural concrete, the German standard specification of chloride content is quite reasonable because chloride attack is a major problem in concrete, since chloride-induced corrosion causes severe damage of the reinforced steel and therefore reduces the strength capacity of the element. However, for less demanding uses of reinforced concrete, chlorides are not a threat. In the case of type 3 and 4 aggregates, which cannot be used in structural concrete, the specified limit is quite strict in comparison to the Brazilian specification.

Table 7.4 Maximum chloride and sulphate contents in RA (DIN 4226-100)

Properties	Type of aggregate			
	1	2	3	4
Maximum amount of acid soluble chlorides (w/w) (%)	0.04			0.15
Maximum amount of acid soluble sulphates (w/w) (%)	0.8			–

The specified maximum allowed limit for sulphate in type 1–3 aggregates is lower than that specified in the RILEM specification and British Standard BS 8500-2, which usually allow a maximum of 1 % like the Brazilian standard, NBR 15.116.

7.2.2.2 Specification for Concrete with Recycled Aggregates

In August 1998, the German Committee for Structural Concrete (DAfStb) published the specification, "Concrete with recycled aggregate" as a supplementary document to DIN 4226-100, which was based on a 4-year national research project, "Baustoffkreislauf im Massivbau (BIM)". This standard is divided into two parts: the first part describes the use of RA in structural concrete and the second part describes the minimum quality of the RA as well as its control.

Part I: Application in Concrete

This part focuses on the production of RAC and its scope. This specification allowed using RA for concrete production with maximum 28-day cube strength of 37 MPa, which is generally represented by the C30/37 class concrete. To use RAC in interior elements, where both coarse and fine RA are allowed, the maximum allowed replacement amount of NA by RA was kept at 35 % for the C20/25 concrete class. However, the RA collected from known origin can only be used for the preparation of concrete intended for use in external environments and RA should be classified as insensitive to alkali–silica reaction. Table 7.5 shows the maximum levels of aggregates present in the mix which depends on the strength of the concrete class and the type of proposed application.

Analysing the applicability of the aggregates by the German specification, the standard seems to have a conservative approach. The RA can only be used to

Table 7.5 Maximum allowed ratio of RA in various types of concrete (DafStB 1998)

Application		Coarse and fine aggregates in concrete, D > 2 mm		Fine aggregate in concrete, D ≤ 2 mm	
		[a] (%)	[b] (%)	[a] (%)	[b] (%)
Interior elements	≤C20/25	35	50	7	20
	C30/37	25	40	7	20
Concrete exposed to the normal environment		20	30	0	0
Impermeable concrete					
Concrete with high resistance to	Freeze–thaw				
	Chemical attack				

[a] With respect to total volume of aggregates
[b] With respect to total volume of coarse or fine aggregates content

prepare concrete up to the C30/37 class and can be considered only for interior elements. The maximum amount of replacement of NA by RA for this class is 25 % by volume which can be considered to be low, taking into account that concrete is preponderant in RA. Compared to this standard, RILEM does not impose any restriction for concrete up to the C50/60 strength class on the replacement amount of NA by RA with quality similar to type 1 aggregates, as described in the German standard. The Dutch Norms, on the other hand, allow using high quality RA as complete replacement of NA in concrete up to the C40/50 strength class. The specification does not allow the use of type 3 and 4 aggregates in structural concrete, which can be considered as a very conservative approach for the use of CDW in construction applications. The RA whose properties are more similar to NA can only be used for very limited purposes and the remaining CDW will have to be disposed of in recycling plants or will need to find other applications.

The specification also recommends the replacement of lower percentages of NA by type 1 and 2 aggregates depending on the severity of natural environmental conditions or physical attacks. Table 7.6 shows the allowed maximum replacement ratio of coarse NA by RA for concrete depending on the external conditions. The maximum allowed replacement ratio also decreases as the exposure conditions become more severe. Such reduction is easily justified, since the RA-based concrete exhibits inferior performance in various durability properties than the equivalent conventional concrete. Thus, decreasing of replacement ratios of NA by RA is inevitable to decrease the severity of external attacks such as freeze–thaw cycles, chlorides and sulphate attacks, and carbonation.

The use of the fine fraction of RA was allowed in the original proposal. Such concrete could be used as interior elements with a replacement of up to 7 % of total aggregate volume by fine RA. However, the use of fine fraction of RA was banned in the preparation of concrete in the DAfStb-2004 document (Solyman 2005). The use of RA was also prohibited in the manufacture of lightweight concrete or pre-stressed concrete (Mueller 2007). The lack of substantial information about the behaviour of RAC in such conditions, e.g. pre-stressed condition and the differences in some properties between the NAC and RAC like shrinkage and creep possibly led to this decision. This limitation is also consistent with the

Table 7.6 Maximum replacement ratio of coarse NA by RA for concrete exposed to various environmental conditions (Mueller 2007)

Environmental class[a]	Exposed condition	Replacement amount (%)	
		Type 1	Type 2
XC1	Carbonation	≤45	≤35
X0	No attack		
XC1 and XC4	Carbonation		
XF1 and XF3	Freeze–thaw without salts	≤35	≤25
XA 1	Chemical attack	≤25	≤25

[a] According to EN 206

standard assumptions for other applications, i.e. the higher the required applications, the smaller the contribution of the RA should be.

In the production of concrete with RA, the specification should be considered to act in the same manner as in the manufacture of concrete with NA. However, because of the higher water absorption capacity of RA as compared to NA, it is necessary to determine the effective water/cement ratio of RAC. For this calculation, it is advised to measure the water absorption at 10 min as it is an expedient method and reaches values of around 90 % of the 24 h water absorption capacity of the RA (Grubl and Ruhl 1998). The inner moisture content of the RA and their water absorption change the workability from the moment of the mix up till the concrete is cast. The standard prevents the addition of water to counteract this phenomenon. This requirement is easily understandable, since this principle is also applied to conventional concrete, due to the detrimental consequences to the properties of concrete which implies namely a decrease of strength and durability. The specification therefore states that the loss of workability of RAC due to the higher water absorption capacity of RA is to be compensated by the addition of superplasticizers. The German specification agrees that structural elements with RAC can be scaled with the same characteristic values used for design of similar elements of NAC. In structures where deflections are to be considered, the properties of the hardened concrete with RA must be evaluated by testing (Grubl and Ruhl 1998).

Although it might seem a conservative specification regarding the use of recycled material, this philosophy allows adjustments of the design parameters in structural analysis, and it is considered that concrete containing RA is equivalent to that containing only NA. This concept is also important for the market image, taking into account that it does not require any additional note, in the delivery form, except that it cannot be used in pre-stressed concrete. In order to make sure that there are no impurities in the RA, each delivery is to be checked visually. To detect differences in the RA the dry density and water absorption at 10 min and at 24 h must be determined weekly (Grubl and Ruhl 1998).

Part II: Aggregates from Waste Concrete

According to Part II of the specification, only aggregates produced from concrete can be used as aggregates in new concrete. In this part of the German guide, considerations are made concerning the properties that the coarse and fine aggregates for concrete shall present, as well as the requirements in terms of processes and technologies used to obtain the material. The specification recommends the use of jaw crushers, followed by rotary crushers, so as to obtain particles with improved particle size distribution and shape. The separation process of wet particles to achieve improved results in the elimination of impurities is also suggested (Grubl and Ruhl 1998).

Grubl and Ruhl (1998) list a number of requirements that the German specification suggests to be considered before using RA in concrete:

- RA must comply with the requirements specified in DIN 4226, which only allow the use of RA of types 1 and 2 specified in DIN 4226-100, in the understanding that their requirements in terms of composition, density and water absorption are inherent to this classification;
- No fixed minimum compressive strength is imposed on the concrete source of the RA, rather being specified that the size of the particles which give rise to the RA must be greater than 32 mm; this measure is to prevent contaminants and fines;
- The sensitivity of the RA to alkali–silica reaction should be assessed prior to use, as well as the possibility of freeze–thaw attack, in which the resistance of the aggregates to these attacks must be checked;
- Tests on RAC shall be in accordance with DIN 1045; some special measures and precautions in the system of quality control are required, such as visual inspections of the supplied RA, more frequent tests for particle size and water absorption and regularly determining the moisture content of RA.

7.2.3 Hong Kong: WBTC No. 12/2002

Specification Works Bureau Technical Circular No. 12/2002, "Specification facilitating the use of recycled aggregates", (2002) has two different parts and intends to outline the use of recycled aggregates in concrete production and the construction of bases and sub-bases of road pavements. Only the first part is discussed in this section.

7.2.3.1 Application

Two alternatives are suggested in this specification for the use of RA in concrete. The complete replacement of NA by RA is prescribed for concrete intended to be used in less demanding structures, such as benches, flowerbeds or cyclopean concrete. On the other hand, only 20 % replacement of NA by RA is allowed in structural concrete with a 28-day compressive strength in the range of 25–35 MPa. The scopes of the RA according to this specification are summarised in Table 7.7.

Table 7.7 Maximum allowed compressive strength of concrete and maximum replacement ratio of coarse NA by RA (WBTC No. 12/2002)

Maximum replacement ratio of coarse NA (%)	Type of concrete	Maximum compressive strength (MPa)
20	Structural concrete	35
100	Less demanding structures	20

In this specification, the mix composition, proposed for concrete prepared by replacing 100 % coarse NA by RA is: Portland cement: 100 kg; fine aggregate: 180 kg; RA with 20 mm size: 180 kg; RA with 10 mm size: 90 kg.

7.2.3.2 Workability

In this specification, only pre-soaked RA is allowed to be used in concrete preparation. This measure seeks to avoid the change of actual water to cement ratio due to the higher water absorption capacity of RA than that of NA, which will affect the cement hydration reaction. It is also recommended that the slump of RAC must be greater or equal to 75 mm at the time of casting.

7.2.3.3 Compressive Strength

This specification also suggested some recommendations for testing the compressive strength of RAC. Accordingly, to determine the strength of concrete with 100 % replacement of coarse NA by RA, at least four concrete cubes should be prepared in the same day to determine the 7- and 28-day compressive strengths; out of these, two should be used to determine the 7-day compressive strength and the other two the 28-day compressive strength. These cubes should have a minimum compressive strength of 14 and 20 MPa at 7 and 28 days respectively. This imposition is unique and deserves to be studied in detail. However, a major drawback is the limitation on the use of large amounts of RAs in the preparation of concrete with CS above 20 MPa.

7.2.3.4 Requirements for RA

The coarse RA should be produced by crushing old concrete and comply with the requirements presented in Table 7.8. This specification prohibits the use of fine RA

Table 7.8 Requirements of coarse RA (WBTC No. 12/2002)

Properties	Limits
Dry density of aggregate (minimum)	2,000 kg/m^3
Water absorption capacity (maximum) (%)	10
Wood or similar material less dense than water (maximum) (%)	0.5
Other contaminants (maximum) (%)	1
Particles finer than 63 μm (maximum) (%)	4
Particles finer than 4 mm (maximum) (%)	5
Sulphate content (maximum) (%)	1
Chloride content (maximum) (%)	0.05

in concrete production. This specification also considers dry density of RA instead of saturated surface dry density that is the choice in some other specifications.

With respect to composition, only RA and NA are allowed to prepare concrete. Therefore, the maximum allowed value of the water absorption of RA should not be a deciding factor in the verification of the criteria of the aggregates, as in the German standard.

The specification requires the measurement of the acid soluble chlorides content (indicated as maximum) instead of the water-soluble chloride content, which results in a more stringent restriction. This limitation does not make the same sense in both cases of application of the RA (Table 7.7). Perhaps there might be some differences of the chloride contents in RA, depending on the stringency of the application of the RAC. In the case of concrete with 100 % RA in less demanding structures, measuring the water-soluble chloride content or allowing an increase of the maximum value of acid soluble chlorides as in German standard should be accepted.

7.2.4 Japan: B.C.S.J

The Building Contractors Society of Japan issued a "Proposed standard for the use of recycled aggregates and RAC" (1977). This document does not limit the use of masonry material and it also establishes a high lower limit for oven-dry density of the RA, which clearly shows that ceramic aggregates cannot be used.

7.2.4.1 Properties of Recycled Aggregates

The aggregates produced from concrete must have normal density and good quality. The separation of the aggregates, depending on the quality of the parent concrete, should be done in different ways. Finishing materials, reinforcement and contaminants must be removed in the best possible way. The aggregates must meet the specifications of Table 7.9 and should not contain contaminants that adversely affect the properties of the new concrete or the reinforcement.

The RA must also comply with the strict maximum allowed levels of impurities, presented in Table 7.10. The determination of the ratio of impurities should be carried out by visual identification and separated into two fractions with

Table 7.9 Quality requirements for recycled aggregates (BCSJ 1977, cited by Hansen 1992)

Properties	Coarse	Fine
Density (kg/m^3)	≥2,200	≥2,000
Water absorption (%)	≤7	≤13
Loss of materials in the washing test (%)	≤1	≤8
Solid volume (%)	≥53	–

Table 7.10 Allowable impurity ratios in RA (BCSJ 1977, by Hansen 1992)

Aggregate type	Coarse (kg/m^3)	Fine (kg/m^3)
Mortars and coatings, with density below 1,950 kg/m^3	10	10
Bituminous materials, plastics, paints, textiles, wood and similar particles retained on 1.2 mm sieve, with density below 1,200 kg/m^3	2	2

densities of 1,200 and 1,950 kg/m^3 by using a heavy liquid separation technique. The method prescribed for determination of levels of impurities, especially heavy liquid separation method, is very swift and gives precise impurities level in RA.

The Japanese standard has no requirements on the source of the RA, e.g. RA may contain a mixture of aggregates generated from demolished concrete and low density ceramic aggregates without any specified ratios. However, one cannot say that this regulation allows the use in concrete of RA with large amounts of masonry or ceramic particles, because there are some restrictions imposed on the limit density and water absorption capacity values of RAs or on the maximum allowed amount of impurities in RA. The standard says nothing about the maximum levels of chloride and sulphate contents in RA, which can be considered as a serious drawback of this standard. This specification presents a conservative approach, not only because it puts a limit on the use of masonry and/or ceramic aggregates in RA but also because it restricts the use of RA in concrete with design strength over 18 MPa.

7.2.4.2 Concrete with Recycled Aggregates

According to this standard, three classes of concrete with limit compressive strengths depending on the fine and coarse RA contents can be prepared. Tables 7.11 and 7.12 show the classifications and applications of concrete made with coarse and fine RA.

The standard also prescribes the use of air-entraining agent or air/water reducing agents in RAC so that the air content can be kept between 3 and 6 % in any case. This measure serves to protect concrete against the effects of freeze–thaw cycles, as water is thus able to expand without damage to the concrete. According to this proposal, the slump of concrete must not exceed 21 cm, the water to cement

Table 7.11 Types of concrete with RA and maximum compressive strength values (BCSJ 1977, cited by Hansen 1992)

Concrete type	Type of aggregate		Maximum compressive strength of concrete (MPa)	
	Coarse	Fine	f_{ck} (design)	f_{cj} (average)
I	RA	NA	18	30
II	RA	RA + NA	15	27
III	RA	RA	12	24

Table 7.12 Types of applications of RAC (BCSJ 1977, cited by Hansen 1992)

Concrete type	Major application
I	Foundations of general buildings, floors of commercial buildings and heavy foundations
II	Foundations for blocks of precast concrete, lightweight non-residential buildings, machine foundations
III	Foundations of wood buildings, fixing fence and gates, foundations of simple machines

ratio must be lower than 0.7 and the cement content should be equal to or greater than 250 kg/m^3. The amount of water used should be kept as low as possible and similarly, the ratio of fine to coarse aggregate ratio should also be kept at a value with which concrete with a minimum workability can be produced.

This specification makes no distinction between the nature and composition of the aggregates to be used in the manufacture of concrete, as it does not prevent the use of the recycled fine fraction. The maximum design strength of RAC is determined by the type of RA used. Thus, RAC made with coarse RA is limited to 18 MPa and RAC made with both coarse and fine RA is limited to 12 MPa. In both cases, concrete is only recommended for non-structural concrete, which is somehow contradictory since the aggregate's requirements are quite demanding. This conservative stance in terms of the use of RA may be due to the experimental state-of-the-art when the document was put forward, in 1977, and a concern with the intense seismic activity registered in Japan.

The coarse fraction in three types of concrete is entirely composed of RA while fine RA can be used in different amounts. However, the application of RA is not conditioning since the standard only considers concrete with very low compressive strength. By comparison with other specifications, this one is conservative in terms of some properties, which can prevent wide-scale application of RA. For example, the RILEM (1994) specification on the use of RA in concrete allows using coarse RA with density of 2,000 kg/m^3 in several high-quality concrete of strength classes up to C50/60, while the Japanese standard only allows relatively high-quality RA (with density of 2,200 kg/m^3) to prepare very low-strength concrete.

There is presently a set of Japanese norms that regulate the use of RA in concrete in a more up-to-date manner than the one mentioned in the previous paragraph: JIS A 5021 (2005) RA for concrete—Class H, JIS A 5022 (2006) RA for concrete—Class M, JIS A 5023 (2007) RA for concrete—Class L.

These documents establish a set of requisites imposed on aggregates, more demanding for class-H concrete and less so for class-L concrete. These demands are concomitant with the conditions of application of the concrete the aggregates are incorporated into. Therefore JIS A 5023 regulates the application of RA in non-structural concrete, JIS A 5022 scope is concrete unaffected by freeze–thaw cycles and JIS A 5021 concerns the use of RA in high-performance concrete.

7.2.5 The RILEM Specification

The RILEM TC 121-DRG recommendation (1994) is a specification that deals with recycled coarse aggregates with minimum size of 4 mm for concrete. As the properties of fine RA fraction are vastly different from those of natural sand, there is no specification for the fine fraction in the RILEM specification. It classes the recycled coarse aggregates and indicates the scope of application for concrete containing these RA classes in terms of acceptable environmental exposure classes and concrete strength classes. Recycled coarse aggregates are classed as follows:

- Type I—aggregates which are implicitly understood to originate primarily from masonry rubble;
- Type II—aggregates which are implicitly understood to originate primarily from concrete rubble;
- Type III—aggregates which are implicitly understood to consist of a blend of RA and NA; the composition shall have at least 80 % NA and up to 10 % type I aggregate.

7.2.5.1 Recommendations for RA

The RILEM specification prohibits the use of RA with more than 15 % of harmful materials which can retard the concrete hardening process. Table 7.13 shows some other requirements for different types of RA. Unlike most of the other specifications, this one imposes several parameters.

The specified minimum density of type III aggregates that also contains a minimum of 80 % NA is 2,400 kg/m^3. If it is presumed that the density of NA is

Table 7.13 Requirements of coarse RA for concrete production (RILEM 1994)

Properties	Type I	Type II	Type III
Density at saturated surface dry (SSD) condition (kg/m^3)	≥1,500	≥2,000	≥2,400
Water absorption (%)	≤20	≤10	≤3
Amount of material with SSD density <2,200 kg/m^3 (%)	–	≤10	≤10
Amount of material with SSD density <1,800 kg/m^3 (%)	≤10	≤1	≤1
Amount of material with SSD density <1,000 kg/m^3 (%)	≤1	≤0.5	≤0.5
Contaminants (bitumen, glass, metals) (%)	≤5	≤1	≤1
Metals (%)	≤1	≤1	≤1
Organic materials (%)	≤1	≤0.5	≤0.5
Filler, <63 μm (%)	≤3	≤2	≤2
Sand (%)	≤5	≤5	≤5
Water soluble sulphite (%)	≤1	≤1	≤1
Natural aggregates (%)	–	–	≥80
Type I aggregate (%)	–	–	≤10

approximately 2,700 kg/m^3, then the density of the remaining amount will have to be only 1,200 kg/m^3. On the other hand, if the recommendation regarding the content of type I aggregates, i.e. 10 %, is considered, then the value for the remaining 10 % RA will be only 900 kg/m^3. If these hypotheses were accepted, there would be disastrous consequences on the properties of concrete. However, the other recommendations proposed in the standard preclude reaching such exceptional conditions.

7.2.5.2 Scope

The coarse RA, as specified in the previous section, can be used in reinforced concrete if the limiting conditions presented in Table 7.14 are satisfied. These recommendations are the most developed of those studied within this survey, since they allow a wide range of applications, as well as the use of a large quantity of recycled coarse aggregates. The replacement of natural coarse aggregates by recycled coarse aggregates can be up to 100 % and the maximum grade strength is C50/50 for type II aggregates, for example. This permissive approach is balanced by the corrective coefficients adopted in the design of the structural elements, to take into account the influence of the aggregates' density on the strength and deformation characteristics of the concrete.

Like for conventional concrete, the document prescribed to use the Eurocode relationships with correction factors. The ones for tensile strength and creep seem to be adequate, since these parameters are usually not much affected by the replacement of NA by RA. Those for modulus of elasticity and shrinkage for different types of RA can also be considered as a good approach due to the differences in quality of these aggregates as this change greatly affects those properties of the resulting concrete. For example, if the basis of comparison is considered concrete with the same compressive strength, the shrinkage of the concrete containing poor quality RA will be higher than that exhibited by concrete containing good quality RA due to the higher cement content that is necessary to achieve the same compressive strength as the concrete with high quality RA. However, this RILEM recommendation ignored the creep behaviour of concrete, as the substitution of NA by RA and the quality of RA can also affect this behaviour of concrete.

If one considers the type of RA use that is prescribed, this specification is far more permissive than the others. The standard also allows preparing a wide range of concrete classes depending on the quality of the RA. It allows the complete replacement of NA by RA, which will also help to increase the used amount of RAs in concrete. However, the situation is slightly tricky for using type III aggregates as it should consist of at least 80 % NA, which rules out the use of large amounts of RA in concrete production.

The fine fraction is also allowed, provided it fulfils the specifications for natural sand, which greatly limits its use. The amount of fine aggregates generated in concrete crushing is typically less than that for coarse aggregates. Thus, it will be

Table 7.14 Conditions for use of RA in concrete (RILEM 1994)

Parameters		Type I	Type II	Type III
Particles greater than ≥4 mm		100 %		
Use of recycled sand		Not recommended		
Maximum authorised strength class		C16/20[a]	C50/60	Unlimited
Exposure classes[b]/additional checks	2a and 4a	ARS testing; not permitted to exposure class 4a	ARS testing	ARS testing
	2b and 4b	No use is allowed	ARS and freeze–thaw resistance tests	ARS and freeze–thaw resistance tests
	3	No use if allowed	ARS and freeze–thaw resistance tests; test with de-icing salts	ARS and freeze–thaw resistance tests; test with de-icing salts
Dimension		Eurocode 2 (EC2) with corrections		
Coefficient for f_{ctm}		1	1	1
Coefficient for modulus of elasticity		0.65	0.80	1
Creep		1	1	1
Shrinkage coefficient		2	1.5	1

[a] In the case of RA with SSD density over 2,000 kg/m^3 , the strength class can be higher than C30/37

[b] Environmental class, ENV 206 1994

[c] Resistance against alkali–silica reaction

hard to find sand-sized RA in accordance with this recommendation in a typical recycling plant where both coarse and fine aggregates with various kinds of compositions are processed.

Additional checks are specified depending on the exposure class proposed for concrete, as seen in Table 7.14. The exposure classes that the standard considers are not in agreement with the present normative documents, since these are prescribed according to ENV 206, instead of the current EN 206-1. The standard considers the use of aggregates in the first exposure classes (dry), 2 (humid environment), 3 (humid environment with ice and de-icing products) and 4 (a marine environment), blocking the use of class 5 (chemically aggressive environment) whatever the type of aggregates. Class 1 is considered suitable for the use of any type of RA. It is also seen that type I aggregates can also be used in humid environments as long as they are not exposed to frost action. Since type II and III aggregates may extend their scope to humid and marine environments, whether or not subjected to frost action, tests to check the resistance of aggregates to freeze–thaw action must be planned for.

7.2.6 United Kingdom

In the United Kingdom, the specification for using RA in concrete production is standard BS 8500-2:2002 "Concrete—Complementary British Standard to BS EN 206-1—Part 2: Specification for constituent materials and concrete". The standard notes that the requirements may not be appropriate under conditions of exposure outside the UK, particularly in hot climates.

The British standard, assuming the same philosophy reflected in other laws of that country, establishes requirements for RA and limits their use, but leaves open the possibility that changes occur to the recommendations, as long as the prescriber takes responsibility for the results that they can bring.

7.2.6.1 Recycled Aggregates

The British Standard specifies requirements for coarse RA only, excluding the use of fine RA for concrete production. These requirements are presented in Table 7.15.

Despite the fact that some requirements are specified for coarse RA, they are insufficient to make an adequate specification. As the potential composition of the aggregates can vary widely, further requirements must be evaluated in each case taking into account the specific composition of the aggregates. In particular, the draft specification for RA should include: maximum content of acid soluble sulphate; a method for determining the content of chlorides; a classification on the alkali-aggregate reactivity; a method for determining the alkali content; any restriction on the use of concrete.

Table 7.15 Requirements for recycled coarse aggregates (BS 8500-2 2002)

Constituents	Maximum amount (% by weight)	
	RCA[a]	RA
Masonry	5	100
Fines	5	3
Lightweight materials[b]	0.5	1.0
Other contaminants[b]	1	1
Acid soluble sulphite (SO_3)	1	–[d]

[a] When the material to be used is obtained by crushing hardened concrete of known composition not contaminated through use, the only requirements are related to the particle size and the maximum content of fines

[b] Glass, plastic, wood, etc.

[c] The measures for RCA must be applied to mixtures of natural coarse aggregates with the constituents listed

[d] The appropriate limit needs to be determined in each case

The standard establishes two types of aggregates, the RCA and the RA, distinguished by composition. While the RCA are aggregates from concrete, in which a small percentage of masonry is allowed, the RA are a less demanding class and may be composed entirely of masonry. The composition of these two types of aggregates is tested in accordance with Annex B of the document and the content of acid soluble sulphates in accordance with BS EN 1744-1 "Tests for chemical properties of aggregates—Part 1: Chemical analysis".

The requirements include mainly the aspects related to the aggregates' composition, lacking requirements concerning density, water absorption or maximum content of particles whose size enables them to act as filler (usually smaller than 63 or 75 μm) aspects that are present in most of the normative documents studied here.

The absence of a requisite imposing a minimum density or absorption of water enables very lightweight materials to be used, such as concrete waste with much paste adhered to the NA. This makes it difficult for the new concrete to reach high strength and modulus of elasticity. The lowest quality of the aggregates implies that larger binder and lower water contents are used, resulting in differences in shrinkage or creep.

No requirement for maximum content of super-fines is provided in this standard and therefore there is the possibility that harmful substances to the mechanical characteristics of the concrete, such as clay particles, arise within the aggregates in higher amounts than appropriate.

The risk of damage due to alkali–silica reactions must be minimised in accordance with a set of conditions. For a prescribed concrete composition or when measures are prescribed to prevent alkali–silica reactions, these conditions must be fulfilled. In these exceptions, the prescriber is responsible for ensuring that the concrete is not subject to any damage due to alkali–silica reactions. The contribution of alkali in RCA must meet the following requirements: Na_2O content must be not more than 0.20 kg/100 kg of RCA; when the composition of the RCA is known (e.g. prefabricated parts surplus, fresh concrete returning to the ready-mixed concrete plant allowed to harden and then be crushed), the alkali content must be calculated for the original concrete.

The risk of damage to alkali–silica reactions can be considered minimum if at least the following conditions are fulfilled:

- Aggregates other than the RCA are not classified as highly reactive;
- The amounts of alkali in blast furnace slag and fly ash do not exceed 1 and 5 % Na_2O equivalent respectively;
- The calculated total content of alkali does not exceed: 3.5 kg/m^3 Na_2O equivalent where the declared average content of alkali in cement or CEM I component of a combination does not exceed 0.75 %; 3.0 kg/m^3 Na_2O equivalent where the stated average content of alkali in CEM I component of a combination is 0.76 % or higher;
- The cement content is not higher than the values in Tables 7.16 and 7.17 depending on the contribution of alkali in the fresh concrete from other components other than cement. The contribution of alkali can be calculated based on the declared mix proportions and the determined alkali content.

7.2.6.2 Scope

The specification authorises the use of RCA in structural concrete, when limited to the strength and exposure classes specified in Table 7.18. The application of RCA in concrete exposed to marine environments, icing salts, harsh environments of freeze–thawing and aggressive soils is generally not allowed. For C20/25 and C40/50 concrete classes, the use of RCA should not exceed 20 % unless the prescriber says so. However, for lower strength classes, RAs can be used to replace all the

Table 7.16 Maximum binders' content, Na_2O equivalent up to 0.20 kg/m^3 of concrete (BS 8500-2 2002)

Binders	Maximum cement content (kg/m^3)											
	Guaranteed amount ≤0.60 % Na_2O equivalent	Declared content of alkali in binders[a], (%)[b]										
		0.50	0.55	0.60	0.65	0.70	0.75	0.80	0.85	0.90	0.95	1.00
CEM I	550	550	550	550	540	500	465	375	355	315	315	300
CEM I plus ≥40 % blast furnace slag	–	550	550	550	550	550	550	550	550	550	525	505
CEM I plus ≥30 % FA		550	550	550	550	550	550	500	470	445	420	400

Table 7.17 Maximum binders' content, Na_2O equivalent up to 0.60 kg/m^3 of concrete (BS 8500-2 2002)

Binders	Maximum cement content (kg/m^3)											
	Guaranteed amount ≤0.60 % Na_2O equivalent	Declared content of alkali in binders[a], (%)[b]										
		0.50	0.55	0.60	0.65	0.70	0.75	0.80	0.85	0.90	0.95	1.00
CEM I	550	550	525	485	445	415	385	300	280	265	255	240
CEM I plus ≥40 % blast furnace slag	–	550	550	550	550	550	550	550	470	445	420	400
CEM I plus ≥30 % FA	–	550	550	550	550	550	470	400	375	355	335	320

[a] Where the alkali content in binders is determined in accordance with a method specified in standard, classification is made based on the declared contents of alkali in CEM I
[b] A linear interpolation for declared medium alkali contents is permitted between 0.50 and 0.75 %, and between 0.76 and 1.00 %
[c] Proportions expressed as a fraction of the cement paste or combinations as appropriate

Table 7.18 Conditions for application of RCA in concrete (BS 8500-2 2002)

Aggregate type	Concrete class	Environmental classes[b]
RCA[a]	C40/50	X0, XC1, XC2, XC3, XC4, XF1, DC-1

[a] The material obtained from crushed hardened concrete and the known composition has not been contaminated through use, can be used in any kind of resistance
[b] Aggregates can be used in other classes of exposure provided that it has been shown that the resulting concrete is suitable for the desired environment (e.g., freeze–thaw resistance, resistance to sulphates among others)

NA. RA is limited to use in concrete with a maximum strength class of C16/20 and only in the mildest exposure conditions.

The aggregates are also limited to a maximum of 0.075 % drying shrinkage.

This specification reflects a little on what is found in some other normative documents, i.e. using aggregates of good quality, in particular of concrete, is authorised in concrete with high compressive strength. However, it can be envisaged that the properties of concrete will not be much affected due to the 20 % replacement of NA by RA. Total replacement of NA by RA is only allowed in low demand concrete, since the expected reduction in strength of the concrete due to an increase in the percentage of aggregates does not have to be compensated by an increase in cement content or other measures that would not only make the concrete more expensive but also change other properties.

RA-type aggregates are precluded in structural concrete, which is justified by the absence of a complete specification, which permits large variations in the composition and properties of aggregates. In this respect, the British standard, although with shortcomings in terms of requirements, is comparable to the RILEM recommendation or the German standard, which specify an aggregate composition mostly composed of masonry whose application is authorised only in not very demanding solutions.

7.2.7 The Netherlands

The Dutch centre CUR (Commissie voor van Uitvoering Research) has developed specifications for the use of RA. In 1984, a specification was released for the use of aggregates from crushing concrete (CUR 1984). In 1986, CUR developed a specification for the use of RA generated from masonry (CUR 1986). Subsequently, another specification was developed for the use of crushed mortar as aggregate (CUR 1994). Hansen (1992) and Schulz and Hendricks (1992) give a summary of the most interesting Dutch rules, providing most of the standard data from which this analysis is performed.

7.2.7.1 Composition of RA

The standard defines concrete waste as that from hydraulic cement concrete and a density greater than 2,100 kg/m^3. According to Loo (1998), cited by Leite (2001), the Dutch standard for concrete aggregates has been adapted in June 1997, including only RA with a specific gravity greater than 2,000 kg/m^3 in the production of concrete.

The Dutch specification proposes two types of RA depending on the amount of concrete and masonry: concrete aggregate and masonry aggregate (Table 7.19). To classify the aggregate, the standard advises analysing the particles retained on the sieve of 8 mm, considering this a representative sample of the whole waste. Various materials present in waste such as conventional concrete, lightweight concrete, bricks, cement and lime mortar, ceramic, natural stone, glass, wood, plastics, bituminous material and other materials should be identified through visual analysis and weighing.

Table 7.19 Composition of RA according to Dutch standards

Material	Concrete aggregate (%)	Masonry aggregate
Concrete	>95	n.d. (%)
Masonry	<5	>65
Lightweight concrete		<20
Ceramic products		
Natural stone		
Cellular concrete		<10
Mortars		<25

The RA, according to the definition in Dutch specification, is composed primarily of waste concrete with high density (2,100 kg/m^3); thus these aggregates become comparable with the best kinds of aggregates defined in other specifications. Besides the tough requirements in terms of density of materials for concrete, a low content of other materials that affect the general characteristics of the aggregate and thereby of concrete is also required. It is expected therefore that, given the requirements assigned to the various groups, these may have an equally demanding use. However, these requirements are likely to hinder the production of large quantities of concrete with this type of aggregate, since no other type of aggregate with less stringent requirements are used in less demanding solutions.

7.2.7.2 Particle Size

The Dutch specification recommends different size requirements for concrete aggregate and masonry aggregate, and these are presented in Tables 7.20 and 7.21.

Tables 7.20 and 7.21 show that, for the same sized fractions, the specification allows the presence of higher ratios of fines in RA generated from concrete than those in masonry aggregates. This is possibly due to the following reasons: the masonry aggregates, due to increased variability in the composition, are prone to the presence of materials which affect the properties of concrete; it is difficult to identify/remove the smaller particles.

Table 7.20 Size requirements for concrete RA in the Dutch specifications (Hansen 1992)

Size fraction (mm)	Percentage retained by the sieve (mm)							
	>31.5	>16	>8	>4	>2	>1	>0.25	>0.063
0–4	–	–	–	0–10	25–31	50–62	80–87	99–100
4–16	–	0–5	55–57	85–100	95–100	–	–	99–100
4–31.5	0–5	32–44	70–75	90–100	–	–	–	99–100

Table 7.21 Size requirements for masonry RA in the Dutch specifications (Hansen 1992)

Size fraction (mm)	Percentage retained by the sieve (mm)							
	>31.5	>22.4	>16	>8	>4	>2	>1	>0.25
0–4	–		–	0	2–10	–	15–50	80–100
4–16	–	0	0–5	35–70	85–100	95–100	96–100	–
4–32	0–2	5–30	25–55	60–85	90–100	–	96–100	–
4–8	–	–	0	0–10	80–100	98–100	–	–
8–16	–	0	0–10	80–100	98–100	–	–	–
16–32	0–10	–	80–100	98–100	–	–	–	–

7.2.7.3 Contaminants

Table 7.22 shows the requirements that the Dutch standards advocate for aggregates generated from concrete and masonry. All the parameters are expressed as maximum allowed ratios of aggregates. The chloride and sulphate contents represent the maximum water-soluble amounts.

By comparing the requirements shown in Table 7.22 with those of other standards studied, the Dutch specifications are very much in agreement. The maximum allowed chloride contents decrease depending on the requirement level of application of concrete and are lower for larger sized fractions. The difference becomes most prominent with respect to the amount of non-mineral particles, whose maximum allowed level is 0.1 % for the fraction of coarse aggregate concrete, when the British standard, the RILEM recommendation, the specification of Hong Kong and the German standard limit is 1 % for aggregate types comparable in terms of composition and application range.

7.2.7.4 Application

The Dutch norm foresees the use of RA in plain concrete, reinforced or prestressed concrete. It states that as long as the replacement ratio of NA by RA, either coarse or fine, is below 20 % the resulting concrete must be considered as if it were conventional. Therefore the measures prescribed are applied to concrete where this replacement ratio is higher than 20 % and may even reach 100 %, which is not common in standards, not even the most recent ones, as seen so far in this study. The Dutch standard is thus a permissible one that, because of the good quality demanded from the aggregates, allows high-strength concrete and therefore encourages the use of large quantities of RA produced and purchased in Holland.

Table 7.22 Requirements for recycled aggregates according to Dutch specifications

Requirements		Concrete aggregate (%)		Masonry aggregate (%)	
		<4 mm	>4 mm	<4 mm	>4 mm
Dispersible fines		–		4	2
Chlorides	Plain concrete	–		1	1
	Reinforced concrete	0.1	0.05	0.1	0.05
	Pre-stressed concrete	0.015	0.007	0.015	0.007
Sulphates		1		1	
Non-mineral components[a]		0.5	0.1	1	
Calcium carbonate		25	10	–	
Lamellar particles		30		30	
Light particles		0.1		–	
Materials other than rocks[b]		1		1	

[a] Wood, vegetable, paper, cloths, etc.
[b] Bitumen, rubber, metal, glass

In order to counterbalance the lower modulus of elasticity and greater creep of concrete with RA, the standard prescribes that, for use in structural concrete in which the design of the structural elements is made in terms of the maximum allowed deformation, the height or thickness of the elements should be increased by 10 %, to ensure their stiffness. This is convenient for pre-design of the structure since, given the greater deformability of the elements with concrete aggregates, the criterion of maximum allowed deflection becomes more important than when conventional concrete is applied.

According to the Dutch standard, it is required to specify clearly the source of aggregate including the name of supplier, type of aggregate, characteristic size, requirements of the standard that the aggregate must meet and origin. The freeze–thaw resistance of the aggregates must be such that, when it is subjected to freeze–thaw cycles, the weight loss should not exceed 3 %.

To compensate for differences in water absorption between NA and RA, the document states that the RA should be subjected to the following methods: adding pre-saturated aggregates to the mix and thus decreasing in the mixing water ratio that is absorbed by the aggregates; or compensation of mixing water, i.e. adding more water to the mix, corresponding to the amount of water absorbed by the aggregates.

Regarding the use of fine fractions of RA in concrete, CUR states that the use of fine RA (<4 mm) and conventional coarse aggregates does not harm concrete. The coefficients proposed in the Dutch standard to establish the relationship between various parameters of NAC and RAC are presented in Table 7.23.

There are also some inconsistencies in the coefficients proposed in the various parameters. For example, the ratios between the modulus of elasticity of the RAC and that of the NAC are above one for classes C25/30 and C40/50 (for recycled concrete aggregate) and C15/20 and C20/25 (for masonry aggregate). In fact the modulus of elasticity of conventional concrete is generally higher than that of RAC, and this is quite significant for RAC with 100 % RA as assumed by the Dutch document.

The shrinkage correction coefficients presented in the Dutch specification are close to the one recommended in the RILEM specification, i.e. 1.50 for concrete

Table 7.23 Proposed coefficients in Dutch standard to establish relationship between various parameters of NAC and RAC (CUR 1994, cited by Lima 1999)

Parameters	Concrete aggregate		Masonry aggregate
	C15/20 and C20/25	C25/30 and C40/50	C15/20 and C20/25
Coefficient of thermal expansion ($\times 10^{-6}$/°C)	13.6	13.1	14.9
Coefficient for f_{ctm}	1	1	1
Coefficient for modulus of elasticity	0.95	1.55	1.25
Coefficient for shrinkage	1.35	1.55	1.25
Creep coefficient	1.45	1.25	1.15

produced with type II aggregates for C50/60 class of concrete. The Dutch standard also proposes lower coefficients for shrinkages for concrete classes with lower design strength, which is due to the lower binder content required to achieve the target strength. However, the coefficient for shrinkage of concrete produced with masonry aggregates is lower than that for concrete aggregates for the same strength classes, which is not be expected, since the masonry aggregates usually have lower stiffness than the RCA.

For creep, the standard assumes that the concrete produced with RA would exhibit poorer performance than that of NAC, which is the correct trend. However, the above correction coefficient for the RAC with RCA is lower than that for RAC with masonry; again this is not to be expected, due to worst characteristics of the masonry aggregates.

7.2.8 Portugal: E 471: 2006

In May 2006, LNEC (National Laboratory of Civil Engineering) prepared pre-norm, prE 469 "Guide for the use of recycled coarse aggregates in hydraulic binder concrete". Later, in September 2006, the final version of the document was published as E 471 with the same name, which classes the coarse RA covered by NP EN 12620 "Aggregates for concrete" and establishes the minimum requirements they must meet in order to be used in the manufacture of hydraulic binder concrete.

7.2.8.1 Recycled Aggregates

Requirements and their applications are not shown for fine RA since, generally, they have a high percentage of particles with dimensions less than 0.063 mm and greater water absorption capacity, making it difficult to control the workability and impairing the mechanical strength of concrete containing that fraction.

The LNEC specification establishes three classes of aggregates from CDW:

- ARB1 and ARB2—consisting mainly of concrete, whether mixed with unbound aggregates or not;
- ARC—its main constituents are concrete, unbound aggregates and masonry units, with no requirements as to the relative percentages of each.

Table 7.24 shows the required composition of each aggregate type, while their requirements are presented in Table 7.25. In contrast with several other specifications, LNEC specification does not mention chlorides content.

Analysing the requirements listed, it is noted that the document only provides for the use of aggregates with quite good characteristics, which is justified by the requirements on minimum density value or maximum water absorption value. Recycled aggregates classified as ARB1 and ARB2 have the same requirements

Table 7.24 Composition of recycled aggregates (LNEC E 471 2006)

Constituents	ARB1	ARB2	ARC
Concrete (%)	≥90	≥70	≥90
Unbound aggregate (%)			
Masonry (%)	≤10	≤30	
Bituminous materials (%)	≤5	≤5	≤10
Lightweight materials[a] (%)	≤1	≤1	≤1
Other materials[b] (%)	≤0.2	≤0.5	≤1

[a] Material with density less than 1,000 kg/m^3 ; when the lightweight particles are mineral constituents not harmful to concrete or its surface finishing, the limit may reach 3 %
[b] Glass, clay, plastics, rubber, metals and putrescible materials

Table 7.25 Requirements of the recycled aggregates (LNEC E 471 2006)

Properties	ARB1 ARB2	ARC
Density (kg/m^3)	≥2,200 (minimum)	≥2,000 (minimum)
Water absorption (%)	≤7 (maximum)	≤7 (maximum)
Fines content (%)	≤4 (maximum)	≤3 (maximum)
Acid soluble sulphates (%)	≤0.8 (maximum)	≤0.8 (maximum)

for density and water absorption, so that, given the differences in composition between the two, it will be expected that the ARB1 class will meet these requirements easily. However, the ARB2 are expected to have difficulty in meeting these requirements only if either the concrete and masonry have in its constituents a significant quantity of mortar since, in a standard situation, in which the density of the concrete aggregate is around 2,300 kg/m^3 and masonry 2,000 kg/m^3, any of these types of RA will comply with the set requirements.

Taking into account its proposed use, the requirements for the ARC aggregate class are quite demanding. However, there will be difficulties in complying with the density limit of 2,000 kg/m^3 if the composition of RA consists exclusively of masonry. It is further noted that the ARC and ARB have the same maximum level of water absorption. For the ARC, it is more difficult to meet the requirement since for the minimum density prescribed for this class of aggregates, it is expected that the aggregates have water absorption of around 10 %. Therefore, when different limits are set for density, it would be expected that different maximum allowed water absorption capacities were set between the various classes of RA as density and water absorption capacity are interrelated. A similar situation is found in the German standard requirements where the minimum densities of type 1 and 2 aggregates are the same but the maximum allowed water absorption capacities are different. The maximum content set for the acid soluble sulphates is consistent with that required by the various standards studied.

The composition and the requirements of the three classes of aggregates proposed by the specification show that there is a great similarity between the ARB1 and ARB2 classes. This is because, despite the differences which may exist with regard to composition, the same requirements are demanded for the properties

Table 7.26 Allowed strength and environmental exposure classes (LNEC E 471 2006)

Aggregate class	Concrete class	Replacement ratio (%)	Environmental classes
ARB1	C 40/50	25	X0, XC1, XC2, XC3, XC4, XS1, XA1[a]
ARB2	C 35/45	20	

[a] In foundations

listed in Table 7.25. Therefore, concrete produced with ARB1 and ARB2 with the same specification as in Table 7.25 will have similar properties although their composition may have some differences.

7.2.8.2 Scope

The aggregates of classes ARB1 (maximum 25 %) and ARB2 (maximum 20 %) can be used in the production of structural concrete but they can be used as full replacement of NA as long in unreinforced, filling or screed concrete in non-aggressive environments. The ARC class aggregates are only allowed for use in non-structural concrete (i.e. filling or screed concrete) that can only be applied in non-aggressive environments. Table 7.26 shows some requirements proposed to use the ARB1 and ARB2 classes along with their application in terms of environmental classes. It is also necessary to mention that the complete percentage of substitution is possible for preparation of concrete that intends to apply in non-aggressive conditions such as plain concrete, fill or compliance.

This specification allows using RCA in structural concrete with fairly high-strength classes and can cover a large share of the construction market, since the produced concrete can be used in a wide variety of environmental exposure classes. However, the allowed ratios of replacement of NA by RA for concrete production are relatively low, which will hinder the use of a high volume of RA in concrete.

The specification states that the inclusion of aggregates from CDW mostly composed of concrete has shown best results and lower dispersion than masonry aggregates, which is why the latter is confined to the use of the concrete without great demands. The replacement ratio of NA is limited by the document in order to avoid great variations of the modulus of elasticity, creep, shrinkage and durability. However, it is allowed to use higher percentages of RA, provided that specific studies are undertaken to assess its influence on the properties relevant to the application considered.

Despite having the same requirements, the differences regarding the concrete classes and maximum replacement ratios proposed for the ARB1 and ARB2 use can be justified by considering the greater percentage of masonry allowed in ARB2. The LNEC specification also presents a conservative stance, since it only considers the use of aggregates in the production of concrete structures under conditions that do not significantly influence the characteristics of the concrete, when compared to conventional concrete.

From a business standpoint, it follows a path that can enhance the use of RA in the concrete construction market, because the use of such waste aggregate is not usual in Portugal due to the lesser advantages they present in this country, when compared to other European countries. However, from an academic standpoint, the specification can be considered limited by comparison with some other specifications, e.g. the RILEM recommendation. Hence, a more detailed specification is necessary that allows dealing with differences in several properties (modulus of elasticity, shrinkage, creep), arising from the use of higher replacement ratios of NA by RA.

7.2.9 Belgium

In Belgium, a recommendation for the use of RA in concrete was compiled by a working group in 1990. Vincke and Rousseau (1994) presented a series of information on the document, which was cited by Roos (2002) and Vazquez et al. (2006). This recommendation is divided into three parts: the first contains the requirements of the RA, the second regulates the scope and the third is related to the calculations of the coefficients and their characteristics.

This standard defines two types of RA, GBSB-I and GBSB-II, only for the coarse fraction. It excludes the fine fraction as in several other specifications. The GBSB-I class is similar to the type I aggregates described in the RILEM recommendation, composed mainly of masonry, while the GBSB-II class is comparable to the type II aggregates in the RILEM specification, i.e. it mostly consists of concrete. The corresponding requirements for these types of aggregates are presented in Table 7.27.

Table 7.27 Requirements of recycled aggregates in the Belgium standard (Roos 2002)

Properties	GBSB-I	GBSB-II
Density of the dry aggregate (kg/m^3)	>1,600 kg/m^3	>2,100 kg/m^3
24 h water absorption (%)	<18	<9
Aggregate with density <2,100 kg/m^3 (%)	–	<10
Aggregate with density <1,600 kg/m^3 (%)	<10	<1
Aggregate with density <1,000 kg/m^3 (%)	<1	<0.5
Total mineral components (%)	>95	
Non-mineral components (%)	<1	
Organic materials (%)	<0.5	
Fine (<80 μm) content (%)	<5	<3
Chlorides content (%)	<0.06	
Sulphates content (%)	<1.0	

Table 7.28 Conditions for applications of concrete with RA (Vincke and Rousseau 1994)

Type of aggregate	GBSB-I	GBSB-II
Concrete strength class	C 16/20	C 30/37
Environmental class	Inside building; dry environment Non-aggressive elements in soil and/or no contact with freeze–thaw conditions	Inside building; dry environment Non-aggressive elements in soil and/or no contact with water

Table 7.29 Corrective factors to determine the characteristics values (Vincke and Rousseau 1994)

Properties	GBSB-I	GBSB-II
f_{ctm}	1	1
Modulus of elasticity	0.65	0.8
Creep	1.5	1.25
Shrinkage	2	1.5

The conditions of application of concrete produced with these RA are presented in Table 7.28 and they are similar to those presented by the RILEM recommendation.

The corrective factors adopted for calculating the characteristics values according to the Belgian recommendation are similar to those of the RILEM recommendation and are presented in Table 7.29. The same discussion presented in Sect. 7.2.5.2 can also be applied to these values.

7.2.9.1 PTV 406: Technical Prescriptions: "Recycled Aggregates from Construction and Demolition Waste"

In Belgium, the standard PTV 406 Technical Prescription "Recycled aggregates from construction and demolition waste" (2003) regulates the composition of the RA that may be used in concrete production, establishing three RA classes: RCA, recycled masonry aggregates (RMA) and a mixture of concrete and masonry aggregates.

According to this specification, the aggregates can be derived from construction, rehabilitation, demolition of buildings or even leftovers of NA. The composition of the various classes of RA is presented in Table 7.30. The specification notes that the requirements must be met individually.

The RA should not contain materials whose nature, shape, size and quantity can be harmful to their use, such as clay particles, vegetable matter or refractory bricks.

Particle size and its tolerances are provided by the manufacturer according to the application. The limits assigned to water absorption, density and maximum

Table 7.30 Composition of recycled aggregate generated from CDW (PTV 406 2003)

Constituents	Concrete aggregates	Mixed (concrete + masonry) aggregates		Masonry aggregates	
Concrete and natural materials (%)	>90	>40		<40	
Masonry (%)	<10	>10		>60	
Other types of mineral materials[a] (%)	<5	–	<10	–	<10
Bituminous materials (%)	<5	<5		<5	
Non-mineral materials[b] (%)	<0.5	<1	<1		
Organic materials (%)	<0.5	<0.5	<0.5		

[a] Mosaic, shingles, tiles, ceramics, cellular concrete, lightweight concrete among others
[b] Gypsum plaster, gypsum, rubber, plastics, glass, metals

level of chlorides are not specified by the standard, being included in the technical specifications.

Requirements for resistance to freeze–thaw conditions, amounts of acid soluble sulphates and sulphur and volume stability according to the classes established by the standard may be imposed by the manufacturer.

7.2.10 Switzerland: OT70085

The Swiss documents published in 2006, Objective Technique OT 70085 "Instruction technique. Utilisation de matériaux de construction minéraux secondaires dans la construction d'abris", creates a wide range of applications for RA, with different approaches depending on user demands. This application is regulated together with the standard SIA 162/4, 1994 "Béton de recyclage". The document establishes requirements to be met by RA as well as their application conditions.

7.2.10.1 Recycled Aggregates and Application Conditions

The standard differentiates between two types of concrete with RA, depending on the degree of demand of their applications:

- Classified concrete: this type of concrete is made with RCA; the RA can be used as complete replacement of NA; the specification of the concrete is carried out similar to concrete with NA; quality requirements must comply with the standard, "SIA 162: Ouvrages en béton";
- Unclassified concrete: this type of concrete can use RA from recycled concrete and/or masonry; RA can be used to replace 100 % of NA; the application is limited to plain concrete and the concrete specification must contain the information about the aggregate used and cement content applied; concrete aggregates

Table 7.31 Requirements of recycled aggregates for various classes of concrete (Roos 2002)

Requirements	Classified concrete	Unclassified concrete
Contaminant (wood, plastic, plaster) contents	≤1 % (volume) or ≤0.3 % (mass) without metals	≤2 % (volume) or ≤0.5 % (mass) without metal
Wood content	0 % when the concrete is used outdoors	See previous line
Mixed material content	≤3.0 %	No limitation
Bituminous material content	No bituminous materials	≤7.0 %
Sulphates content (%)	≤1.0	≤1.0
Chlorides content in un-reinforced concrete (%)	≤0.12	≤0.12
Chlorides content in reinforced concrete	≤0.03 %	Application not allowed

are intended for concrete with cement content higher than 150 kg/m^3, while masonry aggregate may be used in screed concrete and poor cement content concrete.

According to the standard, the application of RA in pre-stressed concrete must be coupled with additional tests. The requirements for the RA used in both types of concrete are shown in Table 7.31 and application conditions in Table 7.32.

It is stated herein that, as long as economic profitability is assured, secondary materials should be used before any other type of material, and a cost 5–10 % higher for the secondary materials should still be considered as economically profitable.

Design involving concrete with RA should be performed according to standard SIA 162/4. The standard states that, in general, concrete with RA can be applied in every building, with varying levels of demand within the same site.

In this specification, three scenarios are defined for use of concrete aggregates:

- Scenario A: the incorporation ratios of RA shall be defined by the supplier in order to meet the requirements of performance and workability of the concrete;
- Scenario B: the maximum incorporation ratios are set according to the performance requirements of the concrete; the given limits ensure that concrete made

Table 7.32 Conditions for application of recycled aggregates (OT 70085 2006)

Application of concrete			Application of recycled aggregates		
			Scenario A	Scenario B	Scenario C
Classified concrete	Outdoor element	≥C25/30	20 % coarse; 20 % fine	20 % coarse	–
	Indoor element	C30/37		25 % coarse; 20 % fine	100 % coarse
		C20/30		35 % coarse; 20 % fine	
	Minor component	C15/20		100 %	
Unclassified concrete	Cement content 150–230 kg/m^3		100 % concrete aggregates		
	Cement content <150 kg/m^3		100 % mixed aggregates		

Table 7.33 Differences in the properties of concrete with RA and NA (OT 70085 2006)

Composition of aggregates	0–4 mm:	100 % natural	0–4 mm	100 % recycled
	4–32 mm:	100 % recycled	4–32 mm:	100 % recycled
Compressive strength (%)	−10		−30	
Modulus of elasticity (%)	−20		−30	
Shrinkage (%)	+10		+100	
Creep	+30 to 40		+100	

with RA has characteristics similar to those of conventional concrete, both in the fresh and hardened states; this approach ensures that no changes are needed in the design of structures;

- Scenario C: the use of this type of concrete must be, for the time being, limited to solutions not so demanding; the design and implementation of the structure should take into account the differences between the concrete with recycled and conventional aggregates, both in the fresh and hardened states; preliminary tests are necessary according to standard SIA 162; when the concrete is produced with 100 % of RA, the standard requires that it must take into account the specific characteristics of the concrete produced, in particular the differences in terms of modulus of elasticity, shrinkage, creep and carbonation (shown in Table 7.33), and testing should be undertaken in accordance with the specifications of SIA 162, SIA162/1 and SIA 162/4.

The value for shrinkage when 100 % of the coarse NA are replaced by coarse RA is below expectations, namely taking into account the other specifications analysed. A value of 50 % would be more fitting. Given the differences in workability between concrete with RA and conventional concrete, it is recommended that a maximum size of 16 mm is imposed on the RA for concrete elements whose minimum dimension is below 30 cm.

7.2.10.2 Final Assessment

The Swiss standard has a set of solutions that allow a high production of concrete with RA. First, it is allows the production of concrete strength classes corresponding to the current market needs of the construction, capable of being used in most buildings. Second, different application scenarios are created: replacement of low amount of NA by RA without effecting any change in the design of structural elements and also replacement of high amounts of NA by AR for which the design of the structure must be corrected taking into account the differences in the properties of concrete caused by the presence of those amounts of RA. By using the Swiss standard, it is also possible to produce concrete with RA that can be used for less demanding purposes.

7.2.11 Other Specifications

In addition to the various specifications studied in this chapter, the references to documents from other countries not sufficient to yield a relevant analysis will only be listed, providing information about the features that stand out most significantly in each document. This information is also useful in that it provides more data for analysis.

7.2.11.1 Denmark

The Danish Concrete Association published a guide for use of RA concrete in 1990, "Recommendation for the use of RA for concrete in passive environmental class". In a further development, an amendment was made to this document in 1995, the "Addition to Danish Concrete Association No-34 for the use of recycled aggregates for concrete in passive environmental class". Roos (2002) gives a description of the contents of the standard, which are summarised in Tables 7.34 and 7.35.

The standard classified RA into three groups according to their properties:

- GP1 tested are aggregates with density greater than 2,200 kg/m^3, comprising concrete aggregate from recycling plants with a 0–32 mm particle size; if they come from other sources, where the quality control of the aggregates is not done efficiently, only the 4–32 mm fraction is authorised;
- GP1 untested are identical to the previous type of aggregates, with the difference that these aggregates from other sources may have a 0–32 mm particle size;
- GP2 are aggregates with density greater than 1,800 kg/m^3.

7.2.11.2 Japan

In 2005, a new standard was adopted for class RA for use in high-strength concrete, prepared by the Japanese organisation BCSJ. This document is called JIS A 5021, "Recycled aggregate for concrete—H class". The standard establishes requirements for aggregates to be used according to JIS A 5308, "Ready-mixed concrete", and these are presented in Tables 7.36 and 7.37. This is the extension of a previous specification for RA which was discussed in Sect. 7.2.4.

Table 7.34 Requirements for recycled aggregates (Roos 2002)

Properties	GP1 tested	GP1 not tested	GP2
Composition	≥95 % concrete	≥95 % concrete	≥95 % concrete and/or masonry
Dry density	≥2,200 kg/m^3	≥2,200 kg/m^3	≥1,800 kg/m^3

Table 7.35 Requirements for concrete with recycled aggregates (Roos 2002)

Type of concrete/exposure condition	GP1 tested	GP1 not tested	GP2
Concrete with recycled aggregate >4 mm	Up to 100 %		
Concrete with recycled aggregate <4 mm	Up to 20 %		
Allowed maximum strength class	40 MPa		20 MPa
Exposure conditions	Low to moderate aggressive conditions		Low aggressive conditions

Table 7.36 Requirements for recycled aggregates to be used in high-strength concrete (JIS A 5021 2005, cited by Mueller 2007)

Properties	Coarse recycle aggregates	Fine recycle aggregates
Dry density (kg/m^3)	$\geq$2,500	$\geq$2,500
Water absorption (%)	$\leq$3.0	$\leq$3.5
Abrasion (%)	$\leq$35	n.d.
Materials <75 μm (%)	$\leq$1.0	$\leq$7.0
Solid volume to evaluate the particle shape (%)	$\geq$55	$\geq$53
Chlorides content (%)	$\leq$0.04	

Table 7.37 Acceptable maximum contaminants content in recycled aggregates to be used in high-strength concrete (JIS A 5021 2005, cited by Mueller 2007)

Category	Contaminants	Limits (% by weight)
A	Roof tiles, bricks, ceramics, concrete, asphalt	2.0
B	Glass	0.5
C	Gypsum plaster	0.1
D	Other inorganic materials	0.5
E	Plastics	0.5
F	Wood, papers, bituminous materials	0.1
Total		3.0

7.2.11.3 Russia

In 1984, the former Soviet Union introduced a specification, developed by a scientific research institute, for the use of RA in plain and reinforced concrete.

Regarding the scope of the standard, it specifies that the replacement ratio of NA by RA can reach 100 %, if the concrete is used in foundations or reinforced concrete with strength below 15 MPa. If the replacement ratio is not more than 50 %, it can be applied in concrete structures with strength over 20 MPa. The use of RA in pre-stressed concrete is not allowed by the document, due to the high shrinkage and creep values induced by the use of RA (Roos 2002).

7.2.11.4 Norway

In Norway, a specification was created for the use of RA, resulting of the research project RESIBA and completed in 2002. The project's objective was to formulate a recommendation for the use of RA, specifying requirements for their properties, depending on the application area. Table 7.38 shows the requirements concerning the composition of RA, stipulated by this recommendation.

7.2.11.5 Spain

Vazquez et al. (2006) presented a proposal, "Recommendations for the use of recycled aggregates", which specifies the requirements for RA to be used in concrete with incorporation up to 20 %. For higher percentages, it is recommended to perform specific studies and complementary experiments in accordance with the application. This document mentions only those considerations that complement the requirements included in the articles of EHE, "Instruction of Structural concrete". The RA can be used only in concrete with compressive strength not exceeding 40 MPa but not in pre-stressed concrete.

This recommendation only allows the use of coarse RA that mainly consist of waste concrete, excluding the possibility of using RA containing substances such as ceramics, asphalt and those coming from concrete structures deteriorated due to

Table 7.38 Composition of recycled aggregate for use in concrete (Karlsen et al. 2002)

Contaminants/Properties of aggregate		Type I Crushed concrete	Type 2 Materials mixture
Principal constituents (%)	Crushed concrete and/or natural aggregates	>94	–
	Crushed concrete, crushed brick	–	>90[a]
Other granular materials (%)	Crushed brick	<5	–
	Crushed bituminous materials	<1	<1
Non-mineral materials (%, by volume)	Total non-mineral[b]	<1	<2.5
	Organic material	<0.1%	<0.5
	Lightweight material	<0.1	<0.5
Density (kg/m^3)	Oven-dry density	>2,000	>1,500
	Saturated surface dry density	>2,100	>1,800
Water absorption (%)		<10	<20

[a] It is recommended that the proportion of concrete and natural aggregate is higher than 80 % for uses where there are requirements for other properties than the materials composition

[b] Wood, paper, metal, plastics, glass, rubber, others

Table 7.39 Maximum content of impurities allowed (Vazquez et al. 2006)

Impurities	Maximum allowed ratio (%)
Ceramic materials	5
Lightweight materials	1
Bituminous materials	1
Other materials[a]	1

[a] Glass, plastics, metals, among others

harsh conditions such as alkali–silica attack, sulphate attack, fire, among others and some special concrete with fibres or polymers.

Recycled Aggregates

This recommendation allows the use of recycle aggregates with size ≥4 mm and the replacement of up to 5 % of coarse NA by RA. The content of clay particles should not exceed 0.6 % in the aggregates and 0.15 % in the NA. The maximum levels of impurities allowed in the RA are presented in Table 7.39.

Regarding chlorides content, this recommendation suggests considering the total chlorides content instead of the water-soluble content as recommended in various other specifications and also in EHE. However, the maximum value of chlorides allowed must be the same as suggested in EHE. This choice is due to the fact that there may be combined chlorides that, in certain circumstances, may be reactive chlorides and attack the reinforcement. Table 7.40 shows the requirements for aggregates according to the EHE standard.

Regarding the allowed maximum water absorption, it is limited to 7 % for RA and 4.5 % for natural coarse aggregates when the replacement ratio of NA by RA is no more than 20 %. For higher replacement ratios, the combination of natural and recycled coarse aggregates must have water absorption below 5 %. To estimate the water absorption capacity, it is recommended to perform tests to determine the water absorption at 10 min, which should be less than 5.5 % for replacement rates up to 20 %. The wear resistance of the RA should be similar to that of the NA, i.e. a coefficient of Los Angeles not exceeding 40 %.

Table 7.40 Specification for aggregates in the Spanish standard (Vazquez et al. 2006)

Properties	Specified amount
Total fine (<63 μm) content (%)	≤1
Shape index	≥0.2
Water absorption (%)	≤5
Los Angeles coefficient (%)	≤40
Water soluble chlorides (%)	≤0.05
Acid soluble sulphates (%)	≤0.08
Lightweight particles (%)	≤1
Clay content (%)	≤0.25

According to this recommendation, the following information should be recorded for each batch of RA:

- The material;
- Recycling centre and carrier material;
- Impurities (wood, ceramics, bituminous materials);
- Details on their origin;
- Any other relevant information such as the cause of demolition, a possible contamination of concrete by chlorides or concrete affected by alkali–silica reactions.

Concrete with Recycled Aggregates

It is recommended to use RA in concrete whose strength is limited to 40 MPa. The following values are prescribed for the relationship between some properties of a concrete containing RA as 100 % replacement of NA and a conventional concrete containing natural aggregate only: 0.8 for the modulus of elasticity; 1.5 for shrinkage; 1.25 for creep.

However, due to the scatter in the quality of the RA, there can be a wide variability in the values of the properties mentioned, and that is why the execution of specific tests is recommended. For replacement ratios less than 20 % it is considered, for all practical purposes, that there are no significant differences between the concrete with RA and conventional concrete.

7.2.11.6 European Standard

The technical committee, CEN/TC 154, "Aggregates" has developed an amendment, currently known as EN 12620:2002/PRA1: 2006. This standard is to be a European regulation, changing the current EN 12620:2002 and its national versions.

The standard is to establish requirements for the composition of the coarse RA, besides their water absorption and density. Also, the maximum amount of chlorides and sulphates is to be controlled. The acid soluble chlorides content is measured according to prEN 1744-5e. This test, although overestimating the chlorides content in concrete, is considered to allow for an additional safety margin. The prEN 1744-5e also suggests measuring the water soluble sulphates, which are considered as potentially reactive.

This amendment also includes a clause reserved for alkali–silica reactions, establishing that all aggregates should be classed as potentially reactive unless it is specified that they are not reactive. The resistance to freeze–thaw cycles and alkali–silica reactions, volume stability, in particular the maximum drying shrinkage, will be taken into account.

Table 7.41 Categories for the constituents of coarse recycled aggregates (prEN 933-11 2006)

Constituents	Limiting content (% by mass)	Category	Constituents	Limiting content (% by mass)	Category
R_C	≥90	R_C90	R_B	≤10	R_B10
	≥70	R_C70		≤30	R_B30
	<70	R_C declared		≤50	R_B50
	No requirements	R_CNR		>50	R_B declared
$R_C + R_U$	≥90	$R_{CU}90$		No requirement	R_BNR
	≥70	$R_{CU}70$	R_A	≤1	R_A1-
	≥50	$R_{CU}50$		≤5	R_A5-
	<50	R_{CU} declared		≤10	R_B10-
	No requirements	$R_{CU}NR$	FL_{NS}	≤0,01	$FL_{NS}0.01$
$FL_S + FL_{NS}$	≤1	$FL_{total}1$		≤0,05	$FL_{NS}0.05$
	≤3	$FL_{total}3$		≤0,1	$FL_{NS}0.1$
$X + R_G$	≤0.2	$X\ R_{G0.2}$			
	≤0.5	$X\ R_{G0.5}$			
	≤1	$X\ R_{G1}$			

Constituents (according to prEN 933-11): R_C concrete, concrete products and mortars; R_U unbound aggregates, natural stones and aggregates treated with hydraulic binders; R_A bituminous materials; R_B masonry units of clay-based materials (bricks, tiles, etc.), masonry unit of calcium silicate-based materials, non-floating cellular concrete; R_G glass; FL_s floating stone materials; FL_{NS} stony materials that do not float; X deleterious materials, cohesive materials, plastics, rubber, ferrous and non-ferrous metals, putrescible materials

Table 7.41 presents the classes defined for the constituents of coarse RA. These classes are defined according to prEN 933-11, "Tests for geometrical properties of aggregates-Part 11: Classification test for the constituents of coarse recycled aggregates". This standard specifies a method in order to estimate the relative proportions of the constituent materials in these aggregates. The classification is based on separating particles from a portion of coarse RA into five classes represented by a letter: bituminous mixtures (A); masonry (B); concrete (C); unbound aggregate (U) and others (X).

Although this document may create future prospects for the use of RA, it is not official which are the requirements for recycled aggregate and their conditions of application in concrete, so it is not possible to say whether this amendment will induce the use of recycled aggregates other than in non-structural concrete.

7.3 Documents Overview

Although the standards reflect different trends and viewpoints, it is possible to identify a pattern in the parameters for the main norms. Therefore, they are summarised in three different groups: the first concerns aggregate and their classification; the second presents the most relevant requisites; finally, the third compares the conditions for using RA in concrete.

Table 7.42 Overview of recycled aggregates composition

Specifications	Classification	Composition (Maximum content %)					
		Concrete	Masonry	Organic material	Contaminants	Lightweight material	Filler
Brazil	RCA	>90	–	2	3[c]	n.a	7
	MRA	<90	–	2	3[c]	n.a	10
Germany	RCA	>90	<10	n.a	1[b]	n.a	n.a
	RCA	>70	<30	n.a	1[b]	n.a	n.a
	RMA	<20	>80	n.a	1[b]	n.a	n.a
	MRA	>80	n.a	n.a	n.a	n.a	
Hong Kong	RCA	<100	–	n.a	1	0,5	4
Japan BCSJ	MRA	n.a	12 kg/m^3	n.a			
Japan JIS A 5021	MRA	–	n.a	3[e]	n.a	1 coarse; 7 fine	
RILEM	RMA	–	<100	1	5	1	3
	RCA	<100	–	0,5	1	0,5	2
	RCA + AP[a]	<20	<10	0,5	1	0,5	2
United Kingdom	MRA	n.a	1[b]	1	n.a		
	RCA	>95	<5	1[b]	0,5	n.a	
The Netherlands	RCA	>95	<5	0,1	1[b]	0,1	n.a
	RMA	–	>65	1	1[b]	n.a	n.a
Portugal	RCA	>90	<10	0,2	1	n.a	
	RCA	>70	<30	0,5	1	n.a	
	MRA	>90	2	1	n.a		
Belgium	RCA	>90	<10	0,5	0,5[b]	n.a	n.a
	MRA	>40	>10	0,5	1[b]	n.a	n.a
	RMA	<40	>60	0,5	1[b]	n.a	n.a
Norway	RCA	>94	<5	0,1	1[d]	0,1	n.a
	MRA	>90	0,5	2,5[d]	0,1	n.a	

(continued)

Table 7.42 (continued)

Specifications	Classification	Composition (Maximum content %)					
		Concrete	Masonry	Organic material	Contaminants	Lightweight material	Filler
Switzerland	RCA	<100	–	n.a	1[d]	n.a	n.a
	MRA	<100	–	n.a	1	n.a	n.a
Denmark	RCA with testing	>95	–	n.a	n.a	n.a	n.a
	RCA without testing	>95	–	n.a	n.a	n.a	n.a
	MRA	>95	n.a	n.a	n.a	n.a	

n.a not available

[a] Primary or natural resources

[b] Asphalt is not included

[c] Organic material content included

[d] Asphalt and lightweight material are not included

[e] Bricks, ceramics, glass, plaster, plastics, wood

7.3.1 Recycled Aggregates Classification

Most specifications classify RA in terms of their composition, and each has its own designation. As a consequence, a standard nomenclature has been proposed for the purpose of comparing all the specifications: RCA; mixed concrete and masonry RA (MRA); RMA. Table 7.42 compares the different classes' composition of the normative references studied.

The norms prescribe that there may be one or two types of RCA, for which a minimum concrete requirement of 90 % is normally imposed for the most demanding type, while the second type is less strict, when there is one. Normally there are no limit values associated with RMA, which therefore may consider that there is a more or less significant percentage of concrete, as well as just state that there is masonry in its composition. When aggregates made of a mixture of concrete and masonry (MRA) are envisaged, several situations may occur: a minimum percentage of the sum of these materials (concrete and masonry), as in Portugal, Norway, Denmark and Germany; minimum values for each of them, as in Belgium; no limits imposed, as in Japan and the UK.

The analysis of the limits stipulated for harmful substances shows that organic materials are between 0.5 % and 1 %, with the exception of the Brazilian standard (2 %). The explanation for this is that only non-structural RAC is allowed. In relation to contaminants, in most cases an upper limit of 1 % is imposed but the range of materials considered in the different standards varies considerably, though glass, gypsum, plastics, timber, paper, bituminous material are normally among those included. Some standards establish individual limits for each material, such as in Germany, the UK and Belgium with respect to bituminous material. Other standards consider some materials other than contaminants, such as timber, which is included in light materials, i.e. lighter than water (density 1,000 kg/m^3). For RCA, the maximum values allowed for these contaminants are less than or equal to 0.5 %, while for MRA and RMA this value rises to 1 %. Portugal is an exception, since the LNEC specification allows a maximum of 1 % for all types of RA. Finally, even though not mentioned in every standard, it was considered relevant to include the filler in this comparison, since it can significantly impair the mechanical properties of concrete, because it tends to include materials such as clay particles. When it is mentioned, the maximum filler content is smaller the stricter the requirements of the application for which the aggregates are proposed.

7.3.2 Recycled Aggregate Requirements

Besides the composition requirements (described above), there are other specifications that can change some of the concrete properties. Four requisites were identified, of which two are fundamental to mechanical demands while the other two mostly influence the durability of concrete. Table 7.43 compares the standards

Table 7.43 Overview of recycled aggregate requirements

Specifications	Classification	Minimum density (kg/m^3)	Maximum water absorption (%)	Maximum chloride content (%)	Maximum sulphate content (%)
Brazil	RCA	n.a	7	1[a]	1[a]
	MRA	n.a	12	1[a]	1[a]
Germany	RCA	2,000	10	0.04	0.8
	RCA	2,000	15	0.04	0.8
	RMA	1,800	20	0.04	0.8
	MRA	1,500	n.a	0,15	n.a
Hong Kong	RCA	2,000	10	0.05	1
Japan BCSJ	MRA	2,200	7	n.a	n.a
Japan JIS A 5021	MRA	2,500 coarse and fine	3 coarse; 3.5 fine	0.04	n.a
Japan JIS A 5022	MRA	2,300 coarse; 2,200 fine	5 coarse; 7 fine	n.a	n.a
Japan JIS A 5023	MRA	n.a.	7 coarse; 13 fine	n.a	n.a
RILEM	RCA + AP	2,400	3	a.i.	1[a]
	RCA	2,000	10	a.i.	1[a]
	RMA	1,500	20	a.i.	1[a]
United Kingdom	RCA	n.a	n.a	n.a	1
	MRA	n.a	n.a	n.a	n.a
The Netherlands	RCA	2,000	n.a	0.05[b]	1
	RMA	2,000	n.a	0.05[b]	1
Portugal	RCA	2,200	7	a.i.	0.8
	RCA	2,200	7	a.i.	0.8
	MRA	2,000	7	a.i.	0.8
Norway	RCA	2,000	10	n.a	n.a
	MRA	1,500	20	n.a	n.a
Switzerland	RCA	n.a	n.a	0.03	1
	MRA	n.a	n.a	n.a	1
Denmark	RCA with testing	2,200	n.a	n.a	n.a
	RCA without testing	2,200	n.a	n.a	n.a
	MRA	1,800	n.a	n.a	n.a

n.a not available; *a.i.* additional information provided by the prescribing entity

[a] Water soluble

[b] For reinforced concrete, there are different values for plain and pre-stressed concrete

analysed under these four requisites. The density values presented refer to oven-dry density and the maximum content of chlorides and sulphates are those soluble in acid.

The minimum density allowed shows a pattern that is common to all the standards, in varying degrees of strictness. For RCA the minimum value demanded is between 2,000 and 2,200 kg/m^3, while for MRA and RMA it ranges from 1,500 to 1,800 kg/m^3, with a greater dependency than the RCA on the composition and level of performance target level of the future concrete.

Water absorption of the aggregates is a function of their density. For the standards that consider both requisites, this relationship is valid up to an acceptable degree, and therefore the maximum water absorption values fall between 7 and 10 % for RCA and from 10 to 20 % for MRA and RMA.

Neither the British nor the Swiss standard contemplates any limitation on either the density or the water absorption of the aggregates. In these cases it is assumed that the demands made for the aggregates' composition are enough to ensure the quality of the aggregates, bearing in mind the aim proposed.

The maximum content allowed for chlorides and sulphates is usually stated in the normative documents and, if no values are stipulated for any of these parameters, a reference is normally made to the need for the person/entity responsible to specify a value through a case-by-case analysis. The maximum authorised content of sulphates varies from 0.8 and 1.0 % of aggregates (in mass) while for chlorides the range of values is much wider, changing according to the demand level of the use, even within the same standard. For structural concrete, values between 0.03 and 0.05 % are common, usually associated with RCA because they are the most suitable RA for this type of application. The Brazilian standard allows much higher amounts of chlorides because it does not envisage the use of RA in structural concrete.

7.3.3 Field of Application

Taken individually the aggregates' requirements do not reveal the specification's approach. It can only be stated that a regulation is conservative or permissive after the aggregates' requirements and the proposed use of concrete are examined simultaneously.

Therefore, on this point the standards for both the application of concrete made with RA and the conditions in which concrete is produced are compared, i.e. replacement ratios of natural by RA, level of demand of the application, and maximum strength level authorised. Table 7.44 shows the field of application of RAC allowed in the different countries, and the conditions by which this application is governed.

Due to its propensity to contain contaminants that may affect the mechanical properties of concrete, the use of a fine fraction as concrete aggregate is barred in most of the standards. The Brazilian and Japanese (BCSJ) norms are exceptions, explained partly by the low demand for this use (non-structural concrete). The Danish and Swiss norms are even more complacent and allow the use of this fraction in structural concrete.

The coarse fraction provides the best conditions for use in concrete. Nonetheless, it is regarded conservatively by some of the standards. The Spanish norm allows a replacement ratio of natural by RA of only 20 %, even though the

Table 7.44 Overview of recycled aggregate field of application

Specification	Classification	Maximum replacement of natural with recycled aggregates		Use conditions[b]	Maximum strength class
		Coarse	Fine		
Brazil	RCA MRA	100 %	100 %	Non-structural concrete	15 MPa
Germany	RCA RCA	20–35 %, depending on the application	0 %	X0, XC1–XC4, XF1–XF3, XA1; pre-stressed concrete not allowed	C30/37 (20 % replacement); C25/30 (35 % replacement)
	RMA MRA	n.a	n.a	Non-structural concrete	n.a
Hong Kong	RCA	20 or 100 %	0 %	Less demanding solutions or structural concrete, for 100 or 20 % replacement, respectively	20 MPa (100 % replacement); 35 MPa (20 % replacement)
Japan BCSJ	MRA	100 %	Up to 100 %, depending on the application	Foundations and less demanding solutions	18 MPa
Japan JIS 5021	MRA	n.a	n.a	No limitations	45 MPa
Japan JIS 5022	MRA	n.a	n.a	Members not subjected to drying or freezing and thawing action	n.a
Japan JIS 5023	MRA	n.a	n.a	Backfill concrete; blinding concrete; levelling concrete	n.a
RILEM	RCA + AP RCA	100 %	Only if the natural aggregate requirements are met	Dry and wet environment; non-aggressive soils and/or water environment	No limit C50/60
	RMA			Dry and wet environment; non-aggressive soils and/or water not exposed to frost	C16/20
United Kingdom	RCA	20 %	0 %	X0, XC1–XC4, XF1, DC-1	C40/50
	RA	n.a	0 %	Non-structural concrete	n.a

(continued)

Table 7.44 (continued)

Specification	Classification	Maximum replacement of natural with recycled aggregates		Use conditions[b]	Maximum strength class
		Coarse	Fine		
The Netherlands	RCA	100 %	Only if applied with natural coarse aggregates	Non-aggressive environments	C40/50
	RMA				C20/25
Portugal	RCA	25 %	0 %	X0, XC1–XC4, XS1, XA1	C40/50
	RCA	20 %	0 %		C35/45
	MRA	n.a	0 %	Non-structural concrete	n.a
Switzerland	RCA	100 %	100 %	Reinforced concrete; pre-stressed concrete only with additional tests	C30/37
	MRA			Not allowed in reinforced concrete	n.a
Denmark	RCA with testing	100 %	20 %	Non-aggressive environments	40 MPa
	RCA without testing				
	MRA				20 MPa
Russia	MRA	100 %		Not allowed in pre-stressed concrete	15 MPa
		50 %			20 MPa
Spain[a]	RCA	20 %	0 %	Not allowed in pre-stressed concrete	40 MPa

n.a not available

[a] Proposed recommendation not yet being used

[b] conforming with EN 206-1

concrete may reach relatively high strength levels. The same stance is taken by the British, Portuguese and German specifications, which allow high strength classes but have narrow application fields for the RA. On the other hand, even though maintaining a low replacement ratio, the Dutch norm does not stipulate any special care in this type of situation, since RAC is considered equivalent to conventional concrete. The RILEM specification allows high-strength classes for RAC, comparable to the ones in the Spanish, British, Portuguese and German norms, with total replacement of the NA by recycled ones. Strictly speaking, the RILEM specification should not be compared directly with the other documents since it is the only one that presupposes a correction of the properties of concrete, which explains the greater incorporation of RA allowed in the mixes.

In relation to environmental conditions, all standards generally aim at applications subject only to non-aggressive conditions, and they are especially careful in terms of chemical attacks. The RILEM recommendation progressively limits the uses as the aggregates' characteristics worsen. Pre-stressed concrete is outside the scope of most standards. However, this limitation by itself will not stop the use of RA in concrete from playing an important role in sustainable development.

7.3.4 Conclusions

The analysis of these international norms allowed two different ways to be identified.

The first and more simplistic way of regulating the use of RA in concrete production consists of establishing limits, both in the incorporation ratio of RA and in the use of the concrete, to ensure that the resulting concrete basically maintains the properties of conventional concrete. The second viewpoint assumes that there are, in fact, differences between the performance of conventional concrete and that of the corresponding RAC, and that corrective coefficients are provided that allow concretes with the same strength class to be compared so as to adjust the design of structural elements in which RA are used.

Finally, it should be noted that the Technical Committee CEN/TC 154 "Aggregates" has been preparing an amendment, presently designated EN 12620:2002/prA1:2006, which is nearly concluded. When approved, it will modify European norm 12620:2002 to include specifications to make RA a viable alternative in a concrete mix.

References

Angulo S (2005) Characterization of aggregates from construction and demolition waste and their influence on the performance characteristics of concrete (in Portuguese). Ph.D. Thesis in Civil Engineering, Polytechnic School of the University of São Paulo, São Paulo

De Brito J (2005) Recycled aggregates and its influence on concrete properties (in Portuguese). Public lecture within the full professorship in civil engineering pre-admission examination, Instituto Superior Técnico, Technical University of Lisbon, Lisbon

De Brito J, Pereira AS, Correia JR (2005) Mechanical behaviour of non-structural concrete made with recycled ceramic aggregates. Cement Concr Compos 27(1):429–433

Etxeberria M, Vazquez E, Marí A, Barra M (2007) Influence of amount of recycled coarse aggregates and production process on properties of recycled aggregate concrete. Cem Concr Res 37(5):735–742

Gonçalves P (2007) Concrete with recycled aggregates—commented analysis of existing legislation (in Portuguese). Masters Dissertation, Instituto Superior Técnico, Technical University of Lisbon, Lisbon

Grubl P, Ruhl M (1998) German committee for reinforced concrete (DafStb)—code: concrete with recycled aggregate. In: Sustainable construction: proceedings of the international symposium organized by the concrete technology unit, University of Dundee

Hansen T (1992) Recycling of demolished concrete and masonry. London: E & FN SPON, RILEM Report 6

Karlsen J, Petkovic G, Lahus O (2002) A Norwegian certification scheme for recycled aggregate (RCA). In: Third international conference on sustainable building (CD), Oslo

Kou SC, Poon CS (2008) Mechanical properties of 5-years-old concrete, prepared with recycled aggregates obtained from three different sources. Mag Concr Res 60(1):57–64

Leite MB (2001) Evaluation of the mechanical properties of concretes made with recycled aggregates form construction and demolition waste (in Portuguese). Ph.D. thesis in Civil Engineering, School of Engineering of the Federal University of Rio Grande do Sul, Porto Alegre

Lima J (1999) Proposition of guidelines for the production and standardization of recycled construction aggregate and its application on mortars and concrete (in Portuguese). Master Dissertation in Architecture and Urbanism, School of Engineering of São Carlos, University of São Paulo, São Carlos

Limbachiya MC, Dhir RK, Leelawat T (2000) Use of recycled concrete aggregate in high-strength concrete. Mater Struct 33(9):574–580

Loo W (1998) Closing the concrete loop—from reuse to recycling. In: Dhir R, Henderson N, Limbachiya M (eds) Sustainable construction: use of recycled concrete aggregate. Thomas Telford Pub, London, pp 227–237

Mueller A (2007) Closed loop of concrete and masonry rubble. Chair of mineral processing of building materials and reuse, Bauhaus-University, Weimar

Roos F (2002) Ein Beitrag zur Bemessung von Beton mit Zuschlag aus rezyklierter Gesteinskörnung nach DIN 1045-1. Dissertation Technischen Universität München, München

Schulz RR, Hendricks ChF (1992) Recycling of masonry rubble. In: Recycling of demolished concrete and masonry, part 2, RILEM Technical Committee Report No. 6 (3rd edn), E & FN SPON, London, pp 132–236

Solyman M (2005) Classification of recycled sands and their applications as fine aggregates for concrete and bituminous mixtures. Doktor-Ingenieurs Dissertation, Fachbereich Bauingenieurwesen der Universität Kassel, Kassel

Vazquez E, Alaejos P, Sanchez M, Aleza F, Barra M, Buron M, Castilla J, Dapena E, Etxeberria M, Francisco G, Gonzalez B, Martinez F, Martinez I, Parra J, Polanco J, Sanabria M (2006) Utilización de árido reciclado para la fabricación de hormigón estructural. Comisión 2, Grupo de Trabajo 2/5 "Hormigón reciclado" (in Spanish), Monografía M-11 ACHE, Madrid

Vyncke J, Rousseau E (1994) Recycling of construction and demolition waste in Belgium: actual situation and future evolution. In: Lauritzen EK (ed) Demolition and reuse of concrete and masonry, proceedings of the third international RILEM symposium, 1st edn. E&FN Spon, Odense, pp 57–70

Xiao JZ, Li JB, Zhang C (2006) On relationships between the mechanical properties of recycled aggregate concrete—an overview. Mater Struct 39(6):655–664

Normative Documents

Addition to Danish Concrete Association No. 34 for the use of recycled aggregates for concrete in passive environmental class (in Danish). Hrsg.: Danish Concrete Association, Copenhagen, 1995

Instrucción de hormigón estructural (EHE) (in Spanish). Ministerio de Fomento, Spain, 2001

Recommendation for the use of recycled aggregates for concrete in passive environmental class (in Danish). Hrsg.: Danish Concrete Association Publication No. 34 Copenhagen, 1990

BCSJ (1977) Proposed standard for the use of recycled aggregate and recycled aggregate concrete (in Japanese). Committee on Disposal and Reuse of Construction Waste, Building Contractors Society of Japan

BS 8500-2 (2002) Concrete—complementary British Standard to BS EN 206-1, part 2: specification for constituent materials and concrete. British Standards Institution, 2002

BS EN 1744-1 (1998) Tests for chemical properties of aggregates—part 1: chemical analysis. British Standards Institution, 1998

CUR (1984) Betonpuingranulaaten als toeslagsmateriaal vor beton (in Dutch). Aanbeveling 4, CUR-VB, The Netherlands

CUR (1986) Betonpuingranulaaten metselwerkpuins granullat alls toeslagmeterial van beton (in Dutch). Rapport 125, CUR, The Netherlands

CUR (1994) Metselwerkpuingranulaat als Toeslagsmateriaal vor Beton (in Dutch). Aanbeveling 5, CUR-VB, The Netherlands

DEUTSCHER AUSSCHUSS FÜR STAHLBETON (1998) DafStb: Richtlinie "Beton mit rezykliertem zuschlag", German Committee for Reinforced Concrete; DAfStb: Guideline "Concrete with Recycled Aggregates", Germany

DIN 4226-1: 2001-7 (2001) Mineral aggregates for concrete and mortar—part 1: normal and heavy-weight mineral aggregates, Germany

DIN 4226-100: 2002-2 (2002) Aggregates for mortar and concrete, part 100: recycled aggregates, Germany

EN 12620:2002/prA1 (2006) Aggregates for concrete (Amendment to EN 12620:2002), Brussels, Belgium

EN 206-1:2000 (2000) Concrete—part 1: specification, performance, production and conformity, Brussels

EN 1744-1:1998 (1998) Tests for chemical properties of aggregates. Part 1: chemical analysis, Brussels

JIS A 5021:2005 (2005) Recycled aggregate for concrete—class H (in Japanese), Japan

JIS A 5308:1995 (1995) Ready-mixed concrete (in Japanese), Japan

LNEC E 471 (2006) Guide for use of coarse recycled aggregates in hydraulic binder concrete (in Portuguese), National Laboratory of Civil Engineering, Lisbon

NBR 15.116 (2005) Aggregates from construction solid waste: use in pavements and non-structural concrete production—requirements (in Portuguese), Brazil

NBR 7211:2004 (2004) Concrete aggregates—specification (in Portuguese), Brazil

NP EN 12620:2004 (2004) Concrete aggregates (in Portuguese), Portuguese Institute of Quality, Lisbon

NP ENV 206:2003 (2003) Concrete—performance, production, casting and conformity criteria (in Portuguese), Portuguese Institute of Quality, Lisbon

OT (Objectif Technique) 70085 (2006) Instruction technique. Utilisation de matériaux de construction minéraux secondaires dans la construction d'abris (in French), Switzerland

Prescriptions Techniques PTV 406 (2003) Granulats de débris de démolition et de construction recyclés (in French), Brussels

pr EN 933-11:2005 (2005) Tests for chemical properties of aggregates. Part 11: classification test for the constituents of coarse recycled aggregate, Brussels

pr EN 1744-5:2004 (2004) Tests for chemical properties of aggregates. Part 5: determination of acid soluble chloride salts, Brussels

RILEM TC 121-DRG (1994) Specifications for concrete with recycled aggregates. Mater Struct 27:557–559

SIA 162 (1989) Ouvrages en béton (in French), Société suisse des ingénieurs et des architectes, Switzerland

SIA 162/4 (1994) Béton de recyclage (in French), Société suisse des ingénieurs et des architectes, Switzerland

WBTC No. 12/2002 (2002) Specifications facilitating the use of recycled aggregates, Works Bureau Technical Circular, Hong-Kong

Index

J. de Brito and N. Saikia, *Recycled Aggregate in Concrete*,
Green Energy and Technology, DOI: 10.1007/978-1-4471-4540-0,

Printed by Publishers' Graphics LLC
BT20130122.19.24.26